Corporate and Global Standardization Initiatives in Contemporary Society

Kai Jakobs
RWTH Aachen University, Germany

A volume in the Advances in Human and Social Aspects of Technology (AHSAT) Book Series

IGI Global
DISSEMINATOR OF KNOWLEDGE

Published in the United States of America by
IGI Global
Information Science Reference (an imprint of IGI Global)
701 E. Chocolate Avenue
Hershey PA, USA 17033
Tel: 717-533-8845
Fax: 717-533-8661
E-mail: cust@igi-global.com
Web site: http://www.igi-global.com

Library of Congress Cataloging-in-Publication Data

Names: Jakobs, Kai, 1957- editor.
Title: Corporate and global standardization initiatives in contemporary society / Kai Jakobs, editor.
Description: Hershey, PA : Information Science Reference, [2018]
Identifiers: LCCN 2017040009| ISBN 9781522553205 (hardcover) | ISBN 9781522553212 (ebook)
Subjects: LCSH: Standardization. | Information technology--Standards. | Technological innovations--Standards.
Classification: LCC HD62 .C67 2018 | DDC 389/.6--dc23 LC record available at https://lccn.loc.gov/2017040009

This book is published in the IGI Global book series Advances in Human and Social Aspects of Technology (AHSAT) (ISSN: 2328-1316; eISSN: 2328-1324)

British Cataloguing in Publication Data
A Cataloguing in Publication record for this book is available from the British Library.

Advances in Human and Social Aspects of Technology (AHSAT) Book Series

Ashish Dwivedi
The University of Hull, UK

ISSN:2328-1316
EISSN:2328-1324

Mission

In recent years, the societal impact of technology has been noted as we become increasingly more connected and are presented with more digital tools and devices. With the popularity of digital devices such as cell phones and tablets, it is crucial to consider the implications of our digital dependence and the presence of technology in our everyday lives.

The **Advances in Human and Social Aspects of Technology (AHSAT) Book Series** seeks to explore the ways in which society and human beings have been affected by technology and how the technological revolution has changed the way we conduct our lives as well as our behavior. The AHSAT book series aims to publish the most cutting-edge research on human behavior and interaction with technology and the ways in which the digital age is changing society.

Coverage

- Technology and Social Change
- Activism and ICTs
- Computer-Mediated Communication
- Public Access to ICTs
- Digital Identity
- ICTs and social change
- Technology and Freedom of Speech
- Human Rights and Digitization
- Human Development and Technology
- Cyber Bullying

IGI Global is currently accepting manuscripts for publication within this series. To submit a proposal for a volume in this series, please contact our Acquisition Editors at Acquisitions@igi-global.com or visit: http://www.igi-global.com/publish/.

The Advances in Human and Social Aspects of Technology (AHSAT) Book Series (ISSN 2328-1316) is published by IGI Global, 701 E. Chocolate Avenue, Hershey, PA 17033-1240, USA, www.igi-global.com. This series is composed of titles available for purchase individually; each title is edited to be contextually exclusive from any other title within the series. For pricing and ordering information please visit http://www.igi-global.com/book-series/advances-human-social-aspects-technology/37145. Postmaster: Send all address changes to above address.

Titles in this Series

For a list of additional titles in this series, please visit: www.igi-global.com/book-series

Experience-Based Human-Computer Interactions Emerging Research and Opportunities
Petr Sosnin (Ulyanovsk State Technical University, Russia)
Information Science Reference • copyright 2018 • 294pp • H/C (ISBN: 9781522529873) • US $165.00 (our price)

Technology-Enhanced Human Interaction in Modern Society
Francisco Vicente Cipolla-Ficarra (Latin Association of Human-Computer Interaction, Spain & International Association of Interactive Communication, Italy) Maria Valeria Ficarra (Latin Association of Human-Computer Interaction, Spain & International Association of Interactive Communication, Italy) Miguel Cipolla-Ficarra (International Association of Interactive Communication, Italy) Alejandra Quiroga (Universidad Nacional de La Pampa, Argentina) Jacqueline Alma (Electronic Arts – Vancouver, Canada) and Jim Carré (University of The Netherlands Antilles, Curaçao)
Information Science Reference • copyright 2018 • 319pp • H/C (ISBN: 9781522534372) • US $205.00 (our price)

HCI Challenges and Privacy Preservation in Big Data Security
Daphne Lopez (VIT University, India) and M.A. Saleem Durai (VIT University, India)
Information Science Reference • copyright 2018 • 275pp • H/C (ISBN: 9781522528630) • US $215.00 (our price)

Handbook of Research on Human Development in the Digital Age
Valerie C. Bryan (Florida Atlantic University, USA) Ann T. Musgrove (Florida Atlantic University, USA) and Jillian R. Powers (Florida Atlantic University, USA)
Information Science Reference • copyright 2018 • 526pp • H/C (ISBN: 9781522528388) • US $275.00 (our price)

Optimizing Human-Computer Interaction With Emerging Technologies
Francisco Cipolla-Ficarra (Latin Association of Human-Computer Interaction, Spain & International Association of Interactive Communication, Italy)
Information Science Reference • copyright 2018 • 471pp • H/C (ISBN: 9781522526162) • US $345.00 (our price)

Designing for Human-Machine Symbiosis Using the URANOS Model Emerging Research and Opportunities
Benjamin Hadorn (University of Fribourg, Switzerland)
Information Science Reference • copyright 2017 • 170pp • H/C (ISBN: 9781522518884) • US $125.00 (our price)

Research Paradigms and Contemporary Perspectives on Human-Technology Interaction
Anabela Mesquita (School of Accounting and Administration of Porto, Polytechnic Institute of Porto, Portugal & Algorithm Research Centre, Minho University, Portugal)
Information Science Reference • copyright 2017 • 366pp • H/C (ISBN: 9781522518686) • US $195.00 (our price)

IGI Global
DISSEMINATOR OF KNOWLEDGE

701 East Chocolate Avenue, Hershey, PA 17033, USA
Tel: 717-533-8845 x100 • Fax: 717-533-8661
E-Mail: cust@igi-global.com • www.igi-global.com

Editorial Advisory Board

Table of Contents

Section 4
1000 Flowers

Detailed Table of Contents

Standardization is a crucial enabler of global business of information and communications technologies. Convergence of the underlying networking paradigms of licensed mobile communication and license-exempted internet has made progress, but full integration is still far from being complete. For standardization professionals, the unpredictable convergence makes decision making and participation in standardization complicated. This chapter examines collaboration in five closely related standardization organizations working in this field during the years from 2003 to 2008. The results show similarities and differences in collaboration structures and behaviours reflecting the specific scope and context of each standardization organization. Furthermore, this chapter extends the use of social network analysis as a tool to the field of empirical standardization research. The results pave the way towards better collaboration in standardization communities of converging mobile internet and beyond by providing better visibility and new insights to standardization leaders, policy makers, and users.

Innovative companies need to make decisions related to protection and sharing of their new knowledge. This chapter explores knowledge patterns during formal ICT standardization and/or in standards consortia. While previous studies focused on the role of intellectual property rights in standardization decisions, this investigation considers standards and patents as sources of information for innovation. Results show that when patents are important, firms are more likely to be engaged in consortia but not in formal standardization. Size moderates this relationship. Smaller firms are more motivated by the quest for knowledge during standardization while larger firms are more sensitive to the importance of patents. The authors conclude that small firms might participate in ICT standardization in order to anticipate development of the standard, rather than place a patent in the standards.

Issues of dominance in the market place, "standards wars," and "battles for dominance" between large companies are frequently addressed by researchers and the business press alike. The existence of companies that could establish internationally dominant solutions to customers' problems within a few years after their founding is quite unknown and the reasons for their success are hardly investigated so far. Therefore, they are not covered by traditional stage models for the establishment of dominant solutions. Presenting 22 cases and a new success factors model, this chapter shows how young companies can successfully

This chapter comparatively examines the valuation methodologies put forward and accepted in four initial FRAND cases adjudicated in the US district court system in 2013-15 and finds a qualitative link to the business models (i.e., value logics) employed by the actors and the type of standards involved. Additionally, valuation models based on prior license agreements seem to currently produce higher valuations of SEPs, everything else equal. While the CAFC has clearly stated that apportionment is the governing rule in SEP cases, the variety of valuation methods put forward in similar contexts indicates the calculation of the SEPs still lacks consensus on both legal and economic grounds. However, a number of the most recent SEP rulings in district courts show clear accordance with clarifications from the CAFC in Ericsson and CSIRO. Additionally, the studied cases confirm the lack of empirical evidence of systemic patent holdup and royalty stacking and indicate that the contractual/transactional nature of SEP disputes can well be addressed on a case-by-case basis by the courts.

During the last decade, many organizations have adopted information technology service management (ITSM) frameworks such as information technology infrastructure library (ITIL). However, such efforts are challenging and costly and not all have been successful. Keeping this in mind, this chapter seeks to identify critical success factors (CSFs) that impact successful implementation of ITSM frameworks with a focus on ITIL. Hence, based on a mixed method, at first, with the semi-structured interviews and literature review, 22 success factors were identified. The data of 122 valid questionnaires were analyzed using robust exploratory factor analysis (EFA) to validate and categorize identified factors. Finally, 22 CSFs are categorized under five "organizational," "human resources," "project management," "managerial," and "process" main groups. The findings of this study have provided a very useful reference for IT standardization scholars and practitioners to identify the important issues of ITSM frameworks implementation projects in research and practice.

Over the years, the importance of open standards has been acknowledged in EU and national policies. Formal (e.g., ISO) standards are often referred to in software development and procurement. Use of

Preface

ON TWO FUTURE TRENDS IN STANDARDIZATION

Introduction

According to the definition adopted by the International Organization for Standardization (ISO, n.d.) a standard is a document:

Established by consensus and approved by a recognized body, that provides, for common and repeated use, rules, guidelines or characteristics for activities or their results, aimed at the achievement of the optimum degree of order in a given context.

The fact that standards are established 'by consensus' distinguishes them from legislation. Typically, the use of standards is voluntary. However, through legislation they may become mandatory (e.g. many health and safety standards) or 'quasi-mandatory' (e.g. Harmonized European Standards).

In the field of Information and Communication Technologies (ICT) international standards are the major mechanism to ensure interoperability between systems. Frequently, ICT standards also describe a commonly agreed (technical) platform upon which innovations can be based and marketed. Standards are also of interest to policy makers. For instance, they were a major enabler of the European Single Market; globally, they serve to remove barriers to trade an thus support the free cross-border flow of goods and, increasingly, services. Moreover, standards are a valuable means of technology transfer, which makes them also relevant for academia and research institutions. Standardization also provides a legal platform for co-operation with competitors and (potential) customers[1]. Last, but decidedly not least must standards' potential economic implications not be under-estimated. A new standard may be used to extend a market, or even help open up a new one. Intellectual Property Rights (IPR), mostly patents, also play a major role here. That is, standards must also be considered as strategic and economic tools.

Indeed, from a macro-economic perspective, a number of studies calculated the macro-economic impact of standards. For instance, Blind (2013) reports that the contribution of standards to the growth rate in different countries was equivalent to 0.9% in Germany, 0.8% in France and Australia, 0.3% in the UK and 0.2% in Canada. Adopting a micro-economic perspective, ISO (2011, 2012) did a number of studies in companies from different business sectors and countries. Based on Porter's value chain (Porter & Kramer, 2011) these studies show that the implementation of standards can provide economic benefits to firms between 0.5% and 4% of their annual revenues.

Not so long ago standardization and innovation were considered as almost mutually exclusive. This has changed by now. In fact, close links between standardization and innovation may often be identified. Today, standardization is no longer considered an impediment to innovation. However, the unqualified claim that 'standards foster innovation' does not fully reflect reality either. Swann and Lambert (2010) observe that standards do both – enable and constrain innovation – but that the enabling aspect is much more important. Specifically, they note that:

Standardization does constrain activities but in doing so creates an infrastructure to help trade and subsequent innovation. Standardization is not just about limiting variety by defining norms for given technologies in given markets. Standardization helps to achieve credibility, focus and critical mass in markets for new technologies. (p. 370)

What's more:

Standards are not only technical questions. They determine the technology that will implement the Information Society, and consequently the way in which industry, users, consumers and administrations will benefit from it. (CEC, 1996, p. 1)

That is, those that develop ICT standards today will to a not inconsiderable degree shape much of the ICT environment we will deploy in the future. It is thus little wonder that these days participation in standards setting has become a major strategic tool for many firms. Frequently, a firm's aim will be to influence the process for its own benefit. This holds particularly – though by no means exclusively – for large companies. Other potential motivations include e.g. networking, intelligence gathering and increasing credibility.

Standardization research is a comparably young field of study, and still a rather small one (albeit one with increasing importance). Most of the early literature – and much of today's – focuses largely on economic aspects. A broader perspective, incorporating elements form a vast array of disciplines (including, but not limited to, business studies, computer science, economics, information systems, legal studies, management studies and sociology) emerged on a somewhat wider scale only in the late 1990s. A special focus of standardization research has always been on the ICT sector. This is for one due to the fact that developments in this sector are extremely fast, with accordingly short technology life cycles. This imposes special requirements particularly, but not exclusively, on the speed of standards development. Private standards consortia that are responsible for a majority of important standards are another phenomenon virtually unique to ICT. Accordingly, much of the literature on ICT standardization addresses issues that relate to these characteristics (including e.g. the issue of 'speed' in general, the pros and cons of formal standards bodies versus private consortia, the co-ordination of standardization activities in different bodies, or more recently, the implications of the new EU legislation (EU, 2012) on stakeholders, processes and policy).

Moreover, policy makers become increasingly aware of the fact that they – and society – stand to benefit from deeper scientific understanding of the various roles standards may play (e.g. technical, economical, societal) and of the exact nature of the underlying standardization process. Fairly recent evidence of this importance assigned to standardization research is the European 'Joint Initiative on Standardization'[2], a policy-driven multi-stakeholder initiative to strengthen research about and education in the field of standardization. Along similar lines, the US National Institute of Standards and Technology (NIST) has launched an initiative to advance standards education.

Future Topics

Hopefully, the brief introduction has made it sufficiently clear that standard and standardization are relevant (from different perspectives) and interesting research topics. This holds not least for ICT – a particularly fast moving sector whose importance is rapidly growing. Thus, this section will briefly discuss two current trends that may be expected to become particularly important in the nearer future. One is from within the ICT sector ('platformization'), the other is enabled by ICT ('smartness').

Platformization

Platforms are becoming more and more popular. Specifically, a software-based platform has been defined as "the extensible codebase of a software-based system that provides core functionality shared by the modules that interoperate with it and the interfaces through which they interoperate (e.g., Apple's iOS and Mozilla's Firefox browser)" (Tiwana et al., 2010, p. 675). They define a module as an "add-on software subsystem that connects to the platform to add functionality to it" (e.g., iPhone apps and Firefox extensions). The platform and the modules specific to it form that platform's ecosystem (see Figure 1).

Gawer (2011) distinguishes between 'Internal', 'Supply Chain' and 'Industry' platforms. The former refers to those that are used internally by one firm. Supply chain platforms link a firm with a set of business partners and the latter type is used throughout a whole industry. Similarly, Gawer and Cusumano (2014) define internal platforms as a set of assets organized in a common structure from which a company can efficiently develop and produce a stream of derivative products. A supply chain platform's role is limited to the facilitation of information exchange or trade, without the possibility for third parties to innovate. External (industry) platforms are defined as products, services or technologies that provide the foundation upon which outside firms can develop their own complementary products or services.

Figure 1. Platform-centric ecosystems
Source: Tiwana et al. (2010)

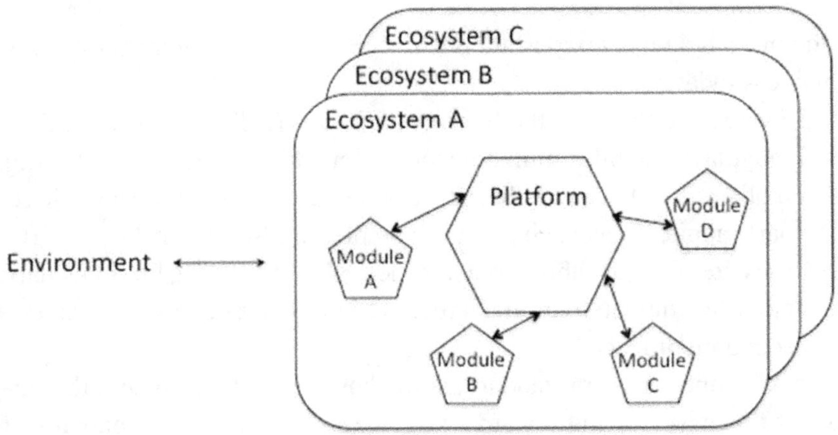

Tiwana et al. (2010) also highlight the importance of technology for a platform's success. They observe that both the technical architecture of a platform and its organizing and governing principles impact its evolution. This observation, in turn, points to the importance of standards as critical elements of any platform. This holds for both internal (to the platform) and, particularly, external standards that link the platform with any complementary modules.

Recently, cloud computing has received considerable attention in both research and the media. Clearly, a 'cloud' is a platform and as such it may serve as an example for platform standardization requirements. NIST (2011) identify three general areas where adequate standards are missing – Interoperability, Portability and Security. Along similar lines, Fischer et al. (2013) identify nine areas for which standards are needed – Compliance, Competition, Portability, Interoperability, Privacy, Information Security, Transparency, Effectiveness and Efficiency. Lewis (2013) uses a slightly different classification, distinguishing between 'Infrastructure as a Service (IaaS)', 'Platform as a Service (PaaS) and 'Software as a Service (SaaS)'. Regarding the need for standards, however, her views differ from those of NIST. She argues that standards are unlikely to emerge in the fields of PaaS and SaaS as such standards would not be in the best interest of the service providers. Focusing on standards for Cloud in the EU, Walden and Gleeson (2014) as well come up with a similar list, including cloud interoperability, data portability and reversibility, data protection standards, cloud security standards, standardized service descriptions and serviced level agreements. Also in Europe, and somewhat uncritically, the European Commission "aims at enabling and facilitating faster adoption of cloud computing throughout all sectors of the economy" (CEC, 2012, p. 2). Their claimed need to cut through the 'jungle of standards' (CEU, 2012), however, is refuted by ETSI's[3] Cloud Standards Coordination group (CSC, 2013). Rather, the authors find that cloud standardization is focused. Perhaps even more importantly, they also highlight the fact that "many standards used for cloud computing are not cloud specific" (p. 28).

Overall and at varying levels of detail the literature identifies three main fields for which standardization activities will be needed – Interoperability, Portability and Security, i.e. those identified by NIST.

Certification, compliance and auditing are other aspects that are at least partially relevant for standards and standardization. This holds particularly for security-related standards that will need to meet legal obligations (CSC, 2013; EC, 2012; NIST, 2011). To this end, Borenstein and Blake (2011) note that evaluative standards, i.e. those that provide a uniform approach to evaluating how well something is being done, can go a long way to reassure (potential) cloud customers.

Smartness

This is perhaps the one single most important (technical) future development the early stages of which we can already witness today.

Cyber-Physical Systems (CPSs) and the Internet of Things (IoT) will add a third dimension to 'anytime' and (almost) 'anyplace' mobile communication – 'anything'. That is, 'smart' objects will be able to interact with each other as well as with their environment, including humans. These capabilities will be deployed by 'smart' applications such as, for example, the Smart Grid or Smart Cities. Although especially Smart Cities are only just about to leave the realm of Science Fiction smart applications are here to stay. They will affect human lives much more than ICT has done so far. And not least, they will have an enormous economic impact.

Indeed, in the past couple of years standardization has seen a development that may be considered an early indicator of this development towards 'smartness' and that may eventually change the whole standardization landscape – the mergers of formerly separate technological sectors. The – almost finalized – integration of (tele)communication and information technology led to ICT. Such mergers of technologies will continue, they are likely to do so on a broad scale and they will not be limited to the ICT sector. The next step (to be) taken will be to instill 'smartness' into 'traditional' applications and the underlying communication infrastructure. And ICT will be the enabling technology of this 'smart' bit.

Most of these emerging new 'smart' applications will integrate hitherto separate disciplines. Examples include, among others, e-health (ICT, medicine), Intelligent Transport Systems (e.g. ICT, transport telematics, traffic engineering), Smart Manufacturing (e.g. ICT, production engineering, robotics, control engineering) and the Smart Grid (e.g. ICT, power engineering). They will deploy an equally 'smart' infrastructure that will be based on Wireless Sensor Networks (WSNs), Machine-to-Machine Communication (M2M), Cyber-Physical Systems and the Internet of Things.

Quite obviously, both the applications and the infrastructure will need to be based on technical standards. Yet, their highly multi-disciplinary nature represents a considerable problem for standards setting. Traditionally, one (or perhaps several) standards setting entity (e.g. a Technical Committee [TC] or a Working Group [WG]) has covered one – potentially broad – field. While close links exist between some such entities, co-operation in general is not overly well developed. Moreover, these emerging new technologies will require co-operation between standardization entities with very different cultures, from equally different backgrounds and with very different technology life cycles. On top of that, ICT in general – and specifically applications like e.g. e-health – need to address issues of information security and safety and of trust. These aspects as well will need to be considered in the standardization process, adding yet another dimension of complexity.

Figure 2 visualizes the development from comparably simple (networked) embedded systems, which have been around for quite some time now via CPSs to an overarching field of application of these technologies, in this case Smart Cities.

Figure 2. From airbags to smart cities
Source: acatech (2011)

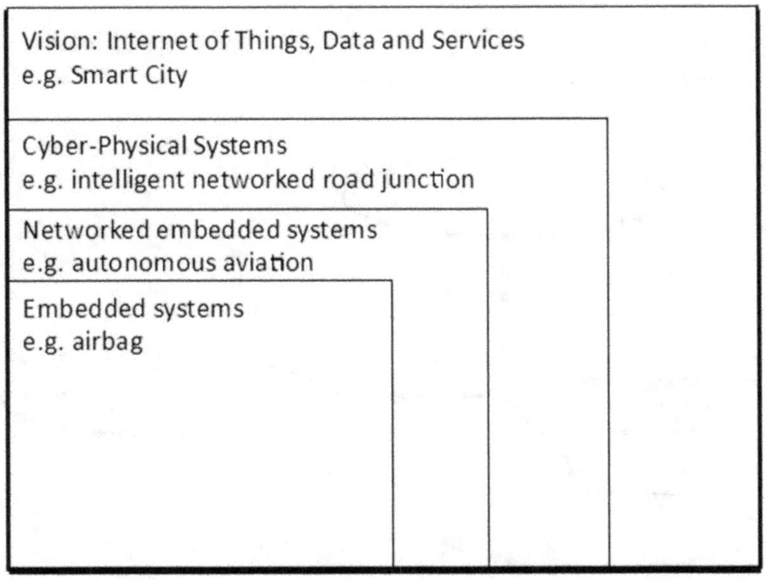

For the European Commission, Cyber-Physical Systems are "integrated systems of computation, networking, and physical processes. Embedded computers and networks in these systems monitor, control and manage the physical processes, with feedback loops where physical processes affect computations and vice versa. Further they can exchange information between them and with human users"[4]. The defining characteristics of a CPS include the integration of both 'physical' and 'virtual' artifacts and the ability to communicate with the outside world, including humans. CPSs are assumed to be part of people's daily lives by 2050 (EP, 2016).

The European Research Cluster on the Internet of Things (IERC) defines the IoT as "A dynamic global network infrastructure with self-configuring capabilities based on standard and interoperable communication protocols where physical and virtual 'things' have identities, physical attributes, and virtual personalities and use intelligent interfaces, and are seamlessly integrated into the information network"[5]. That is, a CPS may be a node of the IoT and may, accordingly, communicate via the IoT. In contrast, not every IoT node is a CPS, as the defining characteristic for an IoT node is ability to uniquely identify each node; it does not need to have a 'physical' component. They also note that the IoT adds the 'anything' dimension to 'anyplace' and 'anytime' (see Figure 3).

Frequently, applications will have hard requirements on the underlying communication system (think of e.g. autonomous driving). Notably, guaranteed levels of latency, resilience, reliability and predictability of the system will become more important than the 'traditional' metrics for (wireless) communication systems like bandwidth and throughput. According to Fettweis et al. (2014), end-to-end latencies of less than 1 ms will be required out of which only 200 μs will be available for the wireless link. This is about one order of magnitude faster than today's LTE[6] networks. Moreover, latency and resilience are diametrically coupled.

As a consequence, application design, communication technology, operating systems and control loops will need to be extremely closely coupled – loosely coupled systems will hardly, if ever be able to meet these requirements. To achieve this, also for the 'smart' infrastructure will close multi-disciplinary co-operation be necessary from the outset. This co-operation will cover various sub-fields of Computer Science, Mechanical/Electrical/Telecommunications Engineering, possibly also Biology and other Natural Sciences plus the respective application area(s).

Figure 3. The three dimensions of IoT communication
Source: ITU (2012)

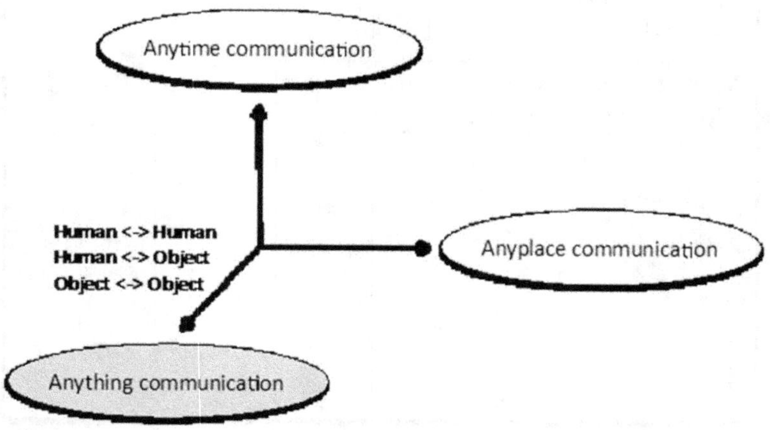

A number of application areas may be identified that are likely to stand to benefit from, or indeed require, the deployment of a smart communication infrastructure. Table 1 summarizes application areas identified in the literature for both the IoT and CPSs.

As note above, each of these application areas is inherently multi-disciplinary. Table 2 shows some of the major disciplines involved in the five most frequently named areas.

Smart applications receive information from and send instructions to the underlying infrastructure, the individual elements of which need to communicate with each other in order to meet the applications' requirements (see Figure 4).

According to the European Commission, Intelligent Transport Systems (ITS) result from:

The integration of information and communication technologies (ICT) with vehicles and transport infrastructure to improve economic performance, safety, mobility and environmental sustainability[7].

In ITS, inter-vehicle communications and road-vehicle communications play an important role in assisting safe (or even autonomous) driving. The quality of such system may, therefore, be a matter of life or death. Accordingly, real-time and robust communication must be secured for ITS.

The idea of 'smart manufacturing' emerged in the late 1980, became more popular as a research topic in the late 1990s and got off the ground with the advent of the German 'Industrie 4.0' initiative in 2013.

Smart Manufacturing has been defined as:

Table 1. Application areas for the IoT and CPSs

CPS (Gunes et al., 2014)	CPS (NIST, 2013)	CPS (excerpt; NIST, 2016)	IoT (AIOTI, 2015)	IoT (IoT-A, 2013)
Smart Manufacturing	Smart Manufacturing and Production	Manufacturing	Manufacturing/ Industry Automation	Productive Business Environment
Health Care and Medicine	Healthcare	Healthcare	Healthcare	Healthcare
Intelligent Transport Systems	Transportation and Mobility	Transportation	Vehicular/ Transportation	Smart Transport Logistics
Critical Infrastructure	Civil Infrastructure & Energy	Infrastructure (communications, power, water)	Energy	Smart Energy
Smart City		Cities	Cities	Smart Cities
	Buildings and Structures	Buildings	Home/Building	Smart Houses
Emergency Response	Emergency response	Emergency response		
			Living Environment	Ambient Assisted Living
		Environmental monitoring	Environment	
		Agriculture	Farming/ Agrifood	
		Supply chain/retail		Retail
	Defense	Defense		
			Wearables	

Source: Jakobs and Wehrle (2017)

Table 2. Disciplines involved in different application areas

Healthcare	Smart Buildings	Smart Manufacturing and Production	Intelligent Transport Systems	Smart Grid	Smart Cities
Medicine Tele-communication Computer Science Mechanical Engineering …..	Architecture Civil Engineering Computer Science Tele-communication …..	Production Engineering Tele-communication Computer Science Robotics Control Engineering ….	Transport Telematics Traffic Engineering Power Engineering Automotive Computer Science, Tele-communication …..	Power Engineering Computer Science Tele-communication …..	The grand total of the others plus City planning Economics Environmental studies …..

Source: Jakobs and Wehrle (2017)

Figure 4. Smart applications and the underlying infrastructure
Source: NIST (2016)

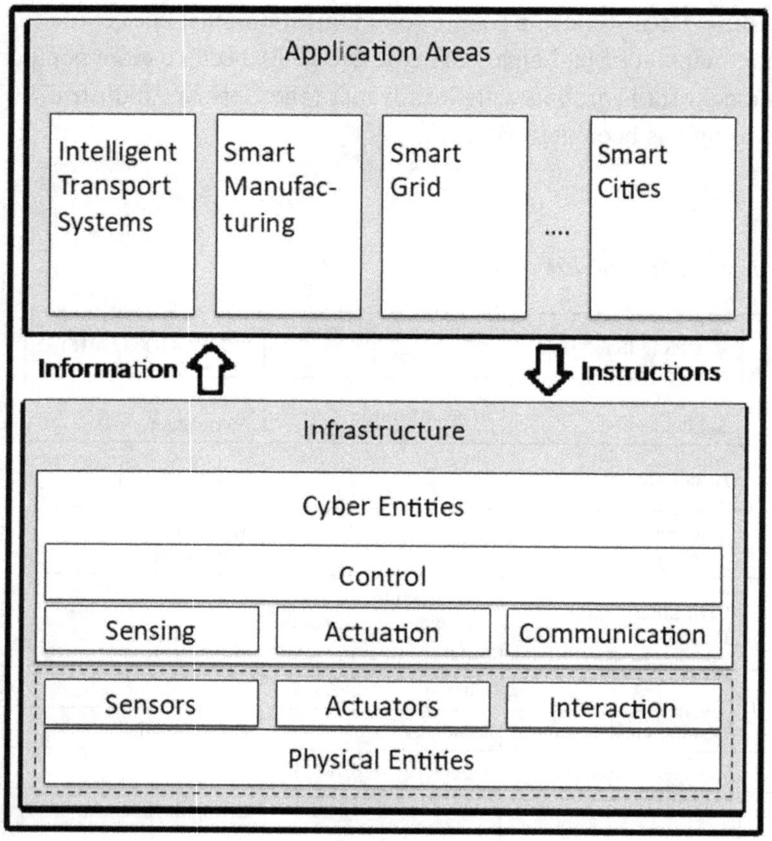

The dramatically intensified application of 'manufacturing intelligence' throughout the manufacturing and supply chain enterprise". It "comprises the real-time understanding, reasoning, planning and management of all aspects of the enterprise manufacturing process and is facilitated by the pervasive use of advanced sensor-based data analytics, modelling, and simulation. (Davis et al., 2012)

From a more practical perspective, Lu et al. (2015) state that smart manufacturing uses:

Information and communication technologies ... to 1) Optimize the use of labor, material, and energy to produce customized, high quality products for on-time delivery. 2) Quickly respond to changes in market demands and supply chains.

The European Commission views smart manufacturing as "a novel architectures for factory automation based on CPS and IoT"[8].

According to Gungor et al. (2010):

The smart grid is a modern electric power-grid infrastructure for improved efficiency, reliability, and safety, with smooth integration of renewable and alternative energy sources, through automated control and modern communication technologies.

More specifically, Farhangi (2010) observed:

The existing electricity grid is unidirectional in nature. It converts only one-third of fuel energy into electricity, without recovering the waste heat. Almost 8% of its output is lost along its transmission lines, while 20% of its generation capacity exists to meet peak demand only (i.e., it is in use only 5% of the time). In addition to that, due to the hierarchical topology of its assets, the existing electricity grid suffers from domino-effect failures". The 'smart grid' is supposed to address these shortcomings. "In essence, the smart grid needs to provide the utility companies with full visibility and pervasive control over their assets and services. The smart grid is required to be self-healing and resilient to system anomalies. And last but not least, the smart grid needs to empower its stakeholders to define and realize new ways of engaging with each other and performing energy transactions across the system.

Notably, the latter refers to the fact that the distinction between energy producer and energy consumer will become blurred. In this context, EC (2006) envisions that "local generation enables each user node in the future network to behave as both sink and source."

'Smart Cities' is arguably the broadest application area. Caragliu et al. (2011) believe:

A city to be smart when investments in human and social capital and traditional (transport) and modern (ICT) communication infrastructure fuel sustainable economic growth and a high quality of life, with a wise management of natural resources, through participatory governance.

In a way, a mart city represents the superset of the above smart applications and then some (like smart buildings, smart government, e-health). It will, therefore, be a particularly complex construct. Nevertheless, a survey by TU Vienna identifies 90 smart cities with 300,000 – 1,000,000 inhabitants and 77 with 100,000 – 300,000 inhabitants[9]. According to Navigant (2011), however, there are many pilot projects and small-scale developments but no examples yet of a smart city on a large scale that implement the model depicted in Figure 5 (which is quite similar to the model used by TU Vienna). That is, numerous comparably simple predecessors of smart cities exist today[10], but a long way remains to be gone to achieve true 'smartness' here.

Figure 5. Smart city model
Source: Navigant (2011)

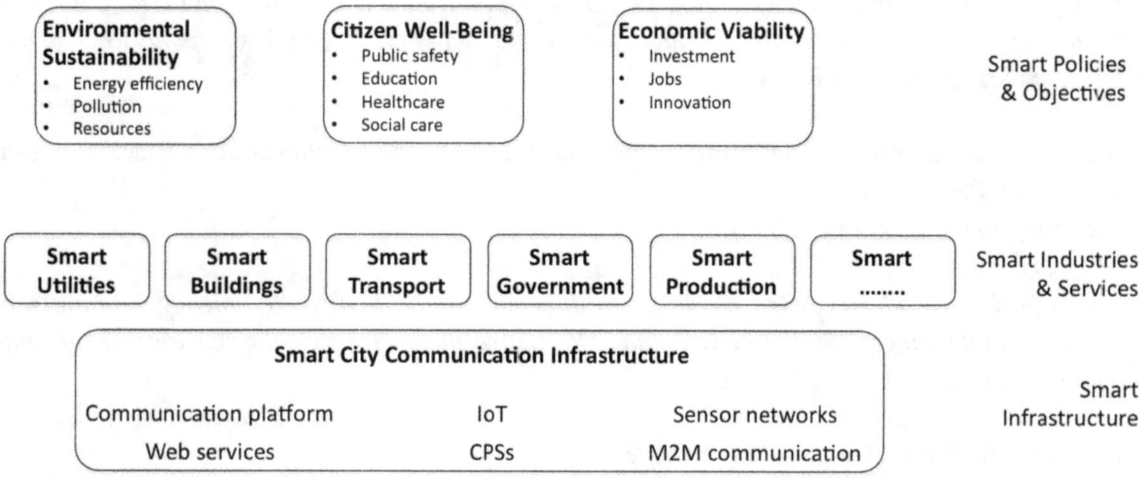

The chapters in this volume do not necessarily address the above topics. But they do help provide a more sound basis for their future analysis. A certain focus is on the development of ICT standards. The resulting insights will still be relevant when it comes to the standardization of platforms and of intelligent infrastructures and smart applications. The same holds for IPR issues, which will not go away in the foreseeable future and for the implementation of standards, where problems are likely to increase with the growing complexity of the standards.

So, over to the papers.

The chapters of the first section look at standards setting.

Cesare Riillo offers answers to the question "Who would like to develop ICT standards?" and discusses "Insights from community survey on ICT usage". He observes that policy makers promote standardization activities to stimulate productivity and growth. To inform policy and to learn about Luxembourgish firms' perception of, and contributions to, ICT standardization part of a national community survey on ICT usage was dedicated to this aspect. To shed some light on standardization phenomenon, the paper explores the features of firms interested in participating in ICT standardization. It turns out the that in Luxembourg a typical firm that is interested in ICT standardization but not currently active in that field belongs to the service industry, uses standards, has a business relationship with public authorities, experiences moderate competition and is not a market leader; the firm's size id not of relevance for its engagement in ICT standards setting.

In the paper "Platform Leaders and Complementors' Strategic Management of Standards: Cases for Complex Products and Systems", Anne Mione and Karim Benmeziane use an example from space industry to show that standards play a crucial coordinating role between different technologies in converging ecosystems. Using Hacklin's strategic profiles they also show that not only technology pioneers, but also ecosystem aggregator and business remodelers can, and do, use standards to implement strategies in a converging industry. These findings indicate that standards are used to manage competitive tensions in combination with strategic alliances

Deploying "Complex Network Perspective on Collaboration in ICT Standardization" Timo Ali-Vehmas' study examines collaboration patterns in five closely related standardization organizations working in the field of mobile communication systems and Internet technologies. Using Social Network Analysis (SNA) it identifies three different types of standardization models in these fields. The collaboration models in 3GPP and OMA are in full alignment, in IETF and in W3C the collaboration networks have strong mutual similarities, and the collaboration in IEEE 802.11 has partly diverging characteristics. The results show that each SSOs collaboration behavior reflects the respective regulatory framework as well as the SSO's specific scope and context. The findings should be of relevance to companies considering engagement in standards setting, for standardization professionals as well as for policy makers.

The paper "Diffusion vs. Knowledge Protection: When Participating in ICT Standardization", by Cesare Riillo and Ivana Mijatovic, explores knowledge patterns in ICT standardization. It considers standards and patents as sources of information for innovation and finds that in modern companies' strategies for managing organizational knowledge may take on different forms that are not mutually exclusive, like safeguarding knowledge through IPR mechanisms and sharing knowledge in standardization processes. Results show that firms are more likely to be engaged in consortia rather than in formal standardization when patents are important. Smaller firms are more motivated by the quest for knowledge compilation during standardization while larger firms are more sensitive to the importance of patents.

Standards wars, typically between large and well-known companies, have frequently been discussed in the literature. In contrast, the paper "Born Global Market Dominators and Implications for the Blockchain Avantgarde", by Knut Blind, Sebastian Fischer and Simone Wurster looks at companies that could establish internationally dominant solutions within a few years after their formation. Based on 22 cases and introducing a new success factors model, this article shows how young companies can successfully establish their technologies as dominant solutions in the global market. Subsequently, the most important success factors identified are discussed in the context of the current Blockchain innovation system. Their respective level of significance for the success of international blockchain innovators is hypothesized.

"A Method for Content Analysis of (Pre-)Standardization Activities" is discussed by Hanah Zoo, Heejin Lee and Vladislav Fomin for "The Case of Cognitive Radio". They developed a document content analysis method to be applied in studies of standardization and technology development. This method integrates two theoretical frameworks: the co-evolutionary technology development framework and the 'D-N-S' (Design, Negotiation, Sense-making) framework for anticipatory standardizing. Applying this method sheds some light on technology development at an early standardization or pre-standardization stage. It also enables an assessment whether or not proposed designs and solutions are described with an adequate level of detail

The chapters in Section 2 discuss an aspect of standards setting that has become more and more important during the past years – Intellectual Property (IP).

Evidence discussed in Marie Barani's chapter "From Patent Hold-Up to Patent Hold-Out?" suggests that there is no patent hold-up. After all, in the mobile phone market, where patent wars have caused considerable upheaval, performance is increasing, prices are decreasing enormously and new companies have no problems entering the market. However, current ruling in the US and changes to a major SSO's bylaws seem to favor patent hold-out, i.e. licensees refusing to pay (F)RAND royalties. The paper argues that this may hamper innovation and competition, thus help reduce or even eliminate standardization and ultimately favor closed proprietary systems.

In his paper "Competing Value Logics in the Apportionment of SEP Damages: A Comparison of FRAND Royalty Methodologies in U.S. Courts 2013-15" Bowman Heiden looks at a related problem. He examines the valuation methodologies accepted in four FRAND cases adjudicated in the US district court system. Based on findings from four case studies, and somewhat contradicting common wisdom, he argues that FRAND commitments are a workable solution to SSOs' IP problems. He claims that the US court system has shown an initial capacity to handle SEP disputes without the need for policy and legislative reform.

The value of an ICT standard depends on whether or not it is actually implemented on a broad scale. Thus, the third Section looks at implementation-related issues.

The chapter by Ahad Zare Ravasan, Mohamma Mohammadi and Homa Hamidi performs "An Investigation Into the Critical Success Factors of Implementing Information Technology Service Management Frameworks: Critical Success Factors of Implementing ITSM Frameworks." It seeks to identify Critical Success Factors (CSFs) that impact on successful implementation of ITSM (Information Technology Service Management) frameworks, with a focus on ITIL. Based on a survey and semi-structured interviews they find 22 CFSs that exert a significant influence on the successful implementation of the ITIL framework. They may be grouped into the categories 'organizational', 'human resources', 'project management', 'managerial' and 'processes'.

Arguing largely from an Open Source (OS) perspective, the paper "On Challenges for Implementing ISO Standards in Software: Can Both Open and Closed Standards Be Implemented in Open Source Software?" by Björn Lundell, Jonas Gamalielsson and Andrew Katz, shows that not all formal standards are open standards and that FRAND commitments (see above) may impose major challenges for the implementation and use of such standards. These include such rather mundane issues like inadequate and inaccurate IPR data bases. A simple yet effective solution is proposed to this. The paper also introduces a conceptual model aimed to clarify the dimensions of 'openness' and 'formality' of standards that may provide an important tool for the analysis and subsequent selection of specific standards. Ultimately, the authors argue that the inclusion of explicit requirements for use of certain formal standards in software procurement may significantly hamper an open and inclusive business friendly ecosystem.

As many companies may hold a different view, the paper by Andrea Fried and Ronny Gey, titled "Metastructuring for Standards: How Organizations Respond to the Multiplicity of Standards", looks at the implementation of process standards in software development organizations. The authors us the concept of technology structuring and apply it to standards as the artifacts to structure. Based on an in-depth single case study the authors identify four areas of metastructuring, four roles for standard mediation and four types of metastructuring activities as ways for an organization to cope with global standard requirements. They also find that metastructuring has a considerable influence on how users perceive, interpret and apply these standards.

The final three chapters fall under the "1000 Flowers" heading – they could not be meaningfully included in any of the other sections.

The first chapter, titled "Impact of Industry Conditions on Innovation: Pre-Existing Standards and Regulations", by Roland Ortt and Tineke Egyedi, highlights the importance of timing when studying the effect of pre-existing standards and regulations on the innovation and diffusion of new high-tech product innovations. The paper looks at the development phases (time interval between invention and first marketable product) and the successive adaptation phases (time interval up to the start of large-scale industrial production) of 50 high-tech innovations since 1850. Results indicate that pre-existing standards and regulations significantly shorten the adaptation phase, but not the development phase.

The former is particularly evident for more radical innovations and for those that are closely linked to a larger technological system.

The second chapter looks at some issues relating to the management of standards and standardization. Entitled "ICT Standards Management: Architectural Challenges and Management Opportunities" and authored by Michael Spring it looks at the historical 'management' of standards, primarily in the US. It finds that, historically, standardization has not been managed so much as it has responded to 'incidents' and 'accidents'. Industrial agreements, cross industry relationships and government control have been incorporated into best practices to enhance the process. The discussion suggests that the current systems for managing standards may not be adequate in areas with many players new to standardization or if standards are appearing very rapidly. In such cases, roadmaps or an architectural framework may be lacking.

Last, but certainly not least, Tapio Levä, Antti Riikonen, Juuso Töyly and Heikki Hämmäinen were active in "Measuring the Deployment of Internet Protocols". The authors note that measuring protocol deployment is important for the identification of any critical factors that may affect a protocol's success. They developed a framework to measure protocol deployment levels and gaps that splits protocol deployment into measurable steps. The deployment process varies depending on the product bundles within which the protocol is included. The framework was applied by measuring how a set of eleven pre-installed protocols spread into handsets in the Finnish mobile market from 2003 to 2012. The illustrative results indicate that protocol deployment is driven by applications, and show the existence of large deployment gaps between protocol possession and its actual usage.

REFERENCES

acatech (Eds.). (2011). *Cyber-Physical Systems – Driving force for innovation in mobility, health, energy and production*. Retrieved from http://www.acatech.de/fileadmin/user_upload/Baumstruktur_nach_Website/Acatech/root/de/Publikationen/Stellungnahmen/acatech_POSITION_CPS_Englisch_WEB.pdf

AIOTI (Ed.). (2015a). *IoT LSP Standard Framework Concepts – Release 2.0*. Retrieved from https://ec.europa.eu/digital-agenda/sites/digital-agenda/files/discussions/aioti_wg3_sdos_alliances_landscape_-_iot_lsp_standard_framework_concepts_-_release_2_0.pdf

Bekkers, R., & West, J. (2009). The limits to IPR standardization policies as evidenced by strategic patenting in UMTS. *Telecommunications Policy*, *33*(1-2), 80–97. doi:10.1016/j.telpol.2008.11.003

Bekkers, R., Bongard, R., & Nuvolari, A. (2011). An empirical study on the determinants of essential patent claims in compatibility standards. *Research Policy*, *40*(7), 1001–1015. doi:10.1016/j.respol.2011.05.004

Blind, K. (2013). *The impact of standardization and standards on innovation*. Retrieved from http://www.innovation-policy.org.uk/share/14_The%20Impact%20of%20Standardization%20and%20Standards%20on%20Innovation.pdf

Borenstein, N., & Blake, J. (2011). Cloud computing standards: Where's the beef? *IEEE Internet Computing*, *15*(3), 74–78. doi:10.1109/MIC.2011.58

Brandenburger, A. M., & Nalebuff, B. J. (2011). *Co-opetition*. Crown Business.

Brismark, G., & Alfalahi, K. (2008). Patent strategies – a fork in the road toward 4G. *Ericsson Business Review*, (3), 62-65.

Caragliu, A., Del Bo, C., & Nijkamp, P. (2011). Smart cities in Europe. *Journal of Urban Technology*, *18*(2), 65–82. doi:10.1080/10630732.2011.601117

CEC. (1996). *Communication from the Commission to the Council and the Parliament on "Standardization and the Global Information Society: The European Approach"*. COM(96) 359 final. Retrieved from http://eur-lex.europa.eu/LexUriServ/LexUriServ.do?uri=COM:1996:0359:FIN:EN:PDF

CEC. (2012). *Unleashing the Potential of Cloud Computing in Europe*. Communication from European Commission COM(2012) 529 final. Retrieved from http://eur-lex.europa.eu/LexUriServ/LexUriServ.do?uri=COM:2012:0529:FIN:EN:PDF

CSC. (2013). *Cloud Standards Coordination*. Final Report. Retrieved from http://ec.europa.eu/information_society/newsroom/cf/dae/document.cfm?doc_id=3988

Davis, J., Edgar, T., Porter, J., Bernaden, J., & Sarli, M. (2012). Smart manufacturing, manufacturing intelligence and demand-dynamic performance. *Computers & Chemical Engineering*, *47*, 145–156. doi:10.1016/j.compchemeng.2012.06.037

EC. (Ed.). (2006). *European Technology Platform SmartGrids. Vision and Strategy for Europe's Electricity Networks of the Future*. Retrieved from https://ec.europa.eu/research/energy/pdf/smartgrids_en.pdf

EP. (Ed.). (2016). *Ethical Aspects of Cyber-Physical Systems – Scientific Foresight study*. Retrieved from http://www.europarl.europa.eu/RegData/etudes/STUD/2016/563501/EPRS_STU(2016)563501_EN.pdf

Ernst, D. (2013). *Standards, innovation, and latecomer economic development: A conceptual framework*. East-West Center Working Paper, Economics Series no. 134. Retrieved from http://scholarspace.manoa.hawaii.edu/bitstream/handle/10125/30723/econwp134.pdf?sequence=1

Estevadeordal, A., & Goodman, L. W. (2017). 21st-century cooperation, regional public goods, and sustainable development. In 21st Century Cooperation: Regional Public Goods, Global Governance, and Sustainable Development. Routledge.

EU. (2012). *Regulation (EU) No 1025/2012 of the European Parliament and of the Council on European standardization*. Retrieved from http://eur-lex.europa.eu/LexUriServ/LexUriServ.do?uri=OJ:L:2012:316:0012:0033:EN:PDF

Farhangi, H. (2010). The path of the smart grid. *IEEE Power & Energy Magazine*, *8*(1), 18–28. doi:10.1109/MPE.2009.934876

Fettweis, G. (2014). *The Tactile Internet*. TU-T Technology Watch Report. Retrieved from http://www.itu.int/dms_pub/itu-t/oth/23/01/T23010000230001PDFE.pdf

Fischer, R., Janiesch, C., Strach, J., Bieber, N., Zink, W., & Tai, S. (2013). Eine Bestandsaufnahme von Standardisierungspotentialen und-lücken im Cloud Computing. *Proc. 11th International Conference on Wirtschaftsinformatik*. Retrieved from http://www.wi2013.de/proceedings/WI2013%20-%20Track%209%20-%20Fischer.pdf

Gawer, A. (2011). *Platforms: Strategy, Leadership and Organization – An overview*. Keynote presented at EuroMOT 2011, Tampere, Finland. Retrieved from https://s3-eu-west-1.amazonaws.com/esrc-files/outputs/T9DpGeKdokaDkureK8C-Fw/oJdW7e8cTUK7qU0Uczx-jw.pdf

Gawer, A., & Cusumano, M. A. (2014). Industry platforms and ecosystem innovation. *Journal of Product Innovation Management*, *31*(3), 417–433. doi:10.1111/jpim.12105

Gunes, V., Peter, S., Givargis, T., & Vahid, F. (2014). A Survey on Concepts, Applications, and Challenges in Cyber-Physical Systems. *Transactions on Internet and Information Systems (Seoul)*, *8*(12), 4242–4268.

Gungor, V. C., Lu, B., & Hancke, G. P. (2010). Opportunities and challenges of wireless sensor networks in smart grid. *Industrial Electronics. IEEE Transactions on*, *57*(10), 3557–3564.

IoT-A. (Ed.). (2013). *Internet of Things – Architecture – Deliverable D1.5 – Final architectural reference model for the IoT v3.0*. Retrieved from http://www.meet-iot.eu/deliverables-IOTA/D1_3.pdf

ISO. (n.d.). *Standards in our world*. Retrieved from http://www.iso.org/sites/ConsumersStandards/1_standards.html

ISO. (Ed.). (2011). *Economic benefits of standards – International case studies – Volume 1*. Retrieved 27 April 2015, from http://www.iso.org/iso/home/store/publication_item.htm?pid=PUB100288

ISO. (Ed.). (2012). *Economic benefits of standards – International case studies – Volume 2*. Retrieved 27 April 2015, from http://www.iso.org/iso/home/store/publication_item.htm?pid=PUB100311)

ITU. (2012). *Recommendation Y.2016 – Overview of the Internet of things: Next Generation Networks – Frameworks and functional architecture model*. Retrieved from https://www.itu.int/rec/dologin_pub.asp?lang=e&id=T-REC-Y.2060-201206-I!!PDF-E&type=items

Iversen, E. J. (2001). Patenting and Voluntary Standards: Tensions Between the Domains of Proprietary Assets and "Public Goods" in the Innovation of Network Technologies. *Science Studies*, *14*(2), 66–82.

Jakobs, K., & Wehrle, K. (2017). *Cyber-Physical Systems – Future Standardization for Europe (CPS-FUSE)*. Unpublished Final Project Report.

Lewis, G. A. (2013). The role of standards in cloud-computing interoperability. In *Proc. 46th Hawaii International Conference on Systems Sciences*. IEEE. Retrieved from http://repository.cmu.edu/cgi/viewcontent.cgi?article=1679&context=sei

Lu, Y., Morris, K. C., & Frechette, S. (2015). Standards landscape and directions for smart manufacturing systems. In *2015 IEEE International Conference on Automation Science and Engineering (CASE)* (pp. 998-1005). IEEE. doi:10.1109/CoASE.2015.7294229

Navigant. (Eds.). (2011). *'Smart City' report*. Navigant Research.

NIST. (2011). *NIST Cloud Computing Standards Roadmap*. Special Publication 500-291. Retrieved from http://www.nist.gov/customcf/get_pdf.cfm?pub_id=909024

NIST. (Ed.). (2013). *Strategic R&D Opportunities for 21st Century Cyber-Physical Systems*. Retrieved from http://www.nist.gov/el/upload/12-Cyber-Physical-Systems020113_final.pdf

NIST. (Ed.). (2016). *Framework for Cyber-Physical Systems, Release 1.0*. Retrieved from https://s3.amazonaws.com/nist-sgcps/cpspwg/files/pwgglobal/CPS_PWG_Framework_for_Cyber_Physical_Systems_Release_1_0Final.pdf

OTA. (1992). *Global Standards: Building Blocks for the Future, TCT-512*. Washington, DC: Congress of the United States, Office of Technology Assessment. Retrieved from http://www.princeton.edu/~ota/disk1/1992/9220/9220.PDF

Porter, M. E., & Kramer, M. R. (2011). Creating shared value. *Harvard Business Review*, *89*(1/2), 62–77.

Swann, G. M. P., & Lambert, R. (2010). Why do Standards Enable and Constrain Innovation? In *Proceedings of the 15th EURAS Conference. EURAS contributions to standardization research* (vol. 2). Mainz Academic Publishers.

Tiemann, M. (2006). An objective definition of open standards. *Computer Standards & Interfaces*, *28*(5), 495–507. doi:10.1016/j.csi.2004.12.003

Tiwana, A., Konsynski, B., & Bush, A. A. (2010). Research commentary-Platform evolution: Coevolution of platform architecture, governance, and environmental dynamics. *Information Systems Research*, *21*(4), 675–687. doi:10.1287/isre.1100.0323

Walden, I., & Gleeson, N. C. (2014). *'It's a Jungle Out There'? Cloud Computing, Standards and the Law*. Retrieved from https://papers.ssrn.com/sol3/Delivery.cfm/SSRN_ID2441182_code1382817.pdf?abstractid=2441182&mirid=1

Yoo, C. S. (2014). *Public Good Economics and Standard Essential Patents*. Faculty Scholarship. Paper 1371. Retrieved from http://scholarship.law.upenn.edu/cgi/viewcontent.cgi?article=2371&context=faculty_scholarship

ENDNOTES

1. 'Co-opetition' (see, e.g., Brandenburger & Nalebuff, 2011).
2. http://ec.europa.eu/growth/tools-databases/newsroom/cf/itemdetail.cfm?item_id=8852.
3. The European Telecommunications Standards Institute.
4. According to a presentation on ICT 30 2015: "Internet of Things and Platforms for Connected Smart Objects", https://ec.europa.eu/digital-agenda/events/cf/info-day-ict4-ict30-fof8-fof9/document.cfm?doc_id=31165 (accessed 22 February 2016).
5. http://www.internet-of-things-research.eu/about_iot.htm.
6. Long Term Evolution.
7. http://ec.europa.eu/research/transport/multimodal/intelligent_transport_systems/index_en.htm.
8. https://ec.europa.eu/digital-single-market/news/h2020-info-day-smart-cyber-physical-systems-digital-automation-smart-anything-everywhere-and.
9. http://www.smart-cities.eu/.
10. See also http://www.smart-cities.eu/.

Acknowledgment

I would like to thank Jan Travers and, particularly, Maria Rhode of IGI Global for their most valuable support throughout the editing process.

Section 1
Setting Standards

Chapter 1
Who Would Like to Develop ICT Standards?
Insights From Community Survey on ICT Usage

Cesare A. F. Riillo
STATEC, Luxembourg & ANEC, Luxembourg

ABSTRACT

Policy makers are promoting standardization activities because standardization can stimulate productivity and growth. To support these policies, a dedicated module of the community survey on ICT usage was designed to investigate the perception of Luxembourgish firms towards ICT standardization. Behaviours and intention of ICT standardization is investigated in two stages. A contingency analysis describes ICT standardization behaviours across industries and size classes. Second, a multivariate analysis identifies the features of firms interested in any area of ICT standardization. The econometric analysis shows that the typical firm that would like to develop ICT standards is active in the service industry, uses standards, is a government supplier, experiences moderate competition, and it is not a leader in its market. Interestingly, large and small firms are equally interested in ICT standardization. Specific policies aiming to provide resources to SMEs that would like but are not active in standardization are particularly advisable.

INTRODUCTION

Standardization, or the process of developing the standards, is recognized by policy makers as a useful tool to foster innovation, competitiveness and growth (e.g. CEN-CENELEC, 2012; EU 2008). the potential impact of standardization is particularly important in the field of Information and Communication Technology (ICT) as standards are necessary allow different items to operate together in a larger system (i.e. interoperability) (Varian 2003).

DOI: 10.4018/978-1-5225-5320-5.ch001

Despite the growingly recognised importance of standards, not many firms are participating in standardization activities. For example, previous research conducted in Luxembourg show that about 2% of all firms with at least ten employees in 2010 are engaged in formal standardization processes (Riillo 2014). Other countries with longer standardization tradition are reporting figures not particularly high (e.g. 11% for Nederland (Blind, Vries, and Mangelsdorf 2012). To better understand the standardization phenomenon, a dedicated module about ICT standardization was designed in collaboration with the "Agence pour la Normalisation et l'Économie de la Connaissance" (ANEC, GIE) and included in the Community survey on ICT usage 2013 for Luxembourg. Policy makers aiming to promote a larger participation in the standardization may be interested to better understand the features of firms more likely to participate in ICT standardization. Firms interested in ICT standardization today should have higher chances to engage in ICT standardization tomorrow compared to firms that are not interested. The current research explores the features of firms not currently engaged in standardisation but that are interested in ICT standardization. Based on the ICT standardization module, firms can be grouped into three categories: (1 those engaged in standardization in any ICT domain, (2 interested but not currently engaged, and (3 not interested.

The research focuses on features of interested firms because they are more likely to engage in standardization compared to firms not interested in any domain of ICT standardization. A dedicated analysis of specific ICT domains is provided. ICT domains are: Cloud computing; Data centre; Telecommunications; Software and system engineering; Security technology; Data management; Electronic signature; E-archiving (ILNAS/ANEC 2012).

This research contributes to ICT standardization research in the following ways. First, it explores both firms' behaviour and perception of ICT standardization on a representative sample of the economy. Second, it describes the features of firms interested in specific domains of ICT standardization on representative sample, at the best of my knowledge.

DATA

The quantitative analysis in this paper is based on a dedicated module in the 2013 wave of the Community survey on ICT usage for Luxembourg (ICT usage 2013). The ICT survey is the official data source for ICT related activities in European Union and Luxembourg. The survey is conducted by the Luxembourgish national statistical office (STATEC). It is representative of the Luxembourgish business economy, including manufacturing and services. The participation to the survey is compulsory. The survey 2013 covers construction industry but does not include financial activities. The ICT usage survey is based on online procedure. The survey is administrated rather on-line or by written questionnaires. Selected firms receive a letter by mail communicating password and login for survey website. Alternatively, the firms can fill and send the questionnaire by mail. The survey collects information on firms' characteristics (e.g. size), perceived market conditions (e.g. competition) and firms' use of ICT. The survey consists of two parts: the first part is common to all European Countries; the second part is country-specific. The ICT 2013 for Luxembourg includes a country specific module about ICT standards and standardization. A firm is defined as "engaged" if it is participating in the elaboration of ICT standards in a recognized body (e.g. ISO, CEN CENELEC, ETSI)" or a consortium (e.g. IETF, W3C, OASIS).[1] (A detailed list of acronyms is provided in the Appendix.)

The exact wording of the question is: "Is your enterprise currently *participating* into the elaboration of ICT standards?" Possible answers are: "a) Recognized body (e.g. ISO, CEN, CENELEC, ETSI) b) Consortia (e.g. IETF, W3C, OASIS)". Additional questions provide information about the interest in standardization in several ICT areas. The exact wording of the question is:

Is your enterprise interested in participating in ICT standardization in any of the following areas?

1. *Cloud computing (e.g. TR 102 997, 103 125)*
2. *Data centre (e.g. EN 50600-1:2012)*
3. *Telecommunications (e.g. ISO/IEC 21481:2012)*
4. *Software and system engineering (e.g. ISO/IEC 20000-1, ISO/IEC 15504)*
5. *Security technology (e.g. ISO/IEC 27001)*
6. *Data management (e.g. ISO/IEC 19503, ISO/IEC 9075-1)*
7. *Electronic signature (e.g. ISO/IEC 7810:2003)*
8. *E-archiving (e.g. ISO 30301)*
9. *Others*

These questions, jointly with the information about firms' characteristics and perception of market conditions, enable the analysis of the standardization attitude and behaviour.

ATTITUDE AND BEHAVIOUR: SOME INSIGHTS

Based on information about participation and interest in ICT standardization, firms are grouped in three categories:

- Engaged (firms participating in ICT standardization recognized bodies or consortia)
- Interested but not engaged (firms not participating but that expressed interest in at least one ICT standardization area)
- Not interested (the rest of the firms)

As shown in Table 1, the approach towards ICT standardization differs significantly among firms. Almost 7% of the firms are engaged, around 13% are interested and 80% are not concerned with ICT standardization issues. We observe that the percentage of firms engaged in standardization raised up to the 7% in 2013 compared to about 2% in 2010 (Riillo 2014). The figure for 2010 is based on the Community Innovation Survey 2010 that includes financial but not the construction industry. Moreover, in the 2010 engagement definition does not include consortia (Riillo 2014). The increasing suggests a growing awareness of Luxembourgish firms towards standardization. Many reasons can explain this trend including the activities conducted by ANEC GIE since 2011. However, there is no evidence to support or discard this explanation. Nevertheless, other countries with a longer standardization tradition show higher percentage. For example, 11% of Dutch firms are engaged in standardization (Blind, Vries, & Mangelsdorf, 2012). Remarkably, at least 445, equal to almost 13% of all firms, report to be interested in at least one ICT standardization domain.

Table 1. ICT Standardization approach of Luxemburghish firms

ICT Standardization	Number of Firms	% of Firms
Engaged	235	7,09
Interested but not engaged	445	13,42
Not interested	2635	79,49
Total	3315	100

Source: Community survey on ICT usage 2013 Luxembourg. Notes: weighted observation; information for 337 firms is missing.

To shed some light on the characteristic of firms engaged in standardisation, Figure 1 shows the breakdown by size. In absolute terms, small firms, with 10-49 employees are the majority of luxembourgish firms (more than 1500) and compared to other size classes, engaged firms are relatively common (almost 80 out of a total of 235 engaged firms). Almost 200 out of 445 of all interested firms are small. With respect to large firms, among almost 300 firms with more than 250 employees, almost 50 are engaged and 40 are interested.

Previous studies (e.g. Blind and Mangelsdorf, 2013) show different behavior of large and small firms. Among small firms almost 5% (80 out of 1500) are engaged while this share is almost 17% (50 out of 300) for large firms. Possible economies of scale or lack of resources among small firms can explain this different likelihood.

While Figure 1 shows the standardization approach towards ICT standardization by size, Figure 2 presents the breakdown by economic activity. Industries are defined according to the European industrial activity classification (NACE code rev. 2). The distribution among industries appears. The highest number of engaged firms are active in trade (55 out of total 235), manufacturing (almost 40 out of 235 total) and ICT industry (almost 36 out of 235). Almost no firm operating in accommodation industry is engaged in standardization activities. The highest number of interested firms is operating in construction and trade industry that are as well the most numerous in absolute terms (both include almost 800 firms). In relative terms, it is worth to point out that almost 16% of ICT firms are active in standardization (about 35 out of about 220), around 13% in manufacturing (almost 40 out of 314) and 13% in support services (25 out of around 200). Other industries have a lower propensity. A dedicated analysis focusing on the propensity of firms interested in standardization is presented in the econometric analysis section.

EXPRESSING INTEREST IN PARTICIPATING: CONTINGENCY ANALYSIS

If the standardization increases the innovation and productivity potential of the economy, the policy makers aiming to achieve these objectives should promote a larger participation in standardization. In this respect, investigating the features of firms not participating but that expressed an interest in ICT standardization is particularly relevant. Therefore, the rest of the investigation focuses exclusively on interested firms which are not currently involved in ICT standardization activities. The following figures show the distribution of interested firms by the size and industry.

As shown in Figure 3 the propensity of a firm of being interested varies little with respect to size. Among firms with 10 to 49 employees, almost 14% declare to be interested in at least one field of ICT standardization. This percentage is almost 17% among firms with more than 250 employees.

Who Would Like to Develop ICT Standards?

Figure 1. Approach towards ICT standardization by size class (number of employees)
Source: ICT usage 2013 Luxembourg. Note: standardization in any ICT area.

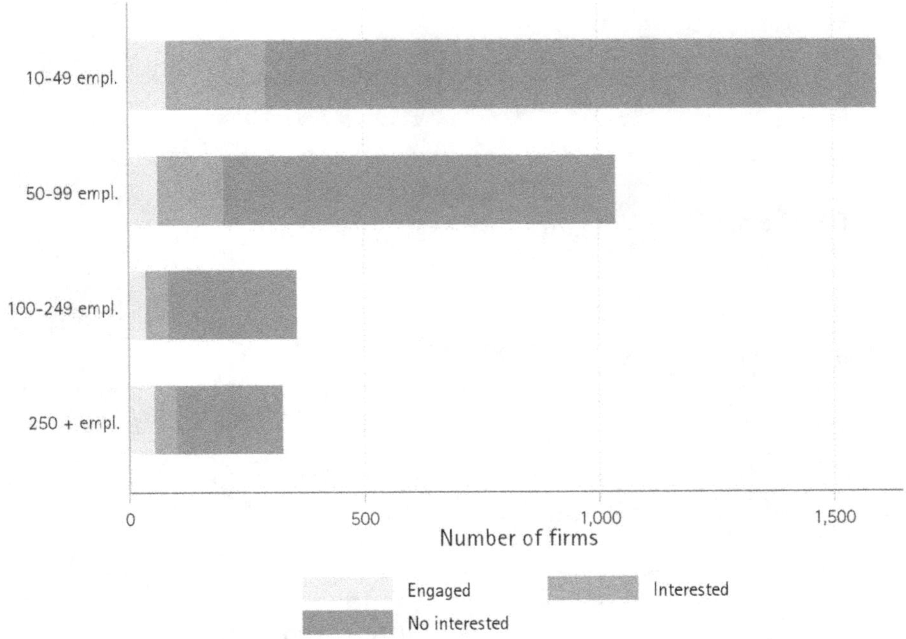

Figure 2. Approach towards ICT standardization by industry
Source: ICT usage 2013 Luxembourg. Notes: Standardization in any ICT area. Manufacturing includes industries C D and E of NACE classification; Professional includes L and M of NACE classification.

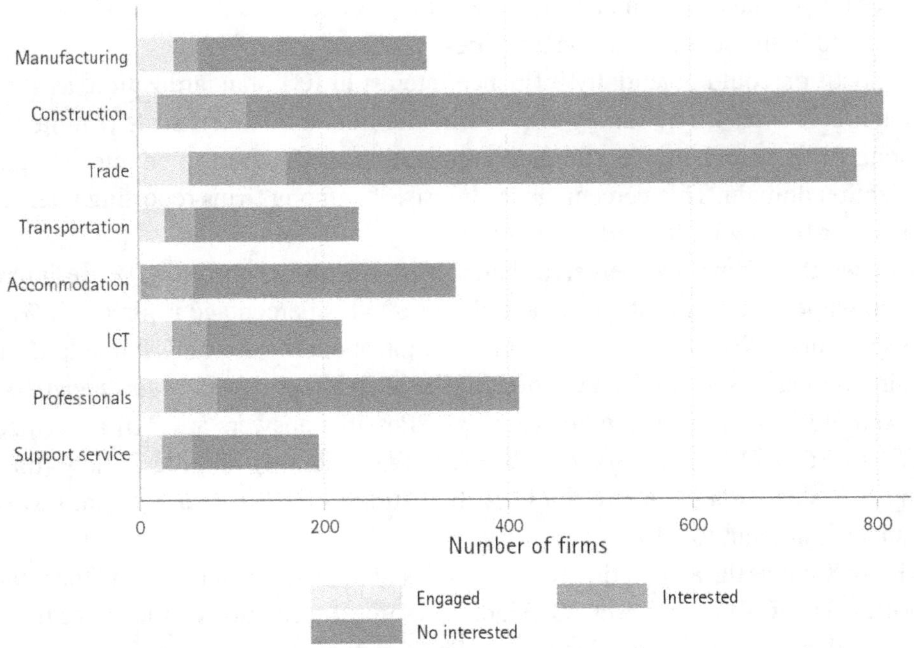

Figure 3. Distribution of firms interested in participating in ICT standardization by size
Source: ICT usage 2013 Luxembourg. Note: standardization in any ICT area. Empl.= employees

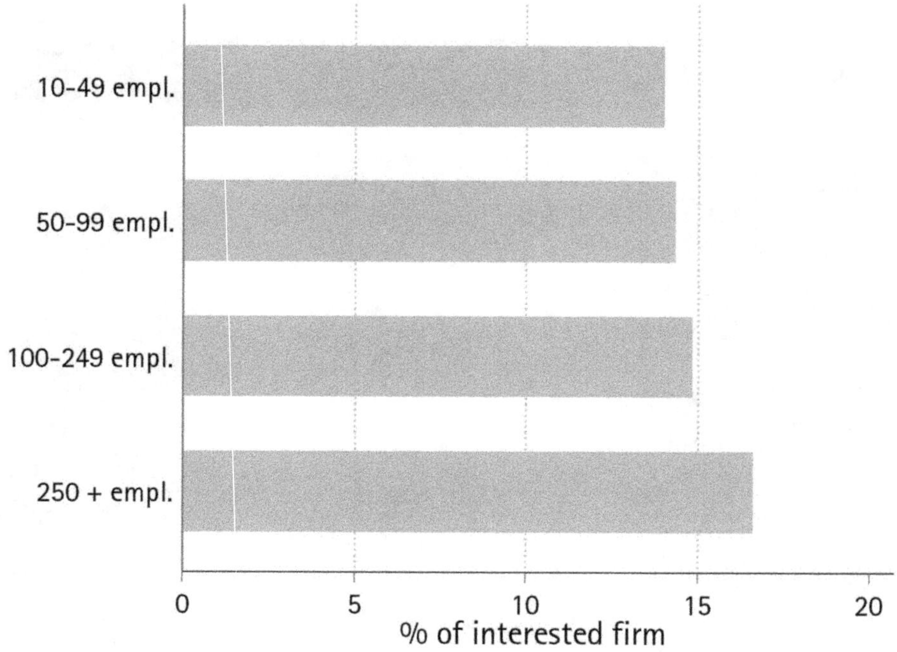

Considering the economic activity of the firm, Figure 4 shows that the percentage of interested firms varies considerably across industries. The share of firms interested in ICT standardization is lowest amongst manufacturing firms (less than 10%) and it is highest among ICT firms (20%) followed by support services (19%) and accommodation (17%). It is worth to note that 13% of manufacturing firms are already engaged in the standardization activities, as shown in Figure 2.

Market competition could potentially influence interest in ICT standardization, as shown in Figure 5. Competition is self-assessed by survey respondent on three-level Likert scale (Limited, intense, very intense). Among firms perceiving very intense competition, 15% report to be interested in at least one ICT standardization domain. This percentage is almost 13% among firms reporting intense competition and around 18% for firms with limited competition.

Figure 6 shows that firms that perceive themselves as leaders, almost 13% are interested in ICT standardization, while amongst challengers and followers, this percentage is almost 15%.

The ICT usage survey 2013 includes some information about the use of ICT standards in the production of its main product or service. The wording of the question is: *"Does your enterprise use any ICT standards producing its main products or services?"* Possible answers are: *"a) recognized body (e.g. ISO, CEN CENELEC, ETSI). b) consortia (e.g. IETF, W3C, OASIS)"* Figure 7 shows that almost 22% of firms using ICT standards are interested in participating in ICT standardization. Firms that are not using ICT standards amount to 13%.

Finally, Figure 8 shows the association between ICT standardization interest and the relationship with public authorities. The ICT usage survey 2013 includes some information on the use of Internet to access documents for tenders to public authorities (i.e. e-Procurement). The exact wording of the question is:

Figure 4. Distribution of firms interested in participating in ICT standardization by industry (NACE)
Source: ICT usage 2013 Luxembourg. Note: Standardization in any ICT area. Manufacturing includes industry C D and E of NACE classification; Professional includes L and M of NACE classification.

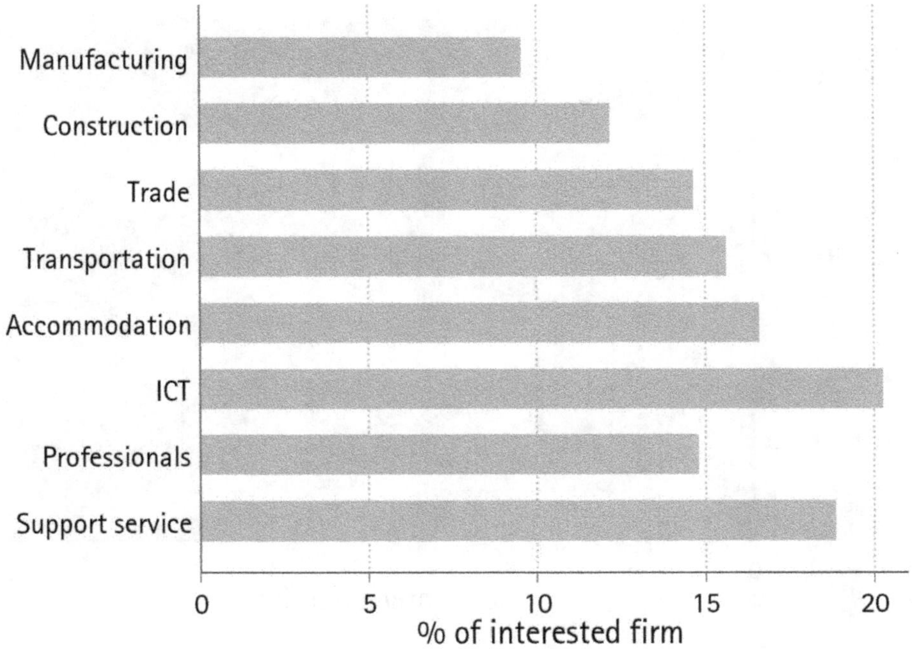

Figure 5. Distribution of firms interested in participating in ICT standardization by competition intensity
Source: ICT usage 2013 Luxembourg. Note: standardization in any ICT area.

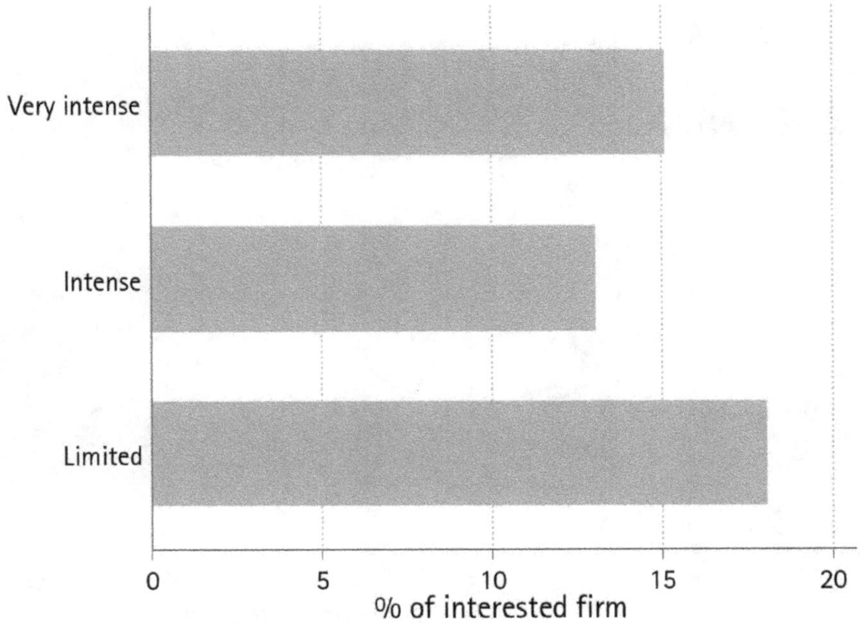

Figure 6. Distribution of firms interested in participating in ICT standardization by market position
Source: ICT usage 2013 Luxembourg. Note: standardization in any ICT area.

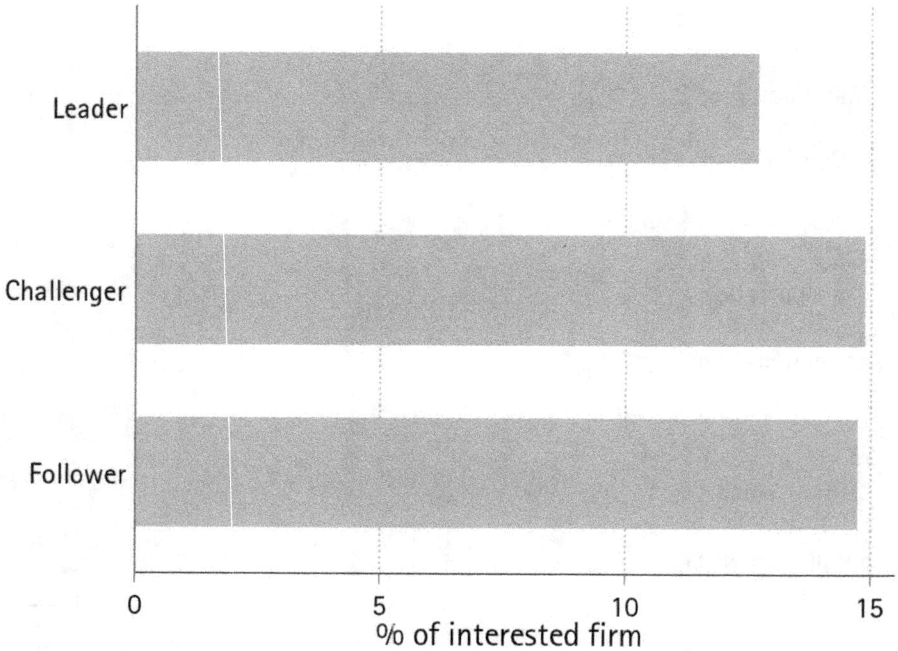

Figure 7. Distribution of firms interested in participating in ICT standardization by use of ICT standards.
Source: ICT usage 2013 Luxembourg. Note: standardization in any ICT area.

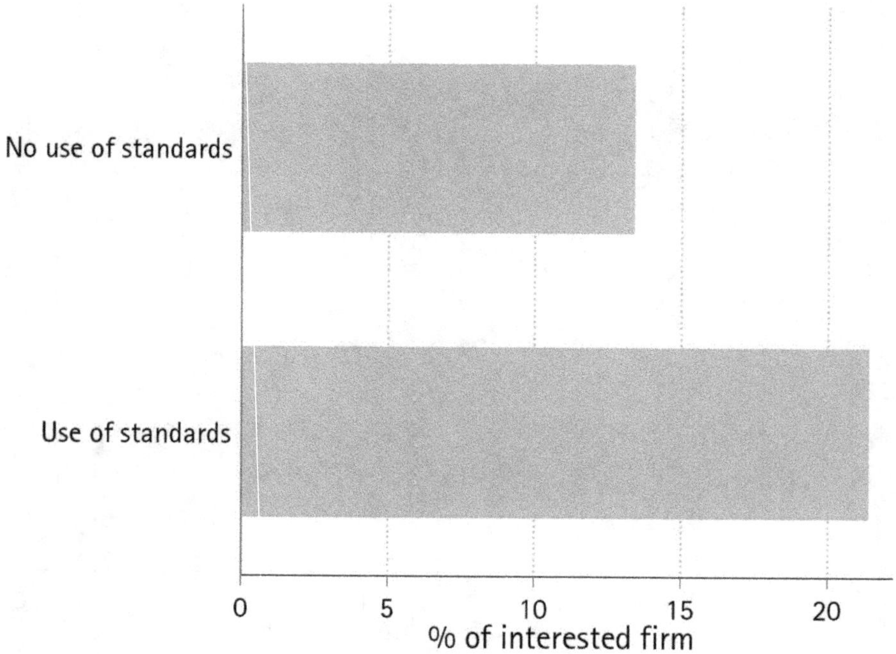

Figure 8. ICT standardization and business relationship with public authorities
Source: ICT usage 2013 Luxembourg. Note: standardization in any ICT area. Rel. = relationship

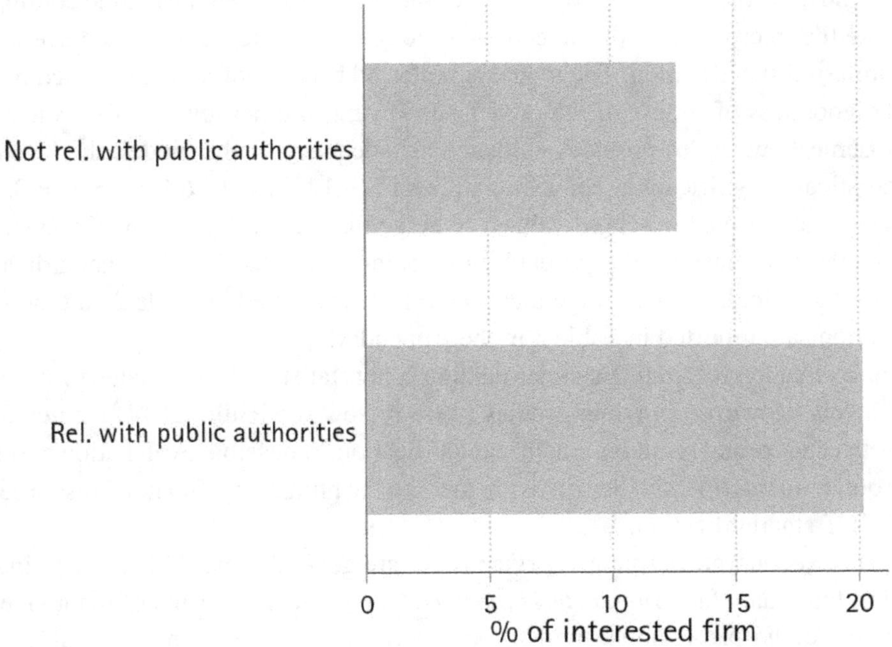

"During 2012, did your enterprise use the Internet for accessing tender documents and specifications in electronic procurement systems of public authorities?". The e-Procurement can be interpreted as proxy variable for identifying firms in a business relationship with public authorities. Looking at Figure 8, it appears interested firms amount to almost 20% amongst firms in a business relationship with public authorities, and 13% otherwise.

ECONOMETRIC ANALYSIS

While the contingency analysis of the previous section inspected the bivariate correlation between different firms features and standardization interest, the econometric analysis accounts for all these features simultaneously. The study explores the probability of the being interested using the logit model (Greene 2008).

Formally, the logit model can be written as:

$$\text{Prob}\left[INT = 1 \middle| X\right] = \frac{e^{\left(X'\beta\right)}}{1 + e^{\left(X'\beta\right)}}$$

INT is the dichotomous dependent variable and it takes the value 1 when a firm is interested but is not currently participating in any ICT standardization and 0 if the firm is not interested. β is the vector of parameters and X is the vector of the features of the firms presented during the contingency analysis.

To facilitate the interpretation of the econometric model, Table 2 reports the average marginal effects of the preferred specification. The model is preferred because of higher adjusted pseudo R-square. Assessing the goodness of fitness of the model, the Hosmer-Lemeshow test for weighted survey data (Archer and Lemeshow, 2006) provides evidence for adequacy of the model. the Hosmer-Lemeshow test is not statistically significant at conventional level $F(9,1473) = 1.26$ Prob $> F = 0.253$ suggesting that estimated values fit well observed values. Average marginal effects show the average impact that the features of the firm have on the probability of being interested in ICT standardization. The logit estimations used to calculate the average marginal effects presented in Table 2 and more parsimonious logit specifications are reported in Table 3 in the Appendix.

Table 2 shows that interest in ICT standardization is not statistically associated to size at conventional significance levels. However, previous studies (for a review see Riillo, 2014) find evidence of strong correlation between size and participation in standardization. A possible explanation of this result is that, despite their potential interest, smaller firms are lacking the human and financial resources to effectively participate in ICT standardization.

With respect to economic activities, service firms are generally more likely to be interested in ICT standardization than manufacturing, *ceteris paribus*. Compared with firms of manufacturing industry, firms in ICT, trade and professional industry are more likely to be interested in ICT standardization (respectively 9, 7 and 6 percentage points).

Considering market conditions, firms that perceive themselves as market leaders are less likely to be interested in ICT standardization than follower firms (5 percentage points). Firms with technological leadership are possibly not willing to cooperate in ICT standardization projects to avoid unintended knowledge spill-overs.

Higher level of competition reduces the likelihood of being interested in standardisation. Firms in competitive markets and very competitive market have respectively 8 and 7 less percentage points than firms experiencing moderate competition. Some considerations may explain why market leaders and firms in high competitive markets are less likely to be interested in ICT standardization. Standardization processes require some degree of trust and coordination among players, firms operating in high competitive markets or firms that are market leaders are probably less likely to cooperate with competitors and then are not willing to engage in the standardization process. However, further research is advisable.

Firms using ICT standards in the production of its main product or service are around 6 percentage points more likely to be interested in ICT standardization than firms not using ICT standards. Remarkably, firms that are in a business relationship with public authorities are more interested in ICT standardization. If public procurement contracts refer to legislation and related standards, then firm in a business relationship with public authorities can have an interest in standardization to anticipate and influence future procurement requirements.

Table 2. Average marginal effects of logit estimates (dependent variable is interest in standardization)

Size	Natural Logarithm of employees	0.0111
		(0.0117)
Industries	Manufacturing	Base
	Construction	0.0230
		(0.0325)
	Trade	0.0726**
		(0.0335)
	Transportation	0.0459
		(0.0433)
	Accommodation	0.0554
		(0.0463)
	ICT	0.0904*
		(0.0506)
	Professionals	0.0625*
		(0.0378)
	Support services	0.0920
		(0.0560)
Market position	Leader	-0.0520*
		(0.0310)
	Challenger	-0.0111
		(0.0277)
	Follower	Base
Competition	Very intense	-0.0677*
		(0.0380)
	Intense	-0.0755**
		(0.0375)
	Limited	Base
Use of standards	Use of standards	0.0612*
		(0.0329)
Relationship with public authorities	Relationship with public authorities	0.0734***
		(0.0254)
Statistics	Observations	1482
	Weighted observation	2921

Source: ICT usage 2013 Luxembourg. Notes: standardization in any ICT area.
Robust standard errors in parentheses;* $p < 0.10$, ** $p < 0.05$, *** $p < 0.01$

CONCLUDING REMARKS

Standardization can be an important tool to promote productivity and growth. Allowing communication and interoperability between different devices, the potential impact of standardization is particularly high in ICT industry. However, not many firms are participating in ICT standardization. To shed some light on standardization phenomenon, the paper explores the features of firms interested in participating in ICT standardization. The investigation is based on a dedicated module of the ICT survey 2013 for Luxembourg. Firms are grouped in three categories: firms engaged in standardization in any ICT area, firms interested but not currently engaged, and not interested firms. The research focuses mainly on interested firms as they may have more chance to participate in standardization than firms with no interest in ICT standardization.

The investigation is performed in two stages. A preliminary analysis describes how firms with different standardization approaches are distributed by industries and size classes. Second, both contingency and econometric analysis is implemented to identify the features of firms interested in any area of ICT standardization. Because of the relevance for policy makers, additional figures show features of interested firms in specific ICT domains.

The econometric analysis shows that a typical firm interested but not currently engaged in ICT standardization is a firm that is active in the service industry, uses standards, is in a business relationship with the public authorities, experiences moderate competition and is not a market leader. It is worth to notice that size does not influence the probability of a firm being interested in standardisation. It is plausible, however, that the reduced resources and not the lack of interest prevent small firms from participating in standardization.

Future research could investigate the obstacles that prevent interested firms from being fully engaged in ICT standardization. Other streams of research could aim to qualify the impact of participation in ICT standardization in terms of productivity and innovation, both at micro and macro level. Future extensions of this research could include a cross country comparison to investigate whether the pattern changes across country.

Despite the explorative nature of the study, some policy implication can be drawn. If policy makers aim to increase the number of firms engaged in ICT standardization, they should keep raising awareness about ICT standardization opportunities and especially target firms with the features identified in the study. Additionally, specific policies aiming to provide resources to interested SMEs are particularly advisable.

ACKNOWLEDGMENT

The author is grateful to Eva Ruckert and STATEC colleagues for useful suggestions. Thanks go to Nicolas Domenjoud, Jean-Philippe Humbert, Hervé Peter and Laurent Wahl for feedbacks and support. The opinions expressed here are those of the author and do not necessarily correspond to those of the institutions of affiliation. Previous version of this work appeared as "ICT standardization: insights from ICT usage survey 2013" in the Cahier Économique STATEC 2014 http://www.statistiques.public.lu/catalogue-publications/cahiers-economiques/2014/PDF-Cahier-118-2014.pdf and in the proceedings of 19th Euras conference 8-10 September 2014, Beograd, Serbia.

REFERENCES

Acemoglu, D., Gancia, G., & Zilibotti, F. (2012). Competing Engines of Growth: Innovation and Standardization. *Journal of Economic Theory*, *147*(2), 570–601.e3. doi:10.1016/j.jet.2010.09.001

Archer, K. J., & Lemeshow, S. (2006). Goodness-of-Fit Test for a Logistic Regression Model Fitted Using Survey Sample Data. *The Stata Journal*, *6*(1), 97–105.

Blind, K., de Vries, H. J., & Mangelsdorf, A. (2012). External Knowledge Sourcing and Involvement in Standardization - Evidence from the Community Innovation Survey. *IEEE International Technology Management Conference*, 1–9. doi:10.1109/ITMC.2012.6306399

Blind, K., & Mangelsdorf, A. (2013). Alliance Formation of SMEs: Empirical Evidence From Standardization Committees. *IEEE Transactions on Engineering Management*, *60*(1), 148–156. doi:10.1109/TEM.2012.2192935

CEN-CENELEC. (2012). *CEN-CENELEC Position Paper on Horizon 2020*. Retrieved September 27, 2017, from http://www.cencenelec.eu/News/Policy_Opinions/PolicyOpinions/PPhorizon2020.pdf

EU. (2008) *Communication from the commission to the council, the European parliament and the European economic and social committee n 133. Towards an Increased Contribution from Standardisation to Innovation in Europe*. Brussels: Commission of the European Communities. Retrieved November 27, 2015, from http://eur-lex.europa.eu/LexUriServ/LexUriServ.do?uri=COM:2008:0133:FIN:EN:PDF

Eurostat. (2009). *High-Technology' and 'Knowledge Based Services' Aggregations Based on NACE Rev. 2*. Retrieved November 27, 2015, from http://epp.eurostat.ec.europa.eu/cache/ITY_SDDS/Annexes/htec_esms_an3.pdf

Greene, W. H. (2008). *Econometric Analysis* (6th ed.). Upper Saddle River, NJ: Prentice Hall.

ILNAS/ANEC. (2012). *Standards analysis ICT sector Luxembourg*. Retrieved November 27, 2015, from http://www.ilnas.public.lu/fr/publications/confiance-numerique/etudes-nationales/standards-analysis-ict-v1-0.pdf

Riillo, C. A. F. (2014). *The Engagement in Standardization Activities A Firm Level Analysis of Formal and Company Standardization*. Retrieved from http://www.statistiques.public.lu/catalogue-publications/economie-statistiques/2014/70-2014.pdf

Varian, H. R. (2003). *Economics of Information Technology* (Unpublished paper). Retrieved September 27, 2017, from http://people.ischool.berkeley.edu/~hal/Papers/mattioli/mattioli.pdf

ENDNOTES

1 Standardization can take place in bodies that are regulated by public authorities (e.g. CEN -European Committee for Standardisation is regulated by European law) or consortia (e.g. W3C -World Wide Web Consortium-) are not regulated by public authorities and are usually they industry organisations.

APPENDIX

Alternative Specifications

Table 3. Model estimates of parsimonious specifications

		(1)	(2)	(3)	(4)	(5)
		Interested	Interested	Interested	Interested	Interested
Size	Ln emp. 2013	0.0984	-0.0836	0.00132	-0.0446	-0.0720
		(0.0773)	(0.290)	(0.299)	(0.308)	(0.327)
Industries	Manufacturing	base	base	base	base	base
	Construction	0.313	-0.301	0.0671	-0.0147	-0.0219
		(0.311)	(1.258)	(1.297)	(1.314)	(1.370)
	Trade	0.538*	-0.457	-0.187	-0.141	-0.0963
		(0.312)	(1.231)	(1.261)	(1.277)	(1.338)
	Transportation	0.555	-1.202	-0.912	-1.007	-1.019
		(0.363)	(1.446)	(1.455)	(1.473)	(1.518)
	Accommodation	0.703*	0.987	0.860	0.523	0.452
		(0.396)	(1.662)	(1.825)	(1.863)	(1.834)
	ICT	0.901**	2.225	2.364	2.062	1.838
		(0.376)	(1.734)	(1.786)	(1.774)	(1.819)
	Professionals	0.558	-0.767	-0.716	-0.738	-0.717
		(0.342)	(1.366)	(1.409)	(1.426)	(1.478)
	Support services	0.788**	-0.179	0.136	0.147	-0.0208
		(0.381)	(1.452)	(1.515)	(1.523)	(1.575)
Size industries interaction	Manufacturing # ln emp.		0.166	0.0784	0.111	0.0805
			(0.332)	(0.340)	(0.347)	(0.364)
	Trade # ln emp.		0.284	0.241	0.230	0.230
			(0.325)	(0.330)	(0.336)	(0.355)
	Transportation # ln emp.		0.461	0.406	0.446	0.444
			(0.367)	(0.369)	(0.376)	(0.392)
	Accommodation # ln emp.		-0.139	-0.125	-0.0286	0.0207
			(0.488)	(0.539)	(0.545)	(0.531)
	ICT # ln emp.		-0.411	-0.419	-0.368	-0.345
			(0.497)	(0.510)	(0.508)	(0.524)
	Professionals # ln emp.		0.391	0.387	0.391	0.387
			(0.368)	(0.376)	(0.382)	(0.396)
	Support services # ln emp.		0.261	0.193	0.211	0.247
			(0.358)	(0.372)	(0.376)	(0.393)

continued on following page

Table 3. Continued

		(1)	(2)	(3)	(4)	(5)
		Interested	Interested	Interested	Interested	Interested
	Leader			-0.358	-0.380	-0.453*
				(0.260)	(0.263)	(0.265)
Market position	Challenger			-0.0311	-0.0583	-0.0899
				(0.217)	(0.220)	(0.224)
	Follower			base	base	base
	Very intense			-0.379	-0.390	-0.528**
				(0.267)	(0.266)	(0.269)
Competition	Intense			-0.446*	-0.479*	-0.593**
				(0.265)	(0.265)	(0.269)
	Limited			base	base	base
Use of standards	Use of standards				0.495**	0.437**
					(0.211)	(0.217)
	Relationship with public authorities					0.573***
						(0.185)
Relationship with public authorities	Constant	-2.612***	-1.945*	-1.833	-1.720	-1.599
		(0.411)	(1.111)	(1.135)	(1.157)	(1.220)
	Observations	1555	1555	1508	1500	1482
	ll_0	-1271.2	-1271.2	-1220.5	-1210.2	-1204.4
	ll	-1260.9	-1253.8	-1198.3	-1183.0	-1166.2

Source: ICT usage 2013 Luxembourg. Notes: standardization in any ICT area; robust standard errors in parentheses; * $p < 0.10$, ** $p < 0.05$, *** $p < 0.01$

List of Acronyms

ANEC: Agence pour la normalisation et l'économie de la connaissance.
CEN: European Committee for Standardisation.
CENELEC: European Committee for Electrotechnical Standardization.
EN: European Standard.
ETSI: European Telecommunications Standards Institute.
ICT: Information Communication Technology.
IEC: International Electrotechnical Commission.
IETF: Internet Engineering Task Force.

ILNAS: Institut luxembourgeois de la normalisation, de l'accréditation, de la sécurité et qualité des produits et services (Luxemburgish institute for standardisation, accreditation, safety and quality of products and services).
ISO: International Organization for Standardization.
OASIS: Organization for the Advancement of Structured Information Standards.
STATEC: Institut national de la statistique et des études économiques du Grand-Duché du Luxembourg.
TR: Technical Report.
W3C: World Wide Web Consortium.

Specific ICT Standardization Domains

While previous sections investigated the firms interested in any ICT domain, this section focuses on the interest in specific domains of ICT to fully explode all available data. The following figures show the percentage of interested firms in each specific ICT domain accounting for size and economic activity. ICT domains are: Cloud computing; Data centre; Telecommunications; Software and system engineering; Security technology; Data management; Electronic signature; E-archiving. For further information please see the "standards analysis ICT sector Luxembourg" (ILNAS/ANEC 2012).

Services are grouped according to knowledge intensity, in line with the aggregation of Eurostat (2009). This classification has the additional advantage of assuring compliance to statistical confidentiality law. The highest percentage of interested firms is reported for E-archiving and E-signature where around 10% of Knowledge Intensive Service –KIS- firms show some interest. Overall, despite some differences across domains, it appears that generally interest increases with size and with level of knowledge intensity.

Figure 9. Distribution of firms interested in participating in E-archiving standardization by size and industry

Source: ICT usage 2013 Luxembourg. Notes: manufacturing is defined as industries C D E F of NACE; LKIS –Low Knowledge Intensive Services- is defined as industries G H I of NACE; KIS -Knowledge Intensive Services- is defined as industries J K L M N of NACE.

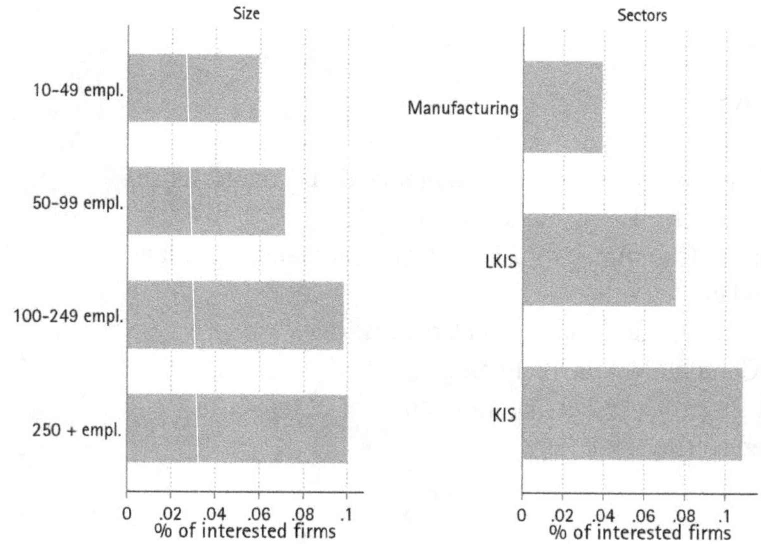

Figure 10. Distribution of firms interested in participating in cloud computing standardization by size and industry
Source: ICT usage 2013 Luxembourg. Notes: manufacturing is defined as industries C D E F of NACE; LKIS –Low Knowledge Intensive Services- is defined as industries G H I of NACE; KIS -Knowledge Intensive Services- is defined as industries J K L M N of NACE.

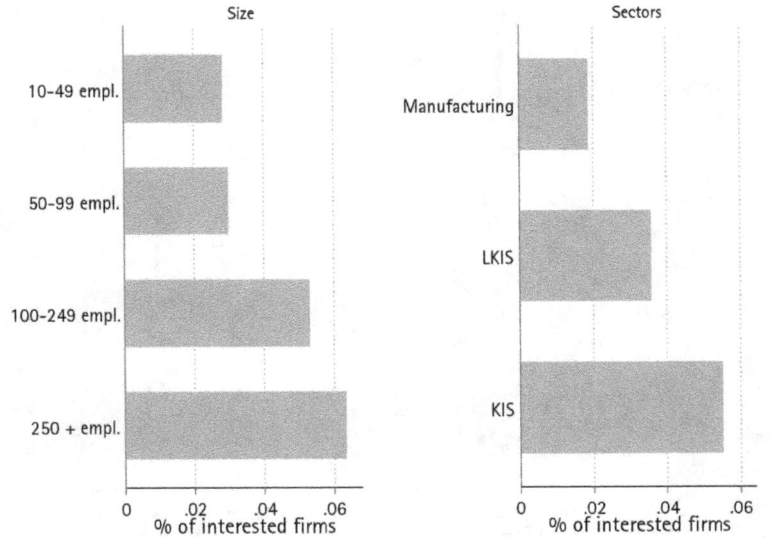

Figure 11. Distribution of firms interested in participating in data centre standardization by size and industry
Source: ICT usage 2013 Luxembourg. Notes: manufacturing is defined as industries C D E F of NACE; LKIS –Low Knowledge Intensive Services- is defined as industries G H I of NACE; KIS -Knowledge Intensive Services- is defined as industries J K L M N of NACE.

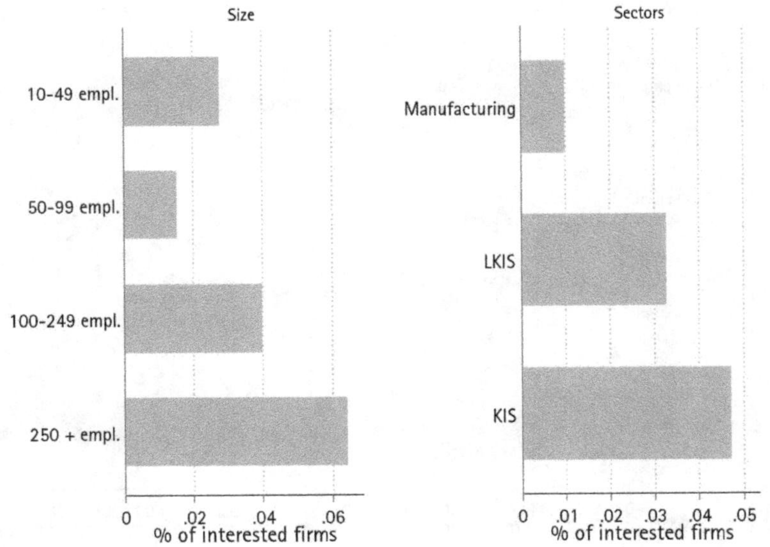

Figure 12. Distribution of firms interested in participating in data management standardization by size and industry

Source: ICT usage 2013 Luxembourg. Notes: manufacturing is defined as industries C D E F of NACE; LKIS –Low Knowledge Intensive Services- is defined as industries G H I of NACE; KIS -Knowledge Intensive Services- is defined as industries J K L M N of NACE.

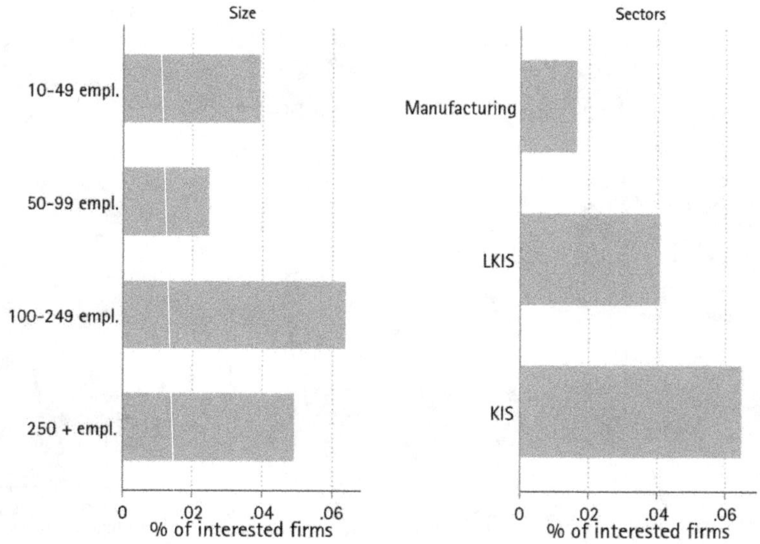

Figure 13. Distribution of firms interested in participating in electronic signature standardization by size and industry

Source: ICT usage 2013 Luxembourg. Notes: manufacturing is defined as industries C D E F of NACE; LKIS –Low Knowledge Intensive Services- is defined as industries G H I of NACE; KIS -Knowledge Intensive Services- is defined as industries J K L M N of NACE.

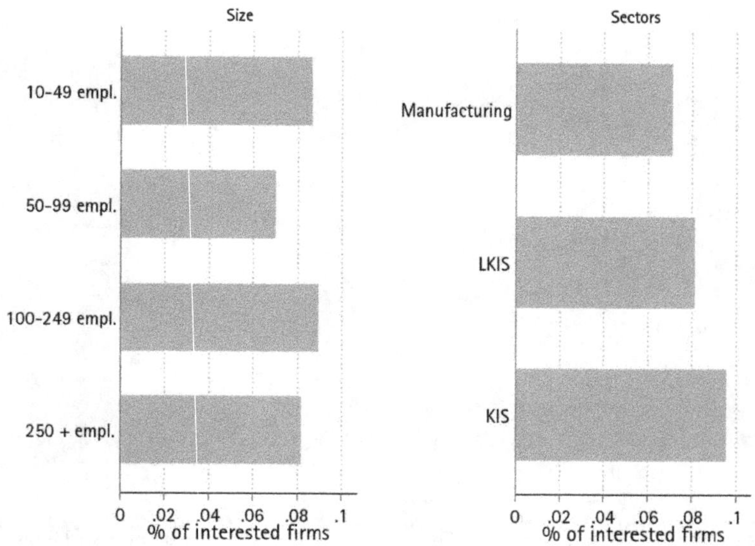

Figure 14. Distribution of firms interested in participating in ICT security standardization by size and industry
Source: ICT usage 2013 Luxembourg. Notes: manufacturing is defined as industries C D E F of NACE; LKIS –Low Knowledge Intensive Services- is defined as industries G H I of NACE; KIS -Knowledge Intensive Services- is defined as industries J K L M N of NACE.

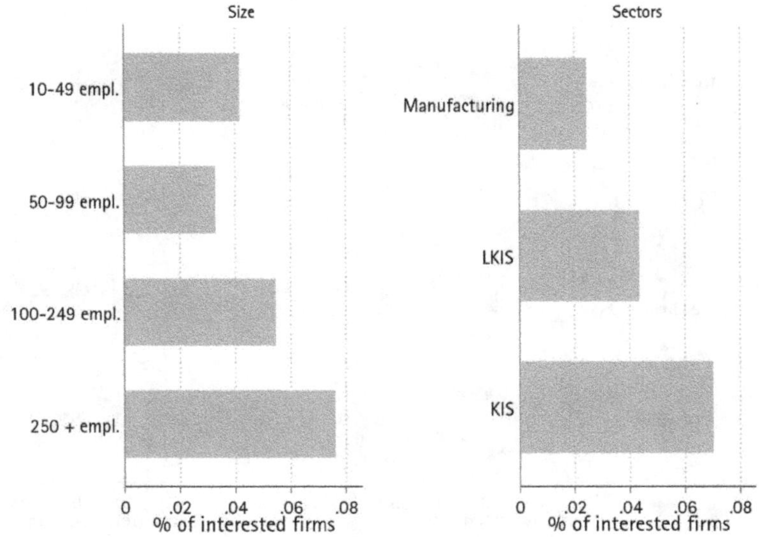

Figure 15. Distribution of firms interested in participating in software and system engineering standardization by size and industry
Source: ICT usage 2013 Luxembourg. Notes: manufacturing is defined as industries C D E F of NACE; LKIS –Low Knowledge Intensive Services- is defined as industries G H I of NACE; KIS -Knowledge Intensive Services- is defined as industries J K L M N of NACE

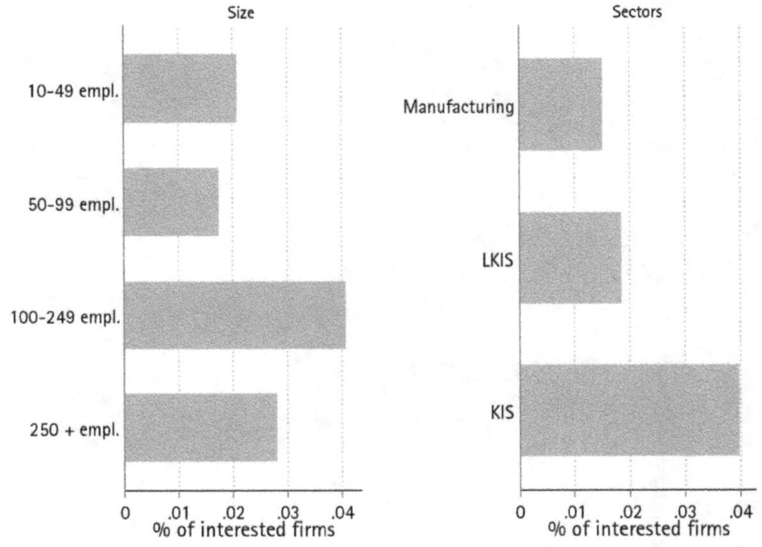

Figure 16. Distribution of firms interested in participating in telecommunication standardization by size and industry

Source: ICT usage 2013 Luxembourg. Notes: manufacturing is defined as industries C D E F of NACE; LKIS –Low Knowledge Intensive Services- is defined as industries G H I of NACE; KIS -Knowledge Intensive Services- is defined as industries J K L M N of NACE.

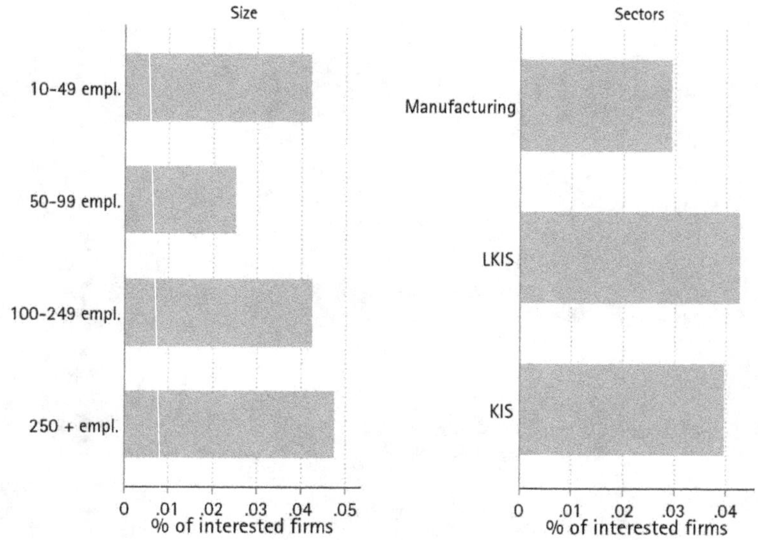

Chapter 2
Platform Leaders and Complementors' Strategic Management of Standards:
Cases for Complex Products and Systems

Karim Benmeziane
Montpellier University, France

Anne Mione
Montpellier University, France

ABSTRACT

In this chapter, the authors describe how companies strategically manage standards in complex products and systems. They explore how standards selection and adoption can leverage the position of a firm as a leader or complementor[1] in a CoPS platform. The authors particularly review the platform leader's management of standards and question whether these standards are used to reinforce its leadership or not. Then, they adopt the complementor's viewpoint and question whether standards adoption and implementation constitute a vehicle for skills acquisition. In other words, the authors address two questions: Does standard selection constitute a strategy to strengthen leadership? Does standard adoption facilitate capability building for a complementor and enable it contesting an installed leadership?

INTRODUCTION

Complex Products and systems (CoPS) are defined as high cost, higlhy customized, engineering-intensive products, systems, networks and constructs which often require several producers to work together simultaneously (Hobday, 1998). In this specific context, standards, and more specifically "local standards", are considered a solution to coordinate the different partners of the project (Steinmuller, 2003). The leader of the project defines and selects a set of specifications, potentially derived from industrial standards to achieve coordination and to ensure complementarity between the components of the complex system. To contribute to the project, the different participants are requested to conform to these

DOI: 10.4018/978-1-5225-5320-5.ch002

standards. To some extent, standards shape the relationships between partners. In this contribution, we aim at exploring this strategic perspective on standards. We refer to Cusumano and Gawer 's description of the platform leader and complementor's positions (Gawer, 2002, 2009, 2010; Gawer and Cusumano, 2002, 2008 2014). Cusumano and Gawer (2002) exposes the strategies from complementors to become leaders or from leaders to maintain their leadership. We explore how standards definition, selection and implementation (de Vries, 1997) can leverage these hierarchical positions. We particularly review the platform leader's management of standards and question whether these standards are used to reinforce its leadership or not. Then, we adopt the complementor's viewpoint and question whether standards adoption and implementation constitute a vehicle for skills acquisition. In other words, does standard adoption facilitate capability building for a supplier or a complementor? How far do these capabilities contribute in the development of new markets or contesting an installed leadership?

Our contribution is based on Complex Products and Systems (CoPS) cases analysis from the space sector. We specifically review space launch development and production activities. The space sector requires converging technologies (between civil and military in the propulsion domain for instance). In such converging ecosystems, Hacklin et al (2013) have described strategic profiles such as technology pioneer, market attacker, ecosystem aggregator or business remodeler. We adopt this typology to refine the role of standards in the description of the strategic profile of platform leaders and complementors.

The first part exposes the strategic management of standards in complex products and systems (Steinmueller, 2003). We expose the different strategic profiles a firm adopts in a high technological converging context and link these profiles to the positions of platform leader and complementor. Then, we explore the decisions made upon standard definition, selection and implementation for platform leaders and complementors. The second part exposes the method. The third part presents the results that will be discussed in the fourth part.

Strategic Management of Local Standards in CoPS

Steinmueller (2003) refers to "local standards" as technical compatibility standards determined through private processes of consultation and through processes of design and problem solving within or between organizations that lead to "privately held" outputs (of the standardization process). These "local standards" are shared within a community whose members are bound by contractual arrangements through their involvement in a particular project. They are particularly important in the context of Complex products and systems. Litterature on CoPS (Steinmueller, 2003, Paoli and Prencipe, 1999; Davies and Brady 2000; Hobday et al. 2000; Davies and Hobday, 2005)) relates to the integration of different components into equipements, equipments into sub-systems, and sub-systems into systems. Behind the integration complexity lie two kinds of issue: the first one, studied by Steinmueller, is how the knowledge supporting the system integration is "acquired, retained and applied". This issue largely relies on the work done on interface standards (David and Greenstein, 1990) which ensure the compatibility of the different levels of artefacts constituting the system. The second issue relates to the decision made during the standard lifecycle as described by De Vries (1997) (e.g. development, selection and implementation stages), especially when those decisions are of strategic nature and leverage the relationships between firms. Our contribution is based on the analysis of the strategic use of standard in the selection and implementation stages in such converging ecosystems. We build on Hacklin et al (2013) definition of strategic profiles such as technology pioneer, market attacker, ecosystem aggregator or business remodeler. We adopt this typology to refine the strategic management of standards of the by platform leaders and complementors.

Applied to CoPS, standards are used as coordination tools and standardization is required at different levels to ensure interoperability of systems in a networked environment (Gerst, 2003). Interface standards particularly play this part in enabling compatibility, complementarity and interoperability within products in the market. Market players influence each other in the standard selection. Actually, when considering compatibility standards in the perspective of coordination, Farrell and Saloner (1988) observed that if an important agent makes a unilateral public commitment to one standard, other agents know that if they follow that lead, they will be compatible at least with the first mover, and plausibly also with later movers. Mione (1994) also noted that the decision to adopt a standard is often made regarding the attitude of the other market participants. The reason why decisions concerning standard selection and implementation are considered relatively to others relates to the compatibility standard literature. Indeed, compatibility induces users to converge around a single technological system rather than support multiple systems, and this convergence forces one technology to become dominant over time (Schilling, 2003). Compatibility can also be seen as embedded in the technological design which enables the product to be used with other products (Suarez, 2004). For these reasons, large, complex technological systems require system-wide compatibility and integration. This strategy is sensitive as a firm is willing to contest a standard and install another one. Doing so, this firm leading a complex system development has to aggregate other firms' skills, at sub-system levels, in order to challenge the supremacy of a competing technology (Soh and Robert, 2003).

This compatibility policy also has a strategic impact on the firm's knowledge and learning process. For firms controlling complex systems, integrating know-how both from outside and within the firm is especially important; the creation of learning, knowledge-sharing, and knowledge integration are critical to business performance (Teece, 2007). Empirically, Schilling (2003) found that continued learning seems to decrease firms' likelihood of lockout. Warner et al. (2006) showed that firms can reduce uncertainty and steer standardizing processes by acquiring relevant knowledge through merger and acquisition. These observations are important to our contribution. Nevertheless, they refer to the leader and question the advantages gained through decisions to enable compatibility and to install a platform. In our perspective, we also take into account the risks generated by this strategy, showing that sharing standards places complementors in a position to challenge the platform leadership.

The Strategic Options of Platforms Leaders and Complementors

Describing converging industries, Hacklin's et al. (2013) identify four profiles: technology pioneer, market attacker, ecosystem aggregator and business remodeler. Technology pioneers are early entrants using their technological advantage as a way to contribute to convergence. Market attackers are also new entrants in the market. They try to turn fragmentation of value chains into commercial and revenues opportunities. The ecosystem aggregators are incumbent. They leverage their competencies and market experience to establish an innovative platform assembling complementary products and services. The business remodelers are also incumbents. They redefine the core business model in well-established networks. Hacklin et al (2013)'s consider defining standards as used by technology pioneers to foster the diffusion on the market (Weitzel et al., 2006). Standards are only mentioned as a tool for technology pioneers to install innovation. They implement standards to demonstrate the potential of their technology on the market and organize the conditions for customer's early adoption. We argue that standards can also constitute strategic tools for the other profiles of the taxonomy. Specifically, we consider business remodeler and ecosystem aggregator as instructive to expose a strategic use of standards.

We particularly consider these profiles as illustrative of a platform leader's role (Cusumano and Gawer, 2002). According to the authors, platform leaders are companies that drive industry-wide innovation for an evolving system of separately developed pieces of technology. Most platform leaders do not have the capabilities or resources to create complete systems by making all the complements themselves. They need to collaborate. The combined efforts of platform leaders and complementary innovators increase the potential size of the pie for everyone (Cusumano and Gawer, 2002). However, the platform leaders face two difficulties: they have to coordinate the contributors to the platform, suppliers and complementors and interact with competitors. Cusumano and Gawer (2002) define four elements of platform leadership specific positions: scope, product technology, relationships with external complementors and internal organization. Product technology management is about decisions made on the architecture of a product. In particular, managers need to decide how much modularity they want, how open their interfaces should be, and how much information about both platform and interfaces to disclose to outsiders who might become complementors or competitors. In our opinion, this type of choice refers to standard selection.

Strategically, platform leaders are challenged by "wanabees" (firms which want to take the leadership). In this situation, Cusumano and Gawer (2002) observe that it is not always clear if today's complementor will become tomorrow's competitor. They also argue that to be effective in the long term, platform leaders need to pursue two objectives simultaneously: first, they must seek consensus among key complementors about what technical specifications and standards will make platforms work with other products. Second, they must influence partners' decisions affecting how well everything works together through new product generations. Finally, the author observes that some companies play multiple roles. In our contribution, we aim at clarifying these elements.

METHODOLOGY

We choose to study the space launcher Ariane 5 through an in-depth case analysis. In order to understand the institutional and economic contexts, we collect secondary and primary data. We review the professional documents such as technical specifications, industry standards, and Space Agencies' documents which are relevant to the space sector. We also collect data on the corporate strategies of the different firms involved, tracing them from the 60s. These documents inform on the firm in charge of the technical specification. We produce a monographic view tracing the evolution of a particular set of standards used for the development and production of the Ariane space launcher family.

A second step requires understanding the strategic objectives of the firms. We realize 10 interviews lead with the major space actors. These interviews are listed in table 1. Our objective is to understand their strategic use of standard in order to identify a specific role according to the different profiles.

Case Description

The cases studies are related to the development and production of the Ariane 5 system, a space launcher that is considered today as the main key of European access to space. In order to qualify the strategic profiles in the European space launch activity, we focus on the main industry actors involved in the development and production of Ariane 5.

Table 1. List of semi-directive interviews

Function	Organization code	Case	Length (min)
Industrial Policy Manager	A	Space launcher and satellites interfaces	98
Structure and mechanisms manager	A	Space propulsion	29
Project Manager	C	Space propulsion	118
Project Manager	C	Space launcher and satellites interfaces	90
Standards and requirements manager	D	Space launcher and satellites interfaces	45
Quality Manager (ex-propulsion expert)	A	Space propulsion	98
VP Quality and General Inspection	A	Space launcher and satellites interfaces	137
VP Quality	B	Space launcher and satellites interfaces	114
Propulsion Manager	A	Space propulsion	127
Director	E	Space launcher and satellites interfaces	60

The Context

After being thought as a human spaceflight launcher in the late 80s as, Ariane 5 was designed to serve the commercial market (e.g. to launch telecommunication satellites for private operators). The development of a new space launcher usually last for 10 years. Therefore, it is only in the late 90s, and after the first launch failure of Ariane 5 G (e.g. a Generic version of Ariane 5) that two other political decisions were made: the first one was to adopt a design to cost approach for developing and producing a second version of Ariane 5 (namely Ariane 5 ECA) and the second one was to give the industry the responsibility of leading the project. This responsibility was given to Aérospatiale (now Airbus Defense and Space, a division of Airbus Group). As the industrial architect, Aérospatiale had to propose the industrial organization in order to develop Ariane 5 ECA to meet the needs of its customers.

Aérospatiale as an Ecosystem Aggregator, Calling for Complementors

In this context, Aérospatiale endorses the role of ecosystem aggregator. According to Hacklin et al (2013), the ecosystem aggregators are incumbent. They leverage their competencies and market experience to establish an innovative platform aimed at complementary products and services. To implement the innovation, Aérospatiale calls for complementors. First, Aérospatiale, as an ecosystem aggregator, does not disrupt the industrial organization, keeping the former one, used for Ariane 5 G and previous launcher development and production. Then, it faces several technology challenges that force it to leverage on complementors capacity to innovate, especially in the propulsion domain. The main technological challenge is to be able to produce enough thrust to orbit a 10 metric tons payload (thrusters were 10 times more powerful than Ariane 4 ones).

Space engine are developed by firms with complementary skills: the engine manufacturers and the propellant producers. Starting with these types of key players, we observe the industrial organization of two solid rocket boosters. The first one is the P230 elaborated during the development of Ariane 5 G. The second one is the P80 mounted on another European space launcher which is VEGA. The first case displays the collaborative relationships between BPD (now AVIO) and SEP (now HERACKLES, a subsidiary of the SAFRAN Group) as complementors. The second case shows the evolution of the status of AVIO from complementor on Ariane 5 P230 to platform leader on VEGA.

SNECMA as a Business Remodeler: The P230 Case

P230 prime contractor is a joint venture, EUROPROPULSION, created in 1985 by two companies, SNIA-BPD (now AVIO) and SEP (which will become SNECMA and then HERACKLES). This engine is the main constituent of Ariane 5 boosters. Those boosters are assembled in Guyana in a dedicated facility (called BIP for the French Bâtiment d'Intégration Propulseur) with the rest of the solid propulsion stage. SNIA-BPD is in charge of delivering the thermal protection, the ignition system and the loading of the front segment of the thruster. The two other segments (rear and medium ones) are left with another joint venture created in 1988 between SNIA-BPD and SNPE (whose propulsion activities are bought by SNECMA later on). In its area, SNECMA is responsible of producing the nozzle and has the responsibility the empty thruster, task that is externalized to a German firm, MAN.

In this particular case, we are interested in understanding the behavior of SNECMA, considered as a business remodeler since they developed thrusters 10 times more powerful than their previous versions, in less quantity for a comparable price. The novelty of the business model is that the solid stage is actually co-developed. SNECMA is a business remodeler since it doesn't not alter the usual business model where the propulsion system is developed by a unique firm. The two business models actually co-exist. We show how they use their relation with AVIO in order to face such a challenge and develop the new business model.

AVIO as Complementor in Ariane Production and Platform Leader of the Launcher VEGA: The P80 Case

P80 prime contractor's role is also left with EUROPROPULSION. The activities are split the same way than in the P230 case with few modifications: the ignition system is delivered by a Belgium company SPE, and the electrical system is done by SABCA, a firm from Netherlands. In addition to its contribution in EUROPROPULSION, AVIO is also the prime contractor for the development of the launcher VEGA through another joint venture, ELV (for European Launch Vehicle), created in 2000 between AVIO and the Italian Space Agency. The joint venture ELV calls for complementors able to produce a series of innovative technologies in order to achieve its primary goal: to develop a low cost and reliable launcher equipped with an innovative propulsion system.

In this particular case, we are interested in understanding the behavior of AVIO, considered at the same time as complementor in the Ariane 5 production and leader in the development of the new launcher VEGA.

Data Analysis

In order to conduct the analysis, we process our data using a multi-nominal coding, identifying commonalities, differences and links between codes (Jorgenson, 1989). To establish the codes, we follow Savall et al. (2008)'s guidelines. First, we use descriptive information and relate them with each code. We used Nvivo 10 to produce the coding. Our findings come from the interpretation of the different types of code (Dumez, 2013).

RESULTS

Our objective is to identify the roles played by different actors in the development and production of CoPS, and study how standards are used by the different actors according to their role in Hacklin's taxonomy and positions as platform leaders and complementors. The result of this qualification is summarized in table 2. In addressing the strategic issues, we are led to refine our definition of standard, according to the specificity of CoPS standardization. Then, we show that standard selection and implementation are drivers for skill building. We also explain how standards reflect leading position when firms try to aggregate ecosystems. The next result is that firms use standards along with alliance strategies in order to manage competitive tensions induced by those alliances. Finally, we show how firms can adopt standards in order to change their position.

Platform Leaders and Complementors as Ecosystem Aggregator and Business Remodeler

Aérospatiale endorses the role of ecosystem aggregator at different levels, as it organizes the coordination of activities based on different technologies. Within these different technologies, we identify two different motors (P230 and P80). Europropulsion is the platform leader for both motors. Europropulsion is a joint venture between AVIO and SNECMA. SNECMA (SEP at that time) endorses the role of business remodeler as it organizes the co-development of the solid stage of Ariane 5 G. Thus, we observe that the market architecture is complex and structured like Russian dolls. In this context, platform leadership can be observed at different levels as shown in table 2.

In the same way that Intel, as chipset manufacturer, is a platform leader in the computer industry (because its technical choices leverage the entire architecture of the computer), the space launcher activity shares a common view based on standardized chemical propulsion design:

Table 2. Role and position of actors according to system and sub-system level of analysis as regards to Ariane 5 G and Vega programs – authors

	Role and position →	Ecosystem Aggregator		Business Remodeler
	Level of analysis ↓	Platform leader	Complementors (for propulsion systems)	-
ARIANE 5 ECA	At system level	AEROSPATIALE	EUROPROPULSION	-
	At sub-system level (P230)	EUROPROPULSION	SNECMA AVIO MAN REGULUS	SNECMA
VEGA	At system level	ELV	EUROPROPULSION	ELV
	At sub-system level (P80)	EUROPROPULSION	SNECMA AVIO REGULUS SPE SAFT SABCA	-

for a launcher system design, you don't have 36 standards. Today, we have chemical space launchers, so I would say, it is the baseline: you choose a chemical substance and you use its potential energy to transform it into kinetic energy.

This shared design can be adapted into different solutions. These solutions differ because the chemical substance can be more or less energetic while delivering the thrust for the space launch operations:

On the first stages of the launcher, we used heavy chemistry: it can be solid rocket booster on Ariane 4, or liquid propulsion using UDMH and L2O4 on Ariane 4. On Soyouz [one of the Russian space launcher], the design is the same but the solution is different as it uses kerosene and oxygen. Then, on the upper stages, you can have liquid oxygen and liquid hydrogen mixes.

This means that propulsion solutions have a direct impact on the architecture and performances of the space launcher, which gives a status of platform leader to the firms responsible in developing or producing space engines. Platform leadership can thus be located at the integrative level which is traditionally devoted to the leader, but also at the sub-system level which corresponds to complementors.

Finally, not only ecosystem aggregators as platform leaders, but also as complementors integrating sub-systems, define, select or implement a set of standards. This may be explained by the specificities of the CoPS standardization.

System Complexity Fosters the Definition of a Technical Standard Portfolio

The CoPS local standards we are observing present some differences with the technical standards usually described in the literature.

Actually, the development and the production of space launchers are complex activities meaning that they are broken down into several activities. Each activity should be reproducible in order to develop know-how within the firm, ensure compatibility with other activities, exchange information with other organizations or simply reduce production costs. For instance, the joint venture EUROPROPULSION ensures the integration of the solid rocket booster of Ariane. It "receives" from their complementors several equipments: an empty metallic structure, a nozzle, a tank filled with solid propellant that will go inside the metallic structure. These firms should also break down their product into several parts: the tank includes three different segments that are produced in different facilities (in French Guyana and in Italy), the metallic structure is done in Germany and is also separated into two parts etc. Technical specifications are used to ensure compatibility between the sub-systems or for the different sub-systems to meet the right level of reliability. Some of these technical specifications are kept within the firm as know-how; others are shared within the supply chain in order for the project to be conducted in a way that all the requirements from the customer can be verified. These shared technical specifications are not de jure or de facto standards and we will refer to as "industry standards". This kind of standards are defined and used as a set. They belong to an intermediary level between internal standards and standards defined by industry consortia. Depending on the activities, the firm will have to define, select and implement a series of industry standards, assembling them in a portfolio which reflects the perimeter of the technical activities they need to share in order to participate to the development and production of a complex system.

In the literature, technical standards are usually studied according to their function (compatibility, interoperability, quality…), the level of applicability (voluntary versus mandatory, used for certification…), their effect on the market (with regards to switching costs, network effects, diversity of products…) and the level of consensus (de facto standards versus de jure standards). When we consider "industry standards", it implies technical specification used by a particular community. The community does not keep the set of standards for its own use (Weber, 1964), but the specifications are not intended for a broad use because of the know-how they contain. The members of the community are tied by contractual arrangements (prescribing normative behavior, Ortmann (2010)) and collaborate to contribute to the development and production of the complex system. This community does not constitute a typical "industry consortia" but an industrial organization that depends on the project they are involved in. The "industry standards" we refer to either provide the necessary thresholds to meet reliability requirements, the definition of the interfaces between subsystems, equipment, units and components, and help formalizing all the necessary information with respect to project management.

An example of these "industry standards" is the specifications of the space launcher Ariane. This set of standards prescribes the requirements to comply with for the management activities (SM) and technical activities (SG). For instance, the management of standards defines the project breakdown structure through which "projects responsabilities and firms relationships are described". It also requires a firm to answer its customer's specification through a management plan explaning the rationale for project implementation (including waiver requests). On the other end, these technical are of two kinds: the first one applies to all the firms involved in the project are related to the qualification of the products. The other kind is more related to data collection and lessons learned for each engineering specialties (e.g. mechincal, thermal, pyrotechnics, and so on…) that are working on the project. The content of such standards are know-how intensive. We argue that this standard portfolio is used by complementors in order to build new skills. We also observe that the entire set of standards (management and technical) is different for the development activities (SM-D, SG-D) and production activities (SM-X, SG-X), the later being derived from the first ones. Such a portfolio of standards, used by the entire Ariane community (around a thousand firms), is a solid indicator that reflects the complexity of a launch system.

Firms Use Industry Standards in Order to Build New Skills

From a strategic standpoint, Avio and Herackles are the result of a complex vertical and horizontal integration aimed at consolidating the solid propulsion activity in Europe. More than 20 acquisition operations, merges and stakes were conducted in a 30 years period to create Avio and Herackles. Those operations, which questioned the industrial organization, induced several managerial issues among which the capacity of the newly created firms to be able to work altogether despite their differences in skills, cultures and workflows. One of the solutions implemented was to share a set of industry standards. The firms discussed and harmonized their standards to foster collaboration while developing and producing their products. Indeed, we observe two situations, one being the consequence of the other: First, AVIO (at that time BPD) is led to use the standard from its French counterpart, HERACKLES (at that time SEP) in order to collaborate within EUROPROPULSION: "*While we gave them these standards, we told them: if you use these standards, you have a good chance to meet the needs*". Then, AVIO implemented the standards in order to gain the necessary skills: "*They had to modify their own internal standards*". Since they were doing so within EUROPROPULSION which was the leading entity for the entire solid rocket booster, they also learned how to properly manage the sub-system development and, after that, production. The main challenge for AVIO was to "adapt their methods of work in order to be able to produce a sub-system that was 10 times bigger than what they used to work on." This result shows that by selecting and implementing the standards of its complementor, a firm can develop new skills that help it to develop and produce an entire and innovative sub-system.

Industry Standards Reflects the Leading Position of Firms

When studying the space launcher activity, we have to differentiate different levels of platform leadership. From a managerial standpoint, the two firms, AVIO and HERACKLES are platform leaders at the propulsion system level (which is one sub-system of the rocket). In collaborating throughout the joint ventures EUROPROPULSION and REGULUS, they have to agree on a how they could share the activities and related responsibilities: *"In all our management specification, we had rules on interfaces management. And in both technical specifications, we defined interface allocations, so each one knows who does what*". This collaboration relies on the resources available when the joint ventures were created (in the late 90s). They define their obligations to each other and to common suppliers. They also specify the product tree and technical interfaces. All these activities are conducted with the agreement of the rocket platform leader, Aérospatiale: "*When we needed to talk with each other we knew how to do it. But often, our industry customer, above us, was doing the interface control*".

Our finding is that the platform leadership for the propulsion sub-system is given to a joint venture. We observe a situation where complementors are also platform leaders (at a sub- level from Aérospatiale viewpoint, the rocket platform leader), not only because of the nature of their activity but also because of the decision made by the two firms (AVIO and HERACKLES) to consolidate solid propulsion activities in Europe. We also notice that the platform leader at rocket level (Aérospatiale) adopts the same behavior with another complementor (Air Liquide) to produce another stage of the rocket (the cryogenic stage) by creating a joint venture named CRYOSPACE. We assume that EUROPROPULSION is not an isolated case. The final result is that the roles of complementor and platform leader are not mutually exclusives, a firm being able to be either one, depending on their resources, skills and alliance strategies. This flexibility is partly based on the fact that firms can rely on a stabilized design (a design not being dominant since there are no eviction from the market) in order to develop their technology, share common industry standards and collaborate with each other. Looking at what is shared and standardized sheds light on different levels of platform leadership.

Firms Use Alliance Strategies and Industry Standards to Manage Competitive Tensions

For geographic return reasons, France and Italy are associated in the development of the solid rocket boosters of Ariane 5. This collaboration leads to the creation of two joint ventures. The first one, REGULUS operates the propellant facility in French Guyana (namely UPG for Usine à Propergol de Guyane). This facility is responsible in producing several parts of the booster. The second joint-venture operates the facility that integrates the solid rocket booster of Ariane 5. REGULUS is a complementor to one of its creator, SNECMA (which produces the nozzle of the solid stage). EUROPROPULSION is the leader of the sub-system platform. These forms of cooperation are deliberate since they allow collaboration, activity partitioning and industrial return:

[The joint venture REGULUS] is formed by two companies that were in competition, namely SNPE [today HERACKLES] and BPD [today AVIO]. The creation of this new company was the consequence of reconciliation of work methods. It can easily be separated from its parents, and there was also a need for partitioning [activities].

This shows that collaboration is not direct, especially when two competitors are affected. It can be done through the creation of joint ventures whose role and position are given by the collaborating firms. This is made possible because the firms reconciliate their working methods by the implementation of industry standards. However, in order to make the joint venture fully operational, the firms have to decide their role and position: complementor in the case of REGULUS, platform leader in the case of EUROPROPULSION. These decisions are made to answer to institutional constraints (geographic return) and to manage competitive tension (activity partitioning).

Industry Standards and Skills Building Help Firms to Challenge the Dominant Position

Ariane 5 boosters' co-development relied on common industry standards used by SNECMA and AVIO. This situation allows AVIO to build new skills, at system level, in order to develop an entire launcher. In doing so, AVIO comes from sub-system platform leader through its participation in EUROPROPULSION to system platform leader through its participation in ELV (European Launch Vehicle) joint-venture with the Italian Space Agency. Competitive aspects firstly relates to the development of system skills within AVIO: "*They won the system skill that would have allowed them to capture the Ariane market*". Then, AVIO uses its skills to modify its own internal standards to gain the necessary capacity to develop an entire sub-system (the P80 engine) and even the entire system through ELV and the VEGA launcher. Doing so, it becomes competitor, at system level, with Airbus Defense and Space. Our last result is that a firm uses standards and skill building to create new market opportunities and challenge leadership position.

DISCUSSION

Our observations entail considerations concerning standards in CoPS. We base our work on Steinmuller (2003) who considers the role of standards in CoPS. Steinmueller argues that technical compatibility based on standards support system integration by allowing a sound division of labour and by defining the interfaces connecting the components and sub-systems of a CoPS. This situation relies on how the coordination between the differents firms is managed and how strong is the coupling between technologies used in the design of the entire system. Considering the "coupling tightness" of the different technologies involved in the system design allowed Steinmueller dig into the problem solving issues that occur between firms participating in the design and development of a complex system, namely the negociation process between integrators and suppliers. Steinmueller further argues that the solutions of inter-organisational issues are somehow distributed and "stored" by systems integrators and component, and subsystem suppliers through a disintegration process. Recalling this accumulated knowledge is, from Steinmueller's viewpoint, a central role of standards.

Our contribution builds on these findings. It confirms the central role of standards. Nevertheless, these issues have been considered from an organisational viewpoint. We argue that considering standards in a CoPS context can help understanding the strategic stakes of compatibility standards. First, coordination is, from our perspective, not only related to the division of labour but also to the influence a firm is able to build on the technological choices made during the system design phases. Secondly, the division of labour in CoPS is not only a vertical issue between integrator and supplier since the firms with system integrator skills can also be suppliers or complementors. In charge of delivering important features of the system, they are also able to leverage the choices made by the integrator, and vice and versa. Thirdly, we argue that problem solving using technical compatibility standards that induce negociation between co-specialized firms not only encouvers solving of the issue but also reveals competitive tensions related to the degree to which know-how is considered in the negociation process. We think that collective strategies are used in order to support the negociation process by partitioning sensitive activities from the tasks that are described in the compatibility standards. Lastly, for our viewpoint, memory is not a static feature and is not only related to struggles for control but also to the decision made, for coordination purposes for retained and use of shared practices. We further argue that this struggle for control is based on hierarchical issues that question the process through which the memory is retained. For these four reasons, we were interested in distinguishing the technological influence that a firm has in the business of integration, by considering the specific context of platforms that answer to CoPS criteria. We were therefore particularly interested in the behavior of platform leaders and complementor's behavior during the standard lifecycle in the context of CoPS development and production.

Specifically, our cases brought questions with regards to the platform leaders and complementors as designed by Cusumano (2002) and Gawer and Cusumano (2012). Adapting these positions to a sector dealing with CoPS, while they were adopted by their authors in a more general way (computers and car industry), drove us to make some observations. We first observe that several platform leaders coexist within the same platform. As Cusumano (2002) observed it, the roles are not so clear and can be reversed. This observation raises the question of the level of platform leadership and complementors. We find that a same firm could be complementor for a system and platform leader in one subsystem. Also, a complementor in one system can be platform leader in a competing system. This reverses the hierarchical idea of the two roles.

Following this line, the status of complementor can also be discussed. Cusumano refers to Nalebuff and Brandenburger (1997)'s definition of complementors as firms adding value to the customer: "A player is a complementor if customers value your product more when they have that player's product than when they have your product alone" (Nalebuff & Brandenburger, 1997, p. 31). Four other definitions are identified by Habets (2012) while he discuss the differences between a supplier and a complementor (Ritter et al., 2004; Adner and Kapoor, 2010; Boudreau, 2008; Yoffie and Kwak, 2006). Those definition are different because they emphasize either on the added-value to the customer (including situations where the customer bundle different products in order to create added-value), or the complementary of products (which is more related to the relationship with the complementor). However, none of these definitions consider the nature of the product, and even less the complexity of the product. Our analysis leads us supporting a more flexible definition to adapt the concept to the context "Firms develop relationships with many other types of firms whose outputs or functions increase the value of their own outputs" (Ritter et al., 2004). This definition used in a complex networks context allows not excluding typical vertical relationships. We thus argue that complementors can be suppliers because the complexity of the products implies that suppliers may also be customers depending on which level of the system is considered as relevant to analyze the relationships.

Our results also entails a discussion about our reference to Hacklin's strategic profiles. We chose this taxonomy because it integrates standard use in strategic profiles designed with no specific focus on standard. Sumer et Bayraktar (2012) reviews twenty one typologies describing firms'competitive behaviors on the market. Only two explicitly state the role of standards (Utterback et Abernaty (1975)) and Miles et al. (1978)). The first one deals with dominant designs in an ecological perspective. Miles and Snow's typology refers to environmental changes and their impact on strategy. To the best of our knowledge, Hacklin et al (2013)'s contribution is the only one that, without a focus on standards, refers to them in describing strategic profiles.

It also combines different strategic levels (business level versus ecosystem level) that enabled to consider a complex standard hierarchy from specifications to dominant designs. Adopting this model helped us to clarify the position in which standards may be used, as a platform leader or complementor, and in which objective notably in aggregating ecosystem or remodeling business. Yet, we met limitations to this use. At first, we respected Hacklin et al (2013)'s taxonomy in selecting incumbents to identify ecosystem aggregators and business remodelers. However, we consider that new entrants could also endorse this role. Secondly, we can discuss the autonomy of the four strategic profiles in Hacklin's typology. We demonstrated that the same firm may adopt different profiles according to the context and we also considered that a firm is able to mix two or more profiles.

Unlike Hacklin et al (2013), we observed not only technology pioneers, but also ecosystem aggregator and business remodelers can use standards to implement strategies in a technology converging industry. However, Hacklin et al (2013)'s model constituted a fruitful matrix to shed light on these specific strategic profiles.

CONCLUSION

In this contribution, we considered a strategic perspective on standards in complex products and services. In this context, standards play a determining coordination role between the different technologies implemented in the projects. They also delineate communities referring to the same standards within the different activities. We referred to Hacklin et al (2013)'s strategic profiles to fathom this architectural complexity. Designing ecosystem aggregators and business remodelers, Hacklin et al (2013)'s model enables us qualifying strategic objectives according to a global or specific approaches of the market. We adopt Cusumano (2002)'s description of platform leaders to specify this status.

We identified firms with the specific role of ecosystem aggregator and business remodeler and showed how they manage a portfolio of standards specifying the requirements for the different activities. As leaders, they impose standards to suppliers and complementors who contribute to this definition. In this regard, standards signal the leading positions of the firms in the market. The positions are however fragiles. The roles and positions can be redistributed. In this dynamic, compliance to standards constitutes a way to gain new skills. The case analysis shows how the compliance to standards helps a complementor to develop new skills that enables it, combined with alliance strategies, to win the leadership status on a new platform. Even in a CoPS context, standards cannot be considered in their restricted role related to organizational issues such as coordination, negociation or memory but also with regards to their impact on strategy: the results particularly show how they are used to manage competitive tensions in combination with strategic alliances.

Finally, we encountered some difficulties with the concept of complementors. Generally, complementors are defined with regards to the added-value they create to their customer. The traditional view considers that the relationship between complementors is horizontal. However, this concept does not seem adapted when dealing with complex system or products. We think that further studies are needed to clarify the concept of complementor with regards to situations where the roles in the value network are versatile.

REFERENCES

Adner, R., & Kapoor, R. (2010). Value creation in innovation ecosystems: How the structure of technological interdependence affects firm performance in new technology generations. *Strategic Management Journal, 31*(3), 306–333. doi:10.1002/smj.821

Alfred, G., & Warner, J. F. F. (2006). Managing Uncertainty in a Formal Standards-Based Industry: A Real Options Perspective on Acquisition Timing. *Journal of Management, 32*(2), 279–298. doi:10.1177/0149206305280108

Barry, A. (2000). Coordinating Technology: Studies in the International Standardization of Telecommunications [review]. *Technology and Culture, 41*(1), 153–154. doi:10.1353/tech.2000.0003

Boudreau, K. (2007). *Too Many Complementors?* Evidence on Software Developers.

Brandenburger, A., & Nalebuff, B. J. (1997). *Co-opetition*. New York: Currency Doubleday.

Cusumano, M. A., & Gawer, A. (2002). The Elements of Platform Leadership. *MIT Sloan Management Review, 43*(3), 51–58.

David, P. A., & Greenstein, S. (1990). The Economics Of Compatibility Standards: An Introduction To Recent Research 1. *Economics of Innovation and New Technology, 1*(1-2), 3–41. doi:10.1080/10438599000000002

Davies, A., & Hobday, M. (2005). *The Business of Projects: Managing Innovation in Complex Products and Systems*. Cambridge University Press. doi:10.1017/CBO9780511493294

de Vries, H. J. (1997). Standardization - What's in a name? *Terminology, 4*(1), 55–83. doi:10.1075/term.4.1.05vri

De Vries, H. J., de Ruijter, J. P. M., & Argam, N. (2011). Dominant design or multiple designs: The flash memory card case. *Technology Analysis and Strategic Management, 23*(3), 249–262. doi:10.1080/09537325.2011.550393

Dumez, H. (2013). *Méthodologie de la recherche qualitative - Les 10 questions clés de la démarche compréhensive*. Paris: VUIBERT.

Farrell, J., & Saloner, G. (1988). Coordination Through Committees and Markets. *The Rand Journal of Economics, 19*(2), 235. doi:10.2307/2555702

Funk, J. L. (2003). Standards, dominant designs and preferential acquisition of complementary assets through slight information advantages. *Research Policy, 32*(8), 1325–1341. doi:10.1016/S0048-7333(02)00121-X

Gawer, A. (2002). *Platform Leadership: How Intel, Microsoft, and Cisco Drive Industry Innovation.* Boston: Harvard Business School Press.

Gawer, A. (2009). Platforms, Markets and Innovation. Cheltenham, UK: Edward Elgar.

Gawer, A. (2010). The Organization of Technological Platforms. *Research in the Sociology of Organizations, 29*, 287–296. doi:10.1108/S0733-558X(2010)0000029022

Gawer, A., & Cusumano, M. (2002). *Platform leardership: How Intel, Microsoft and Cisco drive industry innovation*. Harvard Business School Press.

Gawer, A., & Cusumano, M. (2014). Platforms and ecosystem innovation. *Journal of Product Innovation Management, 31*(3), 417–433. doi:10.1111/jpim.12105

Gawer, A., & Cusumano, M. A. (2008). How Firms Become Platform Leaders. *MIT Sloan Management Review, 49*(2), 28–35.

Gawer, A., & Henderson, R. (2007). Platform Owner Entry and Innovation in Complementary Markets, Evidence from Intel. *Journal of Economics & Management Strategy, 16*(1), 1–34. doi:10.1111/j.1530-9134.2007.00130.x

Gerst, M. H. (2003). The role of standardisation in the context of e-collaboration: a snap shot. *The 3rd Conference on Standardization and Innovation in Information Technology*, 113–119. doi:10.1109/SIIT.2003.1251200

Gnyawali, D. R., & Madhavan, R. (2001). Cooperative Networks and Competitive Dynamics: A Structural Embeddedness Perspective. *Academy of Management Review, 26*(3), 431–445.

Habets, M. J. M. (2012). *Complementors: their effects on new value creation by SMEs in the Dutch printing industry*. Universiteit Twente. doi:10.3990/1.9789036534529

Hacklin, F., Battistini, B., & von Krogh, G. (2013). Strategic Choices in Converging Industries. *MIT Sloan Management Review, 55*(1), 65–73.

Hobday, M. (1998). Product complexity, innovation and industrial organisation. *Research Policy, 26*(6), 689–710. doi:10.1016/S0048-7333(97)00044-9

Miles, R. E., Snow, C. C., Meyer, A. D., & Coleman, H. J. (1978). Organizational strategy, structure, and process. *Academy of Management Review, 3*(3), 546–562. PMID:10238389

Mione, A. (1994). *Les normes, instrument de la stratégie marketing de l'entreprise* (Thèse de doctorat). Université de Montpellier.

Nalebuff, B. J. (1997). *Co-Opetition*. New York: Crown Business.

Paoli, M., & Prencipe, A. (1999). *The role of knowledge in complex product systems; some empirical evidence from the aero engine industry. Journal of Management and Governance.*

Ritter, T., Wilkinson, I. F., & Johnston, W. J. (2004). Managing in complex business networks. *Industrial Marketing Management, 33*(3), 175–183. doi:10.1016/j.indmarman.2003.10.016

Savall, H., Zardet, V., & Bonnet, M. (2008). *Libérer les performances cachées des entreprises par un management socio-économique (2nd ed.)*. Genève: ISEOR.

Schilling, M. A. (n.d.). Technological Leapfrogging: Lessons from the U.S. Video Game Console Industry. *Harvard Business Review*.

Soh, P.-H., & Roberts, E. B. (2003). Networks of innovators: A longitudinal perspective. *Research Policy, 32*(9), 1569–1588. doi:10.1016/S0048-7333(03)00065-9

Steinmueller, W. E. (2003). *The role of technical standards in coordinating the division of labour in complex system industries*. Oxford, UK: Oxford University Press. doi:10.1093/0199263221.003.0008

Suarez, F. F. (2004). Battles for technological dominance: An integrative framework. *Research Policy, 33*(2), 271–286. doi:10.1016/j.respol.2003.07.001

Sumer, K., & Bayraktar, C. A. (2012). Business Strategies and Gaps in Porter's Typology: A Literature Review. *Journal of Management Research, 4*(3). doi:10.5296/jmr.v4i3.1721

Tee, R., & Gawer, A. (2009). Industry architecture as a determinant of successful platform strategies: A case study of the i-mode mobile Internet service. *European Management Review, 6*(4), 217–232. doi:10.1057/emr.2009.22

Teece, D. J. (2007). Explicating dynamic capabilities: The nature and microfoundations of (sustainable) enterprise performance. *Strategic Management Journal, 28*(13), 1319–1350. doi:10.1002/smj.640

Utterbach, J., & Abernathy, W. (1975). A dynamic model of process and product innovation. *Omega, 3*(6), 639–656. doi:10.1016/0305-0483(75)90068-7

Weitzel, T., Beimborn, D., & König, W. (2006). A unified economic model of standard diffusion: The impact of standardization cost, network effects, and network topology. *Management Information Systems Quarterly*, 489–514.

Yoffie, D. B., & Kwak, M. (2006). *With Friends Like These: The Art of Managing Complementors*. Boston: Harvard Business Revie.

ENDNOTES

1. We use the term "complementor" in the sense defined by Brandenburger and Nalebuff (1997), and developed by Gawer and Cusumano (2014) as a shorthand for "the developer of a complementary product" where two products are complements if greater sales of one increase demand for the other. Formally, A and B are complements if the valuation by consumers of A and B together is greater than the sum of the valuation of A alone and of B alone.

Chapter 3
Complex Network Perspective on Collaboration in ICT Standardization

Timo Ali-Vehmas
Aalto University, Finland

ABSTRACT

Standardization is a crucial enabler of global business of information and communications technologies. Convergence of the underlying networking paradigms of licensed mobile communication and license-exempted internet has made progress, but full integration is still far from being complete. For standardization professionals, the unpredictable convergence makes decision making and participation in standardization complicated. This chapter examines collaboration in five closely related standardization organizations working in this field during the years from 2003 to 2008. The results show similarities and differences in collaboration structures and behaviours reflecting the specific scope and context of each standardization organization. Furthermore, this chapter extends the use of social network analysis as a tool to the field of empirical standardization research. The results pave the way towards better collaboration in standardization communities of converging mobile internet and beyond by providing better visibility and new insights to standardization leaders, policy makers, and users.

INTRODUCTION

Standardization (Swann, 2010) and Information and Communications Technologies (ICT) standardization specifically (Shin, Kim, & Hwang, 2015) have been studied extensively. Standards and standardization are major drivers of choice and change. Standards are known to enable value systems where complementary products utilize open interfaces (Katz & Shapiro, 1985). Emerging network effects reflect the strength and type of ties defined by standardized interfaces. Network effects motivate companies to voluntarily contribute their proprietary technologies to open standardization (Economides, 1996) while too strong network effects create undesired technology lock-in as is the case in the QWERTY keyboard (David, 1985). High expectations related to the network effects can make incompatible competition more lucra-

DOI: 10.4018/978-1-5225-5320-5.ch003

tive for leading dominant companies. Therefore, open interoperability through standardization must be favoured by public policy makers (Farrell & Klemperer, 2007).

Linkages between standardization and business models have increased. Scope of standardization has expanded to new areas when focused collaborative consortia have emerged to address weaknesses of the traditional formal standardization (Hawkins & Ballon, 2007), (Blind & Gauch, 2008). Number and volume of different standardization activities has spawned as needs for interoperability, compatibility, scale of economies and for faster innovation diffusion have increased in the globalized markets (Choi, Kim, & Lee, 2010), (Rogers, 1995).

Extensive use of ICT technologies spreads further in our society when the 5G, Internet of Things, Virtual Reality and consumers' data driven applications are emerging. Multiple parallel ICT standardization processes needs more clarity and better structures. Traditional classification of standardization leans to the formal or informal legal status defined by regulation (de jure) or by market actors (de facto). The de jure standards may be promulgated directly by governmental agencies (mandated) or be based on a collaborative work in standards writing organizations (committee) having a formal delegated ("licensed") position. Long time ago governmental organizations alone took care of standards for telecommunications. Today, authorized organizations such as the European Telecommunications Standards Institute (ETSI, 2016) create most of the formal standards for telecommunications. De facto standards may have a dedicated sponsor or owner having interest and full control over the standard (proprietary platform leader controlling publicly available interface specifications) or an "unsponsored" standard is an outcome of a voluntary open collaboration of interested actors. This last model is the most rapidly growing area of standardization, the Bluetooth (Bluetooth, 2016) community being one example. The key difference between the two de facto standardization approaches is the level of openness and control of the standard and the standardization process. (David & Greenstein, 1990), (Funk & Methe, 2001), (Gandal, Salant, & Waverman, 2003). As a summary, Table 1 shows a simplified categorization below:

Standardization is sometimes considered as a competition between different technologies, different business models and between different SSOs. Success of a standardization process depends on a large number of different factors including characteristics of the supporting companies, standardized technology itself and actions of all the stakeholders (Kaa van de & Vries, 2015) as well as on network effects and life cycle dynamics of the standard (Blind, 2011). New needs for research arise related to question on how standardization entities develop over time. When originally small and agile market driven standardization organization gains recognition and develops its processes the initial clear scope may become ambiguous (Pohlmann, 2014). Growing number of market driven standardization consortia parallel to the progressing convergence of the ICT with all other sectors of life have created a need to look at the competition and collaboration of the standardization groups as a larger network of standards and standardization (Jakobs, 2003), (Jakobs, 2008), (Baron, Meniere, & Pohlmann, 2014). Collaboration networks in standardization do not emerge randomly but a number of factors characterises successful

Table 1. Simplified categorization of the system archetypes (Ecosystem Dynamics) of different compatibility seeking approached (Adapted from (Ali-Vehmas & Casey, 2012))

	Mandated/Non-Collaborative	Collaborative
Public de jure	1: Mandated (by government)	2: Delegated to authorized actors
Private de facto	4: Proprietary dominant design	3. Voluntary collaboration

and less successful coalitions. Alliance formation is a strategic not a statistical game (Garas, Tomasello, & Schweitzer, 2014).

Competition and collaboration between companies as well as between SSOs create an invisible fabric of complex dependencies. Flexibility of standard (Ende van den, Kaa van de, Uijl den, & Vries de, 2012) and characteristics of collaborator network (Kaa van de, Ende van den, Vries de, & Heck van, 2011) have been recognized to improve likelihood of success. Distributional conflicts in standards are difficult to agree (Simcoe, 2012). However, high ratio of the same companies participating in both the related industry consortia and the formal standardization project significantly correlates with the success of those companies promoting their technical ideas and solutions in the main standardization project. Like-mindedness of the member companies and their delegates develops in those consortia which helps them to learn from each other both technical and business aspects needed in the standardization (Leiponen, 2008). Like-mindedness is also an organizational element. Inter-organizational learning involving corporatist firms increases significantly when the dominant logic within the arena is also corporatist. When a pluralist logic dominates the arena, corporatist dyads (company pairs where both companies have corporatist background) learn less because firms in the dyad activate a contradictory logic that decouples them from their natural processes for inter-organizational learning (Vasudeva, Alexander, & Jones, 2015).

Standardization is not only a technical but also highly social process.

While standards might aim at the creation of stability and sameness, standardization itself is a highly dynamic phenomenon. Even the stability of standards themselves has to be understood as the result of underlying dynamic processes. (Brunsson, Rasche, & Seidl, 2012)

The multiple dynamical relationships are either competitive, cooperative or simultaneously both, i.e. coopetitive[1] (Majchrzak, Jarvenpaa, & Bagherzadeh, 2015). There is a strong causality between standards, organizational relationships around the standards and the diffusion of the services based on the standards in wireless technologies (Lyytinen & King, 2002). Literature is developing to study broader collaboration aspects of standards setting including information exchange and development of complementing assets (Bar & Leiponen, 2014) and innovations (Delcamp & Leiponen, 2014). Complexity of the collaborative networks, however, is a growing challenge for the future standardized technologies (Katusic, Weber, Bojic, Jezic, & Kusek, 2012).

This study examines collaborative standards setting within five key Standards Setting Organizations[2] (SSO) of Mobile Communications and Internet. The 3rd Generation Partnership Project (3GPP) is authorized to develop specifications for licensed network operators providing public mobile communications service. The Open Mobile Alliance (OMA) was set up to coordinate the development of service enablers for mobile communications. The Internet Engineering Task Force (IETF) has assumed responsibility to develop protocols for the Internet while the focus of the World Wide Web Consortium (W3C) is on standards for web technologies. The Local Area Network (LAN) standards committee for wireless LAN (802.11) in the Institute of Electrical and Electronics Engineers (IEEE) focuses on standards for technologies also known as Wi-Fi. These five forums form the critical core of relevant standardization for global ICT market. Furthermore, the critical interplay of the SSOs address challenges and uncertainties of the multifaceted transition of the traditional mainstream cellular telephony to wireless Internet or alternatively, issues in making the Internet mobile (Tilson & Lyytinen, 2006). The focus period of this research is covering the years 2003 -2008 when the Mobile Internet convergence took its first major steps.

The overall purpose of this research is to reduce uncertainties related to standardization. Standards developers and users want to gain better portfolio decisions (Toppila, Liesiö, & Salo, 2011) and define more successful product plans and platform road maps (Gaynor & Bradner, 2001). Forum shopping is an approach for technology providers (Lerner & Tirole, 2006). Commercial product implementers, however, face different and more severe challenge when integrated product designs have to support multiple competing standards at the same time while the decision-making and leadership of standardization collaboration is a complex multi-forum dilemma.

Chapter 2 defines the research framework based on earlier literature and chapter 3 describes the research area covering the five selected SSOs. Chapter 4 introduces the research methodology and empirical data while chapter 5 presents the main findings of the quantitative analysis. Chapter 6 aims to qualitatively discuss and evaluate the findings. Finally, chapter 7 is a short set of conclusions[3].

RESEARCH FRAMEWORK

For complex problems general systems theory is a source of inspiration (Bertalanffy von, 1972). and concerns (Sterman, 2002) and systems thinking is a mechanism to address the "big things" related also to standardization (Swann, 2010). Research on complex systems utilizes methodologies developed for graph and network theories. Formal quantitative methods such as social network analysis (SNA) open up specific new perspectives on complexities of collaborative networks. Social networks consist of participating collaborators (network node, vertex) and the artefacts of the collaboration (network links, edges). SNA tools provide insightful information on the collaboration beyond simple metrics like the number of standardization contributions per actor. SNA explores relationships between collaborating nodes assessed by e.g. different centrality measures (Wasserman & Faust, 1994). The relational ties between nodes are channels for transfer of resources and knowledge and therefore the networks provide both opportunities and constraints for individual actors. In order to be more central a node has to be close to all other nodes but additionally the node has to be connected to other well-connected nodes in the network (Hanneman & Riddle, 2005). SNA tools have been applied e.g. to business, medicine and computing, for instance to network of hyperlinks on web pages pointing to other web pages of the World Wide Web (Barabási & Bonabeau, 2003).

Social networks typically consist of few nodes, which are highly connected and a long tail of less connected nodes. The node and link distributions of many social networks follow the so-called power law. Power law distributions have been observed also in new business models where products and services are produced by crowds in a decentralized manner and where consumers have become prosumers (Anderson, 2006). Scarcity of a key resource efficiently limits emergence of such a crowd by letting only few players to participate. This is likely to affect the network node distribution also. Furthermore, the mechanism how new nodes are connected to the existing network, such as preferential attachment, influences strongly the network dynamics. For instance, in so-called scale free networks the preference is linearly depending on the number of existing nodes and links in the networks. If the preference linkage is stronger the network grows faster and it is likely to experience the "*Winner take all*" scenario and eventually the network assumes a star topology with one central hub (Barabási & Bonabeau, 2003). In case of regional innovation clusters, hierarchical and assortative characteristics of the collaboration networks indicate their capability to cope with radical changes in their environmental conditions (Crespo,

Suire, & Vicente, 2014). The SNA research on innovation networks of mobile phone industry in Europe have shown similar results (Crespo, Suire, & Vicente, 2016). When analysing very large data sets, like software contributions in open source software projects such as FLOSS community (Free, Libre, Open Source Software) the SNA tools become necessity. The different measures of centrality provide deep insights to the collaboration within the networks including development over time (Martinez-Romo, Robles, Gonzalez-Barahona, & Ortuño-Perez, 2008).

Recently SNA has been utilized also to study different types of consortia in the mobile standard-setting to identify for instance the controlling parties in a collaboration network in China (Kwak, Lee, & Chung, 2012). Similarly, SNA tools have been used to analyse the changes of the interdependency of the RFCs[4] as a measure of collaboration and clustering in IETF (Gençer, 2007). This study focuses on the co-authorship networks as suggested in (Gençer, 2012) and aims to provide a further analysis on collaboration in the five key SSOs of Mobile Communication and Internet.

Social network analysis (SNA) of a large data set derived from documents contributed in these five SSOs provides the quantitative baseline while qualitative observations based on selected expert interviews build bridges between the SNA results and practical work in the standardization. One additional aim is to promote SNA methodology as a tool for standardization research. This study aims to focus on networking and peaceful collaboration rather than fierce competition.

STANDARDIZATION OF INFORMATION AND COMMUNICATIONS TECHNOLOGIES

The 3GPP Release 99 was the first set of specifications for Mobile Communications with full global scope. During the 3G era a serious debate on digital convergence started. New consumers' requirements brought the Internet technologies into the Mobile Communications and separately at the same time, ambitions to make the Internet mobile started to grow. Current integrated multi-mode mobile devices include many standardized functionalities such as 2G, 3G and 4G cellular radio network technologies with Subscriber Identity Module (SIM) based identity as well as local connectivity technologies such as IEEE 802.11 a/b/g/n/. Convergence is strongly visible in the vast range of Internet technologies and World Wide Web (WWW) capabilities as well as in a large amount of vendor specific and platform driven software functionalities and applications running on the mobile devices today. Significant part of the standards commercial products utilise today is a result of collaboration, conceptualization and development effort which took place in the five SSOs during the research period of this study. However, the deep convergence of the standardized technologies is still pending and the developed standards inter-work only in limited fashion. As an example, SIM authentication is still typically not in use in wireless Internet access even if multiple technical specifications have been developed. The different layers of the ICT driven business have different drivers and characteristics (Fransman, 2010). Convergence is a different challenge to each SSO and mutually to all of the SSOs due to their different scope in the overall, layered ICT business domain.

The origin of the standardization groups of this study can be traced back to four different types of genesis: governmental organizations (DARPA for IETF, CERN for W3C), (DARPA, nd; CERN, nd) licensed organizations (ETSI and other similar for 3GPP), combination of commercial and voluntary organizations (OMA) and a voluntary society of professional engineers (IEEE for IEEE 802.11) (IEEE, nd).

The work that later was organized under IETF and W3C had been initiated by governmental organizations. The Defence Advanced Research Project Agency (DARPA) research played a central role in launching the conceptual work, which led to the Internet while the European organization for Nuclear Research (CERN) project ENQUIRE was the first formal step towards World Wide Web. In both cases, the mission was handed over to a number of leading universities such as Massachusetts Institute of Technology (MIT), University of California Los Angeles (UCLA) and Stanford Research Institute (SRI). Evidently, the formal collaborative organizations IETF and W3C were founded to take over the management responsibilities while the technology contributions by a large number of universities continued actively. The role of universities is visible in the collaboration within these SSOs still during the research period of this study (about 10-15% of the data points). The development paths have some historical difference also. IETF emerged among several possible technology initiatives (Campbell-Kelly & Garcia-Swartz, 2013) whereas W3C has a little shorter and less fragmented history of fermentation.

A turning point in the history of IETF took place in 1992 when the Internet Society (ISOC) as a parental legal organization for IETF was established (Internet Society, n.d.). The process was complete by 1996. The change in the governance enabled strong growth of the contributor base including many international contributors. Simultaneously the number of internet users experienced a hockey stick growth curve change in 1997 (Internet World Stats, nd). Today IETF is the only relevant group developing network protocols for the Internet, most notably the Internet Protocol (IPv4/IPv6) and the Transport Control Protocol (TCP) (Crocker, 1993), (Leiner et al., 2009). IETF does not have any specific holistic technical architectural structure but rather the community develops the protocols using a scale free approach in order to support the autonomous re-routing of the data packets. IETF provides communication network layers on the top of the physical link layers, e.g. the functionalities defined by the IEEE 802 group and provides the platform for the content layers defined e.g. in W3C. W3C develops web standards, most notably the Hypertext Markup Language (HTML, currently version 5). The W3C standards define the Open Web platform, which enables developers to create rich interactive web experiences. Multiple browser implementations can exercise the protocols and utilize the Application Programming Interfaces (API) of the mobile devices allowing the web services and web applications to run on almost any hardware. IETF and later W3C have defined the foundations of web architecture including Hypertext transport protocol (HTTP) and Unified Resource Locators (URL). Close co-operation and structural alignment between IETF and W3C has secured high compatibility and interoperability of the protocols developed by these two SSOs (Simcoe, 2015). The first meeting of the W3 Consortium took place in 1995 (W3C, nd).

When setting up the Internet standardization the initial IETF and W3C development environment were not only formally "liberalized" but rather the initial work was simply transferred to a new environment with full, academic freedom. New organizations built their formal structures gradually. They did not redesign the rules of governmental organizations but rather removed the old rules completely and let the new rules to emerge bottom-up.

IEEE 802.11 has its foundation in a voluntary collaboration of professional engineers in their social organization, IEEE and its Standards Association (IEEE-SA). IEEE has a remarkable history of over 100 years and an international position with members in about 100 countries. The IEEE-SA membership is twofold, consisting of about 180 corporate members[5] and the large membership base of individual engineers. The IEEE 802.11 standardization is open for all IEEE members to follow and to participate in the decision-making.

After the liberalization of the specific radio spectrum for Industrial, Scientific and Medical (ISM) use in 1985 IEEE-SA noted a need for a voluntary standard for local area wireless connectivity using this spectrum. Multiple different technologies without any interoperability initially fragmented the field severely. IEEE 802.11 completed its first standard by 1997. This standard utilised earlier technologies, developed in IEEE 802 as the design framework. One key success factor was the networking compatibility with IEEE 802.3 (wired Ethernet). The IEEE 802.11 standards implement the wireless version of the wired Ethernet (Lemstra & Hayes, 2009).

Furthermore, the purpose of the Wi-Fi Alliance (WFA, nd) is to increase the market acceptance of the IEEE 802.11 standards. WFA operates as a post processor prioritizing the IEEE 802.11 standards by creating interoperability certifications for selected subsets of the standards. IEEE-SA and therefore, IEEE 802.11 is a recognized standardization organization by ANSI (ANSI, nd.) while the WFA currently is not ("ANSI Standards", nd).

The roots of 3GPP are deep in the history of telecommunications standardization. A direct continuum from the 2nd generation mobile systems including their promoters, like mobile operators, their product and technology providers as well as the formal standardization organizations in all key markets, Europe, the USA, Japan, China and Korea is evident (Hillebrand, 2013). Related to GSM system evolution the roots of the working processes can be traced back to even further to the 1st generation systems (Haug, 2002). The critical border conditions i.e. operators' license requirements and especially fundamentals of radio spectrum regulations have not changed since the 1990's. For instance, the European Conference of Postal and Telecommunications Administrations (CEPT, nd) still maintains the coordination within Europe. 3GPP has responded to selected regional differences such as different spectrum bands, emergency call numbers and data security settings by providing fully interoperable, adaptive configurations for each region.

The 3GPP membership is open but in practice requires solid financial resources. All the delegates in the meetings represent their affiliations. In order to become a formal standard, the specifications, created in 3GPP require formal approvals by the Organizational Partners, i.e. the regional and national standardization organizations such as ETSI, ATIS and CCSA (ATIS, nd; CCSA, nd).

3GPP covers functionalities on all layers (Fransman, 2010) while the other standardization groups studied in this research cover smaller areas. 3GPP is a system level standardization organization intending to define a large modular architecture with several open internal and external interfaces (Hillebrand, 2013). A specific System Architecture (SA) group takes care of the conceptual development and other dedicated groups develop the necessary interface standards for modules. This approach creates high-level alignment between technical and organization models. 3GPP also collaborates in some key areas with its organizational partners such as for SIM/USIM card specifications with ETSI (ETSI/SCP). The effort on content and application layer is relatively weak in 3GPP. Home location register (HLR/HSS) represents the identity management of the content layer.

The companies working in the 3GPP ecosystem initiated OMA (Grøtnes, 2008). They noted a need to expand the set of contributors and to include the leading Information Technology (IT) companies into the mobile communication ecosystem in order to build stronger content layer functionalities. The need for OMA emerged because the IT companies did not feel comfortable to join directly the 3GPP standardization. The IT companies considered the discussions in 3GPP being telecommunications oriented only (3GPP, 2002). Before OMA, these needs were addressed by a number of small separate consortia of like-minded contributors from IT and Communications industries. The momentum however, was weak and the consolidation intended to improve the situation. Separate industry initiatives of WAP Forum, Wireless Village, SyncML, Mobile Games and Location Interoperability Forum and Mobile Wireless Internet Forum evidently merged into one single organization.

OMA covers networking applications and specifically standards for underlying service enablers. In the early days, 3GPP and OMA agreed to transfer some application layer functionalities like Multimedia Messaging from 3GPP to OMA. The architectural design in OMA leans towards the 3GPP architecture. Several but not all OMA service enabler protocols are reflections of similar protocols developed in IETF and W3C for the Internet. Conceptually, OMA operates on the top of the 3GPP functionalities and enhances the mobile network capabilities beyond the basic voice and short message services. The OMA membership is available for supporting companies only.

The combined effort of IETF, IEEE 802.11, W3C and other related organizations like the SIP Forum (nd) provides a set of standards for the Internet, which forms a comparable set of the combined functionalities of the OMA and 3GPP standards for the Mobile Communications. Originally, these two sets of standards were developed and deployed totally separately. The digital convergence has started to push the systems together forcing the related standardization groups to talk to each other, too. Especially the collaboration between IETF and 3GPP has become important for the support of Internet over Mobile Communications. The Internet Multimedia Subsystem (IMS) of 3GPP is a deliberate action and a major undertaking to seek collaboration between these two SSOs. Furthermore, loosely coupled interoperability has been achieved between the licensed and unlicensed radio technologies (Gunasekaran & Harmantzis, 2008). However, the projected tightly coupled multimode products utilising Wi-Fi and WiMAX[6] with 2G/3G/4G cellular network technologies have not diffused broadly to the market. Figure 1 summarizes historical developments towards a multi-mode mobile device.

Regulatory environment implicitly defines dynamics in competition and collaboration as well as possibilities to build high-level multi-mode functionalities. It is useful to note the difference between two situations. The interplay takes place differently between systems of similar environments like between GSM and CDMA (Cabral & Salant, 2014) and between different types of setups like between 3GPP

Figure 1. Convergence of standards in a Mobile Device (For Ecosystem Dynamics ref to Table 1.)

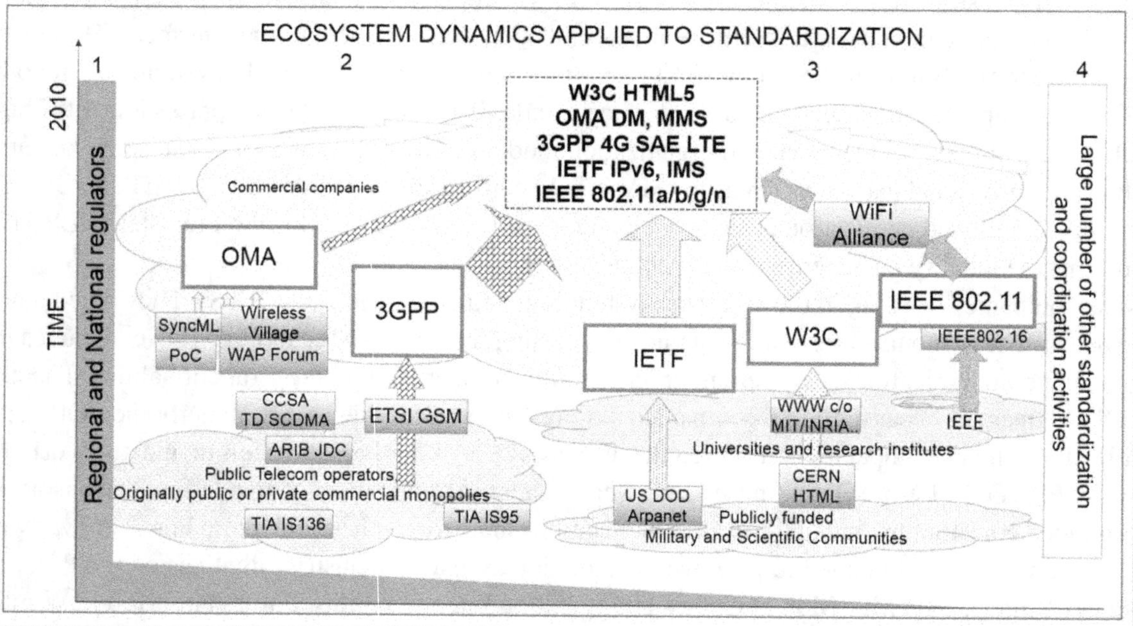

and Wi-Fi (Lemstra & Hayes, 2009), (Trestian, Ormond, & Muntean, 2012). Majority of the earlier research on standardization has been studying aspects of competition between different technologies and standardization organizations. As a complementing step, this research aims to study and compare collaboration in standardization forums based on their internal characteristics. The broader research on collaboration between the standardization organizations is for the future study.

EMPIRICAL RESEARCH METHODOLOGY AND THE DATA

Main approach of this study is quantitative based on social network analysis. However, limited qualitative elements gained through selected expert interviews are included to complement the quantitative findings and to provided deeper insights (Di Minin & Bianchi, 2011). The challenge in using SNA is to interpret the numerical or graphical results beyond the theoretical models. The benefits of SNA however include the possibility to cover large data sets and condense the information into a low number of specific knowledge points or questions. Personal observations, opinions and preferences, on the other hand approach these questions with deep tacit information. The two approaches are usually only partially valid each but by combining the objective results of the SNA with subjective observations of the interviewees there is a possibility to achieve less subjective conclusions and at the same time open up new, more specific research paths for the future.

QUANTITATIVE RESEARCH

Collaboration in writing is an action with an intention to make an impact. Written contributions are considered a meaningful proxy to observe actions of standardization participants (Weiss & Sirbu, 1990). Furthermore, a written document with author's affiliation information typically represent the best interests of the affiliation due to the SSO membership bylaws, processes and traditions as noted in previous studies (Jakobs, Procter, & Williams, 2001), (Isaak, 2006).

There is however, an obvious challenge in modelling standardization collaboration based only on the written documents. A stand-alone contribution may be collaborative in its content. Furthermore, numerous corridor discussions even without any resulting documents are essential in consensus building. Similarly, some single contributions may over time become very valuable, even iconic while the role of other documents is simply to record the process. Still, gaining a public position in standardization between companies in writing always require significant amount of preparations and collaboration and therefore in this research only the written documents are considered quantitatively.

The empirical data collected from the document repositories of these five forums consist of document headers including the names of the contributors with time stamp. All documents provided by working group leaders, formal editors or chairpersons and similar are omitted because these documents are considered neutral related to collaboration and competition. In case when a SSO contributes to some other SSO (liaison statement), contribution by the SSO is included and affiliated to the contributing SSO itself. Furthermore, "*W3C has also its own technical staff able to develop the technology forward*" explains a W3C executive. This dual role as a standardization and research organization makes W3C itself to show up as an active contributor with high number of collaborative contributions. The final set of contributors consists of contributing organizations of these SSOs (companies, universities, research organizations, governmental actors and standardization organizations themselves as well as the similar liaising entities). Some joint ventures, mergers and acquisitions between the contributors took place during the research period, which made manual editing necessary on case-by-case bases.

The final data set forms a bipartite network between the contributors and their contributions. The bipartite networks are converted to simpler projections assuming that the contributors are *connected nodes* in a network if they have collaborated to contribute a document while the joint contributions of multiple contributors form the *links* of the network. Stand-alone contributions do not add any link to the network of collaboration. Therefore, contributors providing only stand-alone contributions form a set of *disconnected nodes*. Similar simplification is used in many earlier social network studies, for instance about collaboration of movie actors (Watts & Strogatz, 1998) and scientists (Newman, 2001).

Comparable data is available on all selected SSOs from 2003 to 2008. Data outside of this period, however, is used selectively. The sampling rate of the data is one calendar year. Table 2 shows the number of documents used in this analysis as well as the number of the contributing organization.

The qualitative interviews of the standardization experts note the key role of the leader in the contribution process. Therefore, each document forms a small elementary start network where the lead contributor is the central node in the elementary network and the collaborating contributors connect to the lead contributor over identical undirected links. The elementary star networks combined result the final undirected but weighted network of collaboration. Link weights indicate the total number of joint contributions between the nodes. Weighted and unweighted networks provide two different perspectives on the collaboration. The weights of the weights, i.e. the specific value of each document are not used[7].

The collaboration network of each SSO forms an adjacency matrix A[8]. The matrix A has N columns and N rows representing the N contributors in the SSO where the elements a_{ij} show the number of collaborative contributions between the contributors i and j. For detailed analysis, a separate adjacency matrix is created for each SSO for each year. Time series of the adjacency matrices enable flexible way to study development of the collaboration over a research period.

QUALITATIVE RESEARCH

The participating people create the standards. Interviews of 13 highly experienced standardization leaders and contributors provide basis for the qualitative findings. All the interviewed experts have a minimum of 5 years of experience in hands-on standardization work in several different standardization organization including at least one of the SSOs studied in this paper. They represent different roles including company lead delegate, document editor, chairperson of a sub group or of a technical plenary, chair or

Table 2. Number of contributors and their contributions in the SSOs. The comparable data is available for the years 2003 – 2008.

	OMA	3GPP	IETF	W3C	IEEE 802.11
Contributions (total)	68066	308967	55292	1243	25761
Contributors (total)	254	623	2879	324	402
Contributions 2003-2008	65959	185789	29728	846	17498
Contributors 2003-2008	248	415	1768	267	294

a member of the board of a SSO as well as director of a SSO with an overall operative or legal or some other executive level responsibility. The qualitative data collection took place between April 2015 and March 2016 in face-to-face and virtual semi-structured interviews. The interviews addressed the whole lifespan of the SSOs with the special focus on the years of the quantitative data (2003 - 2008).

RESULTS OF THE QUANTITATIVE EMPIRICAL RESEARCH

Standardization is a contribution driven process where the contributions are focused in the early phase on new work items with conceptual and architectural requirements while towards the end of the process more detailed comments and change requests dominate (Leiponen, 2008). Many standardization organizations are running several parallel processes and therefore the mixture of contributions in a standards meeting include different types of documents and statements.

COMPETITION, COOPERATION, AND COOPETITION

In this initial analysis, standardization contributors fall into three groups. One group of contributors consists of *Cooperative* actors who never contribute alone. They are considered high collaborative (Collaboration = 1), while another group consists of *Competitive* contributors who never contribute with anybody else (Collaboration = 0). The third group consists of *Coopetitive* contributors who do both (0 < Collaboration < 1) (Majchrzak et al., 2015). With this categorization, the Table 3 shows the numbers of different contributors in each SSO.

The differences between the shapes of distributions shown in the Figure 2 reveal more details. The distribution of different types of contributors forms a collaboration profile for each SSO. The number of contributions each contributor has done is not taken into account in these curves i.e. collaboration profile is unweighted. The total number of all contributors ($N_{SSO)}$ is naturally different for each SSO. Horizontal axis represents all the contributors ranging from 1 to N_{SSO} for each SSO, in other words the axis is scaled to fit the contributors into the same range [0, 100%]. Vertical axis shows the ratio in contributions for each contributor within the range [0, 1] i.e. [0, 100%] collaborative.

Starting from the middle, the shapes of the curves are almost identical for OMA and 3GPP where majority of the contributors (70% for both) is coopetitive. In IEEE 802.11, collaborative only contributors form a small minority (1.7%) while majority (59%) of the contributors are working only alone. W3C is at the opposite mode where half (51%) of the companies are collaborative only. The shape of

Table 3. Number of different types of contributors in the SSOs. Comparable data in this table covers the years 2003 – 2008.

	OMA	3GPP	IETF	W3C	IEEE 802.11
ALL Contributors (N_{SSO})	248	415	1768	267	294
* Competing only	40	83	187	12	172
* Competing and Collaborating	173	291	1038	120	117
* Collaborating only	35	41	543	135	5

the histogram for IETF is similar to W3C with balanced collaborating and competing contributors being the majority (59%).

IT IS A SMALL WORLD

The earlier research on "Small World" and random networks has yielded very basic level observation. The ratio of hierarchy (transitivity, clustering) and communications capabilities (path length) of the network forms a simple starting point for the SNA giving an indication of network topology. Neighbours are neighbours even if only one link connects them together. In other words in this section, weight value of all non-zero link equals 1.

Transitivity in all five SSOs is high when compared to random networks of similar size while the average path length is short. This is an indication of the "Small World" network topology in all SSOs. Transitivity, however in these networks has a trend while the average path length is rather similar for all. The OMA and 3GPP collaboration networks show higher transitivity and lowest average path length. This indicates higher level of hierarchy in collaboration in those organizations than in the other three SSOs. Global transitivity and average path length show similarities between IETF, W3C and IEEE 802.11. This is in particular interesting because the absolute numbers of contributors in IETF, W3C and IEEE 802.11 are one order of magnitude different. The IEEE 802.11 data shows lowest local transitivity and highest path length. Noting that 59% of the IEEE 802.11 contributors are disconnected nodes the longer average path length indicates sparse networking and longer path for information sharing even in the connected part of the SSO network. Figure 3 shows the transitivity and path length values of the SSOs.

Figure 2. Collaboration profiles of the SSOs. Ratio of Collaboration only (Collaboration =1), Competitive only (Collaboration=0) and Coopetitive (0<Collaboration<1) companies is shown separately for each SSO. Data consolidated over the research period 2003-2008.

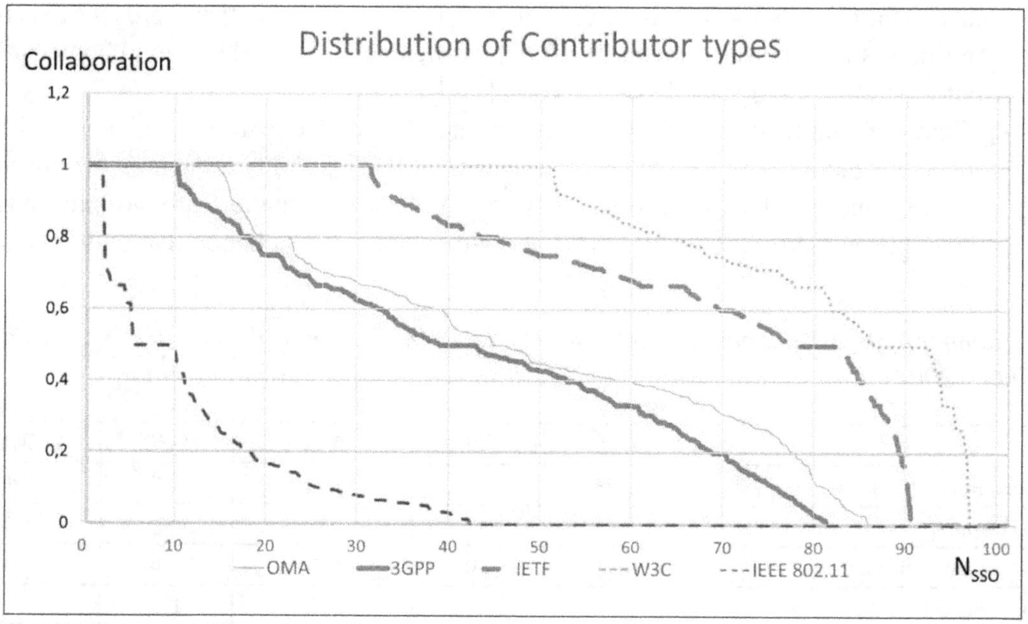

**For a more accurate representation see the electronic version.*

Figure 3. Transitivity and Path length. Only connected contributors included. (2003-2008)

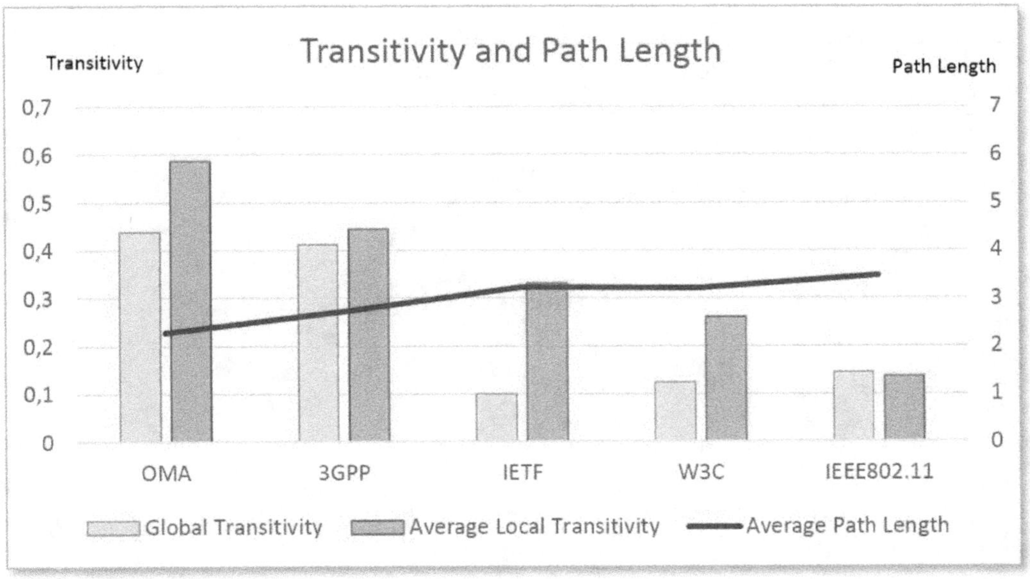

STRUCTURES OF THE COLLABORATION NETWORKS

Analysis of network structure utilises two slightly different approaches: degree centrality and eigenvector centrality. Degree centrality is a measure of local networking showing the number of connected nodes for each network node. Degree centrality is an indicator of the network structure for information sharing. In many collaboration networks, the degree centrality distribution often follows the power law. For structural analysis the number of similar events do not count i.e. all the non-zero link weights equal 1. The contributor ranking is according to their centrality scores.

Distributions of unweighted degree centrality rankings follow two different trends shown in the Figure 4 and Table 4. For 3GPP and OMA, the distribution trend-line is exponential while for IETF, W3C and IEEE 802.11 the trend line follows power law. Tested trend line options were linear, logarithmic and exponential as well as power law. The highest R^2 value indicates best fit. Direct ranking of weighted degrees is used instead of the traditional degree distribution to make the curves comparable to a relevant earlier research (Crespo et al., 2014). Local networking in 3GPP and OMA for information sharing shows rather flat, equally shared distributions especially for the top 40 to 60 contributors while in the other SSOs there are only few strongly networking contributors. Edges of the networks are weakly connected in all SSOs.

Eigenvector centrality extends the analysis to the whole network and as a global centrality metric, it captures the global influential power of collaborating standardization contributors. Eigenvector centrality takes all the unweighted connections into account similarly[9] as the page rank algorithm calculates the centrality of documents in the WWW.

The eigenvector centrality analysis indicates even more clearly that the SSO collaboration styles fall into two different groups as shown in Figure 5. Furthermore, it shows that the global collaboration is broader than local (slope is slower) but also that the cut off is steeper. This is an indication that practices of the information sharing and influencing are different and that they are different between the two groups.

Figure 4. Unweighted Degree Centrality scores of the connected contributors in ranking order in the SSOs. Research period 2003-2008.

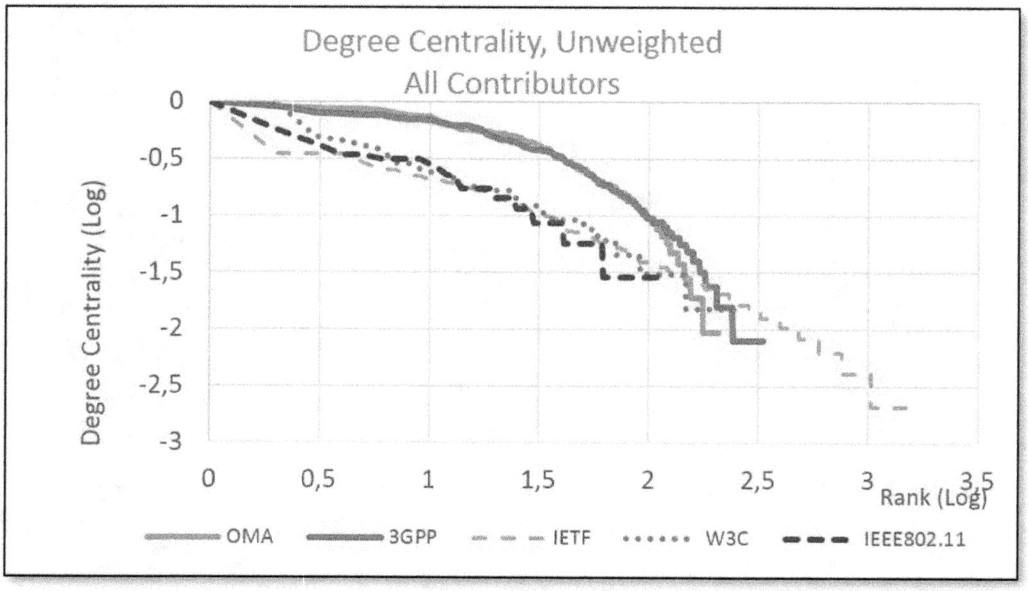

**For a more accurate representation see the electronic version.*

Table 4. Trend line factors, all contributors, Unweighted Degree Centrality.

	OMA	3GPP	IETF	W3C	IEEE 802.11
Power law exponent			**-1,064**	**-0,944**	**-0,891**
Exponential exponent	**-0,023**	**-0,015**			
R^2	0,99	0,97	0,95	0,95	0,94
Number of connected nodes (Log(degree > 0))	208	332	1581	251	122

Influencing for decision-making in mobile communications collaboration is more equally distributed than in the internet-oriented forums. For all SSOs, the long tails of the contributors, i.e. edges of the networks are structurally non-influential in this analysis.

Similar trend matching analysis as for degree centrality is provided in case of the eigenvector centrality distributions in Table 5. Only top 40 contributors are included.

The complete set of eigenvalues of the adjacent matrix forms the spectrum of network. Eigenvector centrality takes into account only the information related to the largest eigenvalue. Therefore, the quality of eigenvector centrality as a characterizing metric depends on the relative size of the first eigenvalue to the other eigenvalues. High relative size indicates that the eigenvector centrality scoring is able to capture a significant amount of collaboration dynamics into the first vector. For OMA and 3GPP distance between the largest and second largest is about 1 to 1/3 while in case of IETF, W3C and IEEE 802.11 the distance is about half of that as shown in the Table 6.

In order to estimate the statistical significance of differences between the ranking scores we can use for instance Kolmogorov Smirnov test to compare two set of samples to see if they follow the same (null

Figure 5. Unweighted Eigenvector centrality scores of the connected contributors in ranking order in the SSOs during the research period 2003-2008.

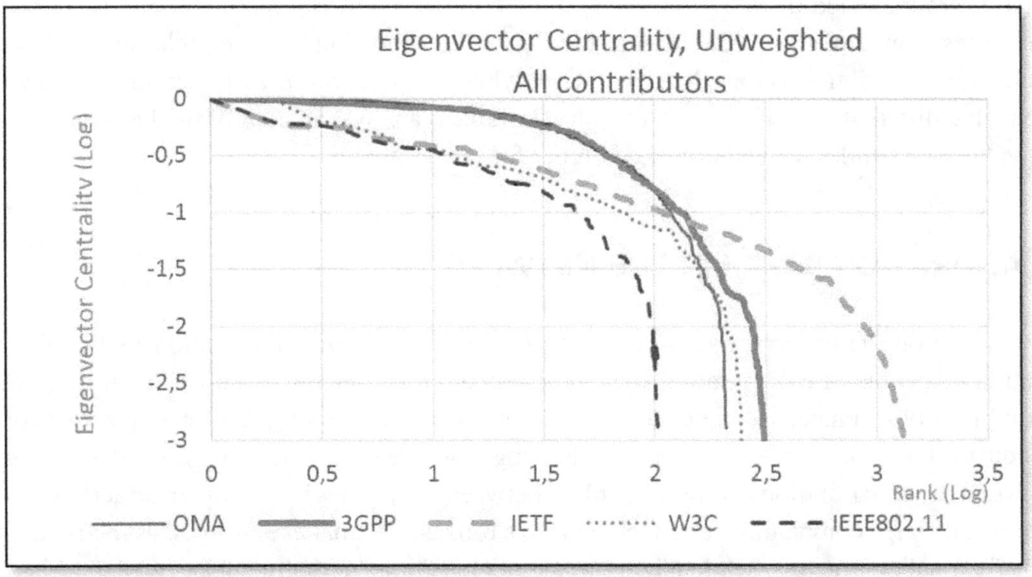

**For a more accurate representation see the electronic version.*

Table 5. Trend line factors, Top 40 contributors, Unweighted Eigenvector Centrality.

	OMA	3GPP	IETF	W3C	IEEE 802.11
Power law exponent			**-0,392**	**-0,517**	**-0,591**
Exponential exponent	**-0,0174**	**-0,0168**			
R^2	0,98	0,99	0,97	0,98	0,97
Number of connected nodes (Log(degree > 0))	40	40	40	40	40

Table 6. Relative top 3 eigenvalues of the connected contributor networks of the SSOs

Relative Eigenvalues	OMA	3GPP	IETF	W3C	IEEE 802.11
# 1	1	1	1	1	1
# 2	0,28	0,31	0,53	0,60	0,62
# 3	0,24	0,28	0,45	0,51	0,53

hypothesis) or different distribution function. The two sample KS test provides an estimate D which is larger than the critical factor c(α)*F(N) if the null hypothesis is not valid. The c(α) for statistical probabilities 0 05 and 0.01 are 1.36 and 1.63, respectively ("Critical Values" nd). Since in this case we want to compare the distributions of the different SSOs related to most active contributors only the factor F(40) is therefore 0.224. The critical values for D are therefore 0.304 and 0.364. Table 7 below shows the results of the KS tests for the eigenvector scores shown in the Figure 5 and related trend lines for the top 40 contributors in Table 5. Significant similarities are found when comparing the distributions, indicated in bold (null hypothesis valid) while in other cases the null hypothesis must be abandoned. Those cases are indicated by *.

The KS test confirms the observation of the Figure 5 by showing the close relationship between the eigenvalue rank distributions for OMA and 3GPP while the other SSOs form another group with similar eigenvalue distributions. The KS test results (D values) are systematically smaller than either of the critical values for similar distributions and higher for other cases.

COLLABORATION STYLE IN THE NETWORKS

Degree correlation, also referred as assortativity provides a behavioural information on the collaboration. Assortativity is a kind of collaboration style indicator and it is used e.g. to assess maturity of industrial organizations. For instance, social networks of scientists and academics often show positive correlation while industrial networks are disassortative, showing negative correlation. Negative degree correlation indicates that the collaboration mainly takes place between companies having opposite activity scores, i.e. active, typically large companies are connected with strong ties to smaller or at least less active companies, typically their subcontractors, while having weaker connections to their equal size peers. Link weights apply when calculating assortativity in order to capture the collaboration behaviour beyond the structure.

In Figure 6, assortativity in IETF shows a long-time evolution from noisy positive scores gradually to stable negative scores. The crossover point in this development matches with the timing of the major organizational change in IETF discussed in section 3. Even if this event is outside of the focus period of this research, it is a useful observation, connecting the assortativity scores to historical events of IETF. Assortativity scores in 3GPP shows similar but much more rapid stabilization. OMA has a short history with no real changes and W3C shows development from rather strong negative towards less negative scores. Finally, IEEE 802.11 shows clearly different behaviour. Instead of rather steady development towards more mature industry collaboration, IEEE 802.11 shows significant cycling with the period of 3-5 years. The collaboration in the IEEE 802.11 cycles strongly between assortative and disassortative scores.

Table 7. Results (D values) of two sample KS test for pairwise comparison for eigenvector centrality distributions

D	OMA	3GPP	IETF	W3C	IEEE802.11
OMA	**0**				
3GPP	**0,125**	**0**			
IETF	0,45*	0,425*	**0**		
W3E	0,525*	0,525*	**0,15**	**0**	
IEEE802.11	0,475*	0,475*	**0,1**	**0,125**	**0**

Figure 6. Assortativity evolution in the SSOs. Comparable data available for the period 2003-2008[10].

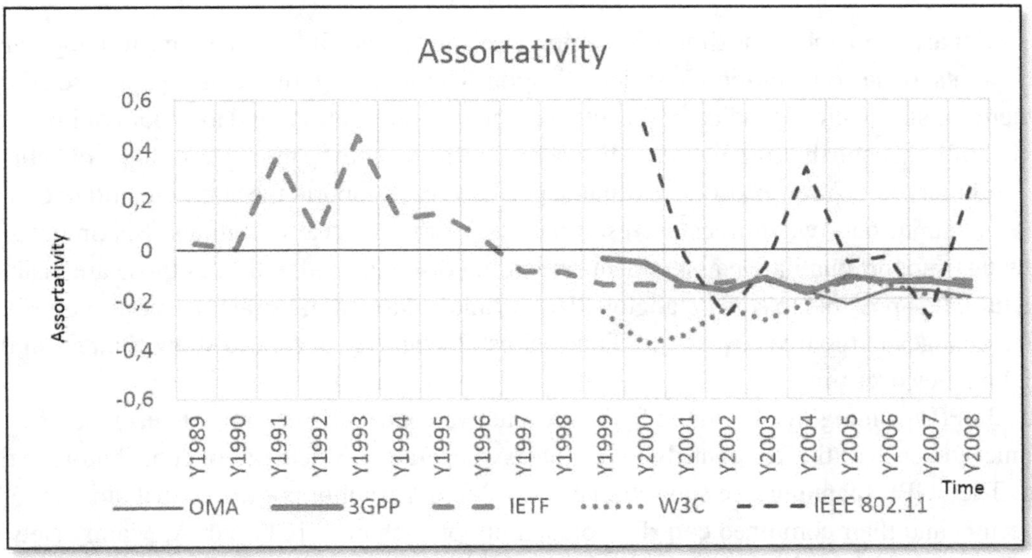

**For a more accurate representation see the electronic version.*

Earlier research has shown that the level of network assortativity gives a good indication of the way knowledge flows between the central and more peripheral nodes (Crespo et al., 2014). Landscape of hierarchy versus assortativity gives a visual indication of resilience in these networks. Figure 7 shows the difference between random, assortative (core-periphery) and resilient network types. All forums show characteristics of resilience by having negative slope in the degree centrality ranking and negative assortativity (note the absolute value for degree centrality |a| in the drawing, positions of the SSOs are illustrative only). The assortativity cycling in IEEE 802.11 makes the SSO to cycle also in this landscape model between the two positions of resilient and core-periphery behaviours.

Figure 7. Resilience of the collaboration modes of the SSOs. (Adapted from (Crespo et al., 2014))

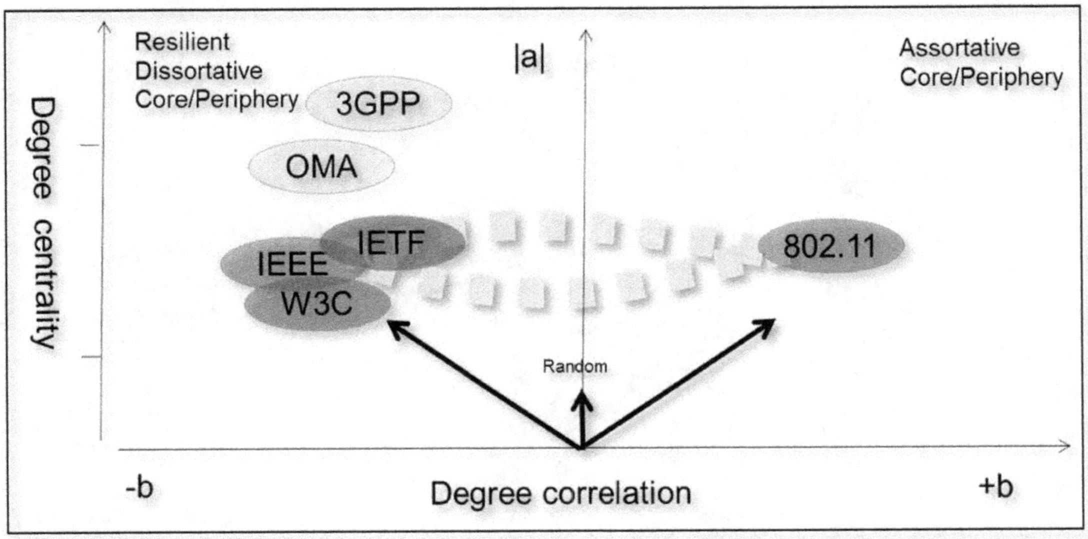

SEMI-QUANTITATIVE VISUALIZATION OF THE COLLABORATION NETWORKS

As the last analysis method, maximum spanning tree shows visually the network topology and gives further insights to the collaboration in standardization. The spanning trees in this paper use all the same scale where the size of circles indicates the total number of contributions and the width of links indicates the number of joint contributions between the nodes. The maximum spanning tree shows only the critical links of collaboration. Additionally, maximum spanning tree supports the semi-quantitative high-level validation of input data when trees are used parallel to the interview findings. Maximum spanning tree drawings support qualitative assessments regarding disconnected nodes, as those are visible in the same picture. It is possible to study e.g. globalization and collaborating strategies of the contributors by observing changes in the trees over time. Time series of spanning trees also show changes in the roles of the companies over time.

In the 3GPP spanning tree in Figure 8, the most active contributors connect to each other with quite strong links and each of the key contributor connects a separate sub-network of contributors to the main network. The 3GPP spanning tree shows rather clear logical grouping of the contributors based on the type of actors and their combined contribution activity of each branch. Group A: Mainly Network operators, Group B: Mainly Korean companies, Group C: Mainly Japanese companies, Group D: Mainly Chinese companies, Group E: Mainly technology and equipment vendors. The stand-alone contributors form a disconnected cloud outside of the connected network.

As another example using the same methodology as in the Figure 8, Figure 9 shows the IEEE 802.11 maximum spanning tree. It is different in many ways. The total number of contributions is much smaller than in 3GPP as is visible in the size of circles. The high number of stand-alone contributors forms a large cloud outside of the main connected network. Branches of the tree include companies without any specific grouping logic. These observations are aligned with other studies related to standardization in telecom and internet oriented standardization communities (Jakobs et al., 2001), (Isaak, 2006).

Figure 8. Maximum spanning tree of the 3GPP collaborative network consolidated over the years 2003-2008.

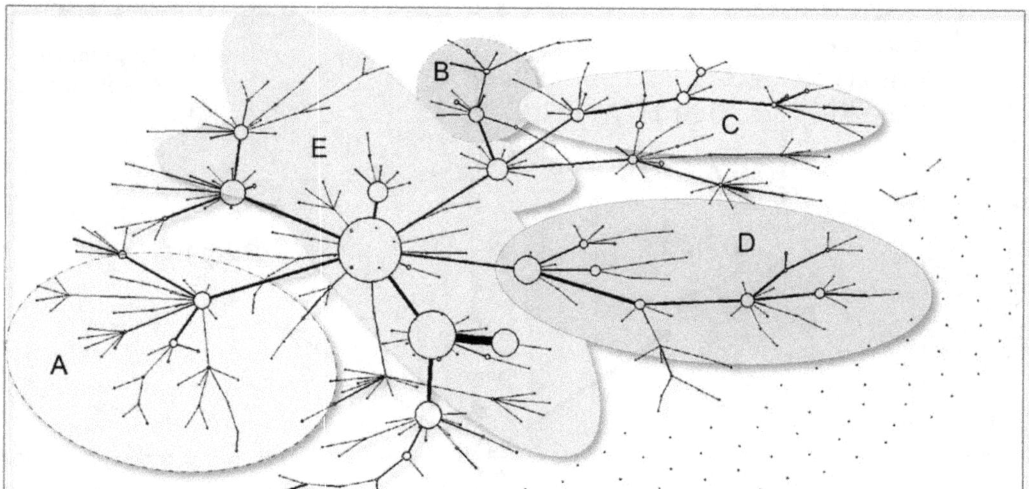

Figure 9. Maximum spanning tree of the IEEE 802.11 collaborative network consolidated over the years 2003-2008.

DISCUSSION

The SNA provides insightful perspective on the internal collaboration in each of the five SSOs. The semi-structured interviews complement the quantitative results by providing views, which in most cases align fully with the SNA findings with some exceptions as discussed below.

3GPP

The maximum spanning tree of the 3GPP collaboration network gives an indication of the importance of historical evolution when companies with the same geographical and regulative origin, China, Korea and Japan show up each in their own branch of the tree. Like-minded companies, like majority of the network operators form up one main branch of the maximum spanning tree while the main body consist of technology and equipment vendors. The interviews note the presence and specific role of licensed network operators in 3GPP. The operators can shape the collaboration in 3GPP individually and as a group through their global and regional collaboration organizations such as the GSM Association (GSMA) and the GSM operators in North America (GSM NA). *"The role of the operators is much more direct and strong than for instance the role of the PCG (Project Coordination Group). ...The GSMA however does not always integrate the opinions of all the operators"* as one of the 3GPP expert explained. Mobile network operators have local natural monopoly due to the licensed radio spectrum their networks utilize. The maximum spanning tree analysis shows the network operators to form their own branch rather than positioning them to the middle. This reflects their role as a significant but not as the most active contributor group.

A specific characteristic discovered in the structure of 3GPP network is that both degree centrality and especially eigenvector centrality show a rather flat distribution of the global information sharing and influence indicating broad and rather equal participation by a large number of contributors to the discussions and decision-making in 3GPP. Weighted degree centrality[11] distribution follows the shape of the unweighted distribution but the slope is steeper indicating that hierarchy in the collaborative actions is higher than in the structure. This means that a smaller group of companies provides large part of the joint documents.

Intuitively, these observations are well in-line with the strong systemic interoperability requirements of the inherited formal standardization environment of 3GPP including regulated, mainly national roles of the network operators. Interviewees noted the heritage of historical roots of the SSOs. One 3GPP expert explained the path dependency from ETSI to 3GPP.

It took several years in ETSI to leave the traditional CEPT mode of operation behind. The situation only changed when the market changed and all participants had to start working according to their new market position. Similar difficulties were experienced when 3GPP was set up. It took several years for participants from the different markets to adopt the new international way of working.

When eigenvector centrality is studied over years the development towards broader collaboration is visible in Figure 10 as the curve becomes more and more flat over the years (from 2003 to 2008).

Speed of technology development in the mobile communications is fast. Lifetime of one generation of technologies is shorter than the lifetime of spectrum license. Large variations of product and service deployments within the operator community together with strong expectation on global roaming have consequently created strong backwards compatibility requirement for the 3GPP technologies

Figure 10. Eigenvalue centrality development over time for period 2003-2008

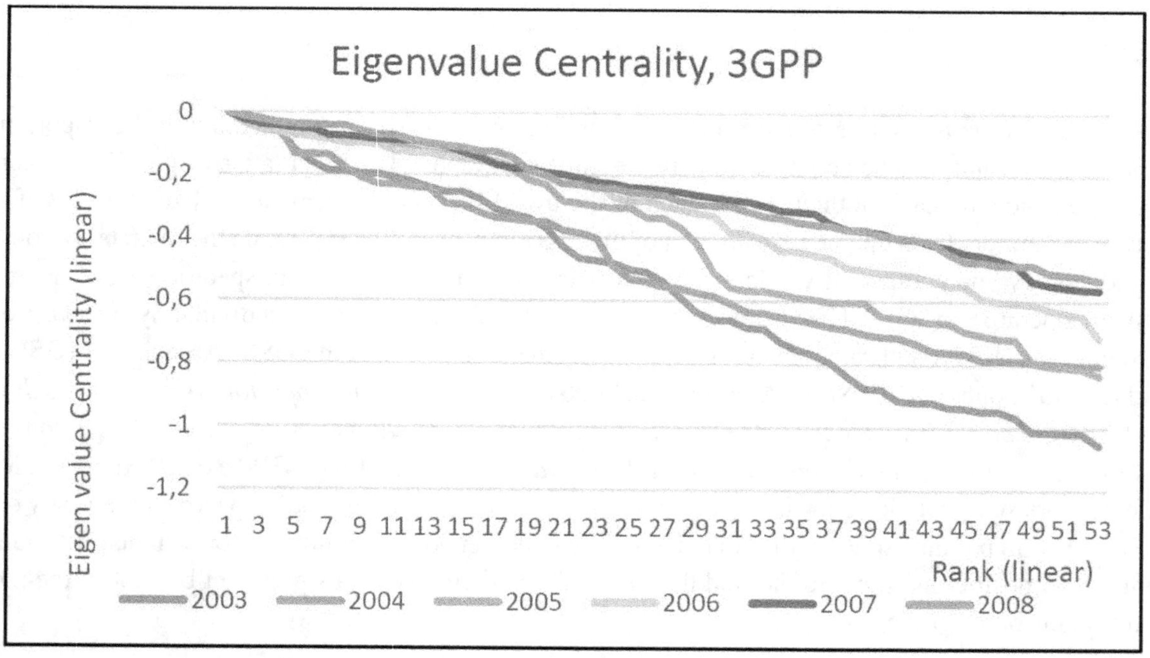

*For a more accurate representation see the electronic version.

and networks. Such a complicated network architecture and a large number of interoperable interfaces obviously benefit from the stability of the collaboration mode. After initial formation period, negative assortativity with rather high clustering and hierarchy indicates stable and resilient operative mode in 3GPP over the timeframe of this study. The initial ambiguous phase of collaboration is weakly visible as the assortativity of the 3GPP contributors stays close to zero until 2001, more than 3 years after the ETSI decision in 1998 to use WCDMA technology for the 3rd generation network was made. One of the 3GPP experts commented:

Japanese delegations did not understand in the beginning how the system worked and there were top level people trying to influence using interpretation (ITU Style) but that was soon taken over by English speaking technical experts and the 3GPP working style was adopted.

OMA

The similarities between OMA and 3GPP are strong in all aspects studied in this research. The fact that the service enablers OMA develops have no license dependent or regulation-oriented requirement shows no reflections in the way the OMA collaboration is structured. As it relates to OMA, the maximum spanning tree constellation provides useful semi-quantitative insight to the structure of collaboration network. Using the same approach as in Figure 8 for 3GPP, Figure 11 shows the following groups. Group A: Mainly network operators, Group B: Mainly equipment vendors, Group C: Mainly IT companies and Group D: Mainly Chinese companies.

Yearly trees show strong network operator influence in the beginning followed by a growing role of equipment vendors and finally by the year 2008 the equipment and technology vendors including also

Figure 11. Maximum spanning tree of the OMA collaborative network consolidated over the years 2003-2008.

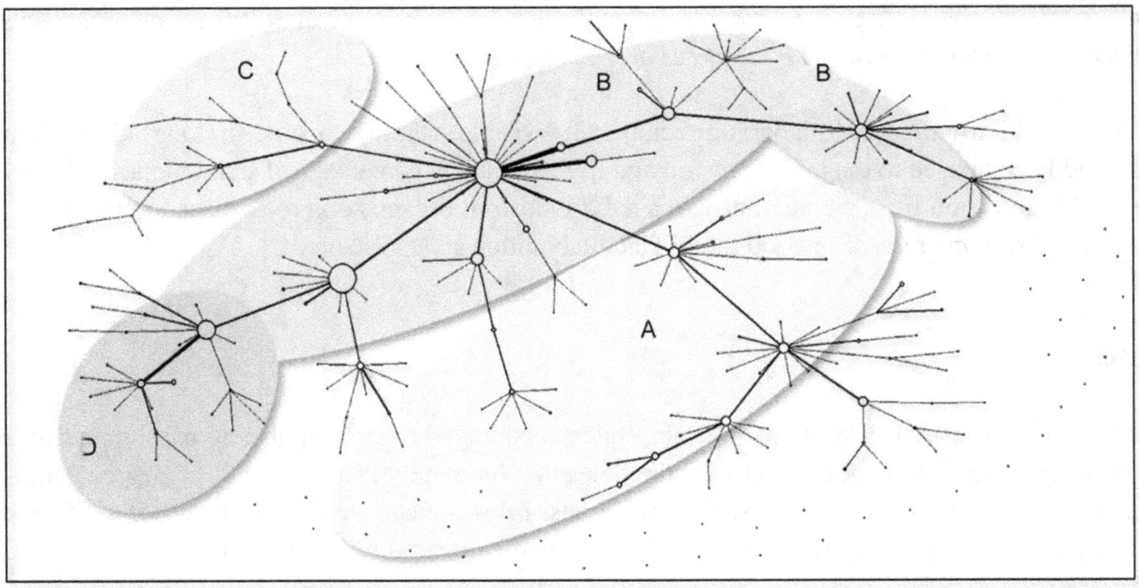

new entrants from Far East, especially from China, dominate the collaboration in OMA. The IT companies tried to seek collaboration with a number of other contributors instead of systematically building their own cluster. The consolidated data shows the IT companies forming a small cluster of their own in the Figure 11.

The role of IT companies was seen stronger in some interviews than what can be observed in the maximum spanning tree. This may reflect the good intention to bring the IT companies into the mobile communications standardization but at the same time, it shows the difficulties on practical level.

In OMA, it was decided in very early on to involve all the key parties, telecom operators, equipment manufactures and IT companies. The Board consisted of these 3 baskets, and the CDMA companies. The board and technical plenary were separated to let the technology decisions be based on technical merits only,

explained one ex-member of the OMA board. The specific role of Chinese companies was not foreseen at all when OMA was set up.

The collaboration style in OMA shows stable industrial behaviour from the beginning. This is in-line with the conscious decision-making in the formation process of OMA.

IETF

The Internet oriented SSOs are working with technologies for businesses with less regulatory constraints. IETF collaboration structure shows power law distribution both in degree and eigenvector centrality. IETF is less hierarchic than 3GPP and OMA but on the other hand, the role of the leading company is significant both locally as a very active contributor and globally as a strong influencer throughout the network. Separately, a 3GPP standardization expert working also in IETF noted the importance of the leading individuals in the consensus building:

The significant role of strong individuals can make big things in IETF. In 3GPP, technical arguments are more powerful while in IETF the gurus are very powerful.

The very high number of collaborating companies is visible in the network, i.e. the long tail contributors are steadily connected to the main network, and IETF has a low percentage of stand-alone contributors. The IETF transition from a community to a solid industrial organization took place around 1997 and therefore during the research period the IETF collaboration style is stable.

W3C

When W3C was established the membership rules encouraged a large number of participants to join. This caused later a rather high churn in the delegate community. The long tail of degree centrality (weighted and unweighted) reflecting the open access rules is clearly visible. The tails of W3C as well as IETF distributions are in relative terms longer when compared to the tails of the OMA and 3GPP distributions. The collaboration profiles and separately the long tail of contributor distributions show similarities between IETF and W3C.

The W3C collaboration shows strong disassortative behaviour from 1999 onwards. The available data covers only the period from 5 years after the formal set up of the organization. The strong negative assortativity is visible in the maximum spanning trees on yearly level also where the trees break down to several separate smaller networks, kind of spanning bushes. This may be an indication of a strong competition between the different browser technologies during those years.

IEEE 802.11

The collaboration in IEEE 802.11 shows structural similarities to other Internet oriented SSOs, IETF and W3C. However, IEEE 802.11 is different from all the other SSOs in two aspects. The collaboration profile shows the high proportion of stand-alone contributors and the assortativity scores show the strong variation over time. Both are unique features of the IEEE 802.11 collaboration requiring further discussion.

The high number of contributors providing only stand-alone contributions leaves the collaborating contributors in minority. The collaboration approach of this minority however, follows the patterns of other Internet oriented SSOs regarding network clustering as well as in degree and eigenvector centrality. By observing the circle sizes in the IEEE 802.11 collaboration spanning tree it is rather clear that the connected contributors are also the most active contributors while the large cloud of disconnected nodes consists of less active contributors. The SNA results reflect the behaviour of the most active contributors in IEEE 802.11, as is the case for other SSOs also. As brought up by one IEEE 802.11 expert during the interview, engineers personally interested in a technology and seeking to promote it for broader use are driving the collaboration in IEEE 802.11. The weak and unstructured maximum spanning tree of IEEE 802.11 therefore is likely to indicate focused technology orientation of the contributors. The traditional "individual" membership makes the difference to the way the SSO is operating as noted in the earlier research also (Lemstra, Hayes, & Groenewegen, 2011).

The oscillating assortativity of the IEEE 802.11 collaboration style is quite an important discovery. All other SSOs show only steady negative assortativity scores during the research period. Therefore, the strong temporary positive scores in the IEEE 802.11 collaboration need special attention. The cycling seems to synchronize with the major undertakings in the IEEE 802.11 community. In the Figure 12, the slopes towards negative peaks in 2002 and 2007 and for the positive peaks in 2000, 2004 (and 2008) align with the technology selection and finalization phases of the IEEE 802.11 releases of 11g, 11n, respectively[12].

When observing the development of the IEEE 802.11 maximum spanning tree parallel with the cycling assortativity it becomes apparent that for each cycle the leading contributors seem to focus first on the concept and system level parameters making the assortativity positive. Later in the process when details are defined, resilient collaboration is needed and assortativity becomes negative. The interviews emphasize the market orientation as a critical success factor for any standardization, including IEEE 802.11. Low presence of end user level customers, like network operators or Wi-Fi access network owners facilitates the specific collaboration characteristic of the IEEE 802.11 group. Furthermore, this observation correlates with a comment by one interviewee that the deployment of the IEEE 802.11 standardized technologies and products require only local decisions. In commercial applications, e.g. in corporations, Wi-Fi networks can be upgraded at the same pace with laptop computer population.

Figure 12. Aligning IEEE 802.11 standards releases and the assortativity of the collaboration

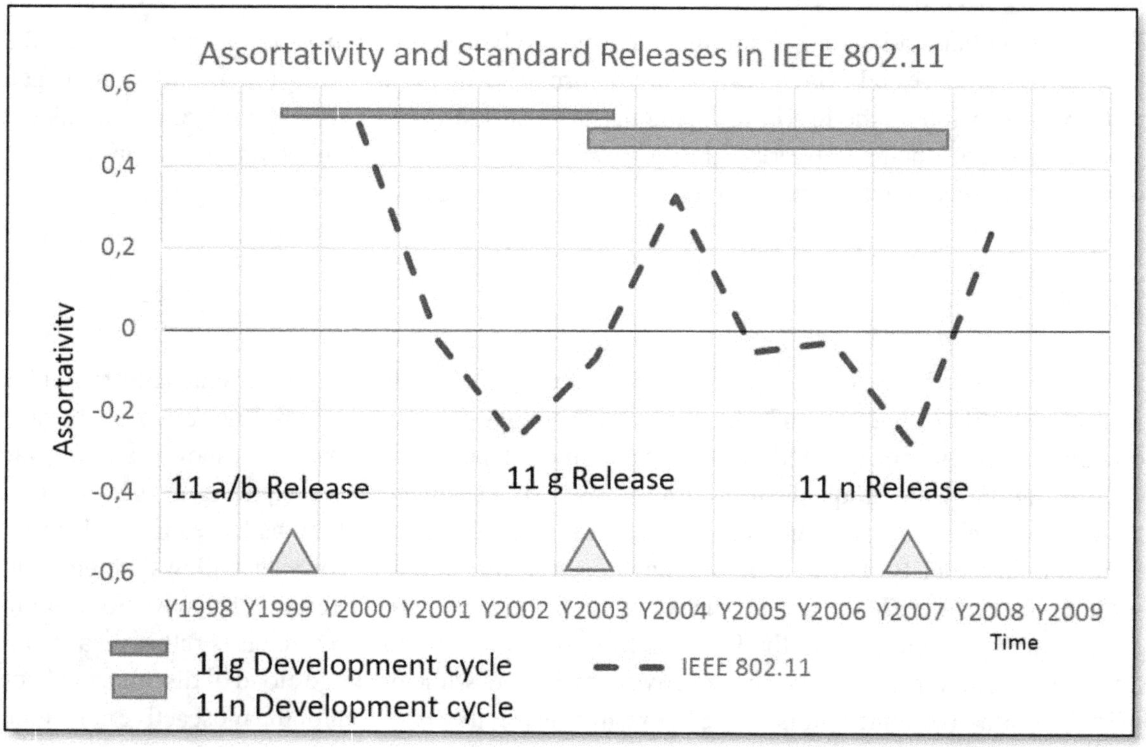

Therefore, each new product renewal cycle is a strong driver in standardization. Backward interoperability and roaming capabilities in Wi-Fi networks have not been very critical features, especially during the years 2003-2008. The Wi-Fi comes as a by-product of the computer industry (Lehr & McKnight, 2003). The less coordinated nature of the IEEE 802.11 process has, however, has caused several hiccups and glitches during the lifetime of the group (Kaa van de & Bruijn de, 2015). Even if there are multiple parallel activities in IEEE 802.11, the main releases of the new Wi-Fi radio technologies set the pace to the overall collaboration.

In the other SSOs similar cycling of collaboration mode is not observed as the groups steadily maintain the negative assortativity. The factors however, behind the observation may be different for each SSO. The interviews related to W3C emphasize the deliberate decision to avoid any specific standard release cycles due to the possibility to download the software implementations of web browsers and servers using the W3C standards as the new versions become available. The same is valid for IETF standards into large extend. Furthermore, the large number of working groups each with their own schedules makes the assortativity to show the average scores only. For 3GPP, where the ridged releases are used the assortativity still does not cycle. 3GPP operative mode covers the development of multiple system level releases at the same time with strong backwards and forwards interoperability requirements. There is no possibility to develop any system standard fully release independently from other releases[13]. The standards development work is in a continuous flow. Originally, OMA was also aiming at full system releases but soon it became evident that it was impossible and each OMA work item got a permission to proceed according to their own schedule. All these factors require mind-set of continuous development and therefore assortativity remains stable over the years. An analysis addressing subgroup level collaboration is for further study.

As discussed in section 3 the WFA promotes the deployment of the Wi-Fi products through the certification process and therefore part of the decision-making related to the Wi-Fi standardization is outside of the IEEE 802.11 group. In other words, the Wi-Fi community sets standards using consecutive technology definition gates as indicated by the assortativity analysis and by observations on the process from standards through certification and market making to commercial products[14]. The other SSOs seem to develop standards in a more continuous process and they rely mainly on their internal competences to define the further steps in the process e.g. the certifications related to 3GPP and OMA (Global Certification Forum, 2014)

SIMILAR AND DIFFERENT

The visible collaboration is similar in all standardization organizations studied in this research. All SSOs follow the process where initial ideas, requirements and technology capabilities are converted into solid specifications and standards to make the compliant products to interoperate. The contribution process is similar based on written documents provided by the contributors. Some contributors are more active in collaboration and contributions as becomes visible through quantitative analyses. All the SSOs studied are "Small Worlds" of their own.

The SSOs working in an environment where licensed network operators have strong role have harmonized their collaborations structures and styles almost completely. The needs of the licensed network operators drive the standardization in 3GPP and OMA (Brzezinski, 2007). Similar alignment towards another set of characteristics is visible in the environment where such licenses are not necessary. These observations are aligned with earlier literature indicating high likelihood of synchronization of two complex systems with similar dynamics (Strogatz, 2001).

Another high-level differentiator between the SSOs is in their focused component or broad systems thinking mind-set. 3GPP covers all the layers of the mobile communications requiring system level approach. Software based implementation of the standards enables, and also requires IETF, W3C and also OMA to pay attention to interoperability through downloading. IEEE 802.11 stays alone as s focused physical layer oriented group. The different natures of hardware and software implementations as well as component and system scope motivate the contributing companies with commercial interests in products and services to optimize their strategies to deal with market uncertainties differently in different environments. (Gaynor & Bradner, 2001).

The third area of different behaviours relates to the historical roots and traditions to compete and collaborate within these communities. The level of collaboration falls into three groups where IETF and W3C form one group OMA and 3GPP form another group whereas the collaboration in IEEE 802.11 has its own characteristic. These traditions increase trust and productivity but when some contributors are highly collaborative such a behaviour may also raise questions on free riding. High collaborative attitude however, brings value by increasing stability and consensus. Similarly, some of the stand-alone contributors, after all, may be collaborative even when their contributions in short term may create confrontation.

The key findings of this study are in the table 8 below.

Table 8. Key findings of the SNA complemented with qualitative observations

	Mobile Communications	Web, Internet	and access
	OMA, 3GPP	**IETF, W3C**	**IEEE 802.11**
Collaboration profile	High	Very high	Low
Transitivity/Clustering	Very high	High	High
Local activity Weighted Degree Centrality distribution	Exponential, Medium tail	Exponential Long/Medium tail	Exponential, Short tail
Global structural influencing Unweighted Degree and Eigenvector Centrality distribution	Exponential Medium tail	Power law Long/Medium tail	Power law Short tail
Assortativity	Negative	Negative	Cycling positive and negative
Regulative framework	Licensed (3GPP)	Unlicensed	License Exempted
End customer presence	High	Medium/Low	Low
Interoperability Focus	Very high Global	High Global	Primarily Local
Scope	System	Network	Access/Component

CONCLUSION

This research identified three different types of standardization models for mobile communications and internet technologies. The collaboration models in 3GPP and OMA are in full alignment, in IETF and in W3C, the collaboration networks have strong mutual similarities, and the collaboration in IEEE 802.11 has partially its own character. The results also give strong implications to see the causality between the predefined regulatory framework for the standardization and the implemented approach in each SSO.

The real difference between the SSOs is hiding deep in the collaborative structures and processes, which become visible using the tools like SNA. The SNA and the intuitive observations of the interviews align very well. The specific need of the global structural collaboration, indicated especially by the eigenvector centrality distributions in case of OMA and 3GPP is not easy to measure without SNA tools. Similarly, the fluctuations of collaborative style over time as observed especially in case of IETF and IEEE 802.11 become visible only when using advanced metrics like assortativity. The number of the documents used in the quantitative analyses is rather high and meaningful differences become visible only by reducing redundant information wisely. In standardization, every company is connected to every other company in large meetings and through even larger email reflectors but the challenge is to find out which of the connections are important. The comparable analysis of the five SSOs aims to find relative differences between the forums rather than absolute values or truths.

When building on the top of the quantitative SNA results the qualitative observations become meaningful. Until today, understanding the way of working in standardization may have been difficult in general but when all the documents in the ICT standardization are downloadable it is possible to run SNA almost in real time and observe very early on if some important changes are about to take place. Similarly, for a company considering joining a collaborative effort understanding the invisible structures of information sharing and influencing might be quite useful. Social network analysis has been applied to many network problems in the past. Standardization collaboration networks have not been extensively studied so far using SNA tools. The results of this study are into large extend aligned with previous SNA based research. Observed deviations will require further work and potentially new algorithms especially for analyses of time variant, dense networks with multiple goals, subgroups and priorities. The SNA is a very promising research approach to study complex collaborative networks.

Standardization professionals might find most of the findings intuitively plausible and encouraging for further collaboration in standardization with holistic and systemic mind-set. Results of this research invite to explore collaboration between SSOs. While visible actions of the contributors in information sharing may be rather similar in all SSOs, the well-known challenges of multi-forum collaboration may become manageable through a deep tacit knowledge on behaviours in each SSO. If the different long term, less visible aspirations as discovered in the assortativity time series or the different needs to collaborate during the standardization process, as visualized in the network structures are not understood, it is likely that the inter-SSO collaboration is vulnerable to several misunderstandings and unrealistic expectations. As a further study the complexities of the inter SSOs activities such as the IMS standardization will be useful.

Finally, the implications of this research to the policy makers include encouragement to build the bridge between constraints set by the policies and the expected output of the related standardization initiatives. Without systemic set of rules, including competition and collaboration models, standardization is not able to progress efficiently. Different licensing models while liberating the use of scarce resources require complementing interoperability requirements. This equilibrium includes implications related to governance rules for the applicable standardization community. By understanding the fundamental differences between the forums and groups better, a peaceful and fruitful collaboration become achievable. These learnings will be crucial when ICT based solutions, services and products will spread to and shape gradually all sectors of human life. Furthermore, explicit needs to standardize new interfaces on higher layers of abstraction e.g. data and identity will emerge. In the future standardization peaceful collaboration will be at least as useful and fruitful behaviour than war and direct competition.

ACKNOWLEDGMENT

Nokia Corporation supported this research by providing access to the raw empirical data. Otherwise, the author is responsible on all the research and results alone. The author wants to thank specifically the anonymous interviewees who provided valuable insights to the collaboration and competition within and between the SSOs. Finally, the author thanks the anonymous reviewers who provided very valuable comments and constructive criticism during the process.

REFERENCES

3GPP. (2002). *Technical Specification Group Services*. Retrieved from http://www.3gpp.org/ftp/tsg_sa/TSG_SA/TSGS_17/Docs/PDF/SP-020418.pdf

Ali-Vehmas, T., & Casey, T. R. (2012). Evolution of wireless access provisioning: A systems thinking approach. *Competition and Regulation in Network Industries*, *13*(4), 333–361. doi:10.1177/178359171201300402

Anderson, C. (2006). *The long tail: Why the future of business is selling less of more*. New York, NY: Hyperion Books.

ANSI. (n.d.a). *About Ansi*. Retrieved from https://www.ansi.org/about_ansi/overview/overview.aspx?menuid=1

ANSI. (n.d.b). *ANSI Standards*. Retrieved from https://share.ansi.org/Shared%20Documents/Standards%20Activities/American%20National%20Standards/ANSI%20Accredited%20Standards%20Developers/AUG2016ASD.pdf

ATIS. (n.d.). *About our Work*. Retrieved from http://www.atis.org/01_about/our_work.asp

Bar, T., & Leiponen, A. E. (2014). Committee composition and networking in standard setting: The case of wireless telecommunications. *Journal of Economics & Management Strategy*, *23*(1), 1–23. doi:10.1111/jems.12044

Barabási, A.-L., & Bonabeau, E. (2003). Scale-Free Networks. *Scientific American*, (May): 50–59. PMID:12701331

Baron, J., Meniere, Y., & Pohlmann, T. (2014). Standards, consortia, and innovation. *International Journal of Industrial Organization*, *36*, 22–35. doi:10.1016/j.ijindorg.2014.05.004

Blind, K. (2011). An economic analysis of standards competition: The example of the ISO ODF and OOXML standards. *Telecommunications Policy*, *35*(4), 373–381. doi:10.1016/j.telpol.2011.02.007

Blind, K., & Gauch, S. (2008). Trends in ICT standards: The relationship between European standardisation bodies and standards consortia. *Telecommunications Policy*, *32*(7), 503–513. doi:10.1016/j.telpol.2008.05.004

Bluetooth. (n.d.). *Our History*. Retrieved from https://www.bluetooth.com/media/our-history

Brunsson, N., Rasche, A., & Seidl, D. (2012). The Dynamics of Standardization: Three Perspectives on Standards in Organization Studies. *Organization Studies*, *33*(5-6), 613–632. doi:10.1177/0170840612450120

Brzezinski, K. M. (2007). Network Operators' Requirements and the Structure of Telecommunications Standards. *International Journal of IT Standards and Standardization Research*, *5*(1), 103–117. doi:10.4018/jitsr.2007010102

Cabral, L., & Salant, D. (2014). Evolving technologies and standards regulation. *International Journal of Industrial Organization*, *36*, 48–56. doi:10.1016/j.ijindorg.2013.07.006

Campbell-Kelly, M., & Garcia-Swartz, D. D. (2013). The history of the internet: The missing narratives. *Journal of Information Technology*, *28*(1), 18–33. doi:10.1057/jit.2013.4

CCSA. (n.d.). *About*. Retrieved from http://www.ccsa.org.cn/english/about.php

CEPT. (n.d.). Retrieved from http://www.cept.org/

CERN. (n.d.). *About CERN*. Retrieved from http://home.cern/about

Choi, H., Kim, S.-H., & Lee, J. (2010). Role of network structure and network effects in diffusion of innovations. *Industrial Marketing Management*, *39*(1), 170–177. doi:10.1016/j.indmarman.2008.08.006

Crespo, J., Suire, R., & Vicente, J. (2014). Lock-in or lock-out? How structural properties of knowledge networks affect regional resilience. *Journal of Economic Geography*, *14*(1), 199–219. doi:10.1093/jeg/lbt006

Crespo, J., Suire, R., & Vicente, J. (2016). Network structural properties for cluster long-run dynamics: Evidence from collaborative R&D networks in the European mobile phone industry. *Industrial and Corporate Change*, *25*(2), 261–282. doi:10.1093/icc/dtv032

Critical Values for the Two-sample Kolmogorov-Smirnov Test. (n.d.). Retrieved from https://www.webdepot.umontreal.ca/Usagers/angers/MonDepotPublic/STT3500H10/Critical_KS.pdf

Crocker, D. (1993). Making standards the IETF way. *StandardView*, *1*(1), 48–56. doi:10.1145/174683.174689

DARPA. (n.d.). *History and Timeline*. Retrieved from http://www.darpa.mil/about-us/darpa-history-and-timeline

David, P. A. (1985). Clio and the Economy of QWERTY. *The American Economic Review*, *75*(2), 332–337.

David, P. A., & Greenstein, S. (1990). The Economics of Compatibility Standards: An Introduction to recent research. *Economics of Innovation and New Technology*, *1*(1–2), 3–41. doi:10.1080/10438599000000002

Delcamp, H. R., & Leiponen, A. E. (2014). Innovating Standards Through Informal Consortia: The Case of Wireless Telecommunications. *International Journal of Industrial Organization*, *36*, 36–47. doi:10.1016/j.ijindorg.2013.07.004

Di Minin, A., & Bianchi, M. (2011). Safe nests in global nets: Internationalization and appropriability of R&D in wireless telecom. *Journal of International Business Studies*, *42*(7), 910–934. doi:10.1057/jibs.2011.16

Economides, N. (1996). Network Externalities, Complementarities and Invitations to Enter. *European Journal of Political Economy*, *12*(2), 211–233. doi:10.1016/0176-2680(95)00014-3

Ende van den, J., Kaa van de, G., Uijl den, S., & Vries de, H. J. (2012). The paradox of standard flexibility: The effects of co-evolution between standard and interorganizational network. *Organization Studies, 33*(5–6), 705–736. doi:10.1177/0170840612443625

ETsI. (n.d.). Retrieved from http://www.etsi.org/about/what-we-are

Farrell, J., & Klemperer, P. (2007). Chapter 31 Coordination and Lock-In: Competition with Switching Costs and Network Effects. Handbook of Industrial Organization. doi:10.1016/S1573-448X(06)03031-7

Fransman, M. (2010). *The New ICT Ecosystem: Implications for Policy and Regulation*. Cambridge, UK: Cambridge University Press. doi:10.1017/CBO9780511676130

Funk, J. L., & Methe, D. T. (2001). Market- and committee-based mechanisms in the creation and diffusion of global industry standards: The case of mobile communication. *Research Policy*, *30*(4), 589–610. doi:10.1016/S0048-7333(00)00095-0

Gandal, N., Salant, D., & Waverman, L. (2003). Standards in wireless telephone networks. *Telecommunications Policy*, *27*(5–6), 325–332. doi:10.1016/S0308-5961(03)00026-0

Garas, A., Tomasello, M. V., & Schweitzer, F. (2014). *Selection rules in alliance formation: strategic decisions or abundance of choice?* Retrieved from http://arxiv.org/abs/1403.3298

Gaynor, M., & Bradner, S. (2001). The real options approach to standardization. *Proceedings of the 34th annual Hawaii International Conference On System Sciences*. doi:10.1109/HICSS.2001.926526

Gençer, M. (2007). Structure and collaborative aspects of internet standards. *Proceedings - IEEE Symposium on Computers and Communications*, 733–739. doi:10.1109/ISCC.2007.4381499

Gençer, M. (2012). The evolution of IETF standards and their production. *International Journal of IT Standards and Standardization Research*, *10*(1), 17–33. doi:10.4018/jitsr.2012010102

Global Certification Forum. (2014). Retrieved from http://www.globalcertificationforum.org/images/downloads/GCF-WP-GCF_Certification_June2014.pdf

Grøtnes, E. (2008). Strategies for Influencing the Standardization Process: Examples from Within. In *16th European Conference on Information Systems*. Association for Information Systems.

Gunasekaran, V., & Harmantzis, F. C. (2008). Towards a Wi-Fi ecosystem: Technology integration and emerging service models. *Telecommunications Policy*, *32*(3–4), 163–181. doi:10.1016/j.telpol.2008.01.002

Hanneman, R. A., & Riddle, M. (2005). *Introduction to social network methods. Riverside, CA: University of California*. Retrieved from http://faculty.ucr.edu/~hanneman/

Haug, T. (2002). A commentary on standardization practices: Lessons from the NMT and GSM mobile telephone standards histories. *Telecommunications Policy*, *26*(3–4), 101–107. doi:10.1016/S0308-5961(02)00003-4

Hawkins, R., & Ballon, P. (2007). When standards become business models: Reinterpreting "failure" in the standardization paradigm. *Info*, *9*(5), 20–30. doi:10.1108/14636690710816426

Hillebrand, F. (2013). The creation of standards for global mobile communication: GSM and UMTS standardization from 1982 to 2000. *IEEE Wireless Communications*, *20*(5), 24–33. doi:10.1109/MWC.2013.6664470

IEEE. (n.d.). *History*. Retrieved from https://www.ieee.org/about/ieee_history.html

Internet Society. (n.d.). *History*. Retrieved from http://www.internetsociety.org/history

Internet World Stats. (n.d.). *Marketing*. Retrieved from http://www.internetworldstats.com/emarketing.htm

Isaak, J. (2006). The Role of Individuals and Social Capital in POSIX Standardization. *International Journal of IT Standards and Standardization Research*, *4*(1), 1–23. doi:10.4018/jitsr.2006010101

Jakobs, K. (2003). Information technology standards, standards setting and standards research. *Cotswolds Conference, Stanhope Center for Communication Policy Research*, 8–9.

Jakobs, K. (2008). ICT standardisation - co-ordinating the diversity. *International Telecommunication Union - Proceedings of the 1st ITU-T Kaleidoscope Academic Conference, Innovations in NGN, K-INGN*. doi:10.1109/KINGN.2008.4542257

Jakobs, K., Procter, R., & Williams, R. (2001). The Making of Standards : Looking Inside the Work Groups. *IEEE Communications Magazine*, *39*(4), 102–107. doi:10.1109/35.917511

Katusic, D., Weber, M., Bojic, I., Jezic, G., & Kusek, M. (2012). Market, standardization, and regulation development in Machine-to-Machine communications. *20th International Conference on Software, Telecommunications and Computer Networks (SoftCOM)*, 1–7.

Katz, M. L., & Shapiro, C. (1985). Network Externalities, Competition, and Compatibility. *The American Economic Review*, *75*(3), 424–440.

Kwak, J. Y., Lee, H., & Chung, D. B. (2012). The evolution of alliance structure in China's mobile telecommunication industry and implications for international standardization. *Telecommunications Policy*, *36*(10–11), 966–976. doi:10.1016/j.telpol.2012.07.017

Lehr, W. H., & McKnight, L. W. (2003). Wireless Internet access: 3G vs. WiFi? *Telecommunications Policy*, *27*(5-6), 351–370. doi:10.1016/S0308-5961(03)00004-1

Leiner, B. M., Cerf, V. G., Clark, D. D., Kahn, R. E., Kleinrock, L., Lynch, D. C., & Wolf, S. et al. (2009). A Brief History of the Internet. *Computer Communication Review*, *39*(5), 22–31. doi:10.1145/1629607.1629613

Leiponen, A. E. (2008). Competing Through Cooperation: The Organization of Standard Setting in Wireless Telecommunications. *Management Science*, *54*(11), 1904–1919. doi:10.1287/mnsc.1080.0912

Lemstra, W., & Hayes, V. (2009). License-exempt: Wi-Fi complement to 3G. *Telematics and Informatics*, *26*(3), 227–239. doi:10.1016/j.tele.2008.11.003

Lemstra, W., Hayes, V., & Groenewegen, J. (2011). *The Innovation Journey of Wi-Fi. The Innovation Journey of Wi-Fi: The Road to Global Success*. Cambridge University Press. doi:10.1017/CBO9780511666995

Lerner, J., & Tirole, J. (2006). A Model of Forum Shopping, with Special Reference to Standard Setting Organizations. *The American Economic Review*, *96*(4), 1091–1113. doi:10.1257/aer.96.4.1091

Lyytinen, K., & King, J. L. (2002). Around the cradle of the wireless revolution: The emergence and evolution of cellular telephony. *Telecommunications Policy*, *26*(3–4), 97–100. doi:10.1016/S0308-5961(02)00002-2

Majchrzak, A., Jarvenpaa, S. L., & Bagherzadeh, M. (2015). A Review of Interorganizational Collaboration Dynamics. *Journal of Management*, *41*(5), 1338–1360. doi:10.1177/0149206314563399

Martinez-Romo, J., Robles, G., Gonzalez-Barahona, J. M., & Ortuño-Perez, M. (2008). Using Social Network Analysis Techniques to Study Collaboration between a FLOSS Community and a Company. *IFIP International Federation for Information Processing*, *275*, 171–186. doi:10.1007/978-0-387-09684-1_14

Newman, M. E. J. (2001). The structure of scientific collaboration networks. *Proceedings of the National Academy of Sciences of the United States of America, 98*(2), 404–409. doi:10.1073/pnas.98.2.404 PMID:11149952

Pohlmann, T. (2014). The Evolution of ICT Standards Consortia. *The DigiWorld Economic Journal,* (95), 17–40.

Rogers, E. M. (1995). *Diffusion of innovations*. New York, NY: Free Press.

Shin, D. H., Kim, H., & Hwang, J. (2015). Standardization revisited: A critical literature review on standards and innovation. *Computer Standards & Interfaces, 38*, 152–157. doi:10.1016/j.csi.2014.09.002

Simcoe, T. (2012). Standard setting committees: Consensus governance for shared technology platforms. *The American Economic Review, 102*(1), 305–336. doi:10.1257/aer.102.1.305

Simcoe, T. (2015). Modularity and the evolution of the Internet. In Economical Analysis of the Digital Economy (pp. 21–54). Chicago: University of Chicago Press.

SIP Forum. (n.d.). Retrieved from http://www.sipforum.org/content/view/13/34/

Sterman, J. D. (2002). All models are wrong: Reflections on becoming a systems scientist. *System Dynamics Review, 18*(4), 501–531. doi:10.1002/sdr.261

Strogatz, S. H. (2001). Exploring complex networks. *Nature, 410*(6825), 268–276. doi:10.1038/35065725 PMID:11258382

Swann, P. G. M. (2010). *The Economics of Standardization: An Update*. Academic Press.

Tilson, D., & Lyytinen, K. (2006). The 3G transition: Changes in the US wireless industry. *Telecommunications Policy, 30*(10–11), 569–586. doi:10.1016/j.telpol.2006.09.002

Toppila, A., Liesiö, J., & Salo, A. (2011). A Resource Allocation Model for R&D Investments: A Case Study in Telecommunication Standardization. In A. Salo, J. Keisler, & A. Morton (Eds.), *Portfolio Decision Analysis: Improved methods for resource allocation* (pp. 241–258). New York, NY: Springer New York; doi:10.1007/978-1-4419-9943-6_11

Trestian, R., Ormond, O., & Muntean, G. (2012). Game Theory — Based Network Selection : Solutions and Challenges. *Communications Survey & Tutorials, 14*(4), 1212–1231. doi:10.1109/SURV.2012.010912.00081

van de Kaa, G., & de Bruijn, H. (2015). Platforms and incentives for consensus building on complex ICT systems: The development of WiFi. *Telecommunications Policy, 39*(7), 580–589. doi:10.1016/j.telpol.2014.12.012

van de Kaa, G., & de Vries, H. J. (2015). Factors for winning format battles: A comparative case study. *Technological Forecasting and Social Change, 91*, 222–235. doi:10.1016/j.techfore.2014.02.019

van de Kaa, G., van den Ende, J., de Vries, H. J., & van Heck, E. (2011). Factors for winning interface format battles: A review and synthesis of the literature. *Technological Forecasting and Social Change, 78*(8), 1397–1411. doi:10.1016/j.techfore.2011.03.011

Vasudeva, G., Alexander, E. A., & Jones, S. L. (2015). Institutional Logics and Interorganizational Learning in Technological Arenas: Evidence from Standard-Setting Organizations in the Mobile Handset Industry. *Organization Science*, *26*(3), 830–846. doi:10.1287/orsc.2014.0940

Von Bertalanffy, L. (1972). The history and status of general systems theory. *Academy of Management Journal*, *15*(4), 407–426. doi:10.2307/255139

W3C. (n.d.). *History*. Retrieved from https://www.w3.org/History.html

Wasserman, S., & Faust, K. (1994). *Social network analysis: Methods and applications*. Cambridge University Press. doi:10.1017/CBO9780511815478

Watts, D. J., & Strogatz, S. (1998). Collective dynamics of "small-world" networks. *Nature*, *393*(6684), 440–442. doi:10.1038/30918 PMID:9623998

Weiss, M. B. H., & Sirbu, M. (1990). Technological choice in voluntary standards committees: An empirical analysis. *Economics of Innovation and New Technology*, *1*(1–2), 111–133. doi:10.1080/10438599000000007

Wi-Fi Alliance. (n.d.). Retrieved from http://www.wi-fi.org/

ENDNOTES

1. In this paper, "Collaboration" refers to "Cooperative and "Coopetitive" behaviors while "Competitive" refers to stand-alone approach. Collaborating contributor aligns with the connected node of a social network while Competing-only contributors refers to disconnected nodes.
2. In this paper, abbreviation "SSO" refers to any kind of standard-setting group or activity including SSOs developing, defining or mandating standards.
3. For a theoretical background of SNA with definitions of the SNA terminology used in this research more details are available in Appendix 1 of the related article in the International Journal of Standardization Research Volume 14 Issue 2 July December 2016.
4. Request For Comments, referring to the documents used in IETF.
5. IEEE-SA claims 183 corporate members
6. Marketing name for IEEE 802.16 based wide area wireless technology.
7. It is for further study to evaluate different structures of the elementary networks by assigning different weights to the different links based on the further interviews of the standardization experts.
8. For more details, see Appendix 1 of the related article in the International Journal of Standardization Research Volume 14 Issue 2 July December 2016.
9. The hyperlinks are directed while the standardization collaboration connections in this research are undirected.
10. Empirical data for each SSO is available differently. Assortativity is visible for all years where sensible amount of data is available noting that the number of documents outside of the research focus area 2003 – 2008 limited.
11. Weighted degree centrality scores can be found in the related article in the International Journal of Standardization Research Volume 14 Issue 2 July December 2016

[12] The IEEE 802.11n standard was technically ready in 2007 but it took two more years to achieve the final acceptance of the work.

[13] There is a number of 3GPP standards developed in the release independent fashion.

[14] The documents contributed in the WFA are not available for this research and therefore the impact of WFA is not taken into account quantitatively in this analysis.

Chapter 4
Diffusion vs. Knowledge Protection When Participating in ICT Standardization

Cesare A. F. Riillo
STATEC, Luxembourg & ANEC, Luxembourg

Ivana Mijatovic
University of Belgrade, Serbia

ABSTRACT

Innovative companies need to make decisions related to protection and sharing of their new knowledge. This chapter explores knowledge patterns during formal ICT standardization and/or in standards consortia. While previous studies focused on the role of intellectual property rights in standardization decisions, this investigation considers standards and patents as sources of information for innovation. Results show that when patents are important, firms are more likely to be engaged in consortia but not in formal standardization. Size moderates this relationship. Smaller firms are more motivated by the quest for knowledge during standardization while larger firms are more sensitive to the importance of patents. The authors conclude that small firms might participate in ICT standardization in order to anticipate development of the standard, rather than place a patent in the standards.

INTRODUCTION

Distinguishing between general knowledge and economically useful knowledge, the creation of economically useful knowledge and its effective conversion into economic activity are essential for economic growth (Carlsson et al. 2009). There are many mechanisms through which economically useful knowledge can lead to economic growth; among them are standardization and patenting. Strategies for managing organizational knowledge in modern companies can have many alternatives which are not mutually exclusive – like safeguarding knowledge through IPR mechanisms and sharing knowledge in standardization processes.

DOI: 10.4018/978-1-5225-5320-5.ch004

Standards are often seen as important and valuable source of knowledge - they preventing "reinventing the wheel" and as base for market access and acceptance of innovations and new technologies (De Vries, 2006). Patents are types of intellectual properties, with main aim to protect invention. A patent is a grant from a government "that permits its owner to prevent others from making using, importing and selling an invention" (Bouchoux, 2009, pp 6.). While both standards and patents aim to promote innovation and market place adoption, there is little else that they have in common (Blind et al., 2011). A company that has invented something can choose between patents and standardization or their combination, depending on which option results in the greatest market success - standards and patents both contain mostly technical solutions - patents are one source of information and standards are another (De Vries, 2006).

The growth of standard has coincided with the increase of intellectual property protection and it is now common place that patents find their way into standardization. They have become a necessary ingredient to implementation and formulation alike. Some incorporated patents are more influential and core to specific standards than others and those that are "core" or "necessary" to effectively implement a standard are called "standard essential patent. (Pierce and Medzmaraishvili, 2015, pp.306)

IPR are an increasingly important part of the technology evaluation process at many standards setting organizations (SSO) and many companies would like to own IPR that are embedded in an industry standard (Rysman and Simcoe, 2008).

According to Blind and Thumm, in area of technological development, the contradiction of intellectual property rights (IPR) and formal standard is based on two aspects: IPR provide knowledge-producers with the temporary right of exclusive exploitation of the benefits deriving from the new knowledge and standards are decisive for the diffusion of new technologies - they make information about new technologies available to everyone (Blind and Thumm, 2004). How standards organizations respond to those who assert IPR is critically important because retaining of IPR in group standards will determine whether a standard is open or closed; who can sell compliant products; it may well influence whether the standard adopted in the market is one chosen by a group or one offered by a single company, on the other side SSO rules governing IP rights will also affect how standards change as technology improves (Lemley, 2002). Competing companies join SSOs and work together to select particular technologies as industry standards and this process promotes price competition among firms, creates demand-side economies of scale, and encourages product innovation by reducing the risks of future research and development efforts (Curran, 2003).

In this research, we are interested about how IPR issues influence company decision to be engaged in standardization. Adopting the "forum shopping model" approach, this research investigates whether IPR concerns have a different influence on the probability of a firm to engage in ICT formal standardization or in consortia. The model considers as well whether firms, who perceive standards as a source of information for their innovation, are more likely to be engaged in standardization activities. Additionally, we investigate the potential moderating role of size of the company.

This paper proceeds as follows. Section II reviews results of previous studies. In Section III we define our research questions. Section IV briefly present the data available for present study, subsequently, in section V we illustrate our methodology, and in section VI we discuss our findings. Finally, section VII summarizes our work and provides some suggestions for future research.

LITERATURE REVIEW

In theory, IPR and standardization, as well as their influence on innovation and technology adoptions are quite well presented. Relation between the strength of patent protection and the rate of innovation was and still is topic of many researches. However, the findings are still quite controversial. Based on data from 19th century, findings of Moser (2005) suggest that patents helped to determine the direction of technical change and that the adoption of patent laws in countries without such laws may alter existing patterns of comparative advantage across countries - the introduction of strong patent laws may trigger changes in the direction of innovative activity in developing countries and initiate significant changes in international patterns of comparative advantage. Lerner (2009) studied 177 of the most significant shifts in patent policy, across 60 countries and during 150 years, shown the lack of a positive impact of strengthening of patent protection on innovation (the author provided some explanation for this results). Lerner (2002) explored the strength of patent protection in 60 countries over a 150-year period in relation to: relative economic strength, political conditions, and legal traditions. His findings suggest that relatively wealthier nations more frequently have patent systems; countries with democratic institutions are more likely to have patent protection and longer awards as well as differences in legal traditions are significant.

Standardization and technology development mutually influence on each other. Research of Jiang et al. (2015) explored the mechanism of these interactions - particularly the effects of technology standardization on technology progress, technology revolution and technology innovation, as well as the effects of technology development level on technology selection, technology standard formulation and technology standard popularization. Based on data from four European countries and 12 sectors, results of Blind and Jungmittag (2008), confirm that both the stock of patents and the stock of technical standards contribute significantly to economic growth in the 1990s - significant differences was observed between the sector models, which indicate that standards are more important for growth in less R&D-intensive industries and patents in R&D-intensive industries.

Motives for company engagements in standardization as well to their choice of suitable standardization organization can be different. Even those differences among formal organization for standardization (e.g. ISO, IEC, ITU); professional and industry associations (e.g. traditional U.S. standardization development organization); business alliances, consortia and fora are quite clear - from the European point of view, many traditional U.S. standardization development organization can be seen as informal (see more in Egyedi, R.Hawkins, 2011). While the economic benefit of standards seems to be widely acknowledged within scientific literature the benefit of taking part in standardization remains a more or less unexplored mystery to date (Wakke et al., 2016). In theory, standard setters not only leverage from the possibility to monitor and influence the upcoming standard but also to access a wide range of knowledge sources in the standards committee (Wakke et al., 2016).

ICT companies have many options to take part in standard development processes. In the ICT area there is a proliferation in number and type of standardization organizations (Iversen et al.,2004) and competition is shifted away from product differentiation toward business aspects like pricing, marketing, and distribution (Mattli, 2001).

The ICT standardization regime has undergone a comprehensive phase of proliferation and internationalization in step with emerging (and merging) information and communication technologies. The landscape of SDOs is global and fragmented, and there is no indication that it develops towards a unitary system. (Iversen et al.,2004)

"Forum shopping" is term used to describe behavior of companies related to phenomena that "firms can increasingly choose from a broad range of standards developing organizations (formal standards bodies, fora, consortia, etc.) and pick that one which suits their interests best (e.g. in terms of the given IPR regime, membership structure or standards preparation speed)" (Egyedi, and Hawkins, 2011). The forum shopping model implicitly assumes that firms have a choice when it comes to formal or informal organization for standardization, but in reality they most often do not - they need to be member of the many relevant standardization organizations in their industry.

Arena of standardization is complex and dynamic and influence or even consequence of forum shopping can be seen in increased level of cooperation among formal and informal organization for standardization as well as changes in formal and informal standardization organizations. On one side, formal organization for standardization introduced "fast track" procedure and specific deliverables (like Publicly Available Specifications PAS, Workshop Agreements WA), and on the other side some of consortia improve their procedures to be more transparent, open and democratic – in that way some consortia became more democratic than formal organizations for standardization.

Lerner and Tirole (2006) analyse general model which can be used in context of the firms (sponsor) which may choose the standardization committee (certifier) that fit their IPR needs ("forum shopping"). According to the authors, forum shopping is shown to exhibit three basic properties: maximal complacency (the sponsor only selects as strong a certifier as is needed), efficient design (the design is optimal for the selected certifier, enabling delegation) and monotonicity (the weaker the sponsor's idea, the more extensive the concessions and the more credible the certifier). Chiao, Lerner, and Tirole (2007) find evidence that *"negative relationship between the extent to which an SSO is oriented to technology sponsors and the concession level required of sponsors"*.

Findings of Blind (2008) based on data gained from 500 German companies, responsible for more than 40% of German application at the European Patent Office, showed that companies involved in formal standardization have lower patent intensity, whereas those engaged in consortia standardization are characterized by a higher patent intensity compared to those not involved in standardization at all. According to Blind (2008), patenting has a significant influence on the likelihood of participation in formal and informal standardization processes. The similar results are provided by Petersen, Blind and Stephan (2012): companies with a large patent portfolio are more active in formal standardization than companies with a small patent portfolio; companies with a high patent intensity are less active in formal standardization than companies with a lower patent intensity and companies with a broad geographical scope of their patent portfolio are more active in international standardization than companies which have a narrower patent portfolio.

According the research of Rauber (2014) the likelihood of entering in the SSO rises with firm size; companies that introduce new products or services into the market are significantly more likely to join formal standardization in the following years, the protection of innovations by patents enhances the likelihood to enter SSOs. According to research of Simcoe (2012) many reasons can influence company willing to participate in SSO: participants favor specific technologies, development lead times, proprietary complements, intellectual property rights, time-to-consensus or agreement as well as amount of concessions (when concessions are costly and new standards create winners and losers, SSO committees will be hampered by the rent-seeking behavior of individual participants). Blind et al. (2012) explored the relationship between firms' external knowledge sourcing and their decision to participate in standardization alliances and claimed that standardization represents a specific form of collaborative knowledge-sharing and knowledge-creating strategy. Their findings suggest that importance of external knowledge is positively correlated with participation in standardization as well as that firms cooperating with different actors are more likely to join standardization.

RESEARCH QUESTIONS

The research work presented in the paper is driven by the following two main research questions (RQs):

RQ1: How do firms' ratings of the importance of patents and their perception of standards as important source of knowledge influence their engagement in formal or informal (consortia) standardization?

RQ2: What is the influence of the firms' size (if any) on their engagement in formal or informal (consortia) standardization, in relation to their perception of importance of patents and standards as source of knowledge?

DATA

The quantitative analysis is based on a dedicate data set of Luxembourgish firms collecting information about importance of patents, standards and other firms characteristics. The data set is the combination of the Community Innovation Survey 2012 –CIS 2012-and the ICT usage survey 2013 –ICT 2013- for Luxembourg. Both surveys are official European surveys and they are made of two parts: the first part is common to all European Countries; the second part is country-specific. Dedicated questions about standards are included in both Luxembourgish questionnaires. CIS 2012 collected information about the importance of patents and standards for innovation activities for firm engaged in standardization. The ICT 2013 retrieves information about ICT standardization engagement in recognized bodies (e.g. International Organization for Standardization) and consortia (e.g. Internet Engineering Task Force).

The surveys are under the responsibility of the Statistical Institute of Luxembourg CIS 2012 and ICT 2013 are performed according to international guidelines and European legislation and they are considered particularly reliable because of the country representativeness and the high response rate (STATEC, 2015). The surveys are cross sectional and include small and large manufacturing and service firms of Luxembourg. Several considerations drive the combination of the two surveys and a dedicate procedure asses the goodness of the final sample used for the analysis. First, we note that despite their names CIS 2012 and ICT 2013, the surveys cover a similar time. Indeed, most information collected by CIS refers to the end 2012 while ICT 2013 collects information referring to January 2013. Second, the CIS 2012 covers different service sectors than ICT 2013 survey. Therefore, the sample was restricted to common sectors only: manufacturing, trade, transport, ICT and professional, scientific and technical activities. It is important to note that both surveys are designed to be representative by sectors. The representativeness is not affected by the restricting the analysis to selected sectors. Third, both surveys include firms with at least 10 employees. Larger firms are included in both surveys and a random sample is drawn for smaller firms. Therefore, there are not major issues for the size. Forth, given the particular structure of the CIS, information about importance of patent is gathered only for firms that are active in technological innovation activities (including ongoing and abandoned). Therefore, similarly other studies on patents (e.g. Brouwer and Kleinknecht, 1999), the analysis focuses on technological innovative active firms.

The number of CIS firms active in innovation and in selected sectors is 301. After the merging these firms with ICT survey, the sample reduces to 210. Merging two different dataset may possibly raise issues of representatives and interpretation of results. To assess if CIS firms retained for the analysis are significantly different from other CIS firms, some tests are performed with respect to size and sector coverage. The average number of employees of merged firm (M= 195; SD= 433) and comparable CIS firms (M=160; SD=409) is not statistically different t (509) = - 0.919, p= 0.358. In terms of sector of economic activities, the Pearson test of independence *X2* (4, N = 511) = 1.405, p= 0.84 shows that there is no significant difference. These results increase our confidence in the data used for the analysis.

The final sample is made of 204 because of same missing information in control variable. Table 1 reports the exact definitions and the descriptive statistics of the variables used in the analysis. Firms engaged in formal standardization are 22% of the sample and 7% are engaged in consortia. Previous study using all observations of ICT 2013 survey (Riillo, 2014) report a lower participation in formal standards development (nearly 6%) because standardization activities are most relevant for firms engaged in innovation activities. Empirical results are qualifying for innovative firms only.

ECONOMETRIC ANALYSIS

The econometric analysis allows investigating the feature that correlate to the decision to engage in standardization activities. In particularly we model the simultaneously decision to participate in formal and informal (consortia) standardization. We are interested to explore whether the importance of patent and the importance of standards as source of knowledge have an impact on the probability of engaging in standardization. We implement a bivariate model to account for the simultaneously decision of consortia and formal standardization. The seemingly unrelated bivariate probit (Greene, 2008), the errors of the equation explaining the consortia engagement are correlated with the errors of the equation explaining the formal standardization. The bivariate probit allows to take in account factors that can simultaneously influence both the decision of formal and consortia standardization. This is modellization is appropriate because both formal and consortia participation variables are dichotomous and information at hands do not allow to include control variables that differ among equations.

In formula:

$$y_1^* = pat.'_i \beta_1 + size_i \# pat.'_i \beta_2 + stand.'_i \beta_3 + size_i \# stand.'_i \beta_4 + \mathbf{controls}'_i \boldsymbol{\beta}_n + \varepsilon_1, \quad y_1 = 1 \text{ if } y_1^* > 0, \quad 0\,otherwise$$

$$y_2^* = pat.'_i \beta_1 + size_i \# pat.'_i \beta_2 + stand.'_i \beta_3 + size_i \# stand.'_i \beta_4 + \mathbf{controls}'_i \boldsymbol{\beta}_n + \varepsilon_2, \quad y_2 = 1 \text{ if } y_2^* > 0, \quad 0\,otherwise$$

$$E\left[\varepsilon_1 \mid \boldsymbol{x}_1, \boldsymbol{x}_2\right] = E\left[\varepsilon_2 \mid \boldsymbol{x}_1, \boldsymbol{x}_2\right] = 0$$

Table 1. Descriptive statistics

	Description	mean	sd
Formal standardization	Participation in elaboration of ICT standards in recognized bodies (e.g. ISO, CEN, CENELEC, ETSI).	0.22	0.41
Standardization in consortia	Participation in elaboration of ICT standards in consortia (e.g. IETF, W3C, OASIS).	0.069	0.25
Importance of patents	Firms reporting that patents are important to protect enterprise's knowledge or innovations (0= Not relevant; 1 = else)	0.37	0.48
Standards source of knowledge	Firms reporting that standards are important information sources for innovation activities (0= Not used or Low importance; 1 = else)	0.37	0.48
Ln emp.	(Ln) Employees 2012 (c)	4.39	1.21
Group	Conglomerate	0.58	0.50
Low Tech	Low Technology (Nace: 10-19; 22-25, excluding 25.4; 30.1; 31-33, excluding 32.5)	0.24	0.43
High Tech.	High Technology (Nace: 20-21;25.4; 26-30 excluding 30.1; 32.5)	0.12	0.33
KIS	KIS Knowledge Intensive Sectors (Eurostat Nace rev.2: 50-51; 58-63; 69-75; 78-80)	0.32	0.47
LKIS	LKIS Less Knowledge Intensive Sectors (Nace: 45-47; 49; 52-53; 55-56; 68; 77; 79; 81-82)	0.32	0.47
Very intense competition	Perceived market competition (1 = Very intense; 0 = else)	0.38	0.49
Intense competition	Perceived market competition (1 = Intense; 0= else)	0.49	0.50
Limited competition	Perceived market competition (1 = Limited or Very limited; 0 = else)	0.14	0.34
Observations		204	

Notes: Statistics computed for firms reporting technological innovation activities; Variables are dummies, unless otherwise indicated; (c) = Continuous variables; Importance of patents and "standards source of knowledge appear with the same marginal distribution only by chance; Sector of activity defined according to Eurostat NACE code revision 2 (EUROSTAT, 2009).

$$\mathrm{Var}\left[\varepsilon_1 \mid \boldsymbol{x}_1, \boldsymbol{x}_2\right] = \mathrm{Var}\left[\varepsilon_2 \mid \boldsymbol{x}_1, \boldsymbol{x}_2\right] = 1$$

$$\mathrm{Cov}\left[\varepsilon_1, \varepsilon_2 \mid \boldsymbol{x}_1, \boldsymbol{x}_2\right] = \rho$$

where $y_1 = 1$ if the firm is engaged in formal standardization, $y_2 = 1$ if the firm is engaged in consortia. The variables of interest are the patent $pat._i' \beta_1$ and its interaction with size $size_i \# pat._i' \beta_2$. and the standards $stand._i' \beta_3$ and its interaction with size $size_i \# pat._i' \beta_2$. ***controls*** are size, conglomerate status, sector of economic activity and perceived competition for both equations. The coefficients of interest and control variables are free to vary among equations.

RESULTS

Table 2 reports the marginal effect and some regression statistics. Only marginal effects, the impact of change of an independent variable on the expected change in the dependent variable, are reported because more easy to interpret due to non-linearity of the model and of the interaction of relevant variables with size. Estimates of different specification are reported in annex. Table 2 shows that, holding other condition constant, firms rating high the importance of patents are more likely to engage in consortia (9 percentage points), but there is no significant effect on formal standardization at conventional level (alpha 10%). Firms that perceive standards as one important source of information for the innovation process have higher chance to participate in formal standardization (13.4 percentage points) but the effect on consortia is no statistically significant. Importantly the rho is positive (.95) and statistically significant suggesting that the participation in consortia is positively correlated with participation in formal standardization. This pattern is in line with the fact that in the ICT standardization arena the border between consortia and official bodies is blurred (Egyedi, R.Hawkins, 2011). A novel result of this study is that size plays an important role as shown in Figure 1 and Figure 2. The full comment of the figures is below. The main result of the analysis is that size moderates the influence of patent and standards on standardization behaviour. The quest for knowledge motivates smaller firms to participate to formal standardization; larger firms are more sensitive to the importance of patents.

As show in left side of Figure 1, the impact of the importance of patents on formal standardization is not statistically different from zero. In other words, patent rating is not influencing formal standardization. However, it appears that smaller firms perceiving patent as important are somehow less likely to join standardization process. Possible explanation is that even if they perceive that patents are important for their business, they prefer rather not to participate to recognized standardization bodies if they are missing resources to properly protect their knowledge.

Table 2. Marginal effects on the probability of engagement

	(1)		(2)	
	Formal Standardization		**Consortia**	
Patents are important	-0.0589	(0.0618)	0.0922***	(0.0347)
Standards are source of knowledge	0.134**	(0.0637)	0.0665	(0.0407)
Ln emp.	0.0155	(0.0257)	0.00731	(0.0140)
Conglomerate	0.111*	(0.0580)	0.0692**	(0.0316)
Low Tech	-0.0691	(0.0770)	0.0590	(0.0393)
High-Tech	-0.00903	(0.104)	0.0426	(0.0419)
KIS	-0.0406	(0.0739)	0.0995**	(0.0427)
LKIS	base	-	base	-
Very intense competition	0.142*	(0.0763)	-0.0318	(0.0727)
Intense competition	0.158**	(0.0732)	-0.0406	(0.0733)
Limited competition	base	-	base	-
Rho	0.948			
X^2	49.76			
p-value X^2	0.000632			
ll_0	-142.9			
LL	-118.0			
Obs.	204			

Standard errors in parentheses
Notes: Estimations include interactions between variables of interest and size; Innovative firms only
$*p<0.10$, $** p<0.05$, $*** p < 0.01$

Figure 1. Impact of patent and standardization importance on formal standardization

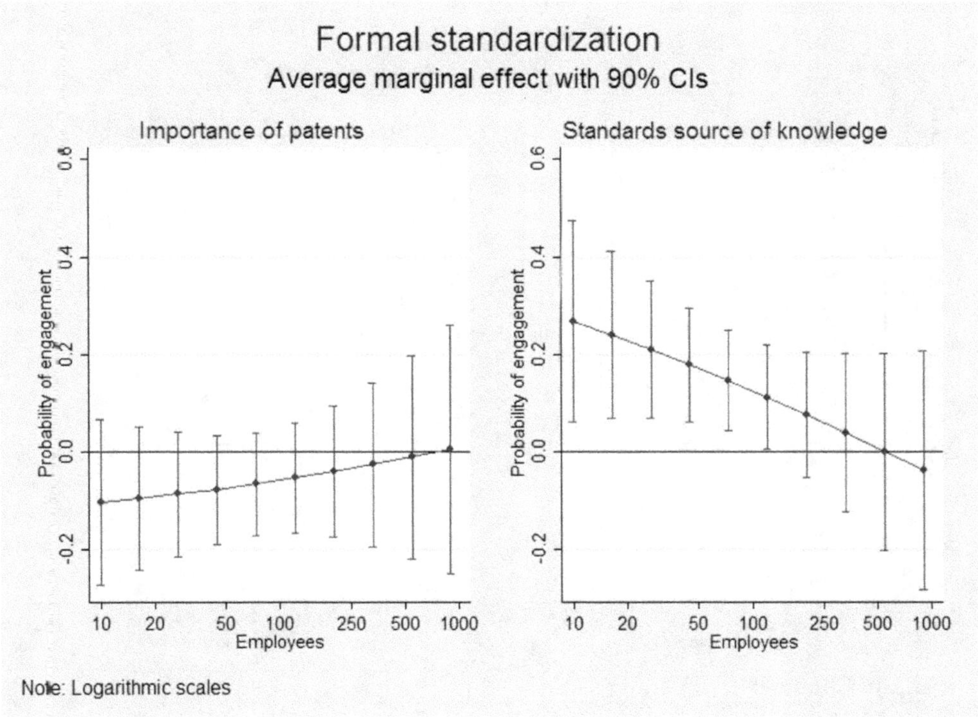

As show in right side of Figure 1, small firms are motivated by the quest for valuable knowledge. Ceteris paribus, firms considering standard an important source of knowledge have more chances to engage in formal standardization (up to 26 percentage points) than firms rating low the informative content of the standards. This effect decreases with size both in magnitude and in statistically significance. The figure reports the impact of standard rating at each level of size and it is not suggesting that participation in standardization is decreasing with size. With respect to importance of standards, after 100 employees, there is no statistical difference. This fact suggests that acquisition of valuable knowledge for innovation may motivate small but not large firms. For example a small firm producing application based on a common platform may be really interested to participate to standardization to anticipate development of the standard even rather than promote own patents in the standards.

When considering consortia, interesting patterns appears as shown in Figure2. The effect of importance of patent is increasing with size. Very large firms (500-1000 employees) have 20% more chances to be engaged in standardization. However, small firms (up to 25 employees) are not statistically different. Small firms perceiving that standards are important source of knowledge have more chances to join consortia (on average 20 percentage points). Larger firms have reversed the pattern. Firms interested in patent are more likely to engage in consortia, but the impact of knowledge from the standards is not statistically different from zero.

The model is estimated varying different specifications to assess robustness of the patterns shown in Figure 1 and Figure 2. The figures showing results of robustness analysis are available upon request. Different formal standardization organizations and consortia have heterogeneous patent policies (Bekkers and Updegrove, 2013) that may affect standardization firm behavior. Particularly important is the

Figure 2. Impact of patent and standardization importance on consortia standardization

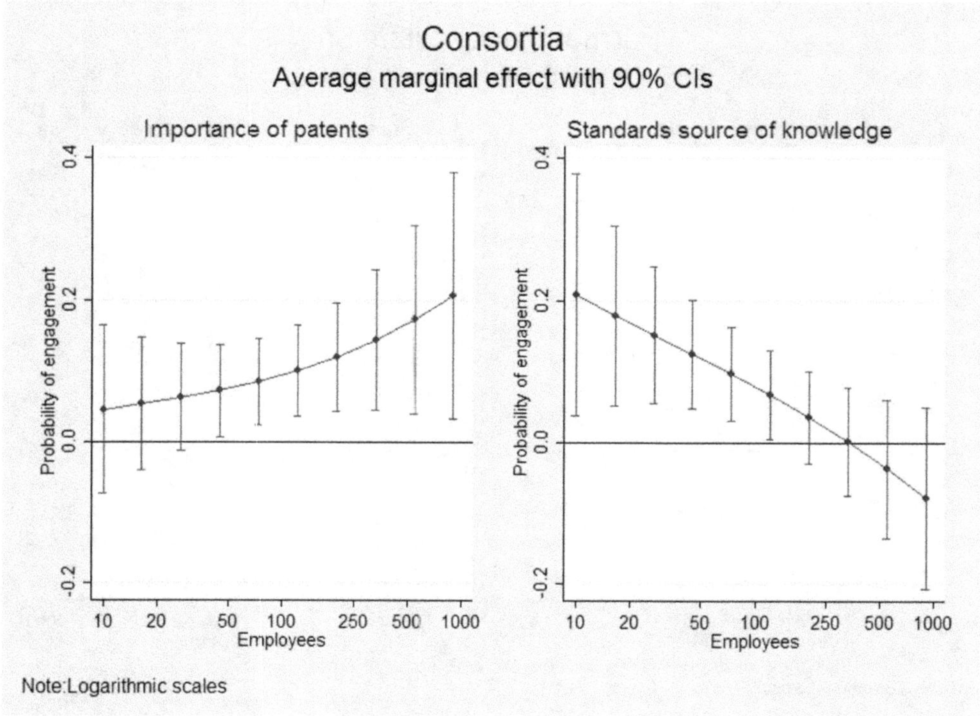

case of W3C and ETSI. W3C obliges patent owners to license their patents for free and ETSI differently from CEN, CENELEC, does require specific patent notifications. For robustness check, the members of these are matched with available the data and the model is estimated once considering ETSI as consortia and once W3C as formal standardization organization. Results are not changing significantly. Including the innovation expenditure, the model is estimated on innovative active firms only, and including the innovation expenditures the results are unaffected. Even when restricting the analysis to ICT sector only (47 observations) the general pattern remains even if precision of estimations decreases considerably because of reduced sample size. Overall, the patterns are considerably robust.

CONCLUSION

Strategies for managing organizational knowledge in modern companies can have many alternatives which are not mutually exclusive – like safeguarding knowledge through IPR mechanisms and sharing knowledge in standardization processes. The main objectives of the research presented in this paper are to explore how perceived importance of patents and of standards as important source of knowledge influence ICT firms' engagement in standardization. Departing from the "forum shopping" approach, the research investigates whether IPR have a different influence on the probability of a firm to engage in ICT formal standardization and/or in standards consortia.

Merging the results of two firm surveys for Luxembourg ICT usage 2013 and CIS 2012, we collected data about perceived importance of patent and of standards, based on representative sample of innovative firms which are engaged or not in standardization. Results showed that firms rating high importance of the patents are more likely to be engaged in consortia standardization. However, our research did not confirm significant effect on formal standardization. Firms that perceive standards as one of the important source for information needed for the innovation process are more likely to participate in formal standardization. A novel result of this study is that size is an important moderator of the reason for participating in standardization activities. The main result related to the size of firms is that smaller firms are more motivated by the quest for knowledge during standardization while larger firms are more sensitive to the importance of patent.

A possible interpretation of our results is that small IT firms might be more interested to participate in standardization in order to anticipate development of the standard, rather than place a patent in the standards. It is important to consider that the analysis is performed on the innovative Luxembourgish firms. Further studies with larger and international dataset are advisable. Investigating whether participation in standardization increases patenting activity would require panel data and is left for our future research.

ACKNOWLEDGMENT

This chapter is largely based on the paper Engagement in ICT Standardization: Pushing the Patents or Questing for Knowledge? *2015 IEEE 9th International Conference on Standardization and Innovation in Information Technology (SIIT)*, Sunnyvale, CA, USA, 6 - 8 October 2015, pp. 1-8; doi: 10.1109/SIIT.2015.7535604.

REFERENCES

Bekkers, R., & Updegrove, A. (2013). *A study of IPR policies and practices of a representative group of Standards Setting Organizations worldwide (updated version)*. Washington, DC: National Academies of Science. Retrieved from http://www.nap.edu/catalog.php?record_id=18510

Blind, K. (2008). The Influence of Companies' Patenting Motives on their Standardisation Strategies. EURAS Proceedings 2008 Aachener Beiträge zur Informatik, 40.

Blind, K., Bekkers, R., Dietrich, Y., Iversen, E., Köhler, F., Müller, B., & Verweijen, J. et al. (2011). *Study on the Interplay between standards and Intellectual Property Rights (IPRs) Final Report. Publications Office of the European Union*. European Union.

Blind, K., de Vries, H. J., & Mangelsdorf, A. (2012). External Knowledge Sourcing and Involvement in Standardization - Evidence from the Community Innovation Survey. *2012 IEEE International Technology Management Conference*. doi:10.1109/ITMC.2012.6306399

Blind, K., & Jungmittag, A. (2008, February). The impact of patents and standards on macroeconomic growth: A panel approach covering four countries and 12 sectors. *Journal of Productivity Analysis*, *29*(1), 51–60. doi:10.1007/s11123-007-0060-8

Blind, K., & Thumm, N. (2004). Interrelation between patenting and standardisation strategies: Empirical evidence and policy implications. *Research Policy, 33*(10), 1583–1598. doi:10.1016/j.respol.2004.08.007

Bouchoux, D. (2009). *Intelectual Property for Paralegals – The Law of Trademarks, Copyrights, Patents and Industry secrets*. Cengage Learning.

Brouwer & Kleinknecht. (1999). Innovative output, and a firm's propensity to patent: An exploration of CIS micro data. *Research Policy, 28*(6), 615-624.

Carlsson, B., Acs, Z. J., Audretsch, D. B., & Braunerhjelm, P. (2009). Knowledge creation, entrepreneurship, and economic growth: A historical review. *Industrial and Corporate Change, 18*(6), 1193–1229. doi:10.1093/icc/dtp043

Chiao, B., Lerner, J., & Tirole, J. (2007). The Rules of Standard-Setting Organizations: An Empirical Analysis. *The Rand Journal of Economics, 38*(4), 905–930. doi:10.1111/j.0741-6261.2007.00118.x

Curran, P. D. (2003, Summer). Standard-Setting Organizations: Patents, Price Fixing and Per Se Legality. *The University of Chicago Law Review. University of Chicago. Law School, 70*(3), 983–1009. doi:10.2307/1600663

De Vries, H. J. (2006). The paradox of standardization and innovation. *ISO Focus, 3*(2).

Egyedi, T. M., & Hawkins, R. (2011). Standards Consortia. In W. Hesser, A. Feilzer, & H. de Vries (Eds.), *Standardization in Companies and Markets. Helmut Schmidt University Germany*.

Greene, W. H. (2008). Econometric Analysis (6th ed.). Prentice Hall.

Iversen, E. J., Vedel, T., & Werle, R. (2004, Summer). Standardization and the Democratic Design of Information and Communication Technology. *Knowledge, Technology & Policy, 17*(2), 104–126. doi:10.1007/s12130-004-1027-y

Jiang, H., Zhao, S., Li, Z., & Chen, Y. (2015). Interaction between technology standardization and technology development: a coupling effect study. *Information Technology and Management*. DOI 10.1007/s10799-015-0215-7

Lemley, M. A. (2002). Intellectual Property Rights and Standard-Setting Organizations. *California Law Review, 90*(6), 1889-1980. Retrieved from http://www.jstor.org/stable/3481437

Lerner, J. (2002). 150 Years of Patent Protection. *The American Economic Review, 92*(2), 221-225. Retrieved from http://www.jstor.org/stable/3083406

Lerner, J. (2009). The Empirical Impact of Intellectual Property Rights on Innovation: Puzzles and Clues. *The American Economic Review, 99*(2), 343-348. Retrieved from http://www.jstor.org/stable/25592422

Lerner, J., & Tirole, J. (2006). Model of Forum Shopping. *The American Economic Review, 96*(4), 1091-111. Retrieved from http://www.jstor.org/stable/30034332

Mattli, W. (2001). The politics and economics of international institutional standards setting: An introduction. *Journal of European Public Policy, 8*(3), 328–344. doi:10.1080/13501760110056004

Moser, P. (2005). How Do Patent Laws Influence Innovation? Evidence from Nineteenth-Century World's Fairs. *The American Economic Review, 95*(4), 1214-1236. Retrieved from http://www.jstor.org/stable/4132712

Petersen, S. S., Blind, K., & Stephan, M. (2012). The Interrelation of Strategic Patenting and Standardization in the Context of the Resource-Based View of the Firm. *EURAS Proceedings 2012 Aachener-BeiträgezurInformatik.*

Pierce, J., & Medzmaraishvili, M. (2015). Standard Essential Patent and Injunctions – A Question of Imbalance? *EURAS Proceedings 2015*, 305-321.

Rauber, J. (2014). Why do Firms join SSOs - Exploring the Influence of Firm Characteristics using Survival Analyses An empirical Analysis for Germany. *EURAS Proceedings 2014*, 197-210.

Riillo, C. A. F. (2014). *ICT Standardization and use of ICT standards: a firm level analysis*. Available at http://mpra.ub.uni-muenchen.de/id/eprint/63436

Rysman, M., & Simcoe, T. (2008). Patents and the Performance of Voluntary Standard-Setting Organizations. *Management Science, 54*(11), 1920–1934. doi:10.1287/mnsc.1080.0919

Simcoe, T. (2012). Standard Setting Committees: Consensus Governance for Shared Technology Platforms. *The American Economic Review, 102*(1), 305–336. doi:10.1257/aer.102.1.305

STATEC. (2015). *Sources and methods*. Available: http://www.statistiques.public.lu/en/methodology/methodes/index.html

Wakke, P., Blind, K., & Ramel, F. (2016, June). The impact of participation within formal standardization on firm performance. *Journal of Productivity Analysis, 45*(3), 317–330. doi:10.1007/s11123-016-0465-3

APPENDIX

Table 3. Biprobit estimates

	(1)		(2)		(3)		(4)	
	No interaction patent-size		Including standards as source of information		Full model		Full model robust SE	
Formal standardization								
Patents are important					-0.724	(0.898)	-0.724	(0.937)
Standards are source of knowledge					1.623*	(0.867)	1.623*	(0.854)
Ln emp.	0.0645	(0.0900)	0.0475	(0.0903)	0.134	(0.151)	0.134	(0.164)
Conglomerate	0.347	(0.225)	0.401*	(0.229)	0.426*	(0.233)	0.426*	(0.231)
Low Tech	-0.276	(0.287)	-0.283	(0.291)	-0.261	(0.296)	-0.261	(0.289)
High-Tech	-0.0806	(0.355)	-0.0277	(0.360)	-0.0315	(0.364)	-0.0315	(0.378)
KIS	-0.188	(0.261)	-0.177	(0.265)	-0.147	(0.268)	-0.147	(0.243)
LKIS	base	-	base	-	base	-	base	-
Very intense competition	0.778*	(0.448)	0.646	(0.454)	0.678	(0.459)	0.678	(0.421)
Intense competition	0.790*	(0.440)	0.699	(0.445)	0.734	(0.449)	0.734*	(0.426)
Limited competition	base	-	base	-	base	-	base	-
Importance of patents	-0.125	(0.241)	-0.208	(0.246)				
Standards source of knowledge			0.471**	(0.214)				
Patents are no important # Ln emp.					base	-	base	-
Patents are important # Ln emp.					0.109	(0.191)	0.109	(0.194)
Standards no source of knowledge # Ln emp.					base	-	base	-
Standards are source of knowledge # Ln emp.					-0.257	(0.188)	-0.257	(0.184)
Constant	-1.831***	(0.564)	-1.859***	(0.565)	-2.302***	(0.810)	-2.302***	(0.770)
Standardization in consortia								
Patents are important					-0.0309	(1.568)	-0.0309	(1.080)
Standards are source of knowledge					3.585***	(1.205)	3.585***	(0.906)
Ln emp.	0.0769	(0.115)	0.0549	(0.117)	0.271	(0.370)	0.271	(0.264)
Conglomerate	0.443	(0.359)	0.624	(0.385)	0.800*	(0.417)	0.800***	(0.278)
Low Tech	0.546	(0.535)	0.677	(0.570)	0.848	(0.593)	0.848	(0.548)
High-Tech	0.366	(0.573)	0.630	(0.617)	0.685	(0.657)	0.685	(0.586)
KIS	0.789	(0.510)	1.021*	(0.557)	1.163**	(0.572)	1.163**	(0.535)

continued on following page

Table 3. Continued

	(1)		(2)		(3)		(4)	
	No interaction patent-size		**Including standards as source of information**		**Full model**		**Full model robust SE**	
LKIS	base	-	base	-	base	-	base	-
Very intense competition	-0.192	(0.528)	-0.376	(0.547)	-0.272	(0.578)	-0.272	(0.449)
Intense competition	-0.301	(0.525)	-0.406	(0.550)	-0.361	(0.593)	-0.361	(0.455)
Limited competition	base	-	base	-	base	-	base	-
Importance of patents	0.967***	(0.364)	0.876**	(0.374)				
Standards source of knowledge			0.682**	(0.308)				
Patents are no important # Ln emp.					base	-	base	-
Patents are important # Ln emp.					0.217	(0.365)	0.217	(0.265)
Standards no source of knowledge # Ln emp.					base	-	base	-
Standards are source of knowledge # Ln emp.					-0.616***	(0.238)	-0.616***	(0.188)
Constant	-3.003***	(0.736)	-3.371***	(0.820)	-4.826***	(1.768)	-4.826***	(1.372)
atanhrho_12								
Constant	1.571***	(0.415)	1.604***	(0.432)	1.816***	(0.556)	1.816***	(0.412)
rho	0.917		0.922		0.948		0.948	
chi2-c	35.84		42.61		49.76		87.71	
p_*c*	0.00305		0.000907		0.000632		8.39e-10	
ll0	-142.9		-142.9		-142.9		.	
ll	-125.0		-121.6		-118.0		-118.0	
n	204		204		204		204	

Standard errors in parentheses
Notes: Innovative firms only
$*p < 0.10$, $** p < 0.05$, $*** p < 0.01$

Chapter 5
Born Global Market Dominators and Implications for the Blockchain Avantgarde

Simone Wurster
Berlin University of Technology, Germany

Moritz Böhmecke-Schwafert
Berlin University of Technology, Germany

Frank Hofmann
Berlin University of Technology, Germany

Knut Blind
Berlin University of Technology, Germany

ABSTRACT

Issues of dominance in the market place, "standards wars," and "battles for dominance" between large companies are frequently addressed by researchers and the business press alike. The existence of companies that could establish internationally dominant solutions to customers' problems within a few years after their founding is quite unknown and the reasons for their success are hardly investigated so far. Therefore, they are not covered by traditional stage models for the establishment of dominant solutions. Presenting 22 cases and a new success factors model, this chapter shows how young companies can successfully establish their technologies as dominant solutions in the global market. Based on the studies' result, the authors then have a look at the groundbreaking IT invention of blockchain that is expected to disrupt many industries. The most prevalent success factors of the study are discussed along with the current blockchain innovation system. Their degree of significance for the success of international blockchain innovators is hypothesised for further empirical analyses.

DOI: 10.4018/978-1-5225-5320-5.ch005

INTRODUCTION

Dominance in the market place is frequently discussed by researchers and the business press alike (see Gallagher, 2007 as well as Suarez, 2004)[1]. According to Gallagher (2007) dominance is an attribute of 'the industry's (favourite) product'. More precisely, it has what he calls the 'dominant design.' Scholars agree that dominant designs are specific technologies that were capable of passing competitive selection processes in the market (Clark, 1985; Suarez & Utterback, 1995; Utterback, 1994; Suarez, 2004; Scott, 1994). The term 'the industry's product' is equivalent to the definition of 'a single winning standard' as a de-facto standard which results from 'the dynamic in the market that gives rise to network effects and switching costs that tip the competition to (the relevant solution)' (Gallagher, 2007). David & Greenstein (1990) distinguish between unsponsored and sponsored standards. Considerations regarding similarities to the dominant design concept concern sponsored standards which are characterized by one or more sponsoring entities having property rights to the standard. According to Lee et al. (1995) and Swann (1987, 2000) the concepts of 'dominant designs' and 'de facto standards' are mutually interchangeable. In contrast, Gallagher (2007) explains in detail the difference between various products and designs and gives an extensive overview of the relevant literature. He argues that compatibility standards in contrast to dominant designs can be controlled by a single firm, which may use them to accrue proprietary rents. More specifically, 'standards can apply to a (single) firm which owns (them)'. Van de Kaa et al. (2007) combine both concepts into the distinctive way of providing a generic service or function' (van de Kaa et al., 2007, p. 6).

In principle, Gallagher's (2007) description of differences between 'dominant designs' and 'de facto standards' focuses more on the ,'design' aspect, while there is no question that a company's product can dominate the market (see e.g. the examples in Suarez, 2004). Therefore, a new dominance concept with a wider object-oriented perspective is needed.

The concepts 'dominant design' and 'de facto standard' refer to hardware and software products like the IBM PC, the Windows operating system; the Playstation etc. (see e.g. Suarez, 2004 and van de Kaa, 2009). Therefore, dominant hardware products and dominant software need to be parts of the new concept. Furthermore, due to Gallagher's (2007) mentioning of compatibility standards such de facto standards as well as interfaces (see van de Kaa et al., 2011 for their relevance in this context) have to be included into the new concept, too. In addition, the new concept has to be extended by cases in which a company grants licenses for a product or solution that reaches dominance in the market (see e.g. Suarez, 2004). We unite all concepts in the umbrella 'dominant solution' (see Figure 1). In a last step, we include dominant technological designs themselves into the concept, too.

Like Anderson & Tushman (1990) and Suarez (2004), who offer a definition for dominant designs, we define a dominant solution as a hardware product, technology, software, interface or design that acquires more than 50% market share for several years and whose history is shaped by one or both of the following events: (a) there is a clear sign that the most closely competing alternative solution has abandoned the active battle, thus acknowledging defeat directly or indirectly; (b) the solution has achieved a clear market share advantage over alternative solutions and recent market trends unanimously suggest that this advantage is increasing. Following van de Kaa et al. (2011) we will make use of literature that uses the terms (de facto) standards or dominant designs for our concept.

In 1995, Netscape Communications, an almost two-year-old technology company, took the world's attention. At the time of its outstanding IPO with a market capitalization of 2.2 billion, the company reached an 80% world market share in the market for internet browsers with the Netscape Navigator. The

Figure 1. Overview of dominant solutions

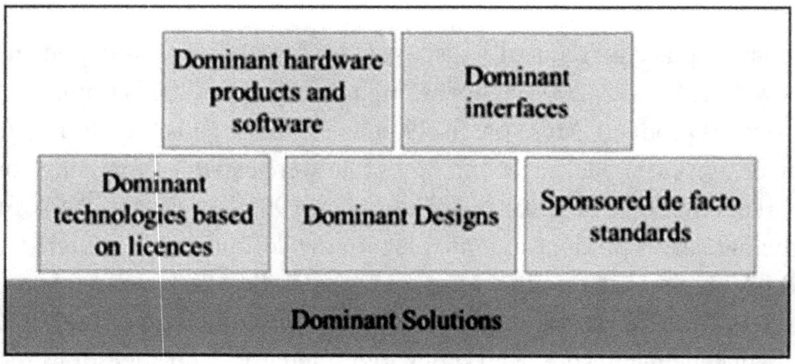

company showed great potential to establish a dominant position in the market. A short time later, Microsoft introduced the Internet Explorer as a competing product and began intense competition measures, which are regarded as the beginning of a standards war. In the following year Netscape continuously lost market share and in 1997 it recorded losses of hundreds of millions U.S. dollars. Shaped by high losses in market share and budget the company was bought by AOL in 1998 (see AOL, 1998). Netscape's example shows the attractiveness, but also risks that establishing dominance imposes on new businesses. This article proposes the establishment of success factors and an integrative framework to better understand the process through which a young company's technology achieves dominance. Scholten et al. (2013) describe the research gap in this area as follows: 'Most of the research on technology dominance focuses on large firms and considers the large resource base of the established firm as important to be successful in early stages of technology dominance.' Blind & Mangelsdorf (2013) analyse the decisions of small- and medium-sized enterprises (SMEs) to participate in official standard setting alliances as well as knowledge spillovers in such co-operation. Although their focus is on small firms, young companies are not specifically considered. According to Scholten et al. (2013), the existence of companies that could establish an internationally dominant solution within a few years after their founding is quite unknown and the reasons for their success are hardly investigated so far: 'Little empirical research has investigated how new entrepreneurial firms develop strategies over time to enter an emerging market and develop market dominance.' Referencing several additional sources, they continue: 'Understanding the factors influencing the predominant design period for start-ups will help practitioners in building more resilient startups' (Scholten et al., 2013). Potential benefits young companies could reap by establishing a dominant product or, more precisely, a de facto standard were also described by Wurster (2011) and Wurster & Blind (2011a, b).

At the time of their founding, a lack of influence and financial resources is characteristic for companies. They have no established position in business networks and are rather unknown to the marketplace. In addition, the literature describes many dominant products that resulted from 'standards wars' and 'battles for dominance' and one would assume that young companies lack the strength for such battles. Companies that internationalise at or near the time of their founding are called 'Born Globals' (BGs) or 'International New Ventures' (INVs) (see e.g. McDougall & Oviatt, 2003 and Knight et al., 2004). We identified a unique class of BGs whose solutions achieved global dominance within a short period of time and analysed the reasons for their success. This article makes a contribution to technology management and international entrepreneurship (IE) research by providing an application of Suarez' (2004) framework for young companies. It also addresses Zettinig & Benson-Rea's (2008) calls for IE research on how companies that internationalise early survive in a sustainable manner as well as for investigations of performance antecedents and outcomes relevant in IE.

We then discuss the implications of our integrative framework along with the emerging technology of blockchain. The technology has the potential to become the new engine for growth in both financial and non-financial industries (Crosby et al., 2015) and is even considered a new 'General Purpose Technology' (Allen 2017). In October 2008, a few weeks after the Emergency Economic Stabilization Act rescued the U.S. financial system from collapse, Satoshi Nakamoto introduced Bitcoin, an electronic cash system based on cryptographic proof, allowing any two willing parties to transact directly with each other without the need for a trusted third party (Catalini & Gans, 2016). Bitcoin cryptocurrency is built with blockchain technology as the underlying infrastructure. This technology allows peer-to-peer transactions, facilitated by decentralized verification protocols, without rent-seeking intermediaries, to go far beyond financial applications (Malinova & Park, 2016). We assume that a substantial share of blockchain-based start-ups internationalise early and are thus considered BGs. We have selected 13 success factors that have proven to be relevant in more than 75% percent of the 22 cases in our sample and draw implications for BG blockchain innovators. We suggest to apply our integrative framework for further empirical analyses of the blockchain pioneers, also called 'avantgarde' here, to draw recommendations on how they could achieve dominance and set de facto standards.

Research Gap

Technology management research builds on industrial economics and strategic management to explain the establishment of dominant designs. Industrial economists focus on the demand side of the market and market-based framework conditions for the relevant players like net-work externalities (Katz & Shapiro, 1985), lock-in effects as well as switching costs (Arthur, 1989; Katz & Shapiro, 1985; Suarez, 2004) in this context. Strategic management research uses constructs like complementary assets (Teece, 1986, 1997) such as a company's reputation, credibility and financial resources to explain the outcome of such processes (see also Scholten et al., 2013 for a summary).

Jones et al. (2011) analysed 323 research papers in IE research, but they did not find re-search on de facto standardization and technological dominance. Vice versa there has been hardly any linkage between dominant design topics on the one hand and the IE research field on the other hand (see Wurster, 2011 and Wurster & Blind, 2011a). An early approach to link entrepreneurship research and research on dominant designs is presented by Scholten et al. (2013). The authors investigate start-up firm networking in the pre-dominance design stage. This article goes a step further. It investigates young firms who a) establish and b) maintain a dominant position themselves. The second aspect addresses an additional research gap: Jones et al. (2011) compiled a comprehensive overview of the state of the art in IE research and identified areas of potential research. One remaining question is related to Zettinig & Benson-Rea's (2008) call for research which was described above. With regard to the time horizon, empirical research on the long-term success of BGs is missing (Oviatt & McDougall, 1997; Autio, 2005; Wurster, 2011 and Wurster & Blind 2011a).

Based on two meta-analyses of 103 and 127 research articles, van de Kaa et al. (2007) and van de Kaa (2009) determined 31 success factors for the establishment of dominant solutions. Important success factors are, for example, financial strength, brand reputation and credibility (Grindley, 1995) technical superiority (Rosenbloom & Cusumano, 1997), entry time (Grind-ley, 1995; Rosenbloom & Cusumano,

1997), pricing (Besen & Farrell, 1994; Katz & Shapiro, 1985; Shapiro & Varian, 1999a, b), alliances (Grindley, 1995; Shapiro & Varian, 1999a), complementary assets (Suarez, 2004), network effects, expectations and an appropriate expectation management (Shapiro & Varian, 1999a), lock-in effects as well as switching costs (Arthur, 1989; Katz & Shapiro, 1985; Suarez, 2004). Suarez (2004) relates success factors to specific stages in the establishment process of a dominant solution. He proposes an integrative framework to describe the process by which a technology achieves dominance. The framework consists of five key phases in the process and related key firm- and environment-level factors that affect the outcome of each stage, see Table 1.

A firm's 'strategic manoeuvring' and 'characteristics in the technological field' are meta factors. The first one comprises the factors entry timing, pricing, licensing and relationships with complementors as well as marketing & PR to manage expectations. Characteristics in the technological field include the number of actors and the level of cooperation versus competition. Traditionally, large, established companies have a head start in these processes. Due to their reputation and market impact, it is quite easy for large companies and their technology concept to be perceived as an attractive partner. In contrast, young companies face three challenges: the 'liability of newness', the 'liability of smallness' (Stinchcombe, 1965) and the liability of foreignness (Zahra, 2005). Zahra (2005) describes this cluster of obstacles as follows: '(BGs) usually experience three types of liability. The first relates to their newness and inexperience, which limits their access to resources and existing networks. Newness raises questions in the minds of other stakeholders about (their credibility and potential viability. The second liability stems from their size, as many (BGs) are small. This limits the (…) resources of BGs and, as a result, their ability to withstand the challenges of internationalisation. The third and final liability arises from the foreignness of (BGs), which means that they have to work hard to over-come barriers to entry, build links to their customers and suppliers, and gain the acceptance of potential customers. Any of these liabilities can increase the risk of (BGs') potential failure.'

The relevance of these problems regarding the fast establishment of a dominant position in the market is emphasized by Wurster (2011) and Scholten et al. (2013). Reflecting the privileged role of large, established companies regarding their opportunities to establish dominant solutions in the market, a detailed comparison of the state of the art in IE and BG research and the research on de facto standards and technological dominance unveils many differences (see Figure 2). For example, the success factors financial strength, brand reputation and credibility as well as (penetration) pricing are not covered in BG literature so far.

Table 1. Success factors and stages in Suarez (2004)

Factor typ	Dominance Factor	Stage 1 (R&D Build Up)	Stage 2 (Technical Feasibility)	Stage 3 (Creating the Market)	Stage 4 (Decisive Battle)	Stage 5 (Post Dominance)
Firm-level	Technological superiority		***			
	Credibility/complementary assets	***			***	
	Installed base				***	***
	Strategic maneuvering			***		
Environ-mental level	Regulation		***			
	Network effects & switching costs				***	***
	Regime of appropriability	***				
	Characteristics of the technological field	***				

Figure 2. Success factors in Born Global research and research on technological dominance and de facto standards

Selected Success Factors in Born Global Research and Research on Market Dominance (MD) and De-facto Standards (DfS)			
Factor	Research on BG	Research on MD/DfS	Sources (examples)
Technological Advantages	Important/Success factor	Important/Success factor	BG: Oviatt & McDougall (1994), MD/DfS: Suarez (2004)
Financial strength	Problem	Important/Success factor	BG: Zahra (2005), MD/DfS: Grindley (1995), van de Kaa et al. (2007)
Brand reputation/ credibility	Problem	Important/Success factor	BG: Zahra (2005), MD/DfS: Grindley (1995), van de Kaa et al. (2007)
Markt power	Problem	Important/Success factor	BG: Zahra (2005); MD/DfS: Suarez (2004)
Networks, network effects	Important/Success factor	Important/Success factor	BG: Oviatt & McDougall (1995), MD/DfS: Shaprio & Varian (1999a)
Alliances, number of allies	Problem	Important/Success factor	BG: Zahra (2005), MD/DfS: Shapiro & Varian (1999a)
Penetration pricing	Problem	Important/Success factor	BG: Cavusgil & Knight (2009) – use of differentiation strategies instead, MD/DfS: Katz & Shapiro (1985)
Internationalisation	Important/Success factor	Research gap	BG: Autio (2005): necessary condition for survival, MD/DfS: -
Key persons	Important/Success factor	Research gap	BG: Oviatt & McDougall (1995), MD/DfS: -
Early market entry	Important/Success factor	Variation	BG: Oviatt &McDougall (1995), MD/DfS: Schewe (1992)
Market niches	Important/Success factor	Variation	BG: Cavusgil & Knight (2009), MD/DfS: Suarez (2004): (small) new market as frame condition

Legend: Important/Success factor; Problem; Variation; Research gap

Figure 2 shows that the recent findings of the two research streams do not offer appropriate explanations for the establishment of dominant solutions or de facto standards by BGs (Wurster, 2011; Wurster & Blind, 2011a). Therefore, specific research is needed.

In addition, blockchain technology has gained wide public recognition and generated great research interest in recent years. The cumulative number of peer-reviewed academic papers increased from 2 in 2013 to 114 at the end of 2016. However, there is neither a comprehensive overview and analysis of the innovation phases of blockchain technology nor an overview of attributes of blockchain that affect the diffusion. Existing research focuses on revealing and improving limitations of blockchain from specific efficiency, legal, and security perspectives (Zhao et al., 2016). We do not seek to fulfil these gaps; however we want to stimulate further entrepreneurial and innovation economics research on the young research field of blockchain technology. Based on our integrative framework we will discuss potential success factors for achieving dominance of blockchain-based BGs. We hypothesise the relevance of the discussed success factors and recommend further empirical research based on this in order to provide blockchain entrepreneurs with important knowledge to develop success strategies.

Research Questions and Design

The previous chapter showed the research gap regarding the establishment of dominant solution by BGs in detail. To close this gap, a qualitative, case-based research approach was chosen. Qualitative research methodologies and case studies are very important to explore new research fields (Yin, 2003). Our article is built on three research questions. As shown before, IE research and the literature on technological dominance both do not consider successful young international market dominators and corresponding success factors specifically. Therefore, question 1 is: Which success factors characterize the successful establishment of a dominant solution by BGs?

Van de Kaa et al. (2007) and van de Kaa (2009) identified research gaps that are related to the establishment of dominant solutions regarding the characteristics of the competition. In ad-dition, relevant authors like Suarez (2004) do not analyse the role of the demand side in the domination process. In particular, initial market characteristics have been almost not investigated. Zahra (2005) shows that research gaps regarding external success factors exist in the BG research, too. With regard to the establishment and maintenance of a dominant position these BGs contextual elements have particular importance, since both activities require the achieving and maintaining of high market shares. Therefore, question 2 can be deduced: What characteristics does a suitable market have?

When considering dominance, not only the immediate success is relevant, but also the long-term preservation. So far long-term success is rarely discussed throughout IE research (Oviatt & McDougall, 1997; Autio, 2005 & Zettinig & Benson-Rea, 2008). Therefore, question 3 refers to internal and external contributions for long-term success: What is important to maintain dominance and to stay successful?

As shown before, literature on dominant designs and de facto standards on the one hand and literature on BGs on the other hand provide two sets of factors which contradict each other in several ways. According to technology management theories, it is nearly impossible for young companies to dominate the market in a very short time. To unveil the specific reason for the success of young market dominators, Grounded Theory (GT) principles were chosen. Based on qualitative text analyses, similarities in the companies' histories and the interviewees' descriptions were identified and enabled the formulation of success factors.

According to Strauss & Corbin (1996) a GT consists of five elements (causal conditions, phenomenon, context, action or interaction strategies and consequences. To answer the research questions an application of the model was developed (see Figure 3).

The stage of causal condition (CC) refers to the time when the idea for a technology or product emerges which later reaches dominance. Concerning young market dominators this incident is often the foundation for the establishment of the company. The stage has similarities with Suarez' first stage, but has a broader focus.

The phenomenon represents the domination process itself while the context consists of the market, the competition and other relevant actors and characteristics of the environment. Based on stage two to four of Suarez' (2004) framework, the phenomenon is divided into three stages: The R&D stage (R&D), the stage of becoming the market leader (BML) and the stage of reaching dominance (SDC). Compared to Suarez' fourth stage, it has a broader focus and does not necessarily require the involvement of the young company in a battle.

According to Suarez (2004) and Christensen et al. (1998) a market share above 50% for three years is a criteria for a design to be considered as dominant (Suarez, 2004). In addition, Suarez (2004) claims that an above 50% market share is only a valid measure if on the one hand the competing design has lost or abandoned the 'standard battle' and on the other hand market trends suggest that the winning design is further advancing (Suarez, 2004). 'Consequences' (CONS) describes the nature of the dominance and its effects on the company and its environment. After reaching dominance, specific strategies are necessary to keep on sustaining competitive advantages. Similar to Suarez' fifth stage, the stage of maintaining

Figure 3. Application of the Grounded Theory model for the study

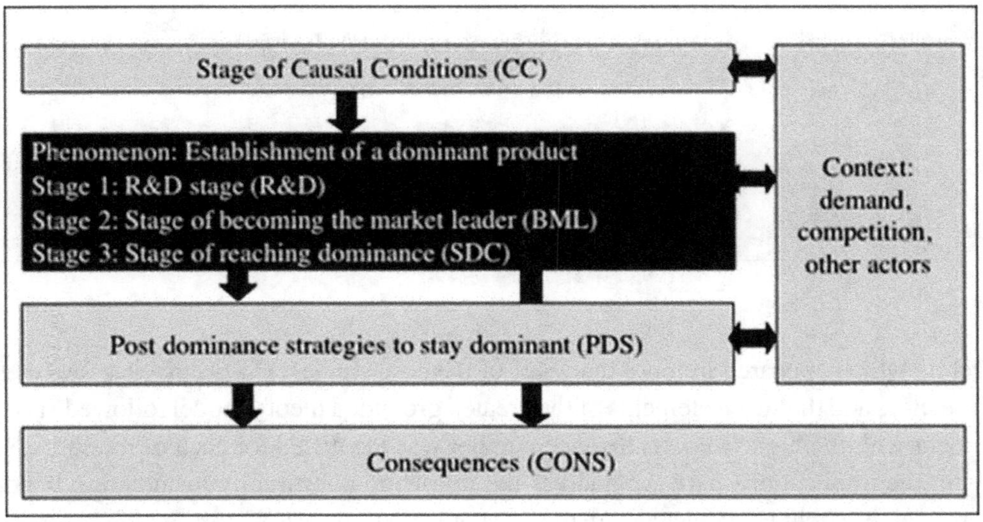

dominance (PDS) ends with the establishment of a new dominating product that replaces the old one. Case studies were completed to collect data. Three criteria were used to identify a BG:

- First international activities within three years after the foundation of the company (Rennie, 1993; McDougall & Oviatt, 2000),
- Entry into a second international market less than three years later (Lindqvist, 1991; Autio et al. 2000; McNaughton, 2000; Stray et al., 2001; Holtbruegge & Wessely, 2009),
- Activities in at least five countries and two cultural/global clusters (Holtbruegge & Wessely, 2009).

The focus was on BGs which established a dominant solution not later than by the age of 18 years. The requirement was operationalized by a market share of at least 50% for at least three years, reflecting the criteria from Suarez (2004). Sources to identify potential cases were the company lists of German and U.S. stock exchanges and several international trade fairs. More companies were identified by screening the literature on technological dominance, de facto standardisation and BG literature, by following advice from industry experts and additional internet research. Altogether, sixty-two companies were identified. One hundred and one per-sons in different leadership positions and promoters were contacted via e-mail. According to Table 2 22 companies (13 from North America, 8 from Europe, 1 from Asia) remained and were analysed by case study research. Thirty-seven people agreed on interviews: 17 CEOs and ex-CEOs, 3 Senior Vice Presidents of Corporate Marketing/heads of marketing, 17 other (particularly co-founders, 1 Executive VP, 1 Vice President, 1 COO, 1 CFO, 1 CRO, 1 CTO as well as an advisor and a co-developer of the relevant software).

According to the principles of the grounded theory (Strauss & Corbin, 1996) data collection and data analysis were linked. Additional literature review enriched the process. All interviews were transcribed and coded with Atlas.TI. The coding procedure consisted of two activities: coding success factors and coding process stages based on the model which was defined be-forehand. A table showing the success factors' frequency in the case studies was produced. The next step was the identification of dominant

Table 2. Overview of the case study firms

Company	Country	Industry	Company	Country	Industry	Company	Country	Industry	Company	Country	Industry
1	USA	S	7	USA	CH	13	USA	S	18	D	B
2	USA	S	8	D	B	14	USA	S	19	USA	I
3	USA	S	9	USA	I	15	D	B	20	FIN	I
4	AUS / USA	B	10	D	B	16	USA	S	21	USA	CH, S
5	D	B	11	CH	B	17	D	B	22	F/USA	I
6	IL	I	12	USA	S						

Industries: B = biotechnology, I = internet, CH = computer hardware, S = software

success factors which appeared in more than half of the case studies. Co-occurrence analyses between the success factors and the seven elements of the created grounded theory model followed. The grounded theory element with the highest co-occurrence number was identified for each of the success factors.

Finally, in the implications part, we address the following question by formulating hypotheses for further empirical research: How relevant are the most prevalent success factors, which we have identified for more than 75% of the cases in our sample, for the successful establishment of dominant solutions by blockchain-based BGs?

RESULTS OF THE FIRST RESEARCH PHASE AND INTERPRETATION

Due to the large number of analysed cases in our first research phase, merely one shall be presented here. Autodesk (company 2) was chosen for that. Its history is presented in Table 3. Although Autodesk realized $ 117 Million in sales in its 7th year of business it was not the fastest growing company in our study. Five companies reached sales revenues of more than $ 400 Million in their seventh year of operation.

For Autodesk, the stage of causal conditions began in the year 1981 when the IBM PC was introduced to the international market and the need for IBM-compatible PC software was aroused tremendously. Just one year later, Autodesk was founded in the U.S. and seized the opportunity to present a prototype of the first CAD program for the PC at a domestic trade show. After this R&D driven phase, Autodesk set off on its way to become a market leader over the next four years and launched their product AutoCAD.

Starting with a promotional feature in a well-known print magazine, Autodesk began to deliver a platform for other companies to implement their own CAD solutions. This was the focal point for the whole AutoCAD ecosystem.

In 1986 Autodesk's market share exceeded 50% and over the next years Autodesk's market value rose to exceed the value of the former CAD market dominator Computer Vision by 90 Million US$ in 1997. Thereby, Autodesk was able to finally win the ‚battle for dominance'. The dominance of Autodesk is reinforced by Microsoft's failed attempt to replace AutoCAD with its own solution Visio. The core reason of the failed attempt was the missing eco-system around Visio. AutoCAD is now the core of an ecosystem dispersed all over the world with countless implementations and applying companies.

As described earlier, 'Autodesk' is part of a series of 22 case studies. Based on the case analyses, 189 interview transcripts and external sources were collected and analysed. 28 firm-specific and 10 context-specific factors were important in at least 50% of the samples (see Figure 4). The results of the analysis of the co-occurrence of the success factors and the fundamental elements of the grounded theory are also shown in Figure 4. The factors 'high added value' and 'international product/solution' occurred in all 22 case studies. Together with an international market, they are regarded as the basis of the model. The results also show the importance of a strong need in the market, technical superiority based on outstanding technical assets of the founders, sales and marketing cooperation as well as an early positioning in lead markets. To illustrate the importance of a strong need in the market as a key pre-condition for early market dominance an interviewee from company 3 says:

I doubt that you would ever have a quick standard in an industry unless there was a long unmet need (company 3).

The success factor 'outstanding technical assets of the founders' is reflected mainly in their outstanding technical expertise. A corresponding reputation and a related scientific degree can represent additional characteristics. High technical expertise means for example that a member of the founding team of company 8 received a Nobel Prize and that the founder of company 5 became a professor. The founder of company 1 and 12 came from Xerox PARC and MIT. Overall, the founders of nine companies have doctorates. For several companies, the doctoral thesis of the founders formed the basis of the company's activities. However, technical expertise is not necessarily reflected by a degree. The co-founder of company 22 developed the relevant product at the age of 17.

Technical expertise is related to another success factor which becomes relevant later: technical superiority. In many cases the considered products are based on disruptive technologies. The co-founder of company 1 describes the importance of this aspect as follows:

(It) was a - if you will a disruptive technology, so I think that is a key component that it provided capabilities that there just were not there before.

He also explains the role of complementary products:

There were three or four other companies out there that had similar ideas at just about the same time and then the majority of those failed and I could have predicted it. (…) (I)f we tried to do it all our-selves, in other words build all the applications (and) all the hardware (…), we would have failed.

Many companies have a specific penetration orientation. An interviewee from company 14 describes the concept as follows:

(M)y opinion is (that) establishing growth (is) more important than profit. I was saying that gaining market shares and market recognition would be of greater value going forward because the market place is very huge (…). We had still only sold a fraction of it. So, by building our reputation and dominance in the market place, we have time to make money later. (…)

He continued by giving an example of another company:

For many years of its early growth didn't make any money at all. But they grew. (…) They grew like crazy. (…) They were sure that that's the only way to be able to establish market dominance. There will be time to make money later.

The following quotation illustrates the importance of specific communication activities:

We went on this massive market education saying we are the market leaders, and we were successful in doing that, and then we signed up channel partners to sell our products, and we were able to (…) hand them this IDC study saying we are the market leader, so (…) we are the safest way to go (…) because you also want to buy the leading product, right? So, we are the leading product. So, that is really the first thing we did, and it took us about at least a half a year to a year to really establish recognition that we were the market leader (company 6).

In addition, an appropriate growth management and IPOs are important for young market dominators. Interviewees from company 2 und 3 describe this in detail.

Table 3. Dominance process of Autodesk

Stage	Description
CC	1981: The introduction of the IBM PC leads to the development of a market for complementary products. A great need for IBM-compatible PC software arises which Mr. Walker, one of Autodesk's future founders recognizes.
R&D	1982: Autodesk is founded in the U.S. by 16 people from four countries. Autodesk presents a prototype of its software AutoCAD at the trade show Comdex. It is the world's first CAD program for the PC. Computer manufacturers and distributors crowd Autodesk's booth to learn more about the software.
BML	12/1982: AutoCAD is launched. From the beginning the product has a wide focus. Autodesk starts collaboration with hundreds of merchants. It works closely with the Press. AutoCAD appears on the cover of the well-known magazine 'PC World'. During this early period, 40% of sales are made in Europe. Without these, Autodesk would not have survived. In combination with third-party products, individual solutions are offered. From the start, AutoCAD is the market leader in the PC-based CAD market. Already the first license includes AutoCAD's DXF format. It allows extensions and data exchange with other programs. Potential channel partners contact Autodesk of their own accord. 1985: Autodesk makes its' initial public offering. It positions itself as a provider of a platform on which other companies can implement their solutions. Thereby the platform becomes the center of an entire ecosystem. 1986: A programming function is added to AutoCAD's toolkit. Initially, the competition consists primarily of software companies. In the first two years, there are two main competitors. One of them offers a superior solution but uses a relatively antiquated operating system. The second competitor is late in recognizing the importance of the new demand and launches new software 18 months after the introduction of AutoCAD. Both competitors soon become negligible. One competitor is even acquired and its CEO is hired by Autodesk.
SDC	1986: AutoCAD's market share exceeds 50%. Several major CAD companies, primarily in the hardware field, introduce similar products, but they are reluctant to cannibalize their previous products. 1990: The software allows the direct development of programs using the programming language C. This is an important basis for the future success of AutoCAD. 1992: At a lower price, the competitive software Visio is introduced by the company Shapeware. 1997: The market value of Autodesk exceeds the value of the company Computer Vision (which had dominated the CAD market previous to Autodesk for twelve years) by 90 Million US$. Seven of the ten largest CAD companies are not profitable any more, some exit the market. 1998: Computer Vision is acquired by Parametric Technology Corp. External market participants develop complementary products for AutoCAD.
PDC	2000: Microsoft buys Visio. An attempt to displace Autodesk fails because Visio offers no ecosystem of companies offering corresponding solutions. 2012++: Autodesk has 7,500 employees and generates $ 2.2 billion in sales. AutoCAD is the core of an ecosystem with uncountable applications provided by several hundred companies for corporate customers in 60 industries. Key success factors are: the continuous provision of innovative products, sensitivity regarding potential competitors who enter the market with new business models as well as an appropriate use of innovative distribution models.

If you are looking at products that become industry standards early – I think you probably going to be looking at high growth companies (company 3).

'Our biggest worry (was) that eventually all of the big vendors were going to end up realizing that they had to have a product, that we were taking away the low end of the market (...) and so the primary rea-son that we went public was because we knew that we needed to have the money that we believed would be necessary to respond when the big companies jumped into our market' (company 2).

An important factor with regard to competition was the long lead time. The interviews yielded very lively descriptions.

How lucky we have been! I can think of almost no parallel in the history of the microcomputer business where one company had the only product in a major market for over a year. Had we entered a market like word processing, BASIC compilers, or spread sheets, we would have long ago been blown away by better organized and capitalised competition (company 2).

In addition, the interviewees described differences to the company Netscape which was mentioned in the beginning:

If you compare for example Netscape versus (our company), in the case of Netscape, they had a well-funded competitor in Microsoft (…) that was very able to understand what makes a good browser (company 3).

We did not become successful as the result of understanding competition and figuring out how to beat the competition because there really effectively wasn't a lot of competition (company 12).

Do you know how long it took for Microsoft to get caught up to what (our company) was doing? Ten years (company 16).

Many success factors for the introduction of the products into the market can only be taken into ac-count if they have already been developed in the R&D phase. It is particularly important to implement a technically superior product with a high customer value and to ensure international applicability. The important realization of a pioneering role is also influenced by an appropriate execution of the R&D phase and the speed of related activities. The relevance of the most prevalent success factors for blockchain-based BGs will be discussed in the next section after a short introduction of the emerging technology of blockchain.

IMPLICATIONS FOR BLOCKCHAIN-BASED BORN GLOBALS

This chapter discusses the previous findings in the context of the emerging blockchain technology. Based on the results of our integrative framework in the previous section, we focus on the most prevalent 13 success factors that we identified for more than 75% of the BGs in our sample and draw implications for blockchain-based BGs. We suggest further empirical analyses of the specific blockchain innovation in order to provide future entrepreneurs insights into the sector-specific success factors.

Figure 4. Success factors and their linkage to the Grounded Theory elements

Area/SF*	Company in General																
All cases N=22	Early Internationalisation	Early in Lead Market	Sales & Marketing Cooperation	Outstanding Techn. Assets Founders	Entry Time	Intens. Communication & Market Education	Economic Assets Founders	Market Sensibility	Market Leader	Financial Resources	IPO	Specific Management Know-how BML+**	Price	Growth Management	Intensive Sales Activities	Penetration Orientation	Large Partners
Total	21	20	19	17	17	17	16	16	16	16	15	12	12	12	12	11	11
Ratio	95%	91%	86%	77%	77%	77%	73%	73%	73%	73%	68%	55%	55%	55%	55%	50%	50%
No. Factor	C1	C2	C3	C4	C5	C6	C7	C8	C9	C10	C11	C12	C13	C14	C15	C16	C17

*) Success factor; **) = Specific know-how to become a market leader and to establish dominance

Co-occurrence

Stage/Area	Company																	Prod./Tech.	
factor no.	C4	C7	C8	C2	C6	C5	C3	C1	C13	C11	C12	C10	C15	C14	C17	C16	C9	P1	P2
CC	15	15	12	10	4	2	2	2	1	0	2	2	0	0	0	0	0	8	1
R&D	5	4	5	2	16	4	2	3	5	0	2	3	1	0	1	0	0	19	19
BML	4	3	6	8	9	21	19	16	11	11	10	10	10	9	7	7	0	19	16
SDC	2	0	1	0	2	6	6	4	5	3	3	0	2	4	1	3	13	8	1
CONS	0	0	0	0	0	1	1	0	0	0	0	0	0	0	0	0	3	2	0
PDS	0	0	0	0	0	3	2	1	1	1	0	0	1	1	0	0	5	1	0

Stage/Area	Product/Technology									Demand							Competition			
factor no.	P3	P6	P7	P8	P9	P10	P11	P4	P5	D2	D6	D1	D5	D4	D7	D3	CO3	CO1	CO4	CO2
CC	3	2	1	0	2	0	0	0	0	14	12	1	1	4	4	0	0	3	2	2
R&D	16	3	1	2	3	1	1	1	0	7	2	0	2	1	1	0	10	1	3	1
BML	13	11	9	8	8	5	3	6	2	12	11	12	11	7	7	4	10	11	8	5
SDC	6	6	6	7	1	4	10	10	4	4	4	0	1	3	0	4	5	6	3	4
CONS	0	0	3	1	0	1	3	10	1	0	0	0	0	0	0	6	0	0	0	0
PDS	4	1	3	3	1	0	0	3	7	2	3	1	2	1	1	2	1	1	1	1

Area/SF	Product/Technology											Demand							Competition			
	High Added Value	International Product/ Solution	Technical Superiority/ Leadership	Self-preservation & Lock-in	Advancement	Openness, Open Elements	Complementary Products, Features	Quality	Leading Technical Environment/Platform	Balanced IP Strategy	Optimization IP Strategy	International Market	Strong Need	Trust & Loyalty	New	International Market Pull	Niche Market	Growing Market	Technical Disadvantage	Exit Competitors	Long Lead Time	Innovator's Dilemma
Total*	22	22	17	17	16	15	15	15	13	11	11	20	19	14	12	12	12	11	17	12	12	11
Ratio	100%	100%	77%	77%	73%	68%	68%	68%	59%	50%	50%	91%	86%	64%	55%	55%	55%	50%	77%	55%	55%	50%
No. Factor	P1	P2	P3	P4	P5	P6	P7	P8	P9	P10	P11	D1	D2	D3	D4	D5	D6	D7	CO1	CO2	CO3	CO4

*) All Cases: N=22

Blockchain Technology and Born Globals

Blockchain technology is a decentralized transaction and data management technology. It allows parties to send, receive, and record value or information in transactions through a peer-to-peer network of computer nodes (Yli-Huumo et al., 2016). The emerging technology deals with the needs of people, technology, and organisations, while addressing the issues of trust, sharing, and privacy as part of an economy and society (Zhao et al., 2016). This is realized by encryption mechanisms, inclusivity and openness of the system as well as an immutable history of all transactions in the blockchain (Tapscott & Tapscott, 2017)

Blockchain is considered a new 'General Purpose Technology' with similar importance as the 'Internet' (MacDonald et al., 2016, Allen 2017). Figures at the blockchain news site Coindesk (2017) for investments in blockchain-based start-ups in the second quarter of 2017 alone are overwhelming. As already mentioned, there are numerous use cases for blockchain industry that go beyond the financial industry (Davidson et al., 2016). The technology has the potential to become the new engine of growth in both financial and non-financial industries (Crosby et al., 2015). However, the financial industry is the first and currently most important application area of the technology (Zhao et al., 2016). With different intensity, blockchain start-ups are internationalising very closely to the time of their founding in order to keep pace with the international developments. Moreover, an important attribute of blockchain technology is its global distribution.

Suarez (2004) identified success factors in five different stages of the 'battle of dominance'. With the publication of the Nakamoto (2008) white paper and the following open source development movements, blockchain technologies have already undergone the R&D build-up phase that has similarities with the stage of causal condition (CC) in our integrative framework. The phase of technological feasibility was characterized by the implementation of the alternative electronic cash system 'Bitcoin' that Nakamoto (2008) described. The technological feasibility was demonstrated in 2010 - according to a famous internet fable - by the purchase of two pizzas worth 10,000 bitcoin (Mack 2013). It can be assumed that blockchain technology is in-between the phases of 'technological feasibility' and 'market creation'. Even traditional industries such as the insurance sector started to plan deployments of blockchain prototypes and installed a consortium (B3i) to deploy a working blockchain reinsurance prototype into production in 2018 (Elliott, 2017). Moreover, it can be observed that three main alliances have been formed by 2017 with the intent to develop and promote their approach to blockchain technology with different vision and technological attributes. The three main consortia, Ethereum, Hyperledger and R3 Corda consist of open source associations, start-ups, and small and medium as well as large corporations that compete in the establishment of a dominant design. However, blockchain-based solutions that seek to transform processes and business models are also in competition against other conventional non-blockchain technologies.

Implications of the Success Factors for Young Blockchain Innovators

The previous subchapter briefly introduced the emerging technology of blockchain and the innovation system. Based on our studies' findings we want to suggest an adaption of our framework to analyse the emerging BGs that are engaging in the blockchain innovation system. We have selected 13 success factors that have proven to be relevant in more than 75% percent of the cases in our sample in order to hypothesise its relevance and characteristics for born global blockchain innovators. As the blockchain technology is very young, the argumentation is mostly based on examples and first findings from discussions with practitioners and based on qualitative literature research. We discuss these success factors along the categories 'Company in General', 'Product / Technology', 'Demand' and 'competition' similar to Figure 4.

Company in General

Early internationalisation on the diverse leading markets (see Figure 4) has been observed in 21 of the 22 analysed BG cases. Digital technology solutions such as software is well suited for international usage, as the programming code is normally international. Of course, this applies to blockchain solutions as well. Basic software solutions are created in highly standardised digital code of common programming languages such as Java or C++, which use English nearly exclusively as a base. Bitcoin was, for example, mainly coded in C++ (see Dinh, 2017) and the Ethereum blockchain known for its smart contract capabilities offers the dedicated customisation language Solidity (see Dinh, 2017) for the creation of smart contracts. Language and culture-dependent information can be integrated modularly in software by coding dedicated interfaces for human operators, assuming that the problem solution principles themselves are not language or culture specific. Additionally, the communication in the international online community is typically in English and internationally oriented as can be seen on forums such as at reddit.com in the blockchain sections, by the blockchain conceptual white papers or in the start-up management teams. Furthermore, as a digital software product, blockchain-based solutions can be distributed by copies or downloads with nearly no additional costs. The arguments support the idea, that early internationalisation may be achieved by blockchain technological solutions relatively easy.

However, to position the enterprise internationally may require knowledge and business skills and the blockchain technology may as well require considerable adaption efforts for a specific solution. Moreover, there is the challenge of legal requirements arising from internationally varying regulations of blockchain. For example, the People's Republic of China has for now stopped blockchain-driven cryptocurrency start-up investments (ICOs) and is curbing exchanges for cryptocurrencies (see e. g. Xiaojin, 2017), while the USA seems currently inclined to consider share offerings of blockchains as securities (see SEC, 2017). Legal uncertainties are reported as well as a great concern by young founders and investors in discussions. Past reports on the misuse of Bitcoin and Bitcoin exchange hacks (Peters et al. 2015, Kiviat, 2015) are additional challenges, also reflected by concerns on the side of regulators although the connection between cryptocurrencies and blockchain infrastructure may not be mandatory (see Nakamoto, 2008, Buterin, 2014). The strength and direction of managing these concerns by policy makers may differ widely between authorities. Regulation may prove to be a critical factor for the blockchain technologies.

To sum up the criteria for early internationalisation, while blockchain solutions as nearly completely digitized software solutions may be, in theory, relatively easily addressed towards international growth markets, the success of doing so may depend significantly on other market-related success factors as mentioned in Table 4.

The early entry time into markets has also been identified as a success factor in 77% of the cases in our analysis. It is argued, based on the cases, that this allows image building and advantageous market positioning. This can be exemplarily observed with Bitcoin (see coinmarketcap.com). As the very first blockchain application, the Bitcoin cryptocurrency took the lead and, despite numerous technical short-comings (Yli-Huumo, 2016, Lin and Liao, 2017), may still be leading in popular image and in market share. The term bitcoin still delivers several times more hits than the term blockchain or its major backbone rival Ethereum as a comparison in Google search results between the terms 'bitcoin', 'blockchain' and 'Ethereum' shows, and as well possibly in market capitalization (see coinmarketcap.com). Although originally developed as a digital currency (Nakamoto, 2008), Bitcoin and its blockchain technology are used as well as a multi-purpose backbone solution as argued in the preceding deliberations. The process of adapting a running blockchain solution in a distributed actor network to mend its shortcomings is complex (Zohar, 2015, Narayanan et al., 2016, pp. 72-75). Still, despite technological limitations, Bitcoin has remained so far the most popular cryptocurrency. In this case, the first mover advantage may still be considerable for the original backbone cryptocurrency Bitcoin. According to discussions on popular forums and the theoretical market capitalization on sites such as coinmarketcap.com, Ethereum is the most popular backbone infrastructure to implement smart contract solutions, which are distributed software code that is executed in the network depending on inputs (see platform column on coinmarketcap.com/assets for examples). Ethereum profited from observing Bitcoin's short-comings and clearly distinguishes itself in these points offering more suited solutions for smart contract programming functionality (see Buterin, 2014).

Therefore, for blockchain solution competition, the first mover advantage or early mover advantage may be hypothesised as discussed in depth for Bitcoin. It could be further hypothesised, that, with the first backbone infrastructure being established, the first mover advantage moves to more specific solution areas taking advantage of the previous solution as does Ethereum and its improvements for smart contracts. It is noteworthy, that both blockchains are not focused on any specific product or geographic market, but are instead used internationally and for different purposes including infrastructure. Other more specific solutions have yet to prove themselves on the markets under the initially discussed limitations.

In case of already established market solutions, blockchains would be the latecomers challenging the incumbents. This success factor may be thus of particular interest for blockchain intra-competition or in newly created markets such the use of smart contracts for the Internet of Things that may compete with other solutions for the market.

The success factor early in lead markets was identified for 20 of the 22 analysed BG. It differs in so far as from the early internationalisation, that the analysed companies started very early in specific leading markets for their product or service.

Where the leading blockchain market may be situated, may yet be in the making, although some sites with intensive blockchain and distributed ledger activities have already become visible. For financial technologies, a Deloitte report in the last year named the UK, Singapore, New York, Silicon Valley and Hong Kong as leading hubs (Deloitte 2016). However, this covers only a part of the blockchain technology applications. The social collaboration and activity site meetup.com lists London with 34,080 groups' members before New York with 23,242 members, San Francisco with 9,824 members, Singapore with 6,902 members, Hong Kong with 6,539 members as well as Toronto with 23,395 members, Berlin with 10,930 members, Zurich with 9,783 members and Bangalore with 9,292 members, among many others under the topic 'blockchain' (meetup.com figures from 08.09.2017). While meetup.com may be employed by a larger share of users in western regions than in the East, and may not cover all blockchain-related groups, it does support centers already recognized in the Deloitte report and also demonstrates the world-wide community interest for knowledge exchange. It is still unclear what specific problems can be successfully solved with it, as discussed at the start of the chapter. Even if a solution is digital, it does not follow that it can be applied in all markets without major changes. The initially discussed constraints may however potentially hinder the early presence in local leading markets.

Additionally, the major start-up locations or developing activity hubs are not automatically the site of the most successful blockchain markets, which may depend as well on factors such as available financing, local customer demand, regulative conditions, geographic proximity or cultural closeness to the actual leading market for specific solutions. Regarding the use of blockchain solutions, coinmap.org shows about 150 venues accepting cryptocurrency in New York and roughly about 100 in San Francisco or London. For Singapore or Hong Kong only 30 to 40 venues are listed (see coinmap.org), which may also stem from the higher popularity of the site with US users.

Sales and marketing cooperation of BGs appeared to be a highly relevant success factor in 19 out of 22 (77%) companies in our sample. BG start-ups offering blockchain solutions are facing the challenge of the disruptive and foundational character of the IT innovation that is often seen as a threat by incumbents. A developed and experienced marketing and sales channel is hypothesised to be very important to promote a complex technology such as the blockchain technology and to address the right target audience for the specific solution. Moreover, marketing cooperations have been proven particularly important for the IT industry (Lu and Yang 2004). A large corporation will struggle to adopt an IT innovation such as a blockchain solution that changes their entire process or imposes a cultural shift in the company, if the provider of the solution does not have a reputation on the market and a reliability at stake. The reputation of the founders as discussed with respect to their technical expertise below are of concern here as well. Therefore, it might be beneficial for BGs to engage with large partners (which is a separate success factor in the presented model). Tapscott & Tapscott (2017) recommend that both user and suppliers engage in open blockchain initiatives (such as Ethereum, R3 Corda or Hyperledger) in order to exchange knowledge and experience about different protocols and application scenarios. Hyperledger also has a joint marketing committee, which might be particularly advantageous for rather unknown start-ups or start-ups that do not have enough capacities for a strong sales and marketing channel (Tapscott & Tapscott 2017). The formation of large consortia to develop a common blockchain backbone and derive industry-specific basic solutions may allow them to meet the challenges of missing market knowledge and market suitability discussed in the initial deliberations by the pooled knowledge and resources of their members. Therefore, one may hypothesise, that consortia solutions may be prime candidates for BGs of the blockchain technology. Consequently, sales and marketing cooperations are considered to be at least as important for BGs with blockchain innovations as for our study sample.

Another often recurring success factor was found to be outstanding technical assets of the founders in 17 cases. This success factor means the superior technical knowledge and reputation in comparison to competition. It was also found that this does not necessarily equal academic postgraduate training.

Indeed, for implementation the blockchain technology does require a wide range of technologies from cryptography to online data transfer technology and network considerations, scripting and consensus algorithms and incentives (Narayanan et al. 2016). Adding to this, knowledge from specific problem fields, e. g., smart home applications, would be required for implementing a problem-specific solution for them as initially discussed. Changes to a default backbone solution or the coding of a novel blockchain and the required level of blockchain knowledge would depend on the problem to be solved. Typically, the founding teams in blockchain start-ups are rather small and it falls to the founders to develop the overall concept often in the form of a white paper following the pattern of the Bitcoin blockchain history.

For blockchain-based solution implementations, it maythus be a question of actual hands-on experience with adapting and managing the blockchain technology stack, instead of deep cryptographic or mathematic knowledge. For the application-specific adaption, in-depth knowledge of the problem, the marketed solutions and the market may be required as discussed initially. If the problem to be solved is a complex technical problem such as market creation for computer processing power contracts (see e. g. iEx.ec), it may require expert experience in the field.

The technological and problem-specific knowledge to adapt or improve the blockchain technology for a specific problem solution may represent an important technological asset, even if the technical skills for adaptation itself become more easily accessible.

Reputation in the open developer community is typically built by involvement in popular blockchain projects or by other community contributions such as frequented blogs. Academic knowledge may not be necessary as is also found in the analysis of the 22 cases: According to an expert from the Ethereum community and additional web-based information, Vitalik Buterin, the spokesman for the Ethereum blockchain, wrote Ethereum's White Paper as a student of computer science, studying at the Canadian University of Waterloo. Mr. Buterin dropped out of the university in the same year after receiving a grant for his programming activities. Subsequently, he would attract an international group of IT experts to start Ethereum. Meanwhile, the blockchain architect Richard Gendal Brown graduated from Cambridge and worked for the IT company IBM before becoming the CTO of the large blockchain consortium R3 Corda. Discussions with members of the community revealed that investors or partners without the ability to deeply analyse white papers attach significant importance to the reputation of start ups' founders and developers.

In summary, it can be hypothesised, that the technological assets of the founders may be a relevant success factor for blockchain-based BGs, but that again the quality of the transfer of the general technology to a specific problem solution may be of critical concern. If this holds true, it may be less about general technological blockchain expertise, and more about the transfer of the blockchain technology's potential to a suitable use case.

In 19 out of 22 (77%) BGs in the sample, intensive communication and market education played an important role for the success of the company. The success factor includes the education of the market by young companies as well as an active communication of success in order to raise attention and trust. In the early phase of evolution of blockchains, there is no clear understanding of the technology itself on the market as current solutions have different attributes and properties (such as the different consensus mechanisms Proof of Work, Proof of Stake, Proof of Importance etc.), are complex to grasp and are continuously and quickly evolving (Tapscott & Tapscott 2017). Therefore, blockchain-based start-ups often face the challenge of the complex technical nature of their products, a lack of specialists, and the requirement of extensive knowledge sharing (Knight, 2017). These challenges reflect the initial discussion about knowledge and business skills as well as product suitability for markets.

A fundamental understanding of the technology's major benefits is important for its wide acceptance and this understanding needs to be increased by further enhancing the general awareness (Mougayar, 2017). Moreover, Tapscott & Tapscott (2017) recommend in their report on blockchain technology for the World Economic forum, the development of knowledge networks by institutions and industries in order to 'address urgent need for diverse talent and expertise in the ecosystem' (Tapscott & Tapscott, 2017, p.34). Many start-ups are publishing white papers in order to introduce their blockchain approach transparently (e.g. Snow et al. 2014; Powerledger, 2017; Ambrosus, 2017) to attract talent and knowledge that is currently rare in the new technology field of blockchain (Tapscott & Tapscott, 2017).

In summary, the success factor of intensive communication and market education is considered as even more important for blockchain-based BGs, especially as marketing may be hampered by its complexity. It is hypothesised, that particularly in the early technology phase, BGs should engage actively in market communication and interaction with end users, the open source communities and potential complementary partners in order to establish recognition and increase understanding of their blockchain products.

As argued regarding sales and marketing cooperation, consortia with industry players may find it easier to communicate and educate the market due to the reputations and the market insights of their members.

Technology / Product

One of the most prevalent success factors that we have identified for every single analysed BG company in our sample was the perceived high added-value of their products and services. For blockchain-based BGs this success factor is particularly tied to the success factors of 'technological superiority' and 'strong need'. As mentioned earlier blockchain technology facilitates the decentralization of financial as well as non-financial transactions. The technology reduces the importance of economic and regulatory intermediaries. It allows individuals to transfer a unique piece of digital property or data to others peer-to-peer in a safe, secure, and historically immutable way. In addition to smart contracts, the technology can create, for instance, decentralized marketplaces that aim to operate free from regulation or decentralized communications platforms that will be increasingly hard to wiretap due to the anonymity of the market participants (Wright & De Filippi, 2015).

These attributes are promising a high added-value in many industries as they are disruptive to incumbent technologies: Processes can be made more cost efficient and prices of transactions may decrease as no intermediary is required anymore. For example, the technology could increase traceability and security in supply chain management by storing each transaction of involved parties in an immutable way. Moreover, blockchain-enabled smart contracts can be a significant leverage for the diffusion of the Internet of Things (IoT) technology. They have the potential to ensure the legitimacy and efficiency of data processing and transactions between billions of devices and machines connected to each other on decentralized systems rather than on central servers (Christidis & Devetsikiotis, 2016). Blockchain-based BGs need to identify the potentials where their solutions can improve processes, the final product and processes in comparison to other technological opportunities as initially argued. In addition, blockchain technology can realize new market constellations and cooperations, even if the involved parties do not necessarily trust each other ('trust in the blockchain code'). The added value of blockchains is particularly revealed when it comes to trust as the technology often unfolds its potential in environments that lack trust between the individual market actors (e.g. in regions where institutions are weak and laws are not fully enforced). From the presented arguments and the central blockchain properties, decentralized consensus and veracity assurance, it could be hypothesised, that if issues of trust, decentralization advantages or both are present in a specific market for blockchain based solutions, then blockchain technology-based solutions may provide benefits. Moreover, the idea of the decentralized organisation principle is considered a revolutionary aspect (Davidson et al., 2016; Kane, 2017), for which factors of political or social acceptance might as well be essential. In summary, the success factor high added-value of the company's own products or services is expected to play also a crucial role for blockchain-based BGs.

An international product or solution, which means international applicability of the solution by homogeneous customer requirements in the international and domestic markets, is another success factor and was found in all cases of our sample. Blockchain is a nearly fully digitized technology with the potential localization by programmed customized interfaces. As argued beforehand, the adaptation for a specific problem introduces challenges, which may not be fulfilled: If the legal, regulatory or customer requirements for the solution to the problem differ strongly, then the solution may not be internationally marketed as easily and would require considerable adaptation to local markets. Blockchain technology solutions may require internationally harmonised or at least transparent regulative and legal standards to be able to establish itself as a de facto standard in international markets. Then again, some general solutions such as Bitcoin or Ethereum may become de facto standards, because they did ignore the uncertainty of regulation and did not wait for regulative frameworks to manifest, but set an example instead. It is very difficult to predict the outcome of such ventures due to the uncertainty, and especially regulatory uncertainty, facing them. At investor talks, the desire for regulatory clarity has been emphasised as well and mentioned as maybe the greatest hurdle to clear. Supportive regulatory findings may significantly promote blockchain technology investments from investors, or restrain it as the impact of the latest regulative measures in China on Bitcoin indicates (Neate, 2017).

Still there are industry reports that over $500 million have been invested in 2015 and in 2016 in blockchain start-ups indicating vested interest in the new technology field (see CB Insight, 2017) and demonstrating the belief in its potential as well. Taken together, an international blockchain product may be a general backbone solution or a solution that fulfils a purpose with homogeneous market requirements.

Technical superiority proved to be a success factor in 17 of the analysed cases. The blockchain technology is considered a disruptive technology in many sectors and industries, and assumed to be technologically superior to other incumbent technological approaches (Matilla, 2017). The combination of cryptography and distributed databases is considered a 'revolutionary new computing paradigm' (Swan 2015, p.92) Blockchain transactions are digitally signed and validation of the distributed database is performed by the network of communicating nodes running dedicated software, which replicate the database peer-to-peer, guaranteeing the ledger's integrity (Le Borne et al., 2016). As a result, there is an operational simplification expected by reducing duplicated and manual tasks which leads to increasing efficiency and cutting costs in comparison to prevalent technologies (Andersen, 2016), providing also an example for the success factor high added-value below. Moreover, a trusted and authorized direct intermediary is not necessary anymore due to the attributes of network consensus, immutable data history and document timestamping within the blockchain. Hence, BG blockchain start-ups could address demands that other technologies are not able to fulfil due to their attributes, substitute intermediaries and processes. An example of a business case where blockchain proves to be a superior technological approach to incumbent technology is Stellar, an open platform based on blockchain with the aim to empower the least developed regions in the world with access to low-cost financial systems. Their system proves so far to be more cost-efficient than traditional payment and banking systems in the developing world. Payments within the financial network of Stellar settle in two to five seconds, making the costs very low in comparison to conventional financial systems (Groenfeldt, 2016).

However, the blockchain technology also experienced setbacks such as hacks and security failures. One of the most prominent ones was the hack on the Ethereum-based 'Decentralized Autonomous Organization' causing fundamental discussions on the basic concepts of the technology and jeopardizing trust in this specific blockchain implementation (O'Shields, 2017). In order to be adopted, blockchain solutions have to be considered as technically mature and secure. In summary, technological superiority is contemplated to be of a major importance for blockchain-based BGs such as the example above.

In 17 out of 22 analysed BGs, self-preservation and lock-in effects were identified as a success factor. Blockchain technology is still in an early phase, where it might be difficult to create lock-in effects: On the one hand, blockchain is a technology that is driven by open source projects and consortia with open architectures (such as the Ethereum or Hyperledger). Therefore, it might be difficult for start-ups to create a technological lock-in by developing their blockchain as a full-scale proprietary solution. Moreover, at the current stage of the technology, a vast amount of use cases and application scenarios are offered to the market in different sectors and industries, and companies are looking for further use cases. The perspective of high switching costs at this stage would probably deter potential clients to adopt a blockchain-based product. On the other hand, many blockchain solutions are technical platforms that need to be deeply integrated into the business model or processes and often require a culture shift, from the traditional paradigm of thinking with trusted third parties acting as intermediaries to the new mindset of distributed trust. The cultural shift and the resulting transformation of businesses and processes might lead to inherent dependencies on the underlying blockchain-based platform. As a result, start-ups that offer well-integrated blockchain solutions to their clients could benefit from lock-in effects, as the implemented platforms are more difficult to exchange. In addition to that, as mentioned in the chapter 'Blockchain Technology and Born Globals', there are currently three main consortia with participants from large corporations to open source developers that drive the development of blockchain technology: Ethereum, Hyperledger and R3 Corda. Their technological approaches to blockchain have different attributes and visions (see Valenta & Sander, 2017 for an overview). Companies that seek to implement a blockchain platform to their business might face the challenge of selecting the appropriate technological approach as suggested by the three consortia. Consequently, they need to build up their know-how to work with the technological approach. Hence, companies focusing on, for example, a blockchain solution based on the technological paradigms of Hyperledger might face high switching costs (in terms of IT resources), if they want to change to an Ethereum-based blockchain solution. In the short run, lock-in effects and self-preservation might play a slightly less important role for blockchain start-ups than our study suggested. However, in the mid and long run, this success factor might increase with the maturity of the technology and its different solutions and be of at least similar high importance as our analysis suggested.

Demand

The existence of an international market was identified as a success factor for 20 of our 22 BG cases. It has already been discussed in principle regarding the international solution applicability and the early lead criteria. The markets could be regarded as international assuming that regulation does not deviate strongly nationally and that the solution for the specific problem in question does have international markets, i.e. markets with homogeneous requirements, as discussed before. The answer is therefore expected to vary widely depending on the specific solution. The trend may currently lean toward a more restrained outlook with national regulators becoming active as discussed regarding legal concerns, which may lead to a market segmentation or growth slow-down. International interest is certainly present, as argued in the discussion of the leading market presence factor.

A strong need for the product was a highly relevant success factor in 19 out of 22 (77%) companies in our sample. For blockchain technology, BG start-ups have been developing manifold application scenarios in various sectors and the technology is projected to disrupt entire industries such as the insurance industry (Swan, 2015). Therefore, blockchain-based start-ups are situated in a very different set of industries with different demand structures. This high demand is also leveraged by the increasing relevance of the Internet of Things (IoT) and machine-to-machine applications. As mentioned above, smart contracts can be a main enabler of fully automatized and immutable machine-to-machine interactions. This suggests that there is a strong need for blockchain solutions on the market. Mattila (2017) argues that this high demand and need play a crucial role for the blockchain ecosystem as network effects are a very important factor for the technology, which indicates that a strong need also plays a key role in the success of BGs with blockchain solutions. This is subject to further empirical analysis.

Competition

Another success factor that we identified for 17 cases of the analysed BGs is technical disadvantage of the competition. As highlighted earlier, technological superiority is hypothesised to be a major success factor for blockchain-based BGs. Consequently, the technical disadvantage of the non-blockchain-based competition might also be a highly relevant success factor for blockchain-based BGs due to, e.g. improvements of operational efficiency, cost-efficiency and the serving of new market segments that could not be realised before by former market solutions (Andersen, 2016). Blockchain-based solutions are currently implemented in a very wide field of application sectors (Crosby et al 2015; Swan, 2015; Underwood, 2016). Nearly all solutions besides Bitcoin and Ethereum have yet to gain acclaim through competition. In addition, blockchain technology still faces several technological limitations and challenges. The potential for a solution of these limitations or at least enhancements of the technology can spur competition between blockchain-based BGs. Swan (2015) lists seven major technical limitations that are a severe issue on the markets, namely throughput (transaction speed within the blockchain network), latency, size and bandwidth, security, versioning and hard forks, usability and the issue of wasted resources (Swan, 2015). It is possible for blockchain-based start-ups to differentiate their technological solutions or products by reducing or externalising these limitations and creating a technical advantage against their competitors. On the other side, development of blockchain technology has an open source character and many blockchain architectures are open (Crosby et al., 2015). In summary, technical disadvantages of the competition between direct blockchain competitors might play a significant role for the establishment of de facto standards: however, at least some solutions are rather easily imitable often due to open architecture. The future importance of this success factor regarding the different options to create smart contracts remains an issue for further research.

DISCUSSION AND CONCLUSION

Based on qualitative research, a success factors model for the establishment and maintaining of dominating solutions by BGs was created. The most prevalent success factors were then discussed along with the emerging technology of blockchain based on the discussions and experiences of practitioners and qualitative literature and web research. The study helps to overcome the lack of BG and entrepreneurship topics in technology management research focusing on market dominance and standardization research which was illustrated by Wurster (2011), Wurster & Blind (2011a) and Scholten et al. (2013). It also makes a contribution to closing the gap of studies concerning the long-term success of BGs shown by Oviatt & McDougall (1997) and Autio (2005). Further empirical research is suggested in order to recommend success strategies particularly to blockchain entrepreneurs in order to establish technological dominance.

Suarez (2004) offered an advanced multi-dimensional model to describe processes to reach dominance in the market. However, the methodology to derive the findings remained unclear. The definition of the different success factors was enriched by examples but their development and linkage to the different process stages was not grounded in empirical work. In addition, the conditions of young companies as well as the demand side were not considered. Scholten et al. (2013) analysed the specific behaviour of young companies in the pre-dominance stages. Grounded in rich empirical data, this article offers a model for young companies which pass the next stages successfully, too.

The model shows that the establishment of a dominant position by BGs does not only depend on specific entrepreneurial skills, but that it is strongly influenced by its environment, the behaviour of external actors, mistakes made by competitors and the behaviour of the target group as well. More precisely, the uncovered success factors are related to technology, the management, the demand and the competition. For blockchain-based BGs, the identified success factors in these categories are hypothesised to be of similar importance with slight variations. Technology-related factors include, for example, technical superiority and high added-value, and are suggested to be particularly important in the complex and technology-intensive blockchain sector. Factors related to management comprise outstanding technical assets of the founders, penetration orientation as well as specific management know-how concerning activities after becoming the market leader. These factors are also supposed to be highly relevant for the success of blockchain-based BGs because they are leading to network effects. Important environmental factors include for example a high demand, an international market pull and, with regard to the competitors, a long lead time. These factors are also supposed to be highly relevant for the success of blockchain-based BGs because they are leading to network effects.

As described before, we defined 'dominant designs' as a sub-concept of 'dominant solutions'. This allows a direct comparison with the current state in this field. We will make it by comparing our results with the summary of the state of research described by Suarez (2004). Since we used a new stage model for the entrepreneurship context, this comparison requires specific preparation. For simplicity's sake we summarize the stages of Suarez' process model into four phases: 1.) R&D and Technical Visibility stage which finishes with the availability of a marketable solution ('R&D'), 2.) the emergence of a market leader, 3.) the establishment of a dominant market position and 4.) a post-dominance stage.

Similarly, we combine our stage of causal condition and our R&D stage to a broader R&D stage. Like Suarez' next two stages, our following stages show the emergence of a market leader and the establishment of a dominant market position. Our next stage deals with post dominance. It has similarities with Suarez' last stage, but we focus on specific strategies of the BG. In summary, our comparison focusses four stages: R&D, the emergence of a market leader, the establishment of a dominant market position and post-dominance.

According to Suarez (2004), five factors which include, for example, technical superiority, complementary assets and the characteristics of the technical field are important in the first stage. Based on our findings, technical superiority is key for BG market dominance as well. Additional factors are outstanding technical assets of the founders who should be skilled inventors and a high demand in the market. The technical field is shaped by a long lead time or the absence of relevant competitors.

The stage that leads to the emergence of a market leader in Suarez' model comprises only company-related factors. Important company-related factors in our study include the early internationalisation of the company as well as cooperation in the fields of sales and marketing enabling the use of partners' complementary assets. In addition, the demand-related factor 'international market pull' and the competition-related factor 'long lead time' were important.

Our key factors in the next stage consist of lock-in effects which can be regarded as synonyms to the switching costs in Suarez' (2004) model (see Wurster, 2011). In the current stage of blockchain technology between 'technological feasibility' and 'market creation' this factor is currently less relevant, but might gain significant importance in the future. In addition, the companies profit from the fact that key competitors exited the market in the previous stage and therefore cannot challenge the company any more.

Table 4. Comparison of the findings with Suarez (2004) and Scholten et al. (2013)

Article/Characteristics	Suarez (2004)	Scholten et al. (2013)	This Article
Consideration of young market dominators	0	✓	✓
Empirical work	0	✓	✓
Cases included	parts of several cases	✓ (1)	✓ (22)
Coverage of the whole dominance process	✓	0	✓
Coverage of the demand side	0	0	✓
Application to current technical developments	0	0	✓

With regard to the next stage, Suarez identified the relevance of the installed base and switching cost. Similarly, our study showed the importance of lock-ins and trust. Furthermore, the importance of product advancement to stay competitive in the market became apparent. Table 4 summarizes the contribution. Grounded in empirical work and based on the data of 22 case studies, it specifically considers the domination process of young market dominators. It covers the whole process and unveils new success factors with regard to the demand side. Moreover, this chapter seeks to stimulate further research on the young emerging innovation system of blockchain technology in the academic fields of Entrepreneurship and Innovation Economics. We suggest applying the integrative framework to use cases of blockchain-based BGs in order to gain insight into the process of the establishment of de facto standards and dominance.

REFERENCES

Allen, D. W. E. (2017). *Blockchain Innovation Commons.* Retrieved September 24th, 2017, from https://ssrn.com/abstract=2919170

Ambrosus. (2017). *Ambrosus White paper.* Retrieved September 24th, 2017, from https://ambrosus.com/assets/Ambrosus-White-Paper-V8-1.pdf

Andersen, N. (2016). *Blockchain Technology - A game-changer in accounting?* Retrieved September 24th, 2017, from https://www2.deloitte.com/content/dam/Deloitte/de/Documents/Innovation/Blockchain_A%20game-changer%20in%20accounting.pdf

Anderson, P. C., & Tushman, M. L. (1990). Technological Discontinuities and Dominant Designs: A Cyclical Model of Technological Change. *Administrative Science Quarterly*, *35*(4), 604–633. doi:10.2307/2393511

AOL. (1998). *America Online, Inc. to Acquire Netscape Communications Corporation in Stock Transaction Valued at $4.2 Billion.* Retrieved September 24th, 2017, from http://www.timewarner.com/newsroom/pressreleases/1998/11/America_Online_Inc_to_Acquire_Netscape_Communications_11-24-1998.php

Arthur, W. B. (1989). Competing Technologies, Increasing Returns, and Lock-in by Historical Events. *Economic Journal (London)*, *99*(394), 116–131. doi:10.2307/2234208

Autio, E. (2005). Creative tension: The significance of Ben Oviatt's and Patricia McDougall's article 'toward a theory of international new ventures'. *Journal of International Business Studies*, *36*(1), 9–19. doi:10.1057/palgrave.jibs.8400117

Autio, E., Sapienza, H. J., & Almeida, J. G. (2000). Effects of age at entry, knowledge intensity, and imitability on international growth. *Academy of Management Journal*, *43*(5), 909–924. doi:10.2307/1556419

Besen, S. M., & Farrell, J. (1994). Choosing How to Compete: Strategies and Tactics in Standardization. *Journal of Economic Perspectives*, *8*(2), 117–131.

Blind, K., & Mangelsdorf, A. (2013). Alliance Formation of SMEs: Empirical Evidence from Standardization Committees. *IEEE Transactions on Engineering Management*, *60*(1), 148–156. doi:10.1109/TEM.2012.2192935

Buterin, V. (2014). *A next-generation smart contract and decentralized application platform*. Retrieved September 24, 2017, from http://www.the-blockchain.com/docs/Ethereum_white_paper-a_next_generation_smart_contract_and_decentralized_application_platform-vitalik-buterin.pdf

Catalini, C., & Gans, J. S. (2016). *Some Simple Economics of the Blockchain*. MIT Sloan School of Management. doi:10.3386/w22952

Cavusgil, S. T., & Knight, G. (2009). *Born Global Firms: A New International Enterprise*. doi:10.4128/9781606490136

Christensen, C. M., Suarez, F. F., & Utterback, J. M. (1998). Strategies for Survival in Fast-Changing Industries. *Management Science*, *44*(12), 207–220. doi:10.1287/mnsc.44.12.S207

Coindesk. (2017). *State of blockchain Q2 2017*. Retrieved September 24th, 2017, from https://media.coindesk.com/uploads/2017/09/state_of_blockchain_q2_2017.pdf

Crosby, M., Nachiappan, Pattanayak, P., Verma, S., & Kalyanaraman, V. (2015). *Blockchain Technology: Beyond Bitcoin*. Sutardja Center for Entrepreneurship & Technology Technical Report.

David, P. A., & Greenstein, S. (1990). The economics of compatibility standards: An introduction to recent research. *Economics of Innovation and New Technology*, *1*(1), 3–41. doi:10.1080/10438599000000002

Davidson, S., De Filippi, P., & Potts, J. (2016). *Economics of blockchain*. Retrieved September 24th, 2017, from http://ssrn.com/abstract=2744751

Deloitte. (2016). *Connecting Global FinTech: Hub Review 2016*. Retrieved September 24, 2017, from https://www2.deloitte.com/content/dam/Deloitte/uk/Documents/Innovation/deloitte-uk-connecting-global-fintech-hub-federation-innotribe-innovate-finance.pdf

Dinh, T. T. A., Liu, R., Zhang, M., Chen, G., Ooi, B. C., & Wang, J. (2017). *Untangling Blockchain: A Data Processing View of Blockchain Systems*. arXiv preprint arXiv:1708.05665

Elliott, A. (2017). *B3i moves blockchain to market with working reinsurance prototype*. Retrieved September 24th, 2017, from https://www.zurich.com/en/knowledge/articles/2017/09/b3i-moves-blockchain-to-market-with-working-reinsurance-prototype

Gallagher, S. (2007). The complementary role of dominant designs and industry standards. *IEEE Transactions on Engineering Management*, *54*(2), 371–388. doi:10.1109/TEM.2007.893991

Grindley, P. (1995). *Standards*. New York: Strategy and Policy.

Groenfeldt, T. (2016). *Stellar Makes Sending Money Internationally As Easy As Email?* Retrieved September 24th, 2017, from https://www.forbes.com/sites/tomgroenfeldt/2016/12/07/stellar-makes-sending-money-internationally-as-easy-as-email/#288ad1f65078

Holtbruegge, D., & Wessely, B. (2009). Initiating forces and success factors of born global firms. *European Journal of International Management*, *3*(2), 232–260. doi:10.1504/EJIM.2009.024324

Insight, C. B. (2017). *Deals To Bitcoin & Blockchain Startups Fall Below 2014 Levels*. Retrieved September 24th, 2017, from https://www.cbinsights.com/research/bitcoin-blockchain-startup-funding

Jones, M., Coviello, N., & Tang, Y. K. (2011). International entrepreneurship research (1989–2009): A domain ontology and thematic analysis. *Journal of Business Venturing*, *26*(6), 632–659. doi:10.1016/j.jbusvent.2011.04.001

Kane, E. (2017). *Is Blockchain a General Purpose Technology?* Retrieved September 24th, 2017, https://papers.ssrn.com/sol3/papers.cfm?abstract_id=2932585

Katz, M. L., & Shapiro, C. (1985). Network Externalities, Competition, and Compatibility. *The American Economic Review*, *75*(3), 424–440.

Kiviat, T. I. (2015). Beyond Bitcoin: Issues in Regulating Blockchain Tranactions. *Duke Law Journal*, *65*, 569.

Knight, G., Madsen, T. K., & Servais, P. (2004). An inquiry into born-global firms in Europe and the USA. *International Marketing Review*, *21*(6), 645–665. doi:10.1108/02651330410568060

Knight, W. (2017). The Technology Behind Bitcoin Is Shaking Up Much More Than Money. *MIT Technology Review*. Retrieved September 24th, 2017, from https://www.technologyreview.com/s/604148/the-technology-behind-bitcoin-is-shaking-up-much-more-than-money/

Konstantinos, C., & Devetsikiotis, M. (2016). Blockchains and Smart Contracts for the Internet of Things. *IEEE Access: Practical Innovations, Open Solutions*, *4*, 2292–2303. doi:10.1109/ACCESS.2016.2566339

Le Borne, F., Treat, D., Dimidschstein, F., & Brodersen, C. (2016). *SWIFT on distributed ledger technologies - Delivering an industry- standard platform through community collaboration*. Retrieved September 24th, 2017, from http://www.ameda.org.eg/files/SWIFT_DLTs_position_paper_FINAL1804.pdf

Lee, J., O'Neal, D. E., Pruett, M. W., & Thomas, H. (1995). Planning for Dominance: A Strategic Perspective on the Emergence of a Dominant Design. *R & D Management*, *25*(1), 3–15. doi:10.1111/j.1467-9310.1995.tb00896.x

Lin, I. C., & Liao, T. C. (2017). A Survey of Blockchain Security Issues and Challenges. *International Journal of Network Security*, *19*(5), 653–659.

Lindqvist, M. (1991). *Infant multinationals. The internationalisation of young, technology-based Swedish firms* (Dissertation). Stockholm School of Economics.

Lu, L., & Yang, C. (2004). The R&D and Marketing Cooperation across new Product Development Stages. *Industrial Marketing Management*, *33*(7), 593–605. doi:10.1016/j.indmarman.2003.11.002

MacDonald, T. J., Allen, D. W. E., & Potts, J. (2016). Blockchains and the Boundaries of Self-Organized Economies: Predictions for the Future of Banking. In P. Tasca, T. Aste, L. Pelizzon, & N. Perony (Eds.), *Banking Beyond Banks and Money. New Economic Windows*. Cham: Springer. doi:10.1007/978-3-319-42448-4_14

Mack, E. (2013). *The Bitcoin Pizza Purchase That's Worth $7 Million Today*. Retrieved September 24th, 2017, from https://www.forbes.com/sites/ericmack/2013/12/23/the-bitcoin-pizza-purchase-thats-worth-7-million-today/#24ce03862509

Malinova, K., & Park, A. (2016). *Market Design for Trading with Blockchain Technology*. RMIT University.

Mattila, J. (2016). *The Blockchain Phenomenon – The Disruptive Potential of Distributed Consensus Architectures*. BRIE Working Paper 2016-1.

McDougall, P. P., & Oviatt, B. M. (2000). International entrepreneurship: The intersection of two research paths. *Academy of Management Journal*, *43*(5), 902–906. doi:10.2307/1556418

McDougall, P. P., & Oviatt, B. M. (2003). Some Fundamental Issues in International Entrepreneurship. *Entrepreneurship Theory and Practice*, (July): 2003.

McNaughton, R. B. (2000). Determinants of time-span to foreign market entry. *Journal of Euromarketing*, *9*(2), 99–112. doi:10.1300/J037v09n02_06

Mougayar, W. (2017). *Why the Blockchain Still Lacks Mass Understanding*. Retrieved September 24th, 2017, from http://www.huffingtonpost.com/entry/why-the-blockchain-still-lacks-mass-understanding_us_587132afe4b08052400ee2f8

Nakamoto, S. (2008). *Bitcoin: A peer-to-peer electronic cash system*. Retrieved September 24, 2017, from https://people.eecs.berkeley.edu/~raluca/cs261-f15/readings/bitcoin.pdf

Narayanan, A., Bonneau, J., Felten, E., Miller, A., & Goldfeder, S. (2016). *Bitcoin and Cryptocurrency Technologies: A Comprehensive Introduction*. Princeton University Press.

NASDAQ. (2008). *Company List*. Retrieved September 16, 2013, from http://www.nasdaq.com/asp/symbols.asp?exchange=Q

Neate, R. (2017). *Bitcoin value plummets after China orders trading in currency to cease*. Retrieved September 24th, 2017, from https://www.theguardian.com/technology/2017/sep/15/bitcoin-value-plummets-after-china-orders-trading-in-currency-to-cease

O'Shields, R. (2017). Smart Contracts: Legal Agreements for the Blockchain. *Banking Institute Journal*, *21*(1).

Oviatt, B. M., & McDougall, P. P. (1994). Toward a Theory of International New Ventures. *Journal of International Business Studies*, *25*(1), 45–64. doi:10.1057/palgrave.jibs.8490193

Oviatt, B. M., & McDougall, P. P. (1997). Challenges for internationalisation process theory: The case of international new ventures. *Management International Review*, *2*(Special Issue), 85–99.

Peters, G., Panayi, E., & Chapelle, A. (2015). *Trends in crypto-currencies and blockchain technologies: A monetary theory and regulation perspective.* arXiv:1508.04364v1

Powerledger. (2017). *Powerledger White paper.* Retrieved September 24th, 2017, from https://tge.powerledger.io/media/Power-Ledger-Whitepaper-v3.pdf

Rennie, M. W. (1993). Global competitiveness. Born global. *The McKinsey Quarterly*, *3*, 45–52.

Rosenbloom, R. S., & Cusumano, M. A. (1987). Technological Pioneering and Competitive Advantage: The Birth of the VCR Industry. *California Management Review*, *29*(4), 51–76. doi:10.2307/41162131

Schewe, G. (1992). Imitations management. Academic Press.

Scholten, V., van de Kaa, G., & Trott, P. (2013). Riding the waves: Developing theory on start-up firm networking in the pre dominance design stage. Academic Press.

Scott, W. (1994). Institutional analysis: variance and process theory approaches. In Institutional Environments and Organisations: Structural Complexity and Individualism. Academic Press.

SEC. (2017). *Report of Investigation Pursuant to Section 21(a) of the Securities Exchange Act of 1934: The DAO.* Security and Exchange Commission, Release No. 81207.

Shapiro, C., & Varian, H. R. (1999a). Art of Standard Wars. *California Management Review*, *41*(2), 8–32. doi:10.2307/41165984

Shapiro, C., & Varian, H. R. (1999b). *Information Rules: A Strategic Guide to the Network Economy.* Cambridge.

Snow, P., Deery, B., Lu, J., Johnston, D., & Kirby, P. (2014). *Factom white paper 1.0.* Retrieved September 24th, 2017, from https://www.factom.com/devs/docs/guide/factom-white-paper-1-0

Stinchcombe, A. L. (1965). Social structure and organizations. In J. G. March (Ed.), *Handbook of organizations*. Chicago.

Strauss, A., & Corbin, J. (1996). *Basics of Qualitative Research.* Academic Press.

Stray, S., Bridgewater, S., & Murray, G. (2001). The internationalisation process of small, technology-based firms. Market selection, mode choice and degree of internationalisation. *Journal of Global Marketing*, *25*(1), 7–29. doi:10.1300/J042v15n01_02

Suarez, F. F. (2004). Battles for technological dominance: An integrative framework. *Research Policy*, *33*(2), 271–286. doi:10.1016/j.respol.2003.07.001

Suarez, F. F., & Utterback, J. M. (1995). Dominant Designs and the Survival of the Firms. *Strategic Management Journal*, *16*(6), 415–430. doi:10.1002/smj.4250160602

Swan, M. (2015). *Blockchain: blueprint for a new economy.* Sebastopol, CA: O'Reilly Media.

Swann, G. P. (1987). Industry Standard Microprocessors and the Strategy of Second Source Production. In H. L. Gabel (Ed.), (pp. 239–262). Amsterdam: Product Standardization and Competitive Strategy.

Swann, P. (2000). The Economics of standardization. Final Report for Standards and Technical Regulations Directorate Department of Trade and Industry.

Tapscott, D., & Tapscott, A. (2017). Realizing the Potential of Blockchain A Multistakeholder Approach to the Stewardship of Blockchain and Cryptocurrencies. *World Economic Forum.*

Teece, D. J. (1986). Profiting from technological innovation: Implications for integration, collaboration, licensing, and public policy. *Research Policy, 15*(6), 285–305. doi:10.1016/0048-7333(86)90027-2

Teece, D. J., Pisano, G., & Shuen, A. (1997). Dynamic capabilities and strategic management. *Strategic Management Journal, 18*(7), 509–533. doi:10.1002/(SICI)1097-0266(199708)18:7<509::AID-SMJ882>3.0.CO;2-Z

Underwood, S. (2016). Blockchain beyond bitcoin. *Communications of the ACM, 59*(5), 15–17. doi:10.1145/2994581

Valenta, M., & Sandner, P. (2017). *Comparison of Ethereum, Hyperledger Fabric and Corda.* FSBC Working Paper.

van de Kaa, G. (2009). *Standards Battles for Complex Systems.* Rotterdam: Empirical Research on the Home Network. ERIM PhD Series Research in Management.

van de Kaa, G., de Vries, H. J., van Heck, E., & Van den Ende, J. (2007). The Emergence of Standards: a Meta-analysis. *Proceedings of the 40th Hawaii International Conference on Systems Sciences (HICSS'07).*

van de Kaa, G., Van den Ende, J., de Vries, H. J., & Van Heck, E. (2011). Factors for winning interface format battles: A review and synthesis of the literature. *Technological Forecasting and Social Change, 78*(8), 1397–1411. doi:10.1016/j.techfore.2011.03.011

Wright, A., & De Filippi, P. (2015). *Decentralized Blockchain Technology and the Rise of Lex Cryptographia.* Retrieved September 24th, 2017, https://ssrn.com/abstract=2580664

Wurster, S. (2011). Leaders, Servants, Independence. A Dynamic Success Factors Model for the Establishment of De-facto Standards by Born Global Firms and Technical Clusters. *International Journal on GSTF Business Review, 1*(2), 172–179.

Wurster, S., & Blind, K. (2011a). Towards a Dynamic Success Factor Model for the Establishment of De-facto-Standards by Born Global Firms and Technical Clusters. GSTF Best Paper Award. doi:10.1037/e605342012-005

Wurster, S., & Blind, K. (2011b). The Role of Standardization and Standards in the Context of Entrepreneurship: Conceptional Model and Insights form Case Studies. *Proceedings of the 6th European Conference on Innovation and Entrepreneurship*, 861-874.

Wurster, S., Blind, K., & Fischer, S. (2014). Born Global Market Dominators: Insight into a Unique Class of Young Companies and Their Environment. *International Journal of IT Standards and Standardization Research, 12*(1), 1–16. doi:10.4018/ijitsr.2014010101

Xiaojin, R. (2017). *ICO ban seen helping improve regulation.* Retrieved September 24th, 2017, from http://www.chinadaily.com.cn/business/2017-09/09/content_31760669.htm

Yin, R. K. (2003). Case study research, design and methods (3rd ed.). Academic Press.

Yli-Huumo, J., Ko, D., Choi, S., Park, S., & Smolander, K. (2016). Where Is Current Research on Blockchain Technology? A Systematic Review. *PLoS One*, *11*(10), e0163477. doi:10.1371/journal.pone.0163477 PMID:27695049

Zahra, S. A. (2005). A Theory on International New Ventures: A Decade of Research. *Journal of International Business Studies*, *36*(1), 20–28. doi:10.1057/palgrave.jibs.8400118

Zettinig, P., & Benson-Rea, M. (2008). What becomes of international new ventures? A co-evolutionary approach. *European Management Journal*, *26*(6), 354–365. doi:10.1016/j.emj.2008.09.009

Zhao, L. J., Fan, S., & Yan, J. (2016). Overview of business innovations and research opportunities in blockchain and introduction to the special issue. *Financial Innovation*, *2*(1), 28–25. doi:10.1186/s40854-016-0049-2

Zohar, A. (2015). Bitcoin: Under the hood. *Communications of the ACM*, *58*(9), 104–113. doi:10.1145/2701411

ENDNOTE

1 This chapter is partially based on Wurster et al. (2014).

Chapter 6
A Method for Content Analysis of (Pre-) Standardization Activities:
The Case of Cognitive Radio

Vladislav V. Fomin
Vilnius University, Lithuania

Hanah Zoo
Yonsei University, South Korea

Heejin Lee
Yonsei University, South Korea

ABSTRACT

This chapter is aimed at developing a document content analysis method to be applied in studies of standardization and technology development. The proposed method integrates two theoretical frameworks: the co-evolutionary technology development framework and the "D-N-S" (design, negotiation, sense-making) framework for anticipatory standardization. At the backdrop of the complex and diversified landscape of science and R&D efforts in the technology domain, and the repeated criticism of the weak link between R&D initiatives and standardization, the authors argue that the method offered in this chapter helps better understand the internal dynamics of the technology development process at the early stage of standardization or pre-standardization, which, in turn, can help mobilize and direct the R&D initiatives. To demonstrate the practical usefulness of the proposed method, they conduct a content analysis of the research contributions presented in the COST Action IC0905 "Techno-Economic Regulatory Framework for Radio Spectrum Access for Cognitive Radio/ Software Defined Radio" (COST-TERRA).

DOI: 10.4018/978-1-5225-5320-5.ch006

INTRODUCTION

There is a general understanding and empirical evidence for the importance of standardization to economic development and regional competitiveness (Brykman, 2011). Scholars agree that the (technology) standardization process is a complex phenomenon, and insufficiently understood among businesses and politicians (Bousquet, Fomin, & Drillon, 2009; Delaere & Ballon, 2008; Isaak, 2006; Jakobs, 2003; K. J. Lyytinen, Keil, & Fomin, 2008). Policy makers often call for standardization process improvements (European Commission, 2007), specifically with regard to the repeatedly criticized "weak link" (Jakobs & Blind, 2011, p. 102) between standardization communities and R&D ones, or "research-to-standardization gap" (Sales, Darmois, Papadimitriou, & Bourse, 2012) – the critique, which, among others, points at the lack of direct support by the European Community for early- and pre-standardization activities. The growing complexities of Information and Communication Technologies (ICT) in general (Schneberger & McLean, 2003) and ICT standards in particular (Delaere & Ballon, 2008) aggravate the situation, which is referred to as "blurred ICT standardization landscape" (European Commission, 2007; Jakobs & Blind, 2011). The latter is also attributed to the lack of theoretical and practical understanding of the processes leading to the creation of standards. More often than not, development of a particular technology is advanced by several standards development organizations (SDOs) and/or consortia, whereas areas of standardization activities may be overlapping (Baldini et al., 2013). Thus, while focused, targeted efforts can help mobilize diverse R&D and standardization initiatives, the "blurred landscape" and the lack of tools or methods for "seeing through" makes it difficult for an observer to command the landscape view (Sales et al., 2012) and hence to efficiently contribute to the technology and standards development.

An analysis of R&D activities directed at emerging technology (i.e., technology at an early or pre-standardization stage) can be helpful in informing the policy makers and standardization professionals on the research directions in general, and on "promising findings" of the research community in particular. Seeking to respond to the contemporary academic and practitioner challenges of the standardization domain, in this work we take a closer look at diverse research initiatives contributing to the development of a particular technology. These research initiatives at the early, pre-standardization stage of technology development constitute the *venue for trying and learning* through which immature technology gradually forms into a product(s) which can be successfully standardized and commercialized (Schot & Geels, 2007; Schot, Hoogma, & Elzen, 1994). An improved understanding of the R&D process and its topical content can not only contribute to establishing the links between similar activities in R&D and standardization communities, but also strengthening existing links through the reduction of duplicate efforts – i.e., contributing to a better pre-standardization environment, which lays ground for future standardization of the technology (Sales et al., 2012, p. 22).

This paper makes a contribution by empirically examining discourses at the early stage of technology and standard evolution, which has been rarely done before (Nickerson & zur Muehlen, 2006), and thereby helps better understand the dynamics and the content of the process. To fulfil on our goals, we choose an emerging technology called Cognitive Radio. CR/SDR[1] (Cognitive Radio / Software Defined Radio; hereafter referred to as Cognitive Radio (CR)) – a new generation for wireless communications, which carries a promise to overcome the existing limitations and imperfections of extant technologies in spectrum access and use.

Recognizing the importance of the wireless communications in general, and CR-enabled technologies in particular, many (if not all) major SDOs and relevant regulatory organizations have embarked on developing standards or defining norms and regulation for one or another aspect of CR-related telecommunications (Baldini et al., 2013; Arturas Medeišis & Holland, 2014). Such attention to the emerging technology by a large number of stakeholders contributes to the high level of diversification of development efforts, makes it difficult for researchers to locate and "connect" with relevant standardization organizations, and for the standardizers to "feed on" relevant and promising research projects.

The case examines the development of CR technology through the prism of European Science Foundation's research networking programme "COST" (European Cooperation in Science and Technology), in specific COST Action IC0905 "Techno-Economic Regulatory Framework for Radio Spectrum Access for Cognitive Radio/ Software Defined Radio" (COST-TERRA)[2]. Documentation produced under the COST-TERRA over the period of two years from 2010 to 2012 was collected, systematized and analyzed using a qualitative data analysis software package NViVo. COST is aimed at coordination of nationally-funded research on a European level, encouraging participation of different types of stakeholders – academy, industry, and regulatory. In its current form, the COST Action is an excellent venue for studying technology development during its early stage as it aims to facilitate exchange of research findings, ideas, and problem formulations between the participating members of any particular Action.

The analysis we undertake has several theoretical and practical implications. Our work allows us to get an insight into the specifics of CR technology development, which is increasingly important due to scarcity in radio spectrum and the growing demand for this resource (Arturas Medeišis & Holland, 2014). In analyzing R&D efforts, we demonstrate how different theoretical frameworks can be combined to structure the otherwise complex process into intuitive and understandable dimensions, helping the researchers yield better insights from the analysis. We also suggest that similar analyses should see wider adoption in the European standardization arena, fostering exchange of information within and between research-, regulatory-, and standardization-oriented communities. It will make the European standardization landscape more visible and processes leading to technology standardization more efficient.

The paper is organized as follows. In the following section, two theoretical frameworks are presented. They are used as dimensions through which we classify and organize text data. Section three describes the methods used and justifies the choice of the research object. Section four presents data analysis. Then we discuss the findings in section five. The last section concludes the paper with contributions, limitations and suggestions for further research.

THEORETICAL FRAMEWORKS

During the last decades, the domain of ICT development in general (Schneberger & McLean, 2003), and Cognitive Radio Systems in particular (European Commission, 2009a), has seen substantial growth in complexity. One of the reasons for the growing complexity in standardization of CR is related to the convergence of previously distinct markets with one another, such as telephony, Internet, and TV broadcasting. This convergence "is not just technological in nature, but in fact comprises four different domains" (Delaere & Ballon, 2008, p. 3); they are institutional, technological, functional and infrastructural.

For any particular technology, anticipatory technology standards are very challenging to formulate and enforce as they do not just reflect existing practices (e.g., quality standards), but establish new practices that need to be coordinated (K. J. Lyytinen et al., 2008, p. 3). In the case of the paradigm-changing CR services, efforts aimed at standardization of related technologies must create totally new solutions, and hence require substantial coordination and are prone to failures. By now very little is known about the possible market impacts of design choices made, and stakeholders participating in the standardization process have difficulties in estimating what choices would serve their interest best.

The growing complexity of technology (standards), to be managed and successfully commercialized, requires a good managerial and scholarly understanding of the underlying processes. However, to date, no single theoretical framework has been widely accepted as suitable in studying or predicting standardization processes – different theories are used to study different aspects of standardization processes, producing different pictures of the same "reality" (Fomin & Keil, 2000; Jakobs, 2009). The theories, which proved successful in explaining standardization processes *retrospectively*, cannot cope well with anticipatory technology developments nor with models implying dynamic complex decision making patterns (K. J. Lyytinen et al., 2008).

This very complexity of standardization processes and the growing risks of technology development failure necessitate scholars and policy makers to seek an appropriate lens to study or monitor the relevant processes. In this work we propose a method, which reduces the ambiguity and complexity of the study object – i.e., technology development process – by slicing it into a number of dimensions, or simply, examining the same standardization process from different perspectives or in different light. The proposed method is based on the synthesis of two theoretical frameworks, both of which have originated from research on ICT standardization.

The first framework we use is referred to as co-evolutionary[3] analysis (Bauer, Ha, & Saugstrup, 2007; K. Lyytinen & King, 2002). Using this framework we can examine technology standardization processes through the prism of three dimensions representing three crucial development domains, or groups of stakeholders – technology (T), market (M), and regulation (R) – while focusing on domain-specific and inter-domain drivers and barriers to the development process. According to the co-evolutionary analysis, for each of the identified domains (T, M, R) there are factors, which, being largely under the discretion of that particular group of stakeholders, interact with factors that are in part or fully external, but which may be influenced by a different group of stakeholders (Bauer et al., 2007, p. 626; Fomin, Medeišis, & Vitkutė-Adžgauskienė, 2012, p. 6). At the backdrop of the "blurred landscape" of ICT standardization, co-evolutionary lens give the scholar a possibility to see whether and how the same or different issues are being addressed in different contexts.

The previous works applying the co-evolutionary analysis to studying Cognitive Radio are helpful in identifying macro-level dependencies between the three co-evolutionary domains (Fomin, Medeišis, & Vitkutė-Adžgauskienė, 2011). However, no insights of the standardization process at the micro-level were generated. To figure out how micro-level interactions by each stakeholder group belonging to one of the three domains contribute to the formation of the emerging future technology (i.e., the-standard-to-be), a process-oriented theoretical framework is needed. To this end, we adopted the second theoretical framework – the so-called "D-N-S" process analysis framework (Fomin, Keil, & Lyytinen, 2003; K. J. Lyytinen et al., 2008), which has been specifically developed to study the internal dynamics of anticipatory standardization processes.

The "D-N-S" framework ((Fomin et al., 2003; K. J. Lyytinen et al., 2008) distinguishes three cognitive orientations of standardization activities as the following:

- Design orients itself towards the creation of technical specifications and draws upon engineering principles that lay out design architectures, components, and their specifications; structural, functional and observable behavioral features of the artifacts dominate the reasoning.

- Negotiation orients itself towards reconciling technical solutions with regard to the interests of the involved actors; strategies and interests of actors and anticipations of their behaviors towards other actors dominate the reasoning.
- Sense-making orients itself towards "reading" technical specifications in ways that relate them to the life-worlds associated with the use of produced artifacts and/or shaping institutional contexts that surround specified technology.

Unlike the chrestomathic sequential phase model of technology development, the D-N-S lens show (pre-) standardization processes to have multiple instances of D, N, and S types of activities to run in parallel and recursively, while requiring "closures" on each of the instances for the process to evolve. All these types of events (or cognitive orientations) are necessary to define and produce an agreed upon standard's specification and/or functioning technology. Design events are necessary to yield specifications that can include novel principles which are shown to work. Negotiations are needed to reach agreements between involved stakeholders, while sense-making is needed to contextualize the specification and negotiation events into an unfolding business, institutional, and user context (Fomin et al., 2003).

By intersecting the dimensions of the two frameworks, we can produce a 3 by 3 matrix (such as Table 3) where each analyzed document or a key word is mapped in an appropriate cell. The distribution of analyzed items in the matrix shows which cognitive orientation (e.g., design) dominates in which domain (e.g., technology), thus helping grasp both the macro- and micro-level dynamics of the technology development process.

RESEARCH METHODS

Research Object and Venue

CR/SDR technology clearly represents one of the potential forces that could form a basis of the next largest breakthrough in the development of ubiquitous wireless broadband services. However, for CR to become a global "next big thing" in telecommunications, the three forces of technology, market, and regulation that shape the creation of the new telecommunications paradigm must be aligned (Fomin et al., 2011; Artūras Medeišis, 2014). Success of such alignment is strongly related to the availability of technology standards, which are necessary both for the creation of economies of scale for the global markets, and for regulatory practice with unambiguous reference points (Kwak, Lee, & Fomin, 2011; Stewart, Shen, Wang, & Graham, 2011).

On one hand, the development of CR technology standards has passed its starting point, as Cognitive Radio is already embraced by major European and international SDOs, such as ITU, ETSI, ECC, CEPT[4], IEEE, etc. (Baldini et al., 2013). Such footing of the emerging technology in standards making arenas is expected to create a solid point for leveraging the technology to the commercial success in the future.

On the other hand, the fact that CR is one of a large technical systems (LTS, see Hughes, 1993) gives difficulties in advancing the technology development and commercialization processes. LTS are systems or networks of enormous size and/or complexity, such as electricity grid, the Internet, mobile communications infrastructure, etc. Implementation of CR services requires complex infrastructure to support them and coordination of diverse stakeholders with vested interests to create a business model which can be sustained. The development of LTS is accompanied by messiness, heterogeneity, and complexity

(Schot & Geels, 2007). Therefore multiple streams of R&D efforts deal with, and are often intertwined with, economic, political, and technology factors in their attempt to create and stabilize a network of key stakeholders, technology, and its use scenarios (Fomin & Lyytinen, 2000; Schot & Geels, 2007). Such complexities are also found in the scene of CR development, and the broad coverage of CR/SDR by different SDOs and R&D units makes it extremely difficult to study the interactions of stakeholders. It would literally require the researcher's involvement in many of the on-going development initiatives.

That being the case, the European Science Foundation's COST (European Cooperation in Science and Technology) provides an advantageous ground, to a certain degree, to overcome the current difficulties of studying the development of Cognitive Radio (or any other emerging technology). COST is an intergovernmental framework, allowing the coordination of nationally-funded research at the European level. COST programs help researchers study the development of nascent technologies by bringing together multiple streams of diverse research into a "bottom-up" think-tank forum without biased control of any particular policy entity. As membership in COST Actions is open to diverse entities, COST provides networking opportunities and helps increase the exchange of ideas among different stakeholders – researchers, policy makers, regulatory bodies and national decision makers, as well as private entrepreneurs. The particular COST Action IC0905 "TERRA" we are using to study the development of CR is one of the many COST Actions in the domain of Information and Communication Technologies, which serves as a platform for knitting together threads of research on CR from participating countries (22 countries) and research groups[5].

In addition, with regard to a specific view of early standardization developments of technology, COST TERRA has established connections with the major stakeholders in the CR/SDR standardization arena. Not only has COST TERRA associated many partners from either recognized SDOs or standardization consortia (ETSI, CEPT, IEEE DySpan Standards Committee, IEEE Technical Committee on Cognitive Networks, and Wireless Innovation Forum), but its members are also involved in CR-related initiatives in the aforementioned SDOs. Thus, studying CR development through the prism of the COST-TERRA Action enables a researcher to examine the "threads" weaving the socio-techno-institutional web of the CR innovation.

Data Collection

The COST TERRA project was launched in May 2010 and ran until May 2014. The work of the project was organized through hosting 2-3 Management Committee (MC) meetings per annum, supporting research workshops, conferences, and short term scientific missions (STSMs). MC meetings are the primary "networking" and "idea exchange" venues. Representatives from different countries gather for intensive 2-3 day workshops where new research results, new developments, challenges, etc. are reported. Normally, the MC work is organized around thematic Work Groups (WGs), each focusing on a specific sub-domain of the broadly defined problem of the COST Action. Since its inception in 2010 and by the time of this study in summer 2012, five meetings had been conducted under four WGs. Each of the WGs is dedicated to the discussion of the following sub-themes: WG1 on CR/SDR deployment scenarios, WG2 on CR/SDR coexistence studies overview and current status, WG3 on economic aspects of CR/SDR regulation, and WG4 on impact assessment of CR/SDR regulation.

According to the process theory framework for anticipatory ICT standardizing and the methodological guidelines provided by Lyytinen et al. (2008), each *event* in technology development process accounts either directly or indirectly for an observed outcome, and those outcomes – the content – can be analytically distinguished to be associated with a specific type of, for example, ideas, experiment results, technology artifacts, or policy proposals.

In our work, COST-TERRA WG reports are taken as proxies for technology development events. Those reports present results – i.e., outputs – of different threads of research by COST-TERRA members, where each report is concerned with a specific problem related to CR in general and the problematic of a specific WG in particular. A total of 68 presentation documents were submitted by participants of five COST-TERRA meetings over two years between August 2010 and April 2012. They were collected (Table 1) for keyword and trend analyses. The documents, as outputs of research and development events, were analyzed in three ways as described below.

First, for keyword identification, five to six keywords were identified from individual documents. In documents where the original author(s) of the documents provided a set of keywords, those keywords were adopted. When keywords were not provided by the document authors, the researchers selected the keywords after individually reviewing each document, and then discussing for consensus. In order to avoid generic terms from receiving unnecessary weights in the analysis, terms such as Cognitive Radio, when used alone without combination of other keywords, were removed from the analysis.

Second, classification according to the cognitive orientation was conducted. Following the process analysis framework developed by Lyytinen et al. (2008), for each document there were one or more of the three cognitive orientation identified: *design* (D), *negotiation* (N), or *sense-making* (S). When the documents had more than one cognitive orientation, secondary and tertiary cognitive orientations were also noted.

Third, classification according to the primary domain of contents was conducted at each document level. Following the co-evolutionary analysis framework postulated by Fomin et al. (2011), documents were broken down into the following three domains: *technology* (T), *market* (M), and regulation/*policy* (R). Again, as in the classification on cognitive orientation, when the documents were associated with more than one domain, secondary and tertiary domains were identified on top of their primary domain.

The document-level analysis was conducted by the three authors. The authors divided the pool of entire documents into three parts, and each author became the primary reviewer of one of the three parts to identify document-level keywords and classify the documents by both cognitive orientations and domains. The initial results conducted by each researcher were cross-reviewed by the other two for further verification.

Table 1. COST-TERRA Meeting History

Meetings	Date	Place	Working Groups / No. of Documents				
			1	2	3	4*	Subtotal
Meeting 1	30-31 August 2010	Brussels	5	5	N/A	N/A	10
Meeting 2	19-21 January 2011	Lisbon	7	7	2	N/A	16
Meeting 3	20-22 June 2011	Brussels	4	7	6	N/A	17
Meeting 4	15-17 November 2011	Brussels	3	3	6	3	15
Meeting 5	25-27 April 2012	Vilnius	4	3	3	0**	10
Total							68

* WG4 was created from Meeting 4.

** No document was available from WG4/Meeting 5.

DATA ANALYSIS

Stakeholder Analysis

Identifying the actors involved in technology development is an important requirement of process analysis. As each document we analyzed is essentially a report on a particular research and development initiative, knowing who contributed to that development helps understand the context in which the technology development takes place.

The result of COST-TERRA document review showed that participants from one of four major affiliations contributed to the output documents. They are universities, R&D institutes, industry, and regulatory organizations. First, participants from universities include faculty and researchers affiliated with universities and colleges located mostly in the EU.[6] Participants from university-related research centers are also put under this category. Second, R&D institutes include research institutes and labs that are either independent or government-funded, without a direct affiliation with universities or private companies.[7] Third, industry participants include those either from a private company, or research institutes and labs directly affiliated with a private company.[8] Last, regulatory organizations include departments, ministries, and regulatory arms of national governments as well as EU-level agencies.[9]

While 52 out of the 68 documents (76%) were authored by participant(s) from a single affiliation group, 16 documents (24%) were written by authors from organizations that are under different affiliation groups; for example, collaboration between a university and a company, or one among a private company, a regulatory organization and a university. Table 2 (2nd column) shows the result of document categorization according to the affiliation of author(s). University actors are by far the greatest contributor to the CR discussion documents; university single authoring takes more than half (38 out of 68, 56%) and every collaboration work includes at least one university (Table 2). Those results second the criticism of lack of coordination between different stakeholders in technology standardization (European Commission, 2007; Jakobs & Blind, 2011).

Table 2. Documents by affiliation of the authors

Affiliation Group	Number of Total Documents by Group	Domains Number of Documents			Cognitive Orientations Number of Documents		
		M	R	T	D	N	S
University	38	10	6	22	11	1	26
R&D	5	0	1	4	4	1	0
Regulatory	5	0	3	2	1	1	3
Industry	4	2	1	1	0	0	4
Regulatory + University	4	2	1	1	0	1	3
R&D + University	6	3	0	3	3	0	3
Industry + University	3	1	1	1	0	1	2
Industry + R&D + University	2	1	0	1	1	0	1
Industry + Regulatory + University	1	0	0	1	1	0	0
Total	68	19	13	36	21	5	42

Following the logic of the co-evolutionary analysis, we distinguish the documents into three development domains, namely Technology (T), Market (M), and Policy/Regulation (R) (3rd column, Table 2). By mapping documents into one of the three domains of the co-evolutionary analysis, we can see that documents concerned with the Technology domain clearly dominate in primarily research-oriented organizations including universities (22 out of 38, 58%) and R&D institutes (4 out of 5). On the other hand, documents presented by industry actors mostly fall under the Market domain (2 out of 4), whereas documents presented by regulatory actors mostly discuss the Regulatory domain (3 out of 5). This distribution coincides with each affiliation's main interests.

We can also distinguish the cognitive orientation of documents according to the process theory framework for anticipatory standardization, namely, Design (D), Negotiation (N) and Sense-making (S) (4th column, Table 2). The differences by contributor affiliation are not as significant as shown in the classification by domains. Documents concerned with the Sense-making stage dominate in most of the affiliation groups, including the university (26 out of 38, 68%), industry (4 out of 4), and regulatory (3 out of 5). It means that those stakeholders have not reached an agreement on the usage of the technology concerned and the contour of its institutional arrangement. A notable exception to this pattern is the R&D institutes, where documents are mostly oriented toward Design (4 out of 5). R&D institutes are likely to focus on design aspects with less regard to the real life context of the technology use.

Table 3 shows the matrix of document distribution by domain and cognitive orientation. For the domains of the co-evolutionary analysis, at an aggregated level, more than half of the documents address the Technology domain (36 out of 68, 53%), followed by Market (19 out of 68, 28%) and Regulatory (13 out of 68, 19%) domains. It reflects the dominant contributions made from the universities. As for the cognitive orientation, nearly two-thirds of the documents (42 out of 68, 62%) are primarily oriented toward Sense-making, followed by Design (21 out of 68, 31%), and Negotiation (5 out of 68, 7%). It indicates that at this stage of CR development, while technical specifications are created (Design), discourses still center on how to use it (Sense-making) and do not yet go to the level of each stakeholder's strategy building (Negotiation).

When we look at cognitive orientations by domain, out of the 36 documents that primarily address the Technology domain, 19 are oriented to Design (53%), closely followed by 17 documents on Sense-making (47%) and zero on Negotiation. Out of the 19 Market domain documents, 18 deal with Sense-making (95%), except only one document addressing Design (5%). It is understandable that the documents on the Market domain mainly deal with Sense-making. Under the regulatory domain, 7 out of 13 documents address Sense-making (54%), followed by Negotiation (5, 38%). It is notable that out of the entire collection of documents, those documents addressing the Negotiation (5 out of 68, 7%) are found only in the context of the Regulatory domain. As regulation is supposed to manage and control various stakeholders with vested interests, those documents cover Negotiation from an early stage of technology development.

By and large, it is obvious that contributions of each stakeholder reflect their primary technology development domains according to the co-evolutionary analysis model. Technology development as a creative process is dominantly represented by research institutes like universities, while market application of the CR for profit-making is a primary concern of industry actors. Policy and regulatory issues emerging from the development of new technology are addressed mainly by regulators. Here, the fact that documents on technology domain, mostly contributed by university stakeholders, take up a significant part of the entire discussion, reflects the still emerging nature of the CR in the process of co-evolutionary technology development.[10]

Table 3. Document Distribution by Domain and Cognitive Orientation

Categories		Domain			
		Market	Regulatory	Technology	Sub-Total
Cognitive Orientation	Design	1	1	19	21
	Negotiation	0	5	0	5
	Sense-making	18	7	17	42
	Total	19	13	36	**68**

Combined together, the results (Table 3) provide ground-level evidence that CR is still a concept of which discussions primarily remain in its technological feasibility. Transition to its business application scenario or potential regulatory concerns has yet to come (Fomin, Medeišis, and Vitkutė-Adžgauskienė 2012).

Keyword Analysis

The keyword analysis of CR documents presents an opportunity to examine firsthand what kind of topics in CR Technology have emerged, developed and sometimes disappeared within the local research arena, and how some of those key issues are being discussed in different contexts of technology, market and regulatory concerns.

We draw out top twenty five keywords by their frequency of appearance from the pool of document-level keywords out of 68 documents (Table 4) [11]. Figure 1 is a graphic representation of the keyword analysis. Due to some keywords that have the same number of appearances, the list of top 25 actually includes a total of 34 unique keywords. By far, the most frequently appearing keywords are *spectrum*, followed by *white space*, *scenario*, *access*, and *secondary*.

The word *spectrum* appears 47 times in 28 documents out of the 68 investigated, which amounts to 5.92% of the weighted percentage of all keywords (Table 4). The result reflects that the radio spectrum is a cornerstone of the global wireless industry upon which the development of CR technology takes its root (Fomin, Medeišis, and Vitkutė-Adžgauskienė 2012).

Table 5 shows the break-down of top twelve keywords by domain and cognitive orientations. Although the appearance of *spectrum* in document keywords list is universal in all three domains, it is used dominantly in the Technology domain (27 out of 47, 57%), followed by the Market (11 out of 47, 23%) and Regulatory (9 out of 47, 19%) domains. In terms of cognitive orientation, it is mostly used in either Sense-making (25 out of 47, 53%) or Design (17 out of 47, 36%), followed by Negotiation (5 out of 47, 11%).

Out of the 27 Technology domain-oriented documents where the keyword *spectrum* appears, about half of them fall under the Design (13), while the other half falls under the Sense-making (14). The result indicates that in the Technology domain, *spectrum* is a topic which involves a similar amount of discussions on both *development* (Design) and *evaluation* (Sense-making) of technical specifications such as hardware interface and design architecture of CR. It is discussed under the context of spectrum database, spectrum management optimization, technical solutions for spectrum assessment and simulation. However, apart from creating new approaches and conducting pilot programs to refine and improve

Table 4. Top 25 keywords by frequency of appearance

Rank (No.)	Words	# of Appearance	Weighted %[12]	Rank (No.)	Words	# of Appearance	Weighted %
1	Spectrum	47	5.92	19 (6)	cases	6	0.76
2	white space	17	2.14		communications	6	0.76
3	Scenario	16	2.02		Policy	6	0.76
4	Access	15	1.89		power	6	0.76
5	secondary	14	1.76		sensing	6	0.76
6	network	13	1.64		wireless	6	0.76
7	market	12	1.51	25 (10)	architecture	5	0.63
8	use	11	1.39		indoor	5	0.63
	channel	11	1.39		management	5	0.63
	model	11	1.39		mobile	5	0.63
11 (2)	business	10	1.26		technology	5	0.63
	interference	10	1.26		device	5	0.63
13 (3)	analysis	9	1.13		opportunity	5	0.63
	database	9	1.13		pilot	5	0.63
	m2m	9	1.13		service	5	0.63
16 (3)	dynamic	8	1.01		sharing	5	0.63
	system	8	1.01	**Total appearance of 34 listed keywords**		324	40.85
	geolocation	8	1.01	**Total appearance of entire keywords**		794	100

Figure 1. Keywords by frequency of appearance
**The font size of each keyword is proportional to the frequency of its appearance*

the technical capacity, it does not currently entail specific discussions as to how to reach an agreement about different technical approaches through *negotiation*.

Out of the 11 Market domain-oriented documents where the keyword *spectrum* appears, ten fall under the Sense-making (91%). It mainly focuses on service providers' business proposition on spectrum, for example, business feasibility analysis of the secondary or dynamic spectrum access and sharing. Notably, most of the documents with the keyword *spectrum* under this domain propose a form of business scenario and postulate business opportunities, business models, or growth strategies.

Out of the nine documents in the Regulatory domain where the keyword *spectrum* appears, five fall under the Negotiation (56%). Mostly, the keyword is used in the context of radio spectrum management and governance issues regarding how to initiate industry coordination or to set up a regulatory framework.

By and large, the discussions on the keyword *spectrum* remain yet at the level of expanding the body of knowledge about individual technical specifications of CR and their evaluation in technological and market contexts, rather than at the level of aligning those technology designs with other actors' interest to create a binding network of engagement, which eventually would have to evolve into a more institutionalized agreement (Lyytinen, et al., 2008).

Trend Analysis

In terms of the co-evolutionary analysis, the highest number of documents from all the five meetings falls under the Technology domain. Even though there is no significant trend of change in terms of the proportion of the documents under the three domains, there is a slight decrease in the Technology domain while there is a slight increase under the Regulatory domain (Figure 2).

In terms of the cognitive orientation, documents under the Sense-making category take up the majority in most of the meetings, except for Meeting 5 where half of the documents fall under the Design

Table 5. Top 12 Keywords break-down by Domain and Cognitive Orientation

Cognitive Orientation	Domain		
	Technology	**Market**	**Regulatory**
Design	Spectrum (13) Channel (10) White space (8) Interference (7) Network (5) Scenario (3) Access (2), Use (2)	Market (2) Secondary (1), Spectrum (1)	Spectrum (3) Secondary (2) Market (1), White space (1)
Negotiation	-	-	Spectrum (5) Access (3), White space (2) Interference (1), Scenario (1), Secondary (1)
Sense-making	Spectrum (14) Scenario (5) Access (4), Secondary (4), Use (4) Network (3) Business (2), Interference (2), Market (1)	Spectrum (10) Business (7) Secondary (6) Market (5), Network (5), Scenario (5), Use (5) Access (4), White space (4)	Market (3) Access (2), Scenario (2), White space (2) Business (1), Spectrum (1)

Figure 2. Domain by Meeting

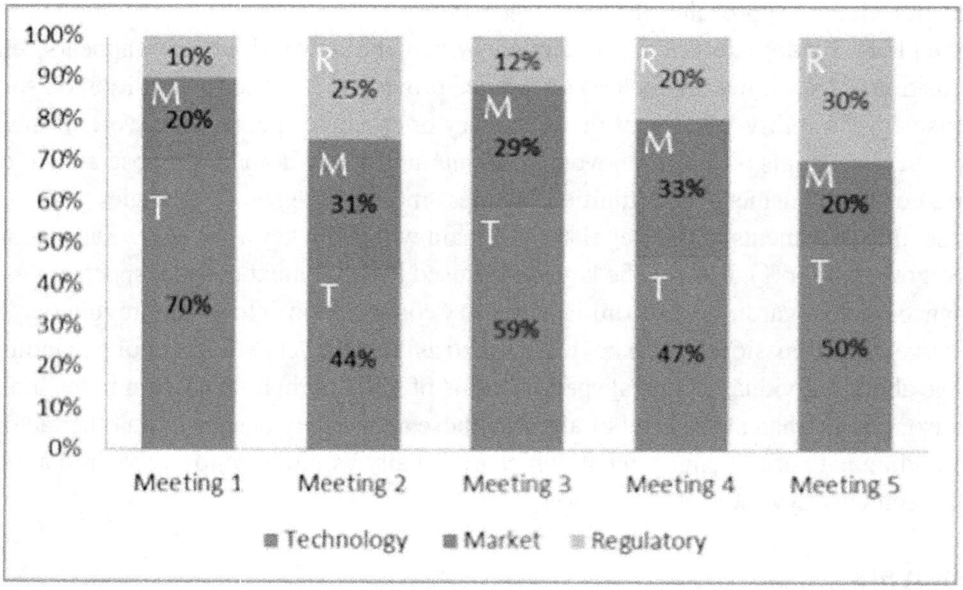

category. There can be no significant trend observed in terms of the proportion of the documents under the three cognitive orientations (Figure 3).

At the keyword level, the keyword spectrum appears most frequently across all the five meetings. Seven keywords appear in all the five meetings: spectrum, white space, access, market, model, interference, and power. Even though there is no significant pattern of change in terms of their weighted percentage

Figure 3. Cognitive Orientation by Meeting

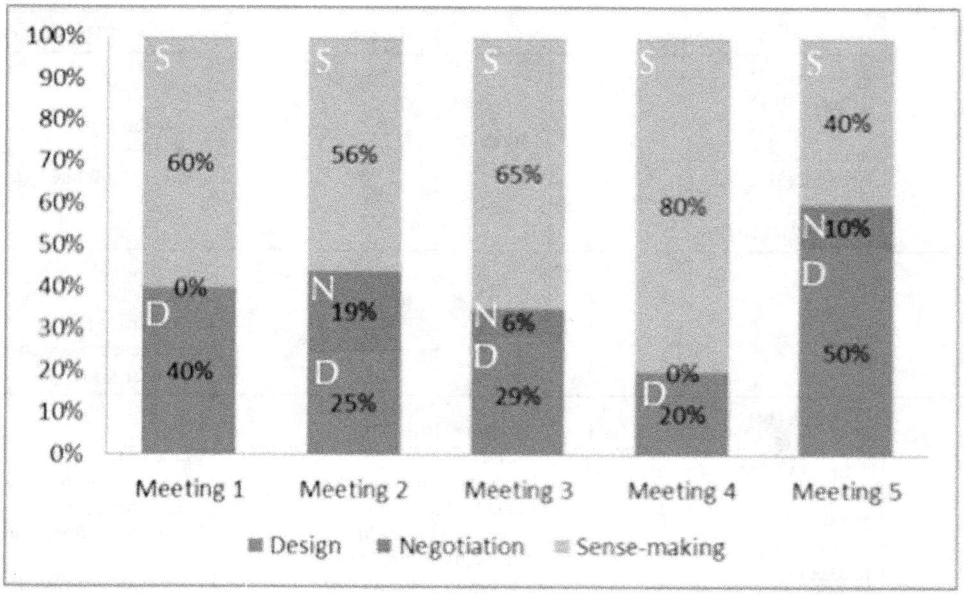

of frequencies, it is notable that two of the most frequently used keywords (spectrum and access) appear less often in later meetings (see Table 6. and Figure 4).

Out of a total of ten documents in meeting 1, 89 keywords appeared more than once. Total counts of all keywords were 123. In Meeting 1, the most frequently appearing keyword was *spectrum* (8 times, 6.35%), closely followed by *network* (5.56%) and *access* (4.76%). Notably, the keyword *network* appeared more frequently in Meeting 1 documents than in other meetings. Likewise, Table 6 exhibits those keywords that appear mostly frequently within each meeting.

DISCUSSION

In this work the document keyword analysis is extended and enriched by the use of the two theoretical frameworks. Below we discuss some findings of significance.

Figure 4. Keyword Trend: Weighted Percentage of Frequency

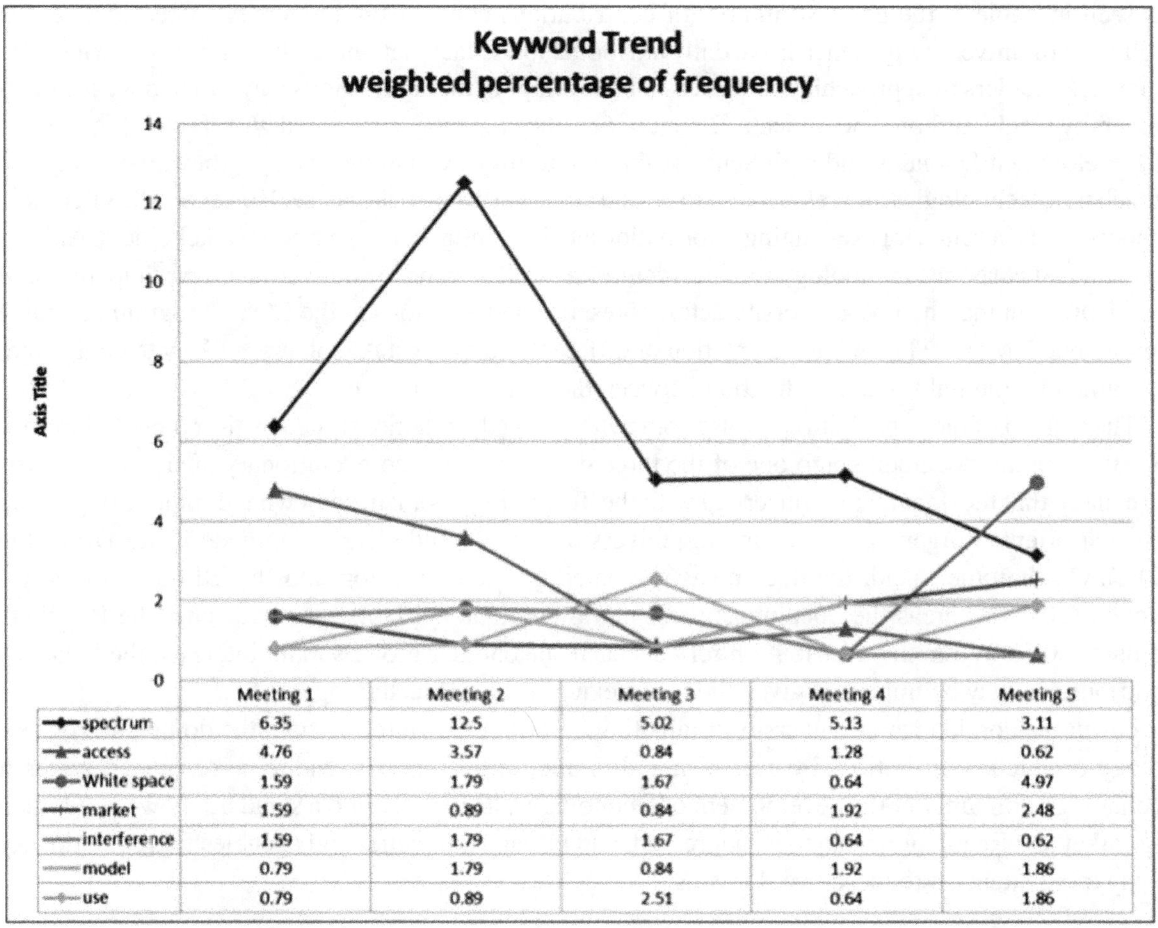

	Meeting 1	Meeting 2	Meeting 3	Meeting 4	Meeting 5
spectrum	6.35	12.5	5.02	5.13	3.11
access	4.76	3.57	0.84	1.28	0.62
White space	1.59	1.79	1.67	0.64	4.97
market	1.59	0.89	0.84	1.92	2.48
interference	1.59	1.79	1.67	0.64	0.62
model	0.79	1.79	0.84	1.92	1.86
use	0.79	0.89	2.51	0.64	1.86

**For a more accurate representation see the electronic version.*

Table 6. Top Keywords by Meeting

Meeting #	# of Documents	# of Unique Keywords	Total Counts of Keywords	Keywords (Weighted Percentage of Frequencies Within Each Meeting)
1	10	89	123	Spectrum (6.35%), Network (5.56%), Access (4.76%), Communications (2.38%), Scenario (2.38%)
2	16	70	109	Spectrum (12.5%), Scenario (5.36%), Sensing (4.46%), Access (3.57%), Secondary (3.57%)
3	17	142	228	Spectrum (5.02%), Use (2.51%), Channel (2.09%), Indoor, Interference, M2M, Secondary, Whitespace (1.67%)
4	15	101	155	Spectrum (5.13%), Business (4.49%), Network (3.12%), Analysis, Policy, System, Technology (2.56%)
5	10	93	155	Whitespace (4.97%), Database, Geolocation (4.35%), M2M, Spectrum (3.11%), Market, Pilot, Secondary (2.48%)

Contributors and Contributions

As seen in Table 2, the greatest number of contributions comes from university representatives. The high level of university presence is partially attributed to the fact that universities are less restricted than other stakeholders in approaching new ideas and developing them in various ways. When we look at the university single-authored documents (38 out of 68), they are more concerned with Technology in terms of Development Domains and with Sense-making in terms of Cognitive. That is, they are willingly approach new technologies and relate them to various life-worlds to make sense. However, COST Action is supposed to be a venue for exchanging information and interacting among various stakeholders. Although national and corporate technology research centres as well as private companies are participating in the COST programme, the non-university actors' presence and activities in the COST programme must be strengthened. More balanced representation of different groups of stakeholders is likely to contribute to strengthening the links and coordination between the different domains.

The content of the contributions is also somewhat skewed, as demonstrated by the co-evolution analysis. By mapping documents into one of the three domains of the co-evolutionary analysis framework, we can see that the documents concerned with the Technology domain somewhat dominate in primarily research-oriented organizations including universities (58%) and clearly dominate in R&D institutes (80%). On the other hand, the documents presented by industry actors mostly fall under the Market domain (50%), whereas the documents presented by regulatory actors mostly discuss the Regulatory domain (60%). While this distribution reflects each stakeholder group's main interests, the balance of contributions may be improved given the complexity of the CR technology.

Given the dominance of university-originating contributions in the Action, the dominance of Technology-oriented contributions by itself is not alarming, as universities tend to be oriented towards the Technology domain. For the development of technology, particularly into a standard, however, domains of Market and Regulation need to be addressed, and private companies and agencies for regulation tend to pay more attention to these two domains.

Cognitive Orientation

Analysis of Cognitive orientation of the contributions allows assessing the completeness or maturity of the proposed technology or business solution. We found that contributions are mostly falling under the Design or Sense-making categories while few keywords are found under the Negotiation. Such distribution indicates that at this stage of CR development, while technical specifications or business models are created (Design), discourses still focus on how to use them (Sense-making) and do not yet go to the level of engaging multi-party (e.g., technology-business, or business-regulation) stakeholder strategy building. This result also echoes the fact that the major part of CR Technology discussions still remains in either the creation of technological solution itself, or in the efforts to develop refined technological applications that relate to the current telecommunication environment.

For example, prevention and management of interference is an important concern of the regulatory domain once the CR technology takes off in the market. In Table 5, however, while interference appears 7 times in the Technology/Design cell and twice in the Technology/Sense-making, it appears only once in the Regulatory/Negotiation. The current discussions on interference mainly come from early-stage studies to understand the interference mechanism itself, and measure and calculate the interference in a given environment. In that respect, the result again confirms the early nature of CR technology in the co-evolutionary process of technology development.

Internal Dynamics

In terms of the keyword dynamics, there is no significant pattern of change in their weighted percentage of frequencies over the studied period. However, it is notable that two of the most frequently used keywords (*spectrum* and *access*) appear less often in later meetings, while "*white space*" gains more weight in the last meeting. Keyword dynamics allows analyzing the "thematic shifts" among contributors.

Categorization of contributions according to one of the three stakeholder domains (Technology, Market, or Regulation) and one of the three cognitive orientations (Design, Negotiation, Sense-Making) can help gain deeper insights on the internal dynamics of the contributions.

With regard to the three domains, there is no significant change in the proportion of the contributions. Yet, we observe that Technology contributions slightly decrease, while those under the Regulatory domain are on the rise. This trend is reflective of the overall goal of the particular COST-TERRA programme, which is aimed at developing "techno-economic *regulatory* framework." Having started the project with the exploration of specific technology solutions, the focus has been gradually shifting to address regulatory issues.

With regard to the cognitive orientations, documents under the Sense-making category take up the majority in most of the meetings, except for Meeting 5 where the majority of documents fall under the Design category. The changes in the proportions between the three categories do not seem to follow any trend, with the exception that the Negotiation category is very weakly represented. Given the *techno-economic* and *regulatory* orientation of the COST-TERRA programme, the under-represented Negotiation and dominating Sense-Making suggest immaturity/ incompleteness of the solutions proposed in the meetings.

Implications for European Standardization

There is a concern in European standardization communities; while inter-firm R&D partnerships are very important (Hagedoorn, 2002a) for technology innovation projects, researchers and standardization practitioners do not know much about one another's activities with regard to specific technology development efforts. That is, R&D and standardization are not linked as much as they are supposed to be (Hagedoorn, 2002a; Sales et al., 2012). This issue was repeatedly identified as one of the major problems of the European standardization arena (Jakobs and Blind 2011; INTEREST 2005). At the same time, new European policy such as *Horizon 2020 Programme*, stresses that "stronger support will be given to the market take-up of innovation, including... a better use of the potential of research infrastructures, as well as setting technical standards…" (EC 2011, p.9).

The method adopted in this paper can take the efforts on mutual orientation of researchers and standardization practitioners to a new level. While one of the identified priorities in improving the European standardization arena is fostering synergy between ICT research, innovation, and standardization (European Commission, 2009b), this action cannot be accomplished with due effectiveness, unless the identified gaps between R&D and standardization are filled. The analysis presented in this paper can help with identifying and filling those gaps. To take the example of CR as conducted in the paper, the Cognitive Radio has been gaining increased popularity in the technology and policy making domain in the telecommunication sector over the last decade. The analysis can help researchers and policy makers better understand the characteristics of the diverse R&D efforts which are contributing to the formation of the CR technology. Understanding the diversity of technology development at an early standardization or pre-standardization stage is important (Hagedoorn, 2002b; Hoel, Hollins, & Pawlowski, 2010; Sales et al., 2012), with some authors arguing that pre-standardization is the most important period in ICT standard-setting process (Lim, 2003, p. 5). Many recognized standards development organizations (SDOs) today are developing CR-related standards, and there are many university- and industry-led research initiatives focused on the topic. This means that there are many concurrent and often uncoordinated R&D and standardization efforts taking place – all aiming to contribute to the emergence of (set of) global CR standard(s). By better knowing who the involved stakeholders are (Hoel et al., 2010), knowing about the issues being tackled by the contributors of the ESF COST Action, we can orient researchers, policy makers, or standards professionals with regard to the general (industry) or specific (project) efforts when expertise or collaboration for the research projects is looked for, or when a link between R&D and standardization is sought to be established.

Once we know what is going on in diverse R&D and standardization activities, as Jakobs & Blind (2011, p. 106) suggest, the Commission can reconsider their funding principles for EU-funded R&D projects. Projects with promising findings and an interest to incorporate them into standards could be granted additional funding for standardization activities. The analysis methods offered in this paper can be used for (more) systematic analysis of pre-standardization activities, which, in turn, can help technology developers and policy makers draw competitive standardization strategies (Sales et al., 2012, p. 19).

Establishing better links within and between standardization- and research- oriented communities is of particular importance in the case of CR-related standardization. There is ambiguity with regard to what exactly consumers and markets will be served by CR standards, and CR application requires horizontal interoperability between (and harmonization of) different technology standards, business practices, and regulatory policies (Delaere & Ballon, 2008; Sukarevičienė & Fomin, 2013). Therefore most actors involved in the development of CR technology standards will require a much higher degree of coordination than before (Delaere & Ballon, 2008, p. 2; Schmidt & Werle, 1998, p. 108). In this regard, practical importance of content-analysis-enabled "matching" process is evident, too.

EC also aims to cut in half the average time it takes to develop standards by 2020 (Nemec, Orviska, & Hudson, 2012, p. 290), that is, to shorten "innovation cycles and the convergence of technologies across the boundaries of the three European standardization organizations" (Brykman, 2011, p. 193). EC has the associated "rewards" policy. It supports European Standardization Organizations (ESOs) conditionally on meeting targets relating to the speed with which standards are produced and to the quality of standards with regard to the stakeholder representation in the development process (Nemec et al., 2012, p. 290). At this backdrop, our work of content analysis presented in this paper, as descriptive and mundane as it may appear, has a potential to improve the effectiveness of the R&D – Standardization – Policy links, and to speed up standardization processes.

CONCLUSION

The multifaceted document analysis of the COST-TERRA Action presented in this paper allows us to shed light on the development of technology at an early standardization or pre-standardization stage. It is when the "global-technology-to-be" is taking shape through diverse R&D initiatives of universities, industry, and regulatory stakeholders, which are taking place concurrently and often overlapping.

The theoretical contribution of this research is two-fold. First, the two theoretical frameworks – the co-evolutionary analysis (Bauer et al., 2007; K. Lyytinen & King, 2002) and the "D-N-S" model of cognitive orientations in anticipatory standards making (K. J. Lyytinen et al., 2008) – were combined to allow technology development process analysis to be conducted at the marco- (industry, stakeholders, topics) and micro- (completeness or maturity of the proposed solution through cognitive orientation of the proposal) levels. The adopted method of data analysis produced deeper insights on the concurrent R&D processes which together contribute to the formation of novel technology standard(s). Specifically, we were able to identify what were discussed in technological, market, or regulatory domains regarding the specific developments. We can also assess whether the proposed designs and solutions are specific enough or are only vaguely defined, and which topics are uniquely discussed under certain domains or categories. While the study is conducted in the area of ICT standardization, the same method can be applied in other areas because standardization, by its nature, involves multiple stakeholders in technology, market and regulation, and passes through the stages of design, sense-making and negotiation. And all those activities by many actors generate a huge amount of documents.

Second, the use of quantitative content analysis combined with the two theoretical frameworks sets a model for future analysis of technology (standardization) development. To our knowledge, this is the first work in which the quantitative text analysis is used for understanding (pre-) standardization processes. No attempts are made to introduce systematic procedures for documents storage and keyword and/or content analysis. However, this is a task which could be easily implemented – a variety of powerful software for lexical and semantic analysis is available today. Monitoring documentation of COST Actions could contribute to forming a better understanding of the topics/themes being discussed. Such simple action can contribute to making the standardization landscape more understandable and transparent to researchers, and help policy makers and standardization practitioners find promising/relevant research projects/ideas with less effort. Document analysis can also help identify links within the researchers' community, and matching between different COST actions can be done. Identification of common interests is crucial in forming research consortia. Consortia, subsequently, are crucial in promoting research findings (e.g., to standardization or to the market) (Jakobs, 2012).

The value of this work is constrained by the amount of documents that we have analyzed. It is partly due to the facts that (1) the studied COST Action was half-way in its four year lifespan at the time when the analysis was conducted, and that (2) there was at least one other COST Action (IC0902) focused on CR/SDR technology development, which has not been included in this analysis. We believe, however, the current work is sufficient to demonstrate the methodological value of the content analysis conducted in this study.

ACKNOWLEDGMENT

This work was supported by the National Research Foundation of Korea Grant funded by the Korean Government (NRF-2011-330- H00002); and the European Social Fund and the Republic of Lithuania (grant number VP1-3.1-ŠMM-10-V-02-025).

The authors are grateful to colleagues from COST Action IC0905 TERRA "Techno-Economic Regulatory framework for Radio spectrum Access for Cognitive Radio/Software Defined Radio" (www.cost-terra.org) for feedback received on earlier versions of this paper.

REFERENCES

Baldini, G., Holland, O., Stavroulaki, V., Tsagkaris, K., Demestichas, P., Polydoros, A., & Allen, D. et al. (2013). The evolution of cognitive radio technology in Europe: Regulatory and standardization aspects. *Telecommunications Policy*, *37*(2-3), 96–107. doi:10.1016/j.telpol.2012.07.003

Bauer, J. M., Ha, I. S., & Saugstrup, D. (2007). Mobile Television: Challenges of Advanced Service Design. *Communications of the Association for Information Systems*, *20*, 621–631.

Bousquet, F., Fomin, V. V., & Drillon, D. (2009). Standardization: a Major Tool for Competitive Intelligence. In *14th EURAS Annual Standardisation Conference "Standardisation and Corporate Intelligence"* (pp. 27–36). Cergy, France: Wissenschaftsverlag Mainz in Aachen.

Brykman, L. (2011). Better Standards to Enhance European Competitiveness. In *Proceedings of the 7th biennial Standardisation and Innovation in Information Technology conference (SIIT 2011)* (pp. 193–194). Berlin: Mainz Publishers.

Delaere, S., & Ballon, P. (2008). Multi-Level Standardization and Business Models for Cognitive Radio: The Case of the Cognitive Pilot Channel. In *3rd IEEE Symposium on New Frontiers in Dynamic Spectrum Access Networks (DySPAN'08)*. Chicago, IL: IEEE. doi:10.1109/DYSPAN.2008.67

European Commission. (2007). *Final report of the study on the specific policy needs for ICT standardization*. European Commission. Retrieved from http://www.epractice.eu/en/library/281182

European Commission. (2009a). *Radio Spectrum Policy Group Report On "Cognitive Technologies". Final Draft* (No. Electronic Communications Policy). Brussels: Information Society and Media Directorate-General. Retrieved from http://rspg.groups.eu.int/_documents/documents/meeting/rspg20/rspg09_299.pdf

European Commission. (2009b). *White Paper: Modernising ICT Standardisation in the EU – The Way Forward.* (White Paper No. COM(2009) 324 final). European Commission. Retrieved from http://eur-lex.europa.eu/LexUriServ/LexUriServ.do?uri=CELEX:52009DC0324:EN:NOT

Fomin, V. V., & Keil, T. (2000). Standardization: Bridging the Gap Between Economic and Social Theory. In *The Twenty First International Conference on Information Systems (ICIS 2000)* (pp. 206–217). Brisbane, Australia: AIS. Retrieved from http://aisel.aisnet.org/icis2000/20

Fomin, V. V., Keil, T., & Lyytinen, K. J. (2003). Theorizing about Standardization: Integrating Fragments of Process Theory in Light of Telecommunication Standardization Wars. *Sprouts: Working Papers on Information Environments, Systems and Organization, 3.*

Fomin, V. V., Medeišis, A., & Vitkutė-Adžgauskienė, D. (2011). The Role of Policy in the Development of Cognitive Radio Systems: Co-Evolutionary Perspective. In *The 7th International Conference on Standardisation and Innovation in Information Technology (SIIT 2011)* (pp. 79–90). Berlin: Mainz Publishing.

Fomin, V. V., Medeišis, A., & Vitkutė-Adžgauskienė, D. (2012). Pre-Standardization of Cognitive Radio Systems. *International Journal of IT Standards and Standardization Research, 10*(1), 1–16. doi:10.4018/jitsr.2012010101

Hagedoorn, J. (2002). Inter-firm R&D partnerships: An overview of major trends and patterns since 1960. *Research Policy, 31*(4), 477–492. doi:10.1016/S0048-7333(01)00120-2

Hoel, T., Hollins, P. A., & Pawlowski, J. M. (2010). On the Status of Learning Technology Specifications and Standards. *International Journal of IT Standards and Standardization Research, 8*(2), i–vi.

Hughes, T. P. (1993). The Evolution of Large Technological Systems. In *The social construction of technological systems: New directions in the sociology and history of technology* (pp. 51–82). Cambridge, MA: MIT Press.

Isaak, J. (2006). The Role of Individuals and Social Capital in POSIX Standardization. *International Journal of IT Standards and Standardization Research, 4*(1), 1–23. doi:10.4018/jitsr.2006010101

ITU. (2009). *Definitions of Software Defined Radio (SDR) and Cognitive Radio System (CRS)* (No. ITU-R SM.2152). ITU-R. Retrieved from http://www.itu.int/dms_pub/itu-r/opb/rep/R-REP-SM.2152-2009-PDF-E.pdf

Jakobs, K. (2003). *Information Technology Standards, Standards Setting and Standards Research: Mapping the Universe*. Retrieved from http://www-i4.informatik.rwth-aachen.de/~jakobs/Papers/stanhope.pdf

Jakobs, K. (Ed.). (2009). *ICT Standardization for E-Business Sectors: Integrating Supply and Demand Factors*. Hershey, PA: IGI Global. doi:10.4018/978-1-60566-320-3

Jakobs, K. (2012). Pitfalls in Standards Making. In *EURAS Proceedings 2012: "Standards and Innovation"* (pp. 211–226). Banska Bystrica, Slovakia: The EURAS Board Series.

Jakobs, K., & Blind, K. (2011). ICT Standardisation Policy in Europe – Recent Past, Presence, and Future(?). In *Proceedings of the 7th biennial Standardisation and Innovation in Information Technology conference (SIIT 2011)* (pp. 101–114). Berlin: Mainz Publishers. doi:10.1109/SIIT.2011.6083611

Kwak, J., Lee, H., & Fomin, V. V. (2011). The governmental coordination of conflicting interests in standardisation: Case studies of indigenous ICT standards in China and South Korea. *Technology Analysis and Strategic Management*, *23*(7), 789–806. doi:10.1080/09537325.2011.592285

Lim, A. S. (2003). *Inter-firm Alliances during Pre-standardization in ICT* (Working paper No. 03.03). Eindhoven Centre for Innovation Studies, Eindhoven University of Technology.

Lyytinen, K., & King, J. L. (2002). Around the cradle of the wireless revolution: The emergece and evolution of cellular telephony. *Telecommunications Policy*, *26*(3-4), 97–100. doi:10.1016/S0308-5961(02)00002-2

Lyytinen, K. J., Keil, T., & Fomin, V. V. (2008). A framework to build process theories of anticipatory Information and Communication Technology (ICT) Standardizing. *International Journal of IT Standards and Standardization Research*, *6*(1), 1–38. doi:10.4018/jitsr.2008010101

Medeišis, A. (2014). European Regulatory Developments related to CR. In Cognitive Radio Policy and Regulation Techno-Economic Studies to Facilitate Development of Cognitive Radio. Academic Press.

Medeišis, A., & Holland, O. (Eds.). (2014). *Cognitive Radio Policy and Regulation Techno-Economic Studies to Facilitate Development of Cognitive Radio*. Springer. doi:10.1007/978-3-319-04022-6

Nemec, J., Orviska, M., & Hudson, J. (2012). Standardization and the European Standards Organizations. In *EURAS Proceedings 2012: "Standards and Innovation"* (pp. 281–294). Banska Bystrica, Slovakia: The EURAS Board Series.

Nickerson, J. V., & zur Muehlen, M. (2006). The Ecology Of Standards Processes: Insights From Internet Standard Making. *MIS Quarterly, 30*, 467–488.

Sales, B., Darmois, E., Papadimitriou, D., & Bourse, D. (2012). A systematic approach for closing the research to standardization gap. *The Future Internet*, 18–29.

Schmidt, S. K., & Werle, R. (1998). Coordinating Technology. Studies in the International Standardization of Tellecommunications. Cambridge, MA: The MIT Press.

Schneberger, S. L., & McLean, E. R. (2003). The Complexity Cross - Implications for Practice. *Communications of the ACM*, *46*(9), 216–225. doi:10.1145/903893.903940

Schot, J., & Geels, F. (2007). Niches in evolutionary theories of technical change. *Journal of Evolutionary Economics*, *17*(5), 605–622. doi:10.1007/s00191-007-0057-5

Schot, J., Hoogma, R., & Elzen, B. (1994). Strategies for shifting technological systems. *Futures*, *26*(10), 1060–1076. doi:10.1016/0016-3287(94)90073-6

Stewart, J., Shen, X., Wang, C., & Graham, I. (2011). From 3G to 4G: Standards and the development of mobile broadband in China. *Technology Analysis and Strategic Management*, *23*(7), 773–788. doi:10.1080/09537325.2011.592284

Sukarevičienė, G., & Fomin, V. V. (2013). *Modeling Dynamic Spectrum Access Infrastructure Development As a Techno-Economic System. Research in progress.* Brussels: EURAS.

ENDNOTES

1 Cognitive Radio System (CRS) is "a radio system employing technology that allows the system to obtain knowledge of its operational and geographical environment, established policies and its internal state; to dynamically and autonomously adjust its operational parameters and protocols according to its obtained knowledge in order to achieve predefined objectives; and to learn from the results obtained" (ITU, 2009).

2 See http://www.cost-terra.org.

3 "Two or more units in such a system are said to co-evolve, if they each have a significant impact on each other to persist" (Bauer, Ha, and Saugstrup 2007, 626)

4 ECC stands for Electronic Communications Committee under CEPT; ECC considers and develops policies on electronic communications activities in European context (see http://www.cept.org/ecc). CEPT stands for European Conference of Postal and Telecommunications Administration and one of the functions is to deal with different aspects of cognitive radio including standardization (see http://www.cept.org/cept).

5 The up-to-date official list of members of the MC for COST Action IC0905 TERRA is available at http://w3.cost.esf.org/index.php?id=177&action_number=IC0905.

6 A small amount of documents are contributed from regions outside of the EU, including Canada, the U.S., China, South Korea, and South Africa.

7 Examples under this category include: VTT Technical Research Centre of Finland; National Institute of Telecommunications, Poland; Communications Research Centre, Canada; Centre Tecnològic de Telecomunicacions de Catalunya, Spain, among others.

8 Examples of research institutes and labs directly affiliated with a private company include: BT Lab and Nokia Research Center.

9 Examples include: Federal Office of Communications, Switzerland; Bundesnetzagentur, Germany; Swedish Post and Telecom Agency; Ministry of EAAI, The Netherlands; and European Communications Office.

10 Over the period of five meetings under the study, the composition of authors' affiliation remained stable in all five meetings, which means that university contributions continued to be the largest.

11 The task, including Figure 1, was conducted using NViVo, a software package for qualitative text analysis.

12 The number denotes the weighted percentage of appearance of the given keyword against the total number of appearances of every keyword identified from 68 documents.

Section 2
IPR Issues

Chapter 7
From Patent Hold-Up to Patent Hold-Out?

Marie Barani
Independent Researcher, UK

ABSTRACT

Patents are an efficient tool for companies contributing to the standardization process to recover investments made in the process and continue participating in future standardization efforts. However, to avoid abusive use of standard-essential patents (SEPs) incorporated in de jure standards, standardization contributors are required to make their SEPs available on fair, reasonable and non-discriminatory terms and conditions (the [F]RAND commitment). This commitment has been created to fairly and adequately reward innovators while at the same time allow access to the standardized technology at (F)RAND terms and conditions; the latter preventing patent hold-up. With the changes in the cellphone market in the last ten years, the content of this (F)RAND commitment has been challenged in courts and in front of antitrust authorities. The question is whether this duty, set up to avoid hold-up, is not used by some implementers to engage in hold-out.

INTRODUCTION

Standardization aims to define standards, which are technical rules. The process[1] is driven by the World Trade Organization ("WTO")'s principles of transparency, openness, impartiality and consensus, effectiveness and relevance, and coherence (TBT Committee 2000). Firms that may compete at a later stage collaborate by pooling human, financial and/or technical resources to have the best technological solution adopted as next standard. Once the standard has been adopted, it is made public and available to anyone wishing to implement it, whether or not they are a member of the standard-development organization ("SDO") that has developed the standard. Besides interoperability and compatibility, even between competing devices (EC regulation No. 1025/2012 on standardization), standards in telecommunications also guarantee high performance (Fraunhofer Study 2011).

DOI: 10.4018/978-1-5225-5320-5.ch007

For the reasons mentioned above, standards are considered as a tool encouraging innovation, technology dissemination and competition. Nonetheless, standardization is closely linked to intellectual property rights ("IPRs") and competition issues. Companies involved in standardization may hold patents. Unlike standards, which are available on an open and non-discriminatory basis, patents are exclusive rights permitting their holders to block third parties from using the patented invention[2]. While both aim to encourage innovation, patents and standards can appear incompatible at first sight. Even more so when patents are qualified as "standard-essential patents" ("SEPs"), as no technical alternative to those patents exists to comply with such standard. Accordingly, each product or service implementing the standard without a license will therefore necessarily infringe the SEPs[3].

As a result, SEPs in principle confer their holders an increased market power by allowing them to control the standardization process and the downstream market (Shapiro 2001, Farrell et al. 2007). Consequently, in the absence of any safeguard, standardization could lead to behaviors impeding competition, as hold-up or royalty-stacking. Patent hold-up refers to the situation where SEP-holders abuse their bargaining power to extract excessive royalty rates from their SEPs, potentially under the threat of an injunction. Such rates, if they are accepted by the implementer, may hamper the diffusion of the standard, as they could be financially burdensome and restrict access to the standard by implementers: the implementer will either have to accept rates "fair in excess of the patent-holder's true contribution", which are similar to a tax on products incorporating the patented technology; or to postpone the sale of products/services incorporating the standard to avoid paying excessive fees or costly litigations (Lemley & Shapiro 2007). Royalty stacking occurs when the cumulative royalty rate for all the patents needed for the standard is excessive: even if each SEP-holder demands a reasonable fee, due to the large number of SEPs to implement, the aggregate royalty rate for all SEPs exploited in the standard may reveal itself excessive and non-bearable (Lemley and Shapiro 2007)[4].

Yet, SEPs play an important role in standardization. Companies holding SEPs are those contributing the most to the development of the standard (on contributors, cf. ABIresearch 2013, on SEP-holders, cf. Fraunhofer Study 2011). Since SDOs seek to "balance between the needs of standardization for public use in the field of telecommunications and the rights of the owners of IPR"[5], they have enacted IPR Policies to avoid anticompetitive behaviors while adequately rewarding SEP-holders for their contribution to the standard. At the same time, these policies should not lead to a counter-abuse on the implementer's side.

One relevant part of the SDO's IPR Policies is the (F)RAND [FRAND, in e.g. Europe, and RAND, in e.g. the US, understood as synonyms] commitment. Through this commitment, SEP-holders agree to share their SEPs accessible on (fair), reasonable and non-discriminatory terms and conditions to any party, instead of keeping their technology covered by such SEPs proprietary. The purpose is to prevent any patent hold-up.

The entrance of new players in the telecommunications market, some of which did not contribute their technologies to standardization but nevertheless manufacture standardized products[6], has been possible thanks to *de jure* standards, as these companies can access to a robust standard and successfully build their business on it (Tapia 2016). The incorporation of stakeholders with new business models has led to worldwide litigations, some of which challenging the suitability of the standardization process as it is currently working.

One concern arising from the so-called "patent war" litigation in the telecommunications sector is whether there is a real risk of "patent hold-up". Inversely, some argue that certain players might be using the (F)RAND commitment to engage in "reverse hold-up" (also called "hold-out"), by refusing or delaying negotiations, in order to impede or reduce the payment for the use of SEPs.

Both questions will be analyzed under IPR Policies in Europe and the United States ("U.S."), as SDOs like the IEEE and ETSI have adopted the most litigated standards (GSM, UMTS, GPRS, Wi-Fi)[7]. Furthermore, this paper will focus on literature, case law and antitrust investigations in both geographic areas. After an explanation about the background of the article, the first section will examine the relationship between the (F)RAND commitment and the academic hold-up theory. The second part will assess how European and American judges and antitrust authorities address the question of hold-up claims. The third one will examine the potential transition from hold-up to hold-out.

BACKGROUND

The research is based on Lemley and Shapiro's article about patent hold-up (Lemley & Shapiro, 2007), related articles, and hold-up claims that have been raised in patent litigations by some standard implementers (among others, cf. Farrell et al., 2007; Hesse, 2012; Prywes & Bell, 2015; EC decisions, *Samsung-Enforcement of UMTS standard essential patents, and Motorola-Enforcement of GPRS standard essential patents*, April 29, 2014; W.D.Wash., *Microsoft Corp. v. Motorola, Inc., et al/ Motorola Mobility, Inc., et al., v. Microsoft Corporation*, April 25, 2013; FTC, *In the Matter of Motorola Mobility LLC, and Google Inc.*, July 24, 2013).

Some authors argue that hold-up allegations are not demonstrated by evidence and that the market may rather face hold-out behavior from some implementers (Galetovic, 2015; Layne-Farrar, 2016; Mallison, 2016).

Implementers alleging that some SEP-holders are holding-up base their claims on a request of an injunction and of excessive licensing fees by the concerned SEP-holder. SEP-holders claiming that some implementers are holding-out try to demonstrate that the concerned implementers have delayed negotiations and/or advocated for a below FRAND royalty rates. Both sides thus ground their allegations on injunction and royalty rates, with different arguments.

The aim is to analyze whether hold-up is supported by evidence or is not rather concealing a hold-out strategy by some implementers.

THE (F)RAND COMMITMENT AND HOLD-UP

The content of the (F)RAND commitment has been challenged in courts (1) among others to mitigate the risk of hold-up as defined by Mark Lemley and Carl Shapiro (2).

The Content of the (F)RAND Commitment

SDOs usually do not interfere in the negotiations of a (F)RAND license, which they consider as a purely commercial matter between companies[8].

While standardization was first driven by firms competing but also cooperating at upstream level, to have the best technology adopted as next standard at downstream level, on the product market, the arrival of smartphones has changed the dynamic of the system: new companies (as Apple, HTC, TCL) have entered the cellphone market, most of them not sharing their technology to the standard development but rather keeping it proprietary. Companies claiming there is a risk of hold-up generally belong to this group of newcomers (Gupta & Snyder 2014). Conversely, some main contributors and SEP-holders (some of them who were initially active both on the upstream development of standards and downstream cellphone market and have divested their handset activity, e.g. Nokia, Motorola) do not sell any smartphone (ABIresearch 2013, Fraunhofer Study 2011)[9]. The main manufacturers seem to rather focus on the development of products and/or services implementing the standards. They may occasionally acquire or develop SEPs, but they are not significantly contributing[10]. When purely manufacturers are involved in litigations, they often use a "FRAND defence", alleging that the injunction a SEP-holder requests on its SEPs beaches the (F)RAND commitment and constitutes an anticompetitive behavior intended to extract excessive licensing fees ("patent hold-up"). Furthermore, manufacturers rather assess "non-SEPs" than SEPs in litigations[11].

The "smartphone patent wars" seem to suggest that upstream competition on the technology market and downstream competition on the end-device market between companies active on both sides have been replaced by a competition between contributors and implementers on a newly defined market, the SEPs market. Litigations illustrate the tension between two different business perspectives: one that tries to reduce as much as possible licensing costs (implementers) and the other one that wants to maximize revenues on SEPs (contributors).

Some implementers claiming the necessity to obtain licenses on SEPs and complaining about a "too ambiguous" (F)RAND regime initiated suits to specify the content of the (F)RAND commitment. Until the *eBay* case (US Supreme Court 2006), injunctions were almost granted automatically in the US, which led to an increase in litigation. In addition to this, some SEP-holders acting in good faith but facing years of unfruitful negotiations, also sued manufacturers.

As confirmed by some courts, the (F)RAND commitment is not a license *per se*[12]. FRAND implies a duty for the SEP holder and for the SEP user to enter into good faith negotiations to determine (F) RAND terms and conditions for a license (*In the Matter of Certain Electronic Devices, Including Wireless Communication Devices, Portable Music and Data Processing Devices, and Tablet Computer,* 2013). Some argue, it has to be interpreted under contractual law, others believe that patent law should be sufficient (Unwired Planet v. Huawei, EWHC [2017] 711). If understood as under contractual law, any contractual negotiation may fail. Under patent law, especially under the (F)RAND framework established in Europe by the Court of Justice of the European Union ("CJEU") in the case *Huawei v. ZTE*, each party will have to follow certain steps to prove their good faith. In both cases, parties can ask an independent third party to determine whether the Huawei requirements have been fulfilled (Europe) or whether the negotiations were fair and reasonable (U.S.) and, if needed, whether the offer made was (F) RAND (Europe and the U.S.).

When negotiations have been unsuccessful but the implementer is still using, hence infringing, the SEPs, and the parties end up in litigation, the SEP-holder often demands to be granted an injunction to stop the unauthorized exploitation of his SEPs. In this case, the implementer generally raises a "FRAND defense".

Courts, antitrust authorities, large part of academia and industry have supported a (F)RAND defense based on patent hold-up concerns. However, it is to be determined whether the so-called patent hold-up has in practice led to too strong restrictions on SEP-holders' rights.

The Academic Hold-Up Theory

The theory of patent hold-up has been put forward by two American academics, Mark Lemley and Carl Shapiro (Lemley & Shapiro 2007). They advise to preclude SEP-holders to be awarded injunctions to thwart the possibility of hold-up and royalty-stacking, especially in sectors where *hundreds or even thousands of patents can read on a single product standard* (Lemley & Shapiro 2007).

However, this study has been criticized on many points.

First, it implies that all the bargaining power lies with the SEP-holder. Nevertheless, some implementers may have a huge countervailing bargaining power due to their position on the downstream market and their financial strength (Geradin et al. 2007). In contractual negotiations, an implementer will probably use these characteristics to successfully lower the rates, as not signing a license with such an implementer would be strategically detrimental to the SEP-holder.

Second, it does not take into account the fact that SEP-holders incur sunk costs they will not recover without a license. Once those costs are invested, there is no guarantee to recoup them (Geradin et al. 2007). And as standardization is a long-term project, SEP-holders are dependent on other players. If they misbehave, their technology will probably not be chosen for the next standard or the next version of the standard and their reputation may be affected (Ginsburg et al. 2015): patent hold-up could induce a potentially negative reputational effect on the concerned SEP-holder, make it more difficult for him to find new allies (Epstein et al. 2012), affect his licensing revue and its position when he has to operate as a licensee (Wright 2013). Therefore, SEP-holders have a clear incentive not to breach their (F)RAND commitment. As explained by Epstein, Krieff and Spulber (2012),

Any participant perceived as behaving "badly" in the licensing context is likely to find few allies and face many difficulties when it seeks to promote its new technologies for inclusion in future standards. In this highly interlaced world, across players and across time, reputational constraints cut deeply. Furthermore, because the same major players show up repeatedly in many different settings, the power of reputation exerts a significant multiplier effect that restraints patentees from acting opportunistically.

Third, it ignores the importance of cross-licensing and portfolio-based licenses (Mallison 2014). These practices seem to have been quite relevant when companies were both active on upstream and downstream markets. Cross-licensing secures freedom to operate as both companies will mutually license their respective patents, and may imply a reduced licensing fee[13]. Portfolio licenses avoid inducing extra-costs linked to the licensing of each patent separately and on a country-per-country basis (in particular in a context where companies are active on a worldwide basis). In the current framework, where participants are less vertically and more horizontally integrated firms, competing on different markets and not on the whole value chain, these practices may have less relevance (Fraunhofer 2011).

Fourth, hold-up is focused is focused on one specific relationship, where SEP-holders and implementers do not compete with each other (Shapiro 2010). This is however not always the case[14].

Fifth, the authors do not provide any explanation of the figures they use and their origin. If we consider that around 75% of all SEPs to 3G are owned by only four firms (Geradin et al. 2007, Mallison 2014) and that the cumulative royalty yield on 2G, 3G and 3G for these four firms has been estimated for 2014 to approximately 2.6% of mobile handset revenues and to around 5.6% of mobile handset revenues for all SEP-holders (Mallison 2015), assumptions of hold-up and royalty stacking may be less pertinent.

Finally, a true patent hold-up would in practice impede the adoption and diffusion of a standard or lead to stagnant innovation or excessive prices for standardized products. Not of this has applied in the telecommunications market. Just the opposite. Standards like 3G have turned out to be fast and worldwide implemented technologies, creating a new market open to a large number of new actors. Data transmission speed is 12,000 times faster from 2G to 4G standard, the network infrastructure cost has decreased 95% from 2005 to 2013, in the same period, the average mobile subscriber cost per megabyte has decreased 99%. Consumers can buy a smartphone for less than $40. Standards related to mobile technology have been widely adopted, have a high penetration rate and bring about a consumer surplus of approximately $16.4 trillion. They rely on core technology developed by firms heavily investing in R&D (21% of revenue, around $1.8 trillion between 2009 and 2013), which enables the creation of a value chain with different business models, from manufacturers of components, devices and infrastructures, to content developers, software companies and start-ups at the end of the value chain. This value chain has generated almost $3.3 trillion in revenues in 2014 (The Boston Consulting Group 2015). Furthermore, prices for SEP-reliant products are decreasing faster than for non-SEP reliant products (Galetovic et al. 2015).

After such impressive (r)evolution in the telecommunications market, one can only question whether implementers really face patent hold-up. For years, patent hold-up has been assumed by different courts, in particular in the U.S.

Nevertheless, as some American judges have pointed out, this has so far not been more than a theoretical principle which has in all the years not been sustained by any evidence (ITC 2014, ITC 2015, Fed. Circ. 2015):

While there may be a hypothetical risk of hold-up, we have evidence that it is not a threat in this case, or in this industry (In the Matter of Certain Wireless Devices With 3G and/or 4G Capabilities and Components Thereof, 2014).

[the party] restates the possibility of hold-up without providing or citing any evidence that has been presented in this case (…).

The ALJ asked Dr. Shampine if he could cite ever one solid example of a hold-up resulting in a non-FRAND contract. Dr. Shampine replied: we do not have a solid example of that occurring yet (...)not a single case of hold-up has been noted (In the Matter of Certain 3G Mobile Handsets and Components Thereof, 2015).

Abstract recitations of royalty stacking theory (expression of collective hold-up) and qualitative testimony that an invention is valuable-without being anchored to a quantitative market valuation- are insufficiently reliable (CSIRO v Cisco, 2015).

An injunction request in itself is not as such a proof of hold-up. Then how do European and American judicial and antitrust authorities assess the situation of patent hold-up?

HOLD-UP UNDER THE SCRUTINY OF COURTS AND ANTITRUST AUTHORITIES IN EUROPE AND THE U.S.

Patent hold-up has been addressed under different grounds in Europe (1) and the U.S (2).

Hold-Up in Europe

Under European law, two different ways of examining patent hold-up have been considered: one more focused on competition law and one on contract and patent law.

The first one is linked to Article 102 TFEU and the abuse of a dominant position. The European Commission ("EC") has investigated Samsung's and Motorola's behavior against Apple on this ground[15]. Those investigations have led to the conclusion that SEP-holders hold a dominant position on the SEPs market, each SEP representing one separate market as such where the SEP-holder holds 100% of market shares. Therefore, a SEP-holder "seeking an enforcement of an injunction for SEPs may result in an abuse of dominance where the SEP-holder has given a FRAND licensing commitment and the implementer expressed willingness to enter into a FRAND license". Nonetheless, the EC did not explain how it comes to the conclusion that each SEP constitute a single market in itself. This conclusion could seem a bit surprising as, like a German Regional Court confirmed in March 2015, not all SEPs automatically confer a dominant position (LG Mannheim,*France Brevets v. HTC,* 2015).

The second one depends on patent and contract law. Injunction is a legally enforceable way recognized to any IPR-holder[16]. Notwithstanding the aforesaid, some implementers believe that a (F)RAND commitment implies that SEP-holders are precluded to be granted an injunction. In case of judicial litigation, some judges have determined whether the injunction request is a breach of the duty to negotiate in good faith, which is a contractual obligation. In the past and before the CJEU ruling in the case *Huawei v. ZTE*, Germany established that, under the *Orange-Book-Standard* case, any patent-holder, even a SEP-holder, was entitled to enforce its exclusive right against an infringer, unless the latter behaved as a true licensee and had made an unconditional (F)RAND offer that the SEP-holder could not reject without being anticompetitive (*Orange-Book-Standard,* 2009). Dutch judges came to a similar position, but they further examined the parties' behavior during the negotiations to assess whether the SEP-holder had breached its (F)RAND commitment. If the implementer had not sought any license from the SEP-holder, the latter was in principle entitled to seek an injunction (*Philips v. SK Kassetten,* 2010). A similar decision was issued in Italy (Genoa, *Koninklijke Philips Electronics N.V. v. Computer Support Italcard s.r.l.* 2004, May 7). In other countries, judges rejected the injunction request, stating that monetary compensation was sufficient to indemnify the SEP-holder (for example *Nokia Corp. v. IPCom GmbH & Co. KG*, [2012] EWHC 1446, *Vringo Infrastructure Inc. and ZTE Limited* [2015] EWHC 214; TGI Paris, *Telefonaktiebolaget LM Ericsson vs. TCT Mobile Europe SAS, TCT Mobile International Limited, TCT Mobile Limited,* 2013, November 29). Most of the European cases did not lead to the determination of a (F)RAND royalty rate mainly for two reasons. First, many cases were under summary proceedings, so the royalty rate would be determined at a later stage of the proceedings (e.g. in the case, *Vringo Infrastructure Inc. and ZTE Limited,* EWHC [2015] 214). Second, some of these cases (involving Apple, Samsung and Motorola/Google) were relinquished (Chowdhry 2014, Rosenblatt et al. 2014).

Nowadays in Europe there is a harmonized position, due to the recent ruling of the CJEU in the Huawei ruling (*Huawei v. ZTE Corp. and ZTE Deutschland GmbH*, 2015). Such ruling, which is binding for the EC as well as for national courts, has established certain steps the SEP-holder and the implementer need to follow in order to respectively obtain or avoid an injunction. These steps have been recently applied and some of the questions left open by the CJEU have been interpreted by national courts, mainly in Germany (cf. Section III.2 below).

Hold-Up Under the Scrutiny of Courts and Antitrust Authorities in the U.S.

In the U.S., the main focus has been put on the injunction request. Although the injunction is also an exclusive prerogative at the disposal of patent-holders[17], it does not mean that an injunction should be granted automatically. Since the *eBay v. MerckExchange* case, the award of a permanent injunction has to follow a four-factor test: irreparable injury for the SEP-holder (1); other remedies inadequate to compensate for this injury (2); remedy in equity warranted considering the balance of hardship between the plaintiff and defendant (3); and the public interest is not disserved (4) *(eBay v. MercExchange,* 2006). Judges usually conclude that granting an injunction against a "willing licensee" would prejudice the public interest and that through the (F)RAND commitment, SEP-holders implicitly acknowledge that monetary compensations are sufficient to indemnify them. The discussion then focuses on the basis to determine a (F)RAND rate and the basis of the royalty, with the risk that a too low royalty rate may enhance hold-out strategies (A). Another factor that may further foster hold-out is the unavailability of injunctions against a "unwilling licensee" (B).

Determining a (F)RAND Royalty and the Risk to Devaluate SEPs

Concerning the determination of a (F)RAND royalty rate, two views oppose each other. On the one hand, the SEP-holder, who has invested in standardization and supported sunk costs, needs both return on its investments("ROI") and revenues to keep investing in the process. On the other hand, the implementer, who mainly gains revenues from the use of the standard and hence, from the SEPs, tries to reduce as much as possible the rate he will have to pay[18].

Therefore, some implementers argue that the basis of the rate should be the smallest patent-practicing unit ("SSPPU") implementing the SEPs. This position, first advocated for non-SEPs, is mostly adhered to by U.S. judges (*LaserDynamics v. Quanta Computer,* 2012; *In re Innovatio IP Ventures LLC,* 2013)[19] to address the situation of juries which are "less equipped" and this may choose a too high percentage if using a higher base (the end device, cf. Sidak 2016). Indeed, the SSPPU Doctrine applies only for some jury trials and does not reflect the commercial practice: in telecommunications, most licenses are concluded on "end-device basis" (Stasik 2010, Teece & Sherry 2016)[20]. And courts never examine whether the standard, thus the SEPs it relies on, drives the demand for the product and confers it properties it would not have otherwise[21] (Teece & Sherry 2016). What if the device would not be able to perform its specific functions without the exploitation of the SEPs? Components are generally incorporated into a product which needs these components to function properly. In case of multicomponent products, how do you calculate on which SSPPU the fee relies (Teece & Sherry 2016, Petit 2016)? In addition to these issues, the SSPPU Doctrine overlooks the fact that the value of a patent depends on and differs following the product it is incorporated into (Teece & Sherry 2016, Petit 2016). It also does not consider the fact that the patented feature will increase the product's value in the eyes of the consumer (Petit 2016) due to properties conferred by the standard[22]. So the exclusion of the end product as a proper royalty basis is questionable.

Second, when we look at the determined royalties, courts cannot agree on a single rule to determine a (F)RAND royalty. How do judges fix a (F)RAND royalty? The application of the *Georgia Pacific* factors, used to award damages in non-SEPs infringement cases, is questioned, because some judges even contest their application to non-SEPs. Judge Robart has adapted those factors to SEPs in the case involving Microsoft and Motorola *(Microsoft Corp. vs. Motorola, Inc.* 2013) but judge Holderman has refused to apply those adapted *Georgia Pacific* factors in the *Re Innovatio* case (*In re Innovatio IP Ventures L.L.C.,* 2013).

Comparable licenses look like the most appropriate comparison mechanism *(Ericsson v. D-Link,* 2014). But they need to be qualified as "comparable", which means related to SEPs, not intended to settle a litigation and not portfolio-based[23]. These criteria make it difficult to consider other licenses. And there seems to be no reason for excluding portfolio licenses, including SEPs and non-SEPs, when apportionment is still possible.

Third, patent pools rates are at first sight an interesting tool to determine (F)RAND fees, when they concern the same standard. Patent pool licenses join together many patents belonging to different holders that are licensed collectively. However, the rate often follows a numeric proportional rule and does not differ according to the SEPs' value or their technological contribution to the standard (Farrar & Lerner 2006). Therefore, if a SEP-holder decides not to join a pool, he may estimate that the proposed rate for his SEPs does not constitute an adequate compensation[24].

Fourth, some courts in the U.S. recall that the value of SEPs should be the value parties would have agreed upon in hypothetical negotiations *ex ante* the incorporation of SEPs into the standard *(Microsoft Corp. v. Motorola Inc.,* 2013; *CSIRO v. Cisco,* 2014). Two issues arise from this " *ex ante* hypothetical negotiation" valuation framework. Before the patents are included in the standard, they are not SEPs. Hence, the patent-holder is free to fix the value he wishes, without being curtailed by any (F)RAND commitment. This impacts the royalty rate on two ways. First, the negotiated value may be higher than once the standard includes the patents declared as SEPs. Furthermore, the patent-holder has not curtailed his exclusive power yet. He consequently remains entitled to refuse to license his patents. The second problem is whether the standard would still be as valuable without the incorporation of the invention covered by the SEPs. If SEPs increase the value of the standard, why could not SEP-holders be fairly and adequately rewarded for them and for the value consumers grant to the standard developed by contributors?

Finally, a (F)RAND determination in court should also take into consideration the investments made by the SEP-holder and his need for a ROI to continue investing in standardization. However, these elements are never analyzed. The risk is that if (F)RAND royalty rates do not provide a sufficient ROI, SEP-holders may decide to stop participating in standardization and rather concentrate on a proprietary business model that may provide them with higher damages and royalty rates in case of infringement. As a result, consumers will face non-interoperable or lower quality devices at a higher price.

If SEP-holders are granted below (F)RAND royalty rates, they are not rewarded for their efforts in standardization. Furthermore, the risk of hold-out is fostered if implementers are aware they can avoid paying a (F)RAND rate through court determination and they have no risk to be prevented from selling infringing products.

The Unavailability of an Injunction and the Risk to Foster Hold-Out

Injunctions requests are examined under two antitrust grounds in the U.S. that are applied to examine the SEP-holders' conduct under the scrutiny of antitrust authorities. First, the Sherman Act and the attempt to monopolize; second, Section 5 of the FTC Act, which addresses antitrust acts.

Section 2 of the Sherman Act has been examined by judges but has never been really endorsed, as decisions mostly reject the injunctive relief on the grounds of *eBay v. MerckExchange* and the public interest factor.

Section 5 of the FTC Act has been scrutinized in FTC investigations. In two cases, the FTC alleged that a SEP-holder under a (F)RAND commitment infringes Section 5 of the FTC by seeking an injunction against a potential "willing licensee", without providing clear guidance as to determining whether a company falls under the "willing licensee" qualification (*In the Matter of Robert Bosch GmbH; In the Matter of Motorola Mobility and Google, Inc.).*

Accordingly, SEP-holders face the risk to engage into long unfruitful negotiations against implementers that will qualify themselves as "willing licensees". Litigating against so-called "willing licensees" in court, SEP-holders will probably not be able to enforce their SEPs through an injunction. Therefore, they may not have any effective enforcement measure against an implementer engaged in a "hold-out" strategy.

Another incentive to hold-out is potentially enhanced by the fact U.S. courts may bar a SEP-holder from enforcing valid injunctions granted abroad (*Microsoft v. Motorola,* 2013; Orzeck 2015). That faculty prevents SEP-holders from benefiting of an enforcement measure considered as valid and legitimate abroad. It potentially fosters a hold-out behavior in case of worldwide litigations because the implementer is aware that he may have the opportunity to have an U.S. court invalidating an injunctive relief legitimately awarded in other jurisdictions. Therefore, it further risks depriving SEP-holders from an effective enforcement measure.

Both in the U.S. and in Europe, antitrust authorities have thus ruled that demanding an injunction against a "willing licensee" constitutes an abusive conduct. But they have not analyzed the behavior of the so-called "willing licensee". As underlined by the current FTC Chairman,

The FTC and other competition agencies responded aggressively, clamping down on perceived efforts by owners of RAND-encumbered SEPs to hold-up standard implementers. They happened upon the rule that such patentees violate antitrust law if they try to enjoin a "willing licensee"—essentially a "no-injunction rule." While that approach has intuitive appeal, it violates core antitrust principles. In America, part of the problem lies in the FTC's reliance on section 5 of the FTC Act to capture conduct that goes beyond the reach of the Sherman Act. (Olhausen 2017).

FROM PATENT HOLD-UP TO PATENT HOLD-OUT?

Some implementers have based their claims against SEP-holders on the hold-up theory. Some SEP-holders retort that certain implementers are engaged in "hold-out" or "reverse patent hold-up" strategy. Analyzing the latter requires first to define hold-out and examine the implementer's conduct (1). The judgement of the CJEU in the case Huawei v. ZTE may also provide for a balance between both sides (2), while the IEEE's decision to amend its Bylaws could further foster hold-out strategies (3).

Patent Hold-Out and Implementers' Behavior

A licensee's willingness is a critical point to evaluate the conduct of the SEP-holder under competition or antitrust law. If a potential licensee strategically delays the negotiations and/or payment in order to avoid any payment or to pay a royalty rate below (F)RAND, it shows not to be a true "willing licensee". This behavior is the opposite of patent hold-up: it is called "patent hold-out" or "reverse hold-up". This, as explained below, can be strategically interesting for the licensee but highly detrimental to the SEP-holder.

First, licensees who do not contribute to the standard development do not bear the R&D costs and risks related thereto.

Second, delaying the conclusion of a license agreement, and further dragging it through litigations and other proceedings postpones the payment of the royalty rate and prejudice the SEP-holder, who will not be remunerated, neither for the investments, nor for the infringement for a long period (up to several years). It also confers a competitive advantage for the "unwilling" manufacturer (versus a competitor that has negotiated in good faith and pays a (F)RAND license), as saving R&D and license costs, the "unwilling" manufacturer can offer unlicensed products cheaper than manufacturers (Sir Jacob 2014). Furthermore, the SEP-holder in order to obtain a (F)RAND compensation will have to incur litigation costs, which for SMEs can lead to bankruptcy (in particular in the U.S. with exorbitant litigation costs).

Third, at worst, some implementers will negotiate a portfolio license but when facing litigation, will request to pay a (F)RAND royalty rate only for a few number of (potentially infringed) patents, i.e. those in suit (cf. Section II.2. above). Such behavior tolerated by some courts in the U.S. has however been considered as "unfair" in Europe (*Unwired Planet v. Huawei,* 2017; *Saint. Lawrence v. Vodafone,*2016)[25].

Finally, if the only risk the implementer faces is to end paying the same amount that it would have paid on the first place, he would be encouraged not to engage in good faith negotiations. As explained by judge Essex from the International Trade Commission ("ITC", administrative court in the U.S.):

but if the patent holder wins they get no more than a FRAND solution, that is, what they should have gotten under the agreement in the first place. There is no risk to the exploiter of the technology in not taking a license before they exhaust their litigation option if the only risk to them violating the agreement is to pay a FRAND based royalty or fee. This put the risk of loss entirely on the side of the patent holder, and encourages patent hold-out, which is as unsettling to a fair solution as any patent hold-up might be (In the Matter of Certain Electronic Devices, Including Wireless Communication Devices, Portable Music and Data Processing Devices, and Tablet Computer, 2014).

Just like hold-up, hold-out needs to be proven. Accordingly, demonstrating a "hold-out" requires examining the "willing licensee's" behavior.

If we examine the behavior of the so-called "willing licensee" in two famous cases in Europe and the U.S., we may come to different conclusions.

The first one is the *Motorola v. Apple* case. In this case, neither the American judges nor the EC have considered Apple's behavior during the negotiations, which have been dragging for three years without success. Not only Motorola's behavior but also Apple's conduct should have been analyzed. Apple, according to German courts, did not make any "Orange-Book -Standard" compliant license offer in Germany till its 6th proposal, despite complaining to the EC about Motorola's abusive behavior. The EC only stated what kind of conduct is not a sign of unwillingness[26]: it deemed the fact that the implementer is "willing to negotiate a FRAND license" and to have a FRAND royalty rate determined by a third party and to be bound with as sufficient (European Commission 2014). In the U.S., the Federal Circuit court dealing with the litigation between both companies applied the *eBay* four-factor test in such a way that it did not grant an injunction against Apple, which was furthermore qualified as a "willing licensee". Such decision, however, was not supported by one of the judges involved in this litigation (Chief-judge Rader dissenting-in-part 2014). Apple itself claimed in the Western District of Wisconsin it would not pay a royalty determined by the court exceeding $1 for any device and retained the right to appeal an award higher than this amount (*Apple v. Motorola*, 2012)[27]. This conduct seems to contradict the above statement from the same company to accept to have the terms of a (F)RAND license determined by a court and be bound with, and which was considered sufficient in Europe to qualify Apple as a "willing licensee". Therefore, relying on a general agreement of the implementer to be willing to enter a (F)RAND license, without assessing his conduct during negotiations, and without comparing his behavior in other related litigations does not seem advisable[28]. Otherwise an implementer may always argue that it is "willing to be licensed SEPs under (F)RAND terms and conditions" (Sir Jacob 2014) but then delay the negotiations in order to avoid or delay payments, eventually paying a royalty below (F)RAND.

The second case involves Apple and Samsung. Apple raised a (F)RAND defence against Samsung, after six months of unsuccessful negotiations. Apple claimed Samsung's requested rates and injunctions were abusive and fostered hold-up. Nevertheless, the ITC granted Samsung an injunction against Apple in 2013 and the Memorandum of Opinion pointed out that the Administrative Law Judge disagreed with Apple's interpretation of (F)RAND (mainly waiving the right to an injunction) and stressed the potential abusive conduct such an interpretation could lead to:

The ALJ also expressed concern that if Apple's position were correct, an infringer of standard-essential patent could make an end-run around a section 337 investigation by merely claiming that the patent owner's offer was not FRAND (In the Matter of Certain Electronic Devices, Including Wireless Communication Devices, Portable Music and Data Processing Services, and Tablet Computers, 2013).

This injunction was nonetheless vetoed by the White House in a letter underlining risks of patent hold-up through an injunction, and the public interest factor (Executive Office of the President, The United States Trade Representative 2013).

Interestingly, in South Korea, dealing with the same parties and same behaviors, the court in Seoul concluded that Apple had not met the requirements to demonstrate it was a "willing licensee" because it had not requested to negotiate a license although it was aware of the existence of the SEPs *(Samsung Electronics Co., Ltd. v. Apple Korea Ltd.*, 2012).

Next to the Korean's findings on (un)willingness from a potential licensee, patent hold-out has been examined and confirmed in other courts outside Europe and the U.S. First in India, in cases opposing Ericsson to local manufacturers (*Telefonaktiebolaget LM Ericsson v. Intex Technologies Limited,* 2015, and *Telefonaktiebolaget LM Ericsson v. M/S Best IT World Private Limited (iBall),* 2015). As the cases evidence it, negotiations lasted between four (for iBall) and six (for Intex) years without any success. Manufacturers tried, according to the court of New Delhi, to delay the negotiations. They filed a complaint to the Indian Competition Authority ("CCI") while negotiations were still ongoing and did not notify Ericsson about the filing of those complaints. They informed the Swedish company about the fact they also challenged the validity of the patents only once Ericsson started assessing its legal rights (*Telefonaktiebolaget LM Ericsson v. Intex Technologies Limited,* 2015; *Telefonaktiebolaget LM Ericsson v. M/S Best IT World Private Limited,* 2015).

The High Court of New Delhi found in the iBall case that the defendant had not taken any step or shown any interest for the purpose of the execution of the (F)RAND Agreement as *on the one hand is alleging that it is not infringing the patents in suit and on the other hand it has filed a complaint before the CCI wherein certain admission of the rights of the plaintiff had been made* (*Telefonaktiebolaget LM Ericsson v. M/S Best IT World Private Limited,* 2015).

The High Court of New Delhi ruled in both cases that the defendants were behaving abusively, trying to delay the negotiations to avoid paying a (F) RAND royalty rate. The court decided thus that the grant of an injunction was justified because in the absence of an injunction, Ericsson would suffer irreparable loss and injury, as the Defendants would continue selling infringing devices without paying a royalty.

A similar situation confronted Ericsson to TCL/TCT in Brazil. Parties had reached a 2G license in 2007. A disagreement arose around the interpretation of TCL's duties related to this license, especially the royalty rate (cf. for example TGI Paris, *Telefonaktiebolaget LM Ericsson vs. TCT Mobile Europe SAS, TCT Mobile International Limited, TCT Mobile Limited,* 2013, November 29). Additionally, parties tried to negotiate the renewal of the (F)RAND license. Negotiations failed and Ericsson initiated proceedings against TCL. The interpretation of the 2G license was dealt by arbitration. The first instance court in Paris rejected in 2013 Ericsson's request to be granted an injunction to prevent this injunction *from skewing the current negotiations between the parties to enable a party, if the need arises, to obtain an unjustified advantage within the framework of the contractual negotiations of a license related to essential patents* (*Telefonaktiebolaget LM Ericsson vs. TCT Mobile Europe SAS, TCT Mobile International Limited, TCT Mobile Limited,* 2013).

Meanwhile TCL filed an antitrust complaint to the Brazilian Antitrust Authority, arguing that Ericsson's injunction request was an anticompetitive conduct used to force TCL entering into unfair terms and conditions.

The Brazilian Antitrust Authority decided in June 2015 that Ericsson's conduct was not anticompetitive, being given the years of unfruitful negotiations (eight and a half years, cf. Charles 2016) and the large number of other license agreements signed by Ericsson. It further points out that antitrust law is not concerned with excessive royalties (contrary to the Chinese Antitrust Authority who limited the fees Qualcomm is entitled to require from Chinese companies on its SEPs, cf. King 2015), but only with the prevention of monopolization. It confirmed that seeking an injunction against an unwilling infringer committing hold-out is reasonable. Furthermore, it stressed that, as Ericsson was not active on TCL's market, it had no interest in not signing any license with TCL. On the contrary, it was in its interest to have such a license agreement (Pereira 2015). But in July 2015, after accepting that a court determines a worldwide portfolio license, TCL obtained from a Californian District Court an anti-suit injunction order prohibiting Ericsson from pursuing any foreign litigation against TCL (Orzeck 2015).

These cases demonstrate that implementers' behavior may be abusive and should be examined to reach a balanced opinion. The CJEU has arrived to a similar conclusion in its ruling *Huawei v. ZTE.*

Balancing Both Sides: The CJEU's Approach

In 2014, the Landgericht (District Court) of Düsseldorf asked a preliminary ruling to the CJEU. The court was indeed confronted with two different positions: on the one hand, the *Orange-Book-Standard* case enabling SEP-holders to enforce their exclusive right against an infringer, unless the defendant makes an unconditional offer that the plaintiff cannot refuse without breaching competition law, and on the other the EC's investigation against Samsung and Motorola, where the EC decided that each SEP confers its holder a dominant position and that it may be abusive in requesting an injunction on SEPs under specific circumstances, i.e. when there is a (F)RAND commitment and a so-called "willing licensee".

The court of Düsseldorf required, among other things, the CJEU to clarify which position should the court follow

Is this sufficient for a successful FRAND defence if the defendant is willing to negotiate a license or does he have to make a binding offer to the SEP-holder on terms the SEP-holder cannot refuse?

Prior to the ruling, the Advocate General Wathelet indicated that none of the positions appeared balanced:

(…) placing reliance only on the alleged infringer's mere "willingness to negotiate" would result in pricing which falls below the true economic value of the SEP. Conversely, placing reliance on the case-law established by the Bundesgerichtshof in Orange-Book-Standard would create the opposite problem, in that the licensed royalties imposed would be very high (though not so high as to constitute a refusal to conclude a contract, in breach of Article 102 TFEU) (Wathelet 2014).

Such conclusion seems to be supported by the CJEU which concluded that the (F)RAND framework applies to both sides, SEPs licensor and licensee. The CJEU explains that there is no abuse *per se* when requiring an injunction, not even for SEPs (Petit 2015), but due to the nature of (F)RAND, a sort of safe harbor was needed for negotiations between a SEP-holder and an implementer, with duties on both parties.

First, before bringing an infringement action, the SEP-holder must inform the implementer it is infringing SEPs and specify how. Second, if the implementer expresses his willingness to conclude a (F)RAND license, the SEP-holder must make a proposal detailing the amount of the royalty and the way it is calculated. If the implementer considers this proposal as not (F)RAND, he has to make a counter-proposal responding promptly and diligently. In case the SEP-holder complies with its duties, and the implementer does not, then the SEP-holder is entitled to seek an injunction without it being at first sight considered abusive.

Although the CJEU provides general guidelines, some questions are left open to be interpreted considering the circumstances of each individual case. For instance, in November 2015, the judge of the Court of Mannheim considered that three months to express willingness to negotiate a (F)RAND license was not diligent (LG Mannheim, *Saint Lawrence v. Deutsche Telekom*, 2015, November 27). Later, in March 2016, the District Court of Düsseldorf underlined that a licensing proposal on one single SEP for one country does not comply with "recognized commercial practices" based on worldwide portfolio licenses, and would lead the SEP-holder to expensive litigations on a country-per-country and patent-per-patent basis if accepted (LG Düsseldorf, *Saint Lawrence v. Vodafone,* 2016, March 31).

Similarly, in the UK, in April 2017, judge Birss considered in April 2017 that (F)RAND licenses are worldwide licenses, and not national licenses (*Unwired Planet v. Huawei,* EWHC [2017] 711). This consideration did not prevent him however from determining distinct rates for China than for Europe and the U.S. Further, in the Netherlands, the Court of The Hague stated in a case that the implementer was not a "willing licensee" because it stated to the SEP-holder that it had to be sued if the SEP-holder wanted to collect more than thousand dollars in royalties. To reject the implementer's claim that the SEP-holder's offer was not (F)RAND, the court of The Hague analyzed comparable licenses concluded by the SEP-holder and comparable rates proposed by a patent pool for similar technologies (Rechtbank's

Gravenhage, *Archos v. Philips,* 2017, February 10). The District Court of Düsseldorf, in *Saint Lawrence v. Vodafone*, considers patent pools however of "limited" significance since *in licensing over a large pool with the property rights of several companies, lower royalties per patent will be paid as a rule than in licensing of the portfolio of only one company, and with a lower total number of property rights* (*St Lawrence v. Vodafone*, 2016).

Not only the CJEU, but also the EC with its Digital Single Market Strategy seems to understand the need for a balanced system between implementers and SEP-holders preventing abuses from implementers and not only SEP-holders:

At the same time, patent holders face unwilling licensees and significant challenges to value the use of their technology as well as their contribution to the elaboration of standards.

(…)

On the other hand, uncertainty in the enforcement framework also undermines the capacity of SEP holders to exert their IP rights due to tactics by the potential infringer of the patent (EC 2017).

While in Europe the CJEU has tried to provide for a balanced solution to limit both hold-up and hold-out, and the EC wants to tackle abuses from both sides, a U.S. SDO, the IEEE, has adopted new Bylaws that could seem unbalanced.

Unbalanced Approach: The new IEEE's Bylaws

WTO's criteria of transparency, openness, impartiality and consensus, effectiveness and relevance, and coherence are followed by SDOs in technical discussions as legal discussions. The compliance with those principles helps ensuring a fair, transparent and balanced process. Balance is important in the legal discussions because there are more companies implementing standards than contributing to their development. Implementers' interest is rather paying a fee below (F)RAND, or nothing at all. If WTO's criteria are not followed in legal discussions, it may lead to an unbalanced situation deserving the interests of one side, and potentially fostering abuses from this side, as it appears to be the case with the new IEEE's Bylaws.

Before the balanced approach introduced by the CJEU as interpreted by national courts, in December 2014, the Board of the IEEE decided to specify the content of the "RAND commitment" in its Bylaws. Besides recommending contributors to determine their SEPs royalty rate on the SSPPU, the new IEEE Bylaws prevents contributors to seek an injunction on their SEPs.

Although IEEE argues that the changes were proposed by "independent experts", these changes are almost identical to those proposals made by some companies (e.g. Apple, Microsoft, Cisco, Intel)[29]. The initial justification for reviewing the policy was some statements made by the U.S. Department of Justice ("DOJ") that there is a risk of patent hold-up, and this, despite no evidence of a systemic patent hold-up. The IEEE and the DOJ claimed that these amendments are crucial to foster innovation and an effective development of standardization (DOJ 2015), even though the DOJ underlines that

If a standard-setting process is biased in favor of one sets of interests, there is a danger of anticompetitive effects and antitrust liability (DOJ 2015).

Notwithstanding these considerations, some main IEEE's contributors have declared they will not license their SEPs under the new IEEE's Bylaws (Decker & King 2015, Lloyd 2015). Those amendments, that seem not to be based on the consensus principle, introduce uncertainty and complexity at three levels. First, for the same standard, SEPs may be licensed under two different IPR Policies (previous Bylaws or new ones), depending on the contributors, or not licensed at all. Second, the same problem may arise for the next generation of standards implementing SEPs of former standards, and licensed under the previous Bylaws, while SEPs only implemented by the new standard will be licensed under the new policy (or the previous, or not licensed) (Katznelson 2015). Third, these new Bylaws may seem incompatible with the ETSI's IPR Policy, which decided in 1994 not to introduce a prohibition of injunction on SEPs (Iversen 1999).

Those changes appear thus contested and challenged, and adopted following a process that may not fit at least the consensus principle of the WTO. Will thet really benefit the IEEE or do they risk lowering and weakening the position and role of this SDO in a context where standardization becomes increasingly important[30]?

Since those changes, it has been reported that submission of Letters of Assurance stating that a member is willing to license its SEPs under RAND terms and conditions has declined (Katznelson 2016).

SOLUTIONS AND RECOMMENDATIONS

Regulators and SDOs trying to amend existing regulations and/or policies or enact new ones to counter hold-up should first assess whether hold-up is concretely affecting the market (which does not seem demonstrated by evidence). The risk of hold-out should also be considered. Antitrust authorities, legislators, judges and SDOs should carefully balance all interests involved when dealing with injunctions on SEPs and hold-up allegations.

FUTURE RESEARCH DIRECTIONS

Hold-up and hold-out claims may keep raising in the near future. Some authorities have already taken measures against hold-up, like the IEEE, with its amended Bylaws. Additionally, the European Commission is working on a communication about SEP licensing (EC 2017). The long term effect of such legal texts, on litigations and licensing, could be later assessed, to analyze whether they have addressed existing issues, created new problematics or strengthened an existing problem.

CONCLUSION

SEPs play an important role in standardization. On the one hand, they represent technical contributions to the standard. On the other hand, they are a tool for their holders to get rewarded for these contributions. The (F)RAND commitment, set up to mitigate abusive conducts as patent hold-up and to provide for a fair, reasonable and adequate compensation to contributors, has been highly successful: standards as the GSM, UMTS, and Wi-Fi have achieved a high and fast penetration rate. However, litigations targeting to challenge the (F)RAND commitment put the equilibrium between contributors and implementers at risk:

33% of litigations between 2000 and 2013 in the U.S. concerned SEPs (Gupta & Snyder 2014). On the other hand, figures need to be put in context. Most of the cases parties sign (F)RAND agreements without being involved in litigations. Looking at the large number of new entrants with new business models, some of them not interested in a long and sustainable standardization system, the disputes seem to be of commercial nature and probably not proportionally higher than when the market only included a few actors. High performance is increasing while prices are decreasing enormously and new companies have no problems entering the cellphone market. Evidence suggests therefore that there is no patent hold-up. It is however not so sure that we can reach the same conclusion with regards to patent hold-out. In fact, the current "no-injunction rule" for SEPs based on a perceived hold-up problem, as recognized by the Acting Chairman of the FTC, Maureen Ohlhausen, "violates core antitrust principles" (Ohlausen 2017) since injunction is only justified when the conduct harms competition. Not surprisingly, despite lack of evidence, firms supporting that there is a risk of hold-up are very often successful companies, extracting high revenues from the use of the standard[31]. The changes inserted in the IEEE's Bylaws seem to be in the interest of (mainly American) companies, which have advocated such positions. These changes have been adopted under the breach of the basic principles of standardization, i.e. openness, transparency, impartiality and non-discrimination, favoring one business model (purely implementers) over the others. For these reasons, as well as for their incompatibility with the CJEU's ruling in the case *Huawei v. ZTE*, such one-sided approach may hinder rather than foster innovation. While the CJEU's ruling provides for safeguards against abusive behaviors from both sides, hold-up and hold-up, these revisions of the IEEE's Bylaws and U.S. court decisions question the fairness of the framework and the possibility to get remedies against "unwilling licensees" engaging in hold-out tactics.

Patent hold-out tactics show two tendencies. On the one hand, a fierce competition between SEP-holders and newcomers through commercial litigations, where it seems that newcomers try to redefine the rules in their advantages, and a shift of bargaining power from contributors to implementers (Froeb & Shor 2015). On the other hand, the threat of a (F)RAND defense used to strategically delay the conclusion of a license and the payment of (F)RAND fees may have a more detrimental effect than the threat of an injunction (which is almost never granted). In a context where new technical challenges (EC 2016) depend on standardization, a balance between contributors and implementers is needed. Automatic injunctions can lead to implementers paying more than what they would have agreed on a good faith negotiation and may make the standard more expensive. Depriving SEP-holders from incentives to innovate and to legally enforce their rights could prejudice innovation, competition and thus reducing or eliminating standardization, favoring closed proprietary systems. The framework established by the CJEU in *Huawei v. ZTE* seems to be sufficient to guarantee such balance. Further intervention, even through soft law, risks to break such a balance.

REFERENCES

ABIresearch. (2013). *Standard leadership within the 3GPP*. Retrieved July 29, 2017, from https://www.abiresearch.com/market-research/product/1016671-standards-leadership-within-the-3gpp/

Apple Inc. and NeXT Software Inc. v. Motorola Inc. and Motorola Mobility, Inc., No. 11-cv-08540 (N.D. Ill. 2012) (2012).

Apple Inc. v. Motorola Inc., No. 11-cv-0017 (W.D.Wisc. 2012) (2012).

Bundesgerichtshof. (2009, May 6). *Orange-Book-Standard.* KZR 39/06.

BylawsIEEE (2015).

Charles, I. (2016). *Ericsson's offer of SEP terms to TCL held to adhere to FRAND standards by US courts.* Retrieved July 29, 2017, from https://spicyip.com/2016/08/ericssons-offer-of-sep-terms-to-tcl-held-to-adhere-to-frand-standards-by-u-s-court.html

Chowdry, A. (2014). Apple And Samsung Drop Patent Disputes Against Each Other Outside Of the U.S. *Forbes.* Retrieved July 29, 2017, from https://www.forbes.com/sites/amitchowdhry/2014/08/06/apple-and-samsung-drop-patent-disputes-against-each-other-outside-of-the-u-s/#6dd91db5418c

TBT Committee. (2000). *Decision on Principles for the Development of International Standards, Guide and Recommendations with Relation to Article 2, 5 and Annex 3 of the TBT Agreement.* Author.

Court of Justice of the European Union (2015, July 16). *Huawei v. ZTE Corp. and ZTE Deutschland GmbH.* C-170/13.

CSIRO v. Cisco (Fed. Circ. 2015) (2015).

Decker, S., & King, I. (2015). Qualcomm Says It Won't Follow New Wi-Fi Rules on Patents. *Bloomberg.* Retrieved July 29, 2017, from https://www.bloomberg.com/news/articles/2015-02-11/qualcomm-says-new-wi-fi-standard-rules-unfair-may-not-take-part

Directive 2004/48 of the European Parliament and the Council of 29 April 2004 on the Enforcement of Intellectual Property Rights.

eBay v. MerckExchange, L.L.C., 547 U.S.388 (2006).

Einhorn, B. (2015). The Cheap Phones Quiety Winning the U.S. *Bloomberg.* Retrieved July 29, 2017, from https://www.bloomberg.com/news/articles/2015-09-03/zte-s-cheap-phones-quietly-winning-the-u-s-

Elmer-DeWitt, P. (2016). How Apple Sucks the Profit Out of Mobile Phones. *Fortune.* Retrieved July 29, 2017, from http://fortune.com/2016/02/14/apple-mobile-profit-2015/

Epstein, R. A., Kieff, F. S., & Spulber, D. F. (2012). The FTC, IP, and SSOs: Hold-Up Replacing Private Coordination. *Journal of Competition Law & Economics, 8*(1), 1–46. doi:10.1093/joclec/nhs002

Ericsson v. D-Link, 773 F.3d (Fed. Circ. 2014) (2014).

European Commission. (2014a, April 29). *Antitrust Procedure, Case AT-39939-Samsung-Enforcement of UMTS Standard Essential Patents.* Author.

European Commission. (2014b, April 29). *Antitrust Procedure, Case AT-39985-Motorola-Enforcement of GPRS Standard Essential Patents.* Author.

European Commission. (2016). *Communication from the Commission to the European Parliament, the Council, the European Economic and Social Committee and the Committee of the Regions/ ICT Standardisation Priorities for the Digital Single Market.* COM (2016) 176 final.

European Commission. (2017). *Roadmap for a Communication for Standard Essential Patents for a European Digitalized Economy.* Author.

Farrell, J., Hayes, J., Shapiro, C., & Sullivan, Th. (2007). Standard Setting, Patents, and Hold-Up. *Antitrust Law Journal, 74*(3), 603–670.

Fraunhofer. (2011). *Study on the Interplay between Patents and Standards.* Retrieved July 29, 2017 from http://www.iplytics.com/download/docs/studies/ipr_study_final_report_en.pdf

Froeb, L., & Shor, M. (2015). Innovators, Implementers, and Two-Sided Hold-Up. *The Antitrust Source, 205*, 1-10.

Galetovic, A., Harber, S., & Levine, R. (2015). An Empirical Examination of Patent Holdup. *Journal of Competition Law & Economics, 11*(3), 549–578. doi:10.1093/joclec/nhv024

Geradin, D., Layne-Farrar, A., & Padilla, A. J. (2007). *Royalty Stacking in High Tech Industries: Separating Myth From Reality*. CEMFI Working Paper No.071. Retried July 29, 2017, from ftp://ftp.cemfi.es/wp/07/0701.pdf

Ginsburg, D.H., Wrong-Ervin, K., & Wright, J.D. (2015). The Troubling Use of Antitrust to Regulate FRAND Licensing. *CPI Antitrust Chronicle, 2015*(1), 2-8.

Gupta, K., & Snyder, M. (2014). *Smart Phone Litigation and Standard Essential Patents*. Hoover IP[2] Working Paper Series No.14006.

High Court of Delhi at New Delhi. (2015, September 2). *Telefonaktiebolaget LM Ericsson v. M/S Best IT World Private Limited (iBall).*

High Court of Delhi at New Delhi. (2015, March 31). *Telefonaktiebolaget LM Ericsson v. Intex Technologies Limited.*

High Court of England and Wales. (2012, May 10). *Nokia Corp. v. IPCom GmbH & Co. KG*, [2012] EWHC 1446 (Pat.).

High Court of England and Wales. (2015, January 30). *Vringo Infrastructure Inc. and ZTE Limited.* EWCH 214 (Pat).

High Court of England and Wales (2017, April 5). *Unwired Planet International Ltd. v. Huawei Technologies Co. Ltd, Huawei Technologies (UK) Co. Ltd v. Unwired Planet LLC.* EWCH 711 (Pat).

In re Innovatio IP Ventures L.L.C, No. 2303, 2013 WL 5593609 (N.D. Ill. 2013) (2013).

Intellectual Property Rights PolicyETSI (2017).

International Trade Commission. (2013c, June 5). *In the Matter of Certain Electronic Devices, Including Wireless Communication Devices, Portable Music and Data Processing Devices, and Tablet Computer.* Inv. 337-TA-794.

International Trade Commission. (2014, June 13). *In the Matter of Certain Wireless Devices With 3G and/or 4G Capabilities, and Components Thereof.* Inv. 337-TA-868, Initial Determination.

International Trade Commission. (2015, April 27). *In the Matter of Certain 3G Mobile Handsets and Components Thereof.* Inv. 337-TA-613.

Iversen, E. (1999). *Standardization and Intellectual Property Rights: ETSI's controversial search for new IPR-procedure*. Paper presented at the IEEE Conference on Standardization and Innovation, Aachen, Germany.

Jacob, R. (2014). *FRAND: A Legal Analysis*. Retrieved July 29, 2017, from https://ec.europa.eu/jrc/sites/jrcsh/files/jrc_is_20141027_-_patentsws-04.frand_.robinjacob.pdf

Katznelson, R. (2015). *Perilous Deviations from FRAND-Harmony-Operational Pitfalls of the 2015 IEEE Patent Policy*. Paper presented at the 9th International Conference on Standardization and Innovation in Information Technology, Sunnyvale, CA. doi:10.1109/SIIT.2015.7535599

Katznelson, R. (2016). *Decline in non-duplicative Letters of Assurance (LOA) from Product/System Companies for IEEE standards*. Retrieved July 29, 2017, from http://www.iam-media.com/files/LOAs-March-2016_stamped.pdf

King, I. (2016). Qualcomm Fined $975 Million in China, Sets Licensing Rates. *Bloomberg*. Retrieved July 29, 2017, from https://www.theglobeandmail.com/report-on-business/international-business/qualcomm-fined-975-million-in-china/article22883526/

Landgericht Düsseldorf. (2015, March 26). *France Brevet vs.* HTC, 4b O 140/13.

Landgericht Düsseldorf. (2016, March 31). *Saint Lawrence v.* Vodafone, 4a O 73/14.

Landgericht Mannheim. (2015, November 27). *Saint Lawrence v. Deutsche Telekom.* 2 O 106/14.

LaserDynamics, Inc., v. Quanta Computer, Inc., 694 F.3d 51, 67 (Fed. Circ. 2012) (2012).

Layne-Farrar, A. (2016, February). Why Patent Holdout is Not Just a Fancy Name for Plain Old Patent Infringement. *CPI,* 1-4.

Lemley, M., & Shapiro, C. (2007). Patent Holdup and Royalty Stacking. *Texas Law Review*, *85*, 1991–2049.

Lerner, J., & Layne-Farrar, A. (2011). To Join or Not to Join: Examining Patent Pool Participation and Rent Sharing Rules. *International Journal of Industrial Organization*, *29*(2), 294–303. doi:10.1016/j.ijindorg.2010.08.006

Lloyd, R. (2015). Ericsson and Nokia the latest to confirm that they will not license under the new IEEE patent policy. *IAM*. Retrieved July 29, 2017, from http://www.iam-media.com/blog/detail.aspx?g=d07d0bde-ebd6-495a-aa72-4eecb9dac67d

Lopez Research LLC. (2015). *An Introduction to the Internet of Things (IoT). Part 1 of the IoT Series.* Retrieved July 29, 2017, from http://www.cisco.com/c/dam/en_us/solutions/trends/iot/introduction_to_IoT_november.pdf

Mallison, K. (2013). *Absurd (F)RAND licensing-rate determinations for SEPs*. Retrieved July 29, 2017, from http://www.ip.finance/2013/11/absurd-frand-licensing-rate.html

Mallison, K. (2014). *Stacking the Deck in Analysis of Smartphone Patent Licensing Costs*. Retrieved July 29, 2017, from http://www.wiseharbor.com/pdfs/Mallinson%20on%20Intel's%20Smartphone%20Royalty%20Stack%2019Sept2014.pdf

Mallison, K. (2015). *Cumulative mobile-SEP royalty payments no more than around 5% of mobile handset revenues*. Retrieved July 29, 2017, from http://www.wiseharbor.com/pdfs/Mallinson%20on%20cumulative%20mobile%20SEP%20royalties%20for%20IP%20Finance%202015Aug19.pdf

Mallison, K. (2016). *"Patent holdup" allegations encourage SEP free-riders*. Retrieved July 29, 2017, from http://www.wiseharbor.com/pdfs/Mallinson%20on%20Holdup%20and%20Holdout%20for%20IP%20Finance%2016%20Aug%202016.pdf

Microsoft, Corp. v. Motorola, Inc., No. 10-1823 (W.D. Wash. 2013) (2013)

Microsoft, Corp. v. Motorola, Inc., No. 14-35393 (9th Circ., 2015) (2015).

Ohlhausen, M. K. (2015). *Antitrust Oversight of Standard-Essential Patents: The Role of Injunctions*. Paper presented at the Antitrust Forum, Beijing, China.

Ohlhausen, M. K. (2017). *The Elusive Role of Competition in the Standard-Setting Antitrust Debate*. Retrieved July 29, 2017, from https://www.ftc.gov/public-statements/2017/07/elusive-role-competition-standard-setting-antitrust-debate

Orzeck, K. (2015). TCL Wins Anti-Suit Injunction in Ericsson Patent Row. *Law360*. Retrieved July 29, 2017, from https://www.law360.com/articles/678680/tcl-wins-anti-suit-injunction-in-ericsson-patent-row

Pereira, A. A., Jr. (2015, Fall). TCL v. Ericsson: Brazil's First Standard Essential Patent Antitrust Investigation. *Newsletter of the Antitrust Section's Intellectual Property Committee,* 18-27.

Petit, N. (2016). *The Smallest Salable Patent Practicing Unit ("SSPPU") Experiment, General Purpose, Technologies and the Coase Theorem*. Retrieved from https://papers.ssrn.com/sol3/papers.cfm?abstract_id=2734245

Petit, N. (2015). Huawei v. ZTE: Judicial Conservatism at the Patent-Antitrust Section. *CPI Antitrust Chronicle, 12*(2).

Rader, R.R. (2014). Chief Judge, dissenting-in-part in U.S. Court of Appeals for the Federal Circuit, *Apple, Inc. and NeXT Software, Inc., v. Motorola Inc. and Motorola Mobility, Inc.*

Rechtbank's Gravenhage. (2010, March 17). *Philips v. SK Kassetten*, HA ZA 08-2522 en HA-ZA 08-2524.

Rechtbank's Gravenhage. (2017, February 10). *Archos v. Koninklijke Philips N.V.*, ECLI:NL:RBDHA:2017:1025.

Regulation (EU) No 1025/2012 of the European Parliament and of the Council of 25 October 2012 on European standardisation, amending Council Directives 89/686/EEC and 93/15/EEC and Directives 94/9/EC, 94/25/EC, 95/16/EC, 97/23/EC, 98/34/EC, 2004/22/EC, 2007/23/EC, 2009/23/EC and 2009/105/EC of the European Parliament and of the Council and repealing Council Decision 87/95/EEC and Decision No 1673/2006/EC of the European Parliament and of the Council.

Rosenblatt, J., Womack, B., & Satariano, A. (2014). *Apple and Google declare cease-fire in legal war*. Retrieved July 29, 2017, from http://www.canada.com/technology/apple+google+declare+cease+fire+legal/9848740/story.html

Shapiro, C. (2001). Setting Compatibility Standards: Cooperation or Collusion? In R. Dreyfuss, D. L. Zimmerman, & H. First (Eds.), *Expanding the Boundaries of Intellectual Property Innovation Policy for the Knowledge Society* (pp. 81–101). Oxford.

Shapiro, C. (2010). Injunctions, Hold-Up, and Patent Royalties. *American Law and Economics Review, 12*(2), 280–317. doi:10.1093/aler/ahq014

Sidak, G. (2015). The Meaning of FRAND, Part II: Injunctions. *Journal of Competition Law & Economics, 11*(1), 1–69.

Stasik, E. (2010, September). Royalty Rates and Licensing Strategies For Essential Patents On LTE (4G) Telecommunication Standards. *les Nouvelles.*

Tapia, C. (2016, January). Securing a competitive future in Europe. *The Patent Lawyer*, 31-35.

Teece, D., & Sherry, E. (2016). *On the "Smallest Salable Patent Practicing Unit": An Economic and Public Policy Analysis*. Tusher Center for the Management of Intellectual Capital Working Paper Series No.11.

The Boston Consulting Group. (2015). *The Mobile Revolution. How Mobile Technologies Drive a Trillion-Dollar Impact.* Retrieved July 29, 2017, from https://www.bcgperspectives.com/content/articles/telecommunications_technology_business_transformation_mobile_revolution/

Treaty on the Functioning of the European Union, Article 102 (2008).

Tribunal de Grande Instance de Paris. (2013, November 29). TGI Paris, *Telefonaktiebolaget LM Ericsson vs. TCT Mobile Europe SAS, TCT Mobile International Limited, TCT Mobile Limited,* 2013, November 29, 12/14922.

Tribunale di Genoa. (2004, May 7). *Koninklijke Philips Electronics N. V. v. Computer Support Italcard s.r.l.*

Tribunale di Milano. (2012, January 5). Sezione Specializzata Proprieta' Industriale ed Intelletuale, *Samsung Electronics Co., Ltd. e Samsung Electronics Italia c. Apple Inc., Apple Italia S.R.L., Apple Retail Italia S.R.L., Apple Sales International,* R.G. 59734/2011.

U.S. Department of Justice. (2015). *Response to the Institute of Electrical and Electronics Engineers, Incorporated.* Retrieved July 29, 2017, from https://www.justice.gov/atr/response-institute-electrical-and-electronics-engineers-incorporated

U.S. Federal Trade Commission. (2013a, April 23). *In the Matter of Robert Bosch GmbH.* FTC File No. 121-0081.

U.S. Federal Trade Commission. (2013b, July 24). *In the Matter of Motorola Mobility and Google Inc.* FTC File No. 121-0120.

Wright, J. (2013). *SSOs, FRAND, and Antitrust: Lessons from the Economics of Incomplete Contracts.* Paper presented at the Inaugural Academic Conference of the Center for the Protection of Intellectual Property on The Commercial Function of Patents in Today's Innovation Economy, Arlington, VA.

KEY TERMS AND DEFINITIONS

FRAND: Commitment made by a SEP-holder to the relevant SDO to license his SEPs under fair, reasonable, and non-discriminatory terms and conditions.

Injunction: Enforcement measure available to intellectual property rights (IPR) holders against an infringer. If granted by a court, an injunction enables an IPR-holder to stop or prevent an infringer from selling and/or manufacturing infringing products.

Patent Hold-Out: Behaviour of a potential licensee who, instead of engaging into good faith negotiations about a FRAND license, tries to delays and postpone such negotiations and/or proposes below FRAND royalty rates.

Patent Hold-Up: Behaviour of a patent-holder who successfully imposes on a license disadvantageous licensing terms and conditions, potentially under the threat of an injunction.

SDOs: Standard-development organisations. A SDO is an organisation where official standards are developed and adopted by members.

SEPs: Standard-essential patents, which means that no technical alternative to those patents. Therefore, any company selling standard-compliant products without a license will necessarily infringe the relevant SEPs.

Standard: A set of technical specifications enabling a product implementing them to perform specific functions. Official standards in telecommunications adopted by SDOs ensure interoperability, even among competing devices.

ENDNOTES

1. At least for *de iure* standards, developed or recognized by official SDOs (while *de facto* standards are developed by one or more companies and gain market acceptance).
2. Cf. Article 28§1 TRIPS:
   A patent shall confer on its owner the following exclusive rights:
   (a) Where the subject matter of a patent is a product, to prevent third parties not having the owner's consent from the acts of: making, using, offering for sale, selling, or importing for these purposes this product.
   (b) Where the subject matter of a patent is a process, to prevent third parties not having the owner's consent from the act of using the process, and from the acts of: using, offering for sale, selling, or importing for these purposes at least the product obtained directly by that process.
3. Cf. Article 6.2 of the IEEE Bylaws that define SEPs as
   any Patent Claim the practice of which was necessary to implement either a mandatory or optional portion of a normative clause of the IEEE Standard when, at the time of the IEEE Standard's approval, there was no commercially and technically feasible non-infringing alternative implementation method for such a mandatory or original portion of the normative clause.
   And Article 15.6 of ETSI IPR Policy stating that
   it is not possible on technical (but not commercial) grounds, taking into account normal technical practice and the state of the art generally available at the time of standardization, to make, sell, lease, otherwise dispose of, repair, use or operate EQUIPMENT or METHODS which comply with a STANDARD without infringing that IPR.

4 Royalty stacking refers to situations in which a single product is potentially infringing many patents, and thus may bear multiple royalty burdens. The term "royalty stacking" reflects the fact that, from the perspective of the firm making the product in question, all of the different claims for royalties must be added or "stacked" together to determine the total royalty burden borne by the product if the firm is to sell that product free of patent litigation. As a matter of simple arithmetic, royalty stacking magnifies the problems associated with injunction threats and hold-up, and greatly so if many patents read on the same product. In this key sense, the problem of injunction threats and royalty stacking are intertwined (Lemley & Shapiro 2007).

5 Cf. Article 3 ETSI IPR Policy.

6 Between 2005 and 2015, the former manufacturers Nokia and Motorola have divested their handset business, and companies like Apple, Samsung, HTC, Huawei, RIM, TCL, ZTE, have entered the cellphone market, where they were previously not active.

7 These standards may also be the cornerstone of the Internet of Things ("IoT"), as the IoT can be described as a system where items in the physical world, and sensors within or attached to these items, are connected to the Internet via wireless and wired Internet connection. These sensors can use various types of local area connections such as RFID, NFC, Wi-Fi, Bluetooth, and Zigbee. Sensors can also have wide area cellular connectivity such as GSM, GPRS, UMTS and LTE (Lopez Research LLC 2013). The IoT may thus rely on (among others) current ETSI' and IEEE's standards and the evolution of current standards. Therefore, it is important to mitigate risks of abusive conducts as hold-up and hold-out.

8 Cf. Sections 2.2 and 4.1 ETSI Guide on Intellectual Property.

9 To the exception of Samsung and Huawei (cf. ABIresearch report 2013, Fraunhofer Study 2011).

10 Cf. acquisition of Nortel's patents by Rockstar consortium, involving among others Apple and Microsoft.

11 Cf. litigations involving Apple or Microsoft. For this paragraph, cf. also 3GPP's website with reports on submissions and accepted contributions.

12 Cf. Article 6.2 IEEE Bylaws stating that
The Submitter [SEP-holder] *and the Applicant* [implementer/potential licensee] ***should engage in good faith negotiations (if sought by either party) without unreasonable delay*** *or may litigate or, with the parties' mutual agreement, arbitrate: over patent validity, enforceability, essentiality, or infringement; Reasonable Rates or other reasonable licensing terms and conditions; compensation for unpaid past royalties or a future royalty rate;(…)* [Emphasis added].
And case law such as Rechtbank's Gravenhage, *Philips v. SK Kassetten,* HA-ZA 08-2522 en HA ZA 08-2524, March 17, 2010; Tribunale di Milano, Sezione Specializzata Proprieta' Industriale ed Intelletuale, *Samsung Electronics Co., Ltd. e Samsung Electronics Italia c. Apple Inc., Apple Italia S.R.L., Apple Retail Italia S.R.L., Apple Sales International,* R.G. 59734/2011, January 5, 2012.

13 The royalties decrease because patents are "exchanged against patents".

14 Some implementers- as, for example, Samsung and Huawei- are contributing to the standard development process (ABIresearch 2013). Furthermore, implementers and contributors may compete on other segments than either the standard development or the smartphone market (e.g. on the network infrastructure, as it is the case between Ericsson, Nokia, Cisco, Huawei, ZTE).

15 European Commission, Antitrust Procedure, Case A.T.39939-Samsung-Enforcement of UMTS Standard Essential Patents, April 29, 2014; European Commission, Antitrust Procedure, Case A.T.39985-Motorola-Enforcement of GPRS Standard Essential Patents, April 29, 2014.

16 Articles 9 and 11 of the Directive 2004/48 of the European Parliament and of the Council of 29 April 2004 on the Enforcement of Intellectual Property Rights.

17 Cf. 35 U.S.C. §283 Injunction, and 19 U.S.C.§1337.

18 It does not prevent implementers from acquiring, developing and asserting SEPs in litigations but many of them have (currently) a less large SEPs portfolio than the main contributors.

19 Cf. Federal Circuit, *LaserDynamic v; Quanta Computer,* August 30, 2012, as "per se rule" for non-SEPs and used in SEPs cases:

Where small elements of multi-components products are accused of infringement, calculating a royalty on the entire product carries a considerable risk that the patentee will be compensated for non-infringing components of that product.

Quoted in the SEPs case, Northern District of Illinois, *In Re Innovatio IP Ventures,L.L.C.* October 3, 2013:

Moreover, calculating royalties based on the end-product would invite error, as those end-products include myriad features that are unrelated to Wi-Fi.

20 Cf. Teece & Sherry 2016:

The reason for the inconsistency is that (with one exception discussed in the next section) we have never seen a license that call for the licensee to pay percentage-based running royalties based on the selling price of an SSPPU. Instead, the most common practice is for percentage-based running royalties to be calculated based on the price of the product as actually sold by the licensee.

21 Cf. Mallison (2013) about the difference of market demand, prices and functions between the iPod and the iPhone to value 3G and 4G standards (consumers are ready to pay more for 3G/4G and Wi-Fi standards).

22 The product will have additional properties thanks to the standard (cf. *supra*, end note 22).

23 On comparable licenses: rejection of proposed "comparable licenses": *In Re Innovatio IP Ventures L.L.C.,* 2013, because not negotiated within a (F)RAND context; same for the rejection of Symbol and Proxim licenses advanced by Motorola, and rejection of the RIM-Motorola license because applied to all Motorola's portfolio and apportionment was difficult, but acceptance of the ARM licenses as comparable whereas ARM business model is based on a license per component (*Microsoft Corp. v. Motorola, Inc.*, 2013); rejection of Motorola license with VTech as not probative of a RAND rate because *part of a broader agreement that settled infringed claims licensed* (*Microsoft Corp. v. Motorola, Inc.*, 2013) while this case also concerned infringement claims and parties seeking to settle a dispute through a license; non rejection *per se* of licenses presented as "comparable" in *Ericsson v. D-Link* (Fed. Circ., 2015):

otherwise comparable licenses are not inadmissible solely because they express the royalty rate as a percentage of total revenues rather than in terms of the smallest sales units.

24 Cf. rejection of Via Licensing Pool in the case *In Re Innovatio IP Ventures LLC,* October 3, 2013; cf. U.S. 9th Circuit, *Microsoft Corp. v. Motorola, Inc.,* July 30, 2015: in pools, patents are *generally licensed at lower rates than might be achieved in a bilateral agreement,* which has been confirmed in Europe by the Landgericht Düsseldorf, Case 4a O 73/14, March 31, 2016.

25 On the use of the "willing licensee" term

that label has at times been applied too broadly to include parties that didn't seem willing to pay a RAND rate (…). Competition agencies should examine the facts of each case before reaching a conclusion. (Olhausen 205).

(…) the following sentence from Apple's submission to the Commission on April 10, 2013, indicates that Apple has no intention of paying Samsung any royalty until after the conclusion of litigation: *(…) and if that judgement were affirmed on appeal, Apple would stand ready to pay FRAND royalties* (ITC 337-TA-794, July 5, 2013)

26 European Commission, Antitrust Procedure, *Case AT.39985-Motorola-Enforcement of GPRS Standard Essential Patents*, April 29, 2014, §443, 454:

Fourth, Apple's conduct during the rate-setting proceedings before the Mannheim District Court is not a sign of its unwillingness to enter into a license agreement for Motorola's SEPs on FRAND terms and conditions.

(…)

Seventh, Apple's refusal to initially acknowledge its past infringement of all SEPs covered by the Settlement Agreement and Motorola's damages claims is not a sign of its unwillingness to enter into a license agreement for Motorola's SEPs on FRAND terms and conditions.

This last sentence is questionable: a refusal to acknowledge past infringement means that the potential infringer does not want to pay for the past exploitation of patents he has benefited from. This situation is quite unfair for the patent-holder and means a free usage of someone's else technologies, which is detrimental to the patent-holder but also discriminatory towards other licensees who are paying for the exploitation of the patents.

27 What if a court has determined a (F)RAND rate higher than this amount?

28 European Commission, Antitrust Procedure, *Case AT.39985-Motorola-Enforcement of GPRS Standard Essential Patents*, April 29, 2014, §111:

Motorola and Apple have been engaged in licensing discussions, on and off, since 2007, following the termination of the Chi Mei Agreement by Motorola. These discussions are not, however, addressed in this section, as they do not affect the Commission's assessment of the abusive nature of Motorola's conduct as established by this Decision.

29 For Apple and Microsoft, cf. their arguments in litigations against Motorola (Apple; Microsoft) and Samsung (Apple). Concerning Cisco and Intel, cf. their responses to the EC's questionnaire on Patents and Standards: A Modern Framework for Standardisation Involving Intellectual Property Rights, available at http://ec.europa.eu/growth/tools-databases/newsroom/cf/itemdetail.cfm?item_id=7833.

30 Cf. for example, Europe Digital Single Market Strategy and the Communication on ICT Standardisation Priorities, where standardization is described as a cornerstone of the Digital Single Market in Europe, and end note 8, *supra,* on the IoT.

31 For example, it has been reported that Apple got in 2015 more than 90% of smartphones profits with only 17% market shares (Elmer-De Wit 2016) and ZTE is gaining market shares (Einhorn 2015).

Chapter 8
Competing Value Logics in the Apportionment of SEP Damages:
A Comparison of FRAND Royalty Methodologies in U.S. Courts 2013–15

Bowman Heiden
Chalmers University of Technology, Sweden

ABSTRACT

This chapter comparatively examines the valuation methodologies put forward and accepted in four initial FRAND cases adjudicated in the US district court system in 2013-15 and finds a qualitative link to the business models (i.e., value logics) employed by the actors and the type of standards involved. Additionally, valuation models based on prior license agreements seem to currently produce higher valuations of SEPs, everything else equal. While the CAFC has clearly stated that apportionment is the governing rule in SEP cases, the variety of valuation methods put forward in similar contexts indicates the calculation of the SEPs still lacks consensus on both legal and economic grounds. However, a number of the most recent SEP rulings in district courts show clear accordance with clarifications from the CAFC in Ericsson and CSIRO. Additionally, the studied cases confirm the lack of empirical evidence of systemic patent holdup and royalty stacking and indicate that the contractual/transactional nature of SEP disputes can well be addressed on a case-by-case basis by the courts.

INTRODUCTION

The convergence of computing, Internet, and telecommunications has created intense competition over intellectual property in recent years. This is due to the fact that multi-technology products, such as smartphones and tablets, include thousands of patents across a broad range of technical functions that are owned by many different actors from disparate sectors, all of which compete to receive a share of the expanding telecommunications market.[1] Additionally, all devices are increasingly becoming

DOI: 10.4018/978-1-5225-5320-5.ch008

telecommunication-enabled through wireless functionality (i.e. smart devices) in what has been labeled as the Internet of Things (IoT), which in turn opens up new opportunities for telecommunication actors in non-traditional markets[2]. Thus technology convergence has created new innovations and markets but also increased competition by actors with different intellectual property and business norms.

One major area of contention regards intellectual property and technology standards[3] in the ICT sector, in particular, the pricing of licenses for patents essential to the implementation of a technology standard (i.e. standard essential patents or SEPs).[4] As each patent essential to the standard could theoretically be used to block or delay the implementation, most standard setting organizations (SSOs) have developed IPR policies to govern the open use of patents essential to the standard on fair, reasonable and non-discriminatory (FRAND) terms.[5] These policies are meant to facilitate a balance between the need for innovation and adoption and diffusion of the standard. In effect, FRAND policies work to incentivize upstream actors to invest in the development of standards and reduce the uncertainty of patent access among downstream implementers who will make irreversible, standard-specific investments before having obtained licenses from all the actors holding standard essential patents (SEPs). Thus FRAND agreements can be seen as an incomplete contract designed to ensure both compensation and access to patented technology but allowing the terms to be resolved through bilateral market negotiations outside of the SSOs, which alleviate antitrust concerns.

In 2013-15, the US federal court system issued several rulings determining the royalty rate of standard essential patents (SEPs) under FRAND commitment to standard setting organizations (SSOs). While not generating as much popular press as the smartphone wars between Apple and Samsung, these cases represent the culmination of a growing battle over the distribution of profits in the telecommunication value chain between technology owners and technology implementers fought through the context of standard essential patents and the FRAND agreements under which they are licensed.[6] From a downstream manufacturing perspective, SEPs are a cost to be minimized, while for an upstream technology provider, SEPs represent the output of its R&D investments from which it looks to maximize its return. This new mode of vertical competition in the value chain has opened up for new business models and new roles for intellectual property that challenge the traditional industrial norms. This has, in turn, generated a growing sentiment that the patent system is broken and no longer supports innovation (e.g. Boldrin and Levine, 2012; Jaffe & Lerner, 2011; Bessen & Meurer, 2008). Thus, these court cases not only have a strong influence on the value of patents in the telecom industry but on the efficacy of the patent system as a whole to support the emergence of new knowledge-based business models and a new division of innovative labor in the knowledge economy.

Theoretical concerns emerged around the turn of the century over the shear increase in the number of patents (i.e. patent thickets) and the increase in the number of different patent holders (i.e. royalty stacking), which created an influential strand of literature around what has been termed, patent holdup (e.g. Lemley and Shapiro, 2007; Farrell et al., 2007; Shapiro, 2001).[7] Proponents of patent holdup claim that patent owners are able to command royalties greater than the value of the contribution of their patents by capturing the value of investments of specific-purpose assets by downstream actors. More specifically, in standardization contexts, patent holdup is interpreted as claiming the economic value of a standard as opposed to the contribution of the patented invention apart from the value associated with its inclusion in the standard. In other words, it is proposed that SEP holders are able to use the market power created by the lock-in effect of the standard to charge royalties greater than what they would have been able to charge in market competition before the standard was set (i.e. ex ante).[8]

In order to balance the interests of standard innovators and implementers, standard setting organizations (SSOs) have for some time now adopted either royalty free (RF) or FRAND commitments in their IPR policies to members of the SSO and third parties as described above. However, despite the adoption of FRAND IPR policies by many SSOs, there has remained a strong concern that (1) FRAND is insufficient to manage the theoretical possibility of patent holdup and (2) courts struggle to properly apportion patent damages in multi-technology products, which has led to proposals to change legislation, alter procedural law, and adopt new regulations and policies regarding the licensing and valuation of SEPs (e.g. Rysman & Simcoe, 2011; Lemley, 2007; Lemley and Shapiro, 2007). Let's take each of these issues in turn.

Firstly, regarding the efficacy of FRAND, there is a lack of empirical evidence of adverse economic impact (neither distributional or efficiency) due to patent holdup and royalty stacking in standards operating under FRAND IPR policies.[9] There are, however, theoretical pre-conditions such as indications of a patent thicket,[10] several specific cases of alleged transactional holdup,[11] and data on increased litigation,[12] but the question is whether these could be said to rise to the level of empirical evidence of market failure in need of policy intervention. Additionally, the theoretical frameworks that suggest the existence of an economic efficiency problem, and posit reasons why intervention would produce a set of better market conditions, have been contested (e.g. Galetovic and Haber, 2017; Heiden and Petit, 2017; Sidak, 2013; Denicolo et al., 2008; Elhauge, 2008; Sidak, 2008; Geradin and Rato, 2007).[13] There does exist, however, general evidence that FRAND-enabled standards have been highly successful[14], such as in cellular standards, where market penetration of mobile phones has grown to include the poorest segment of people in the world[15] and growth of mobile broadband continues to expand aggressively in spite of the increased number of patent disputes and the perceived threat of patent holdup.[16]

Secondly, the landmark FRAND case, *Microsoft v. Motorola*, resulted in a 207-page ruling providing a reasoned methodology for the determination of FRAND royalty rates that even the staunchest critic of the court system would not argue systemically overcompensates the patent holder. In fact, a strong argument could be made that the court sets a precedent that may systemically reduce the value of all SEPs beyond the narrow context of the specifics of the case (Heiden, 2016). In addition, the US court system adjudicated FRAND royalties in several other initial cases handed down in 2013-15 by the federal courts.[17] However, while these rulings suggest that the court system can manage the specific contractual/transactional circumstances of patent holdup without legislative, anti-trust, and policy reform, these rulings create the potential for a large systemic effect on the telecommunication, as well as the broader ICT, industry as business strategy is always shaped in the shadow of the court. Thus, these decisions have repercussions that will impact not only economic performance but also potentially economic efficiency as firms respond to the new rules of the game.

Several authors have commented on the SEP/FRAND court cases under study (e.g. Heiden, 2016; Sidak, 2016; Contreras, 2015; Layne-Farrar and Wong-Ervin, 2015; Sidak 2013), but there has not been an in-depth comparative analysis across all the cases.[18] Thus, this paper will focus on a comparative assessment of the SEP valuation models put forward and evaluated in four of the initial SEP US district court cases from 2013-15. Based on these decisions and the subsequent rulings of the CAFC, the paper will provide a discussion of the systemic implications for industry and policy makers (both SSO and governmental) given the potential impact on economic performance and economic efficiency in the context of the shift from industrial to knowledge-based business models. More recent district court cases from 2015-16 are referenced to provide further validation of the impact of the initial 2013-15 cases.

More specifically, this paper will investigate (1) the impact of evolving knowledge-based market structures and firm positioning on SEP value, (2) the operationalization of valuation principles and norms based on competing business models/value logics and competing theories of patent holdup, royalty stacking, and economic efficiency in the FRAND context, and (3) the level of validation of SEP/FRAND valuation principles by the US federal district courts and appellate courts. These investigations are then formulated into a set of propositions towards a preliminary theoretical framework for an improved understanding of SEP value in the knowledge economy.

Section 2 below will provide an overview of the changing role of intellectual property in the knowledge economy, particularly in relation to the development and implementation of technology standards in telecommunications. Section 3 will provide a short overview of the four initial cases and a summary of a number of key characteristics across the cases. Section 4 will provide a comparative analysis, and Section 5 will discuss the key findings and implications for industry and policy makers towards a better theoretical understanding of SEP value. Section 6 will conclude. This paper does not evaluate the opinions of the court cases from a traditional legal perspective. Instead, information is adduced from the court cases and used to analyze how techno-economic and legal principles are applied by market actors and interpreted by the court to value SEPs and discuss the consequences these rulings could have on economic performance and efficiency. The analysis and conclusions are based on received theory and public data from the four court cases and associated public records on the industry and market actors involved. In addition, qualitative data from industry experts has been sought to improve the validity of the research.[19]

THE CHANGING ROLE OF IP IN THE KNOWLEDGE ECONOMY

The Transition From a Static to a Dynamic IP Logic

The transformation from an industrial to a knowledge economy can be characterized by a fundamental shift in the role intellectual property from a static to a dynamic approach, where IP is not only used to block competition (i.e. static approach) but is increasingly used as objects for commercial transactions (i.e. dynamic approach). In other words, knowledge is not only embedded in physical products, but instead is packaged and transacted as intellectual property through different value propositions such as virtual products, license offers, and knowledge-based services (Petrusson and Heiden, 2008; Petrusson, 2004). Thus one useful way of describing the new roles of intellectual property is through the evolution of value logics and knowledge-based business models. Figure 1 below shows this evolution as four models characterized by the type of entity (i.e. implementing or licensing entities) and origin of intellectual property (i.e. own/internally created or acquired externally)[20]. This model allows for a differentiation between a product market approach, where the role of IP is to protect the sales of physical products, and a technology market approach, where IP itself is transacted for revenue generation (Petrusson and Heiden, 2008; Petrusson, 2004).

1. **Freedom to Operate:** This business model is characterized by the use of IP portfolios to block competitors or cross-license to gain access to the technology required to produce products (i.e. achieve freedom to operate). Historically, the value of patents in the electronics industry has been strongly influenced by the cross licensing between competing manufacturers, which has been a norm

Figure 1. Evolution of value logics

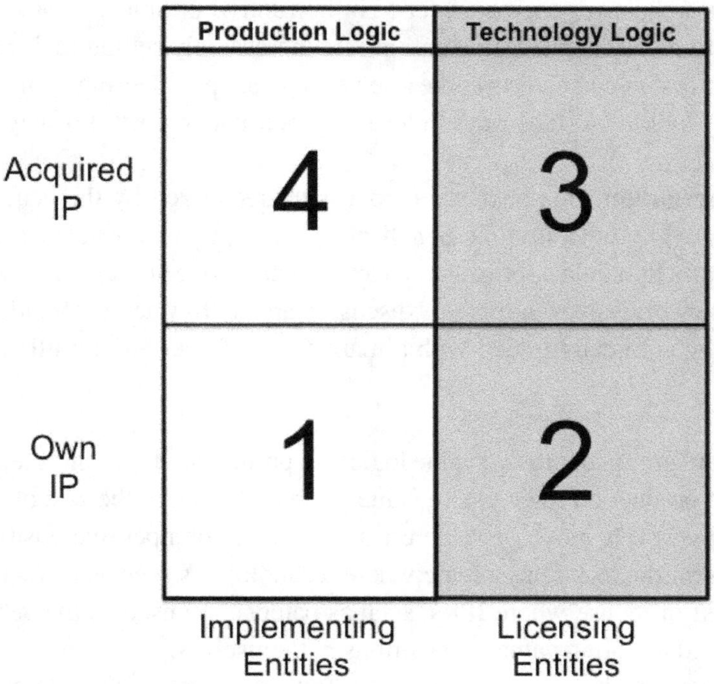

in the industry since the very beginning.[21] More recently, the smartphone war between Apple and Samsung is an example of the static use of patents between competitors vying for market share.[22] Computer and Wifi device manufacturers such as Dell and Cisco are also examples of firms primarily utilizing this model.

2. **Revenue Licensing:** This business model is characterized by the use of IP as a strategic source of revenue generation. In this model implementing firms look to leverage the strength of their patent portfolios on the technology market in addition to competing on the product market. Furthermore, the strength of patents facilitates the opportunity for a new division of innovative labor in the value chain where innovation specialists can generate intellectual property through their own R&D activities but choose to license their technology to others instead of manufacturing products themselves. Historically, in the electronics industry, the relatively free flow of technology and the use of cross-licensing as a means to facilitate freedom-to-operate was augmented by a licensing strategy focused on generating royalty income, which generated hundreds of millions of dollars annually for pioneering firms such as IBM and Texas Instruments in the late 1980s and early 1990s.[23] Today, in the telecommunications industry, companies such as Qualcomm, Ericsson, and Nokia Technologies among others generate hundreds of millions of dollars of revenue each year from their technology licensing programs implemented through differing overall corporate strategies.[24]
3. **Offensive Aggregation:** This business model also focuses on revenue licensing but instead through the acquisition of intellectual property generated by others with the sole purpose of monetizing the IP assets into greater value through licensing and often litigation. This relatively new model has grown rapidly to include both private and public traded companies. Actors employing this model are often referred to pejoratively as patent trolls or Patent Assertion Entities (PAEs) by the firms

they target for litigation. Two of the most successful firms using this model, Intellectual Ventures and Acacia Research, have generated over $3 billion and $1 billion in cumulative royalties to date.[25] It is important to note that practicing entities are increasingly selling their patent portfolios to offensive aggregators, in what has become known as patent privateering, which shows that all actors are increasingly looking to find ways to leverage their patent portfolios beyond the production of physical products.[26]

4. **Defensive Aggregation:** this business model is characterized by the acquisition of intellectual property generated by others on the behalf of operating firms for defensive purposes. Thus this model represents a dynamic approach to enacting a defensive strategy as a response to the growth in IP assertions by offensive aggregators discussed above. Key actors include RPX, which was VC funded as is now publically traded with a market cap of over $850 million, and Allied Security Trust.[27]

The four models are grouped into two value logics – a production logic and a technology logic. Boxes 1 and 4 represent the production logic where value is created through the sale of products and services and IP is used to support this mode of value creation through competitive positioning on the market. Boxes 2 and 3 represent the technology logic where technology is seen as its own value proposition in relation to the context of its use where IPRs, such as patents, are used as the vehicle to transact these value propositions (and are often value propositions in themselves).

The four models can also be seen as stages in the transformation of the role of intellectual property. The first transition is from a static, defensive approach (box 1) to a dynamic, revenue generating approach (box 2) by technology-based firms (i.e. IP generating firms) in the industrial value chain. However, the transition to box 3 represents the creation of a more developed market for intellectual property, where intermediaries such as patent aggregators take on a share of the offensive and defensive activities that were traditionally managed only by the industrial actors. The growth of new actors (i.e. NPEs) that operate in boxes 2-3, which represent both private and public firms[28], have facilitated a large increase in market transactions and litigation, which has driven much of the debate concerning the efficacy of the patent system.[29] Stage 4 represents a dynamic, defensive approach where market transactions are used to build freedom to operate among product manufacturers. This model, while simplified, provides a useful understanding of the evolution of roles for intellectual property, differentiating from an industrial and knowledge-based business perspective. It also illustrates how market actors respond with market solutions to new business behaviors, which can mitigate the need for policy intervention, such as the development of defensive aggregators to counteract the potential negative impact of offensive aggregators.[30]

SEPs, FRAND, and Standardization in a New Division of Labor

The telecommunication sector exemplifies the challenge in balancing the role of patents to incentivize actors to participate in open standards by rewarding inventive activity with the role of standard setting organizations (SSOs) to facilitate the timely development and diffusion of technical standards. Telecommunication standardization processes[31] can involve tens of actors and require access to thousands of patents that are essential to the implementation of a standard.[32] The most complex standardization processes are typically governed formally or by voluntary participation by specific standard setting organizations (SSOs) [33] run by industry consortiums, professional associations, and national or international NGOs.[34] SSOs are solely focused on developing the best technical specifications and formally forbid issues concerning patents and licensing to take place within standard deliberations.[35] However, as the creation of standards generates a lock-in effect on the market[36], holders of patents essential to the execution of the standard potentially possess a much higher degree of market power than would normally be attributed to actors in a competitive market.[37]

For this reason patents and standards have traditionally been depicted at odds with one another, but the division of innovative labor in the knowledge economy makes patents essential (excuse the pun) to the development of most standards, as many knowledge-based firms increasingly compete in the upstream technology market, not only the downstream product market. Figure 2 below shows the creation of a new industrial dynamic within the telecommunication sector, which has resulted in a new division of labor including innovation specialists and implementation specialists together with integrated firms, who all compete in the same value chain with very different strategies and incentives (Schmalensee, 2009; Geradin and Rato, 2007). These actors apply the patent system and anti-trust regulations to the standardization process in very different ways in their search to maximize economic performance. In particular, the division of innovative labor, represented by Firm B in Figure 2, illustrates the transition from a production to a technology logic as discussed above in section 2.1. Theoretically, from a standardization perspective, SEP valuations based on a production logic may be lower than a technology logic as practicing entities use standards to develop markets where they can sell their products while non-practicing entities look for a return on investment for the technology in the standard itself.

The history of FRAND in the telecommunication industry can be dated back to the 1956 AT&T consent decree.[38] However, the modern era of FRAND and the strategic use of IP can be exemplified by the development of digital cellular telecommunication when Motorola altered the structure of the nascent GSM market in the late 1980s and early 1990s by forcing implementing firms to enter into (cross-) licensing agreements in order to produce GSM standard compliant equipment (Bekkers et al.,

Figure 2. New division of innovative labor

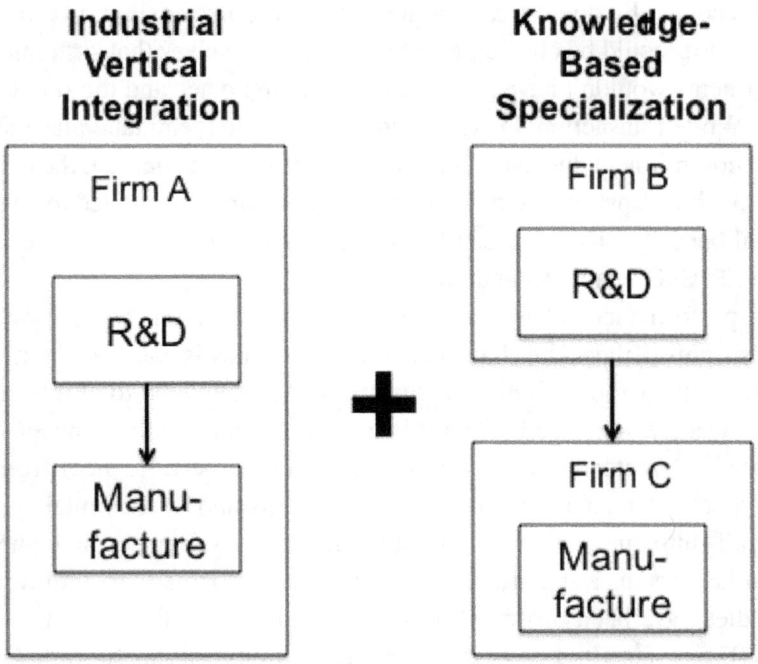

2002). The renewed recognition of the potential power of SEPs generated a movement within ETSI, as well as other SSOs, to develop IPR policies to govern the role of patents in the development and commercialization of standards.[39] In parallel, the fragmentation of the telecommunication value chain and the introduction of IP-based business models created a new competition among firms in the technology market based on the relative strength of SEP portfolios. Thus in response to the new strategic use of patents augmented by a change in the division of labor in the telecommunications value chain, the viability of FRAND-based IPR policies came under increased scrutiny over whether FRAND, specifically designed as an incomplete contract, could manage the challenge of balancing the complex interests of innovators and implementers.[40]

Patent Holdup, Holdout, and Economic Efficiency

FRAND-enabled technology standards are built on voluntary technical contributions that once finalized are made openly available.[41] Thus technology standardization is a process of technology transfer where FRAND licenses are typically negotiated after products have already been put on the market. This process creates the potential for opportunism by both parties, where SEP holders could leveraging their power position through higher than reasonable royalty rates (i.e. patent holdup) and implementing firms can evade and delay payment of royalties while marketing products that include technology developed and owned by others (i.e. patent holdout).[42]

Currently there is no substantial empirical evidence of systemic patent holdup in any industry where standard essential patents are offered for license under FRAND terms (Sidak, 2016; Mallinson, 2016, Galetovic and Gupta, 2016, Galetovic, Haber and Zaretzki, 2016, Galetovic, Haber and Levine, 2015, Layne-Farrar, 2014, and Padilla, Geradin, and Layne-Farrar, 2007.[43,44] For example, the H.264 and 802.11 standards in question in the studied cases have been extremely successful on the market, even though both standards contain thousands of SEPs and as many as 90 SEP holders. One explanation for this is that patent holdup is not based on ex post opportunism, but based on an ex ante FRAND agreement acknowledged and accepted by downstream companies before irreversible investments are made. The success of these standards could be considered evidence in themselves that systemic holdup doesn't exist, otherwise investments wouldn't have been made in the first place and the standards would not have been so successful. While transactional patent holdup and holdout can take place at any time amongst parties to a FRAND commitment due to the incomplete nature of the contract, the market success of ICT standards suggests that both upstream and downstream actors are incentivized to negotiate in good faith for their own mutual benefit.[45] This is possibly another explanation for the lack of empirical evidence of market failure in FRAND-enabled standards.

While economic performance is based, at least in the short run, on the fight over who receives the greater share of the surplus in the value chain, economic efficiency is concerned with the total aggregate surplus. It is only when the (re)distribution of surplus value affects the total that economic efficiency is affected and remedies are warranted. Thus it is critical to separate the concept of patent holdup as a means to negotiate for greater private returns (i.e. economic performance) from the arguments of economic efficiency, both static and dynamic, which are motivated by maximizing social welfare. This distinction further calls into question the nature of holdup as a systemic, ex ante phenomenon blocking the formation of markets versus a specific, transactional ex post phenomenon of rent appropriation[46].

Numerous remedies have been proffered in the name of solving the theoretical problem of patent hold-up. Lemley (2007) advanced ten suggestions, including limiting injunction relief in multi-technology products, ex ante disclosure and negotiation of FRAND royalties within SSOs and the concomitant loosening of anti-trust regulations that would entail, and patent reform focused on redefining reasonable royalty rates in damage calculations[47]. Additional solutions have been put forward such as non-assertion after a specified time (Rysman and Simcoe, 2011), a pseudo patent pool approach (Contreras, 2013), "baseball-style" arbitration (Lemley and Shapiro, 2013), a "contingent ex ante" valuation framework (Siebrasse and Cotter, 2015), a return to a traditional incremental value model applicable to both SEPs and non-SEPs alike (Contreras and Gilbert, 2015), and a "structured price commitments" approach (Lerner and Tirole, 2015). While all of these solutions have a theoretical possibility to reduce patent holdup and increase static economic efficiency, they also have a theoretical possibility to create holdout and reduce dynamic efficiency potentially resulting in an overall reduction in aggregate social welfare. Numerous scholars have questioned the systemic impact of patent holdup and royalty stacking, challenging its existence and operationalization by the scholars, courts, and regulatory actors (e.g. Galetovic and Haber, 2017; Layne-Farrar and Wong-Ervin, 2015; Sidak, 2013; Dorsey and McGuire, 2013; Denicolo et al., 2008; Sidak, 2008, and Geradin and Rato, 2007 among others).

The implementation of the concept of patent holdup as a means to properly apportion the rents associated with SEPs is investigated through the valuation principles and methods put forward and adopted in the court cases discussed and analyzed in section 4 below.

OVERVIEW OF INITIAL FRAND COURT CASES IN THE US

This section provides a brief description of the outcome of the landmark SEP ruling by the Western District of Washington and the Ninth Circuit Court of Appeals (*Microsoft v. Motorola*) and three subsequent FRAND cases in the Eastern District of Texas and the Federal Circuit (*Ericson v. D-Link et al. and CSIRO v. Cisco Systems*) and the Northern District of Illinois (*Innovatio IP Ventures*) together with a comparative overview of the key issues. The next section will then provide a more in-depth comparative analysis of the four initial cases.

Microsoft v. Motorola (April 25, 2013)[48]

In the end of October 2010, Motorola sent Microsoft two letters offering to license their portfolio of essential patents related to the IEEE 802.11 (WiFi) and ITU H.264[49] (video coding) standards at a royalty rate of 2.25% of the end product price for all Microsoft products that comply with the aforementioned standards subject to a grant back license of Microsoft's patents under the standards.[50] On November 9, 2010, Microsoft filed a breach of contract suit against Motorola in the Western District of Washington claiming that the offers breached their RAND commitment to the standard setting organizations (SSOs), IEEE for the 802.11 standard and ITU for the H.264 standard.[51]

On November 13, 2012, after clarifying the basic contractual obligations and staying the pending German injunction and US infringement actions, the bench trial began to determine the FRAND royalty range and the exact FRAND licensing rate for the SEPs in suit, which the court deemed fundamental to addressing both the breach of contract action as well as the impasse over the determination of the actual FRAND rate. The bench trial ended on November 20, 2012, and after receiving several rounds of post-trial briefs, the court issued its decision on April 25, 2013, setting the royalty range and rate of the two SEP portfolios as follows:

- H.264 SEP Portfolio – Royalty Range: $0.00555 – $0.16389 per unit Royalty Rate: $0.00555 per unit
- 802.11 SEP Portfolio – Royalty Range: $0.008 – $0.195 per unit Royalty Rate: $0.03471 per unit

In contrast to the rate of 2.25% of the end product price proposed by Motorola, this is an extreme discount. For matters of comparison, if we assume a low-end price of an Xbox at $200, the anticipated royalty would have been $4.50 per unit, which makes the court ruling a discount of approximately 810 times the expected rate for the H.264 SEP portfolio and approximately 130 times the offered rate for the 802.11 SEP portfolio. As a percentage, this equates to a royalty rate of less than 0.003% and 0.02%, respectively.

This benchmark ruling was the landmark US court case regarding the determination of FRAND royalty rates and thus the value of standard essential patents (SEPs). In September 2013 a jury awarded $14.5 million in damages to Microsoft for Motorola's breach of their FRAND obligation. The decision was affirmed by the Court of Appeals for the Ninth Circuit filed July 30, 2015.[52] It was reported on October 1, 2015, that Google (owner of Motorola's patent portfolio) and Microsoft resolved their patent disputes and dropped their lawsuits.[53]

Ericsson v. D-Link Systems et al. (August 6, 2013)[54]

In 2012 Ericsson brought an infringement action in the Eastern District of Texas against several WiFi and computer manufacturers including D-Link, Netgear, Belkin, Acer, Gateway, Dell, and Toshiba. Intel asked and was granted the right to join the suit as a chipset supplier of the accused parties. The case regarded several of Ericsson's SEPs for the use of the 802.11 (WiFi) standard developed within IEEE. After a Markman hearing to determine the scope of the relevant patent claims, a jury trial was held one year later in June 2013, which returned an infringement verdict on three of the five patents in suit. The court upheld the verdict, and an ongoing future royalty was set during the post-trial motions as follows:

- 802.11 SEP Portfolio – Royalty Rate: $0.15 per unit

The damages and ongoing royalties returned by the jury were in line with those requested by Ericsson based on prior comparable licensing agreements with HP and RIM among others over the same 802.11 SEP portfolio.[55]

The decision was appealed to the Court of Appeals for the Federal Circuit (CAFC), where several organizations and technology firms, such as IEEE, American Anti-Trust Institute (AAI), Qualcomm, and Nokia submitted amicus briefs regarding the case. The CAFC ruled to vacate the jury's damages award and ongoing royalty rate and remanded the case with new jury instructions.[56] In addition, Broadcom petitioned the USPTO to institute an inter partes review of the three central Ericsson patents asserted in the court case to re-evaluate their validity, which resulted in a finding of invalidity for all three patents by the Patent Trial and Appeal Board (PTAB) of the USPTO in March 2015. The SEPs in suit have subsequently been assigned from Ericsson to Wi-Fi One LLC, and the decision of the PTAB is currently under appeal before an en banc panel at the CAFC.

As the case was decided based on evidence provided to the jury, the court did not produce a detailed decision of its own regarding the FRAND royalty determination as in *Microsoft* and *Innovatio*. However, the final ruling on post trial motions by the district court and the CAFC opinion did provide insight into the valuation methodology applied in the case.

Innovatio IP Ventures (October 3, 2013)[57]

In 2011, based on a portfolio of SEPs that they had previously acquired from Broadcom[58], Innovatio IP Ventures began what has been described as an patent assertion campaign targeted towards end users of wireless devices employing the 802.11 standard such as coffee shops, hotels, restaurants, etc[59]. Innovatio sent approximately 13,000 letters asking over $2,000 from each location employing a hotspot.[60] Device manufacturers, including Cisco Systems, Motorola Solutions, Netgear, SonicWALL and Hewlett-Packard intervened on behalf of the end users of their products and the resulting case was consolidated in the Northern District of Illinois. Cisco even went as far as charging Innovatio with racketeering based on their tactics.[61]

Prior to a Markman hearing and determination of validity and infringement, all parties agreed to a bench hearing to rule on the FRAND royalty rate (and subsequent damages) for the portfolio of 23 SEPs (reduced to 19 SEPs during trial) related to the 802.11 standard asserted by Innovatio. The first bench hearing on July 2013 confirmed the essentiality of the Innovatio patent portfolio, and the second bench hearing in September 2013 ruled on the FRAND royalty rate for the manufacturers only (i.e. not the end users). The parties agreed to the use of the methodology put forward in the *Microsoft* case to determine a FRAND royalty rate.

In Oct 2013 the district court issued its ruling on the FRAND royalty rate for the Innovation SEP portfolio using the chipset as the unit of reference as follows:

- 802.11 SEP Portfolio – Royalty Rate: $0.0956 per unit

The court, in this case, drew extensively from the landmark *Microsoft* case and benchmarked its ruling with regard to the *Ericsson* case described above.[62] After the trial Cisco reached a settlement with Innovatio to pay $2.7 million to cover 85 million WiFi devices, equating to $0.032 per device.[63]

CSIRO v. Cisco Systems (July 23, 2014)[64]

In 2011 the Commonwealth Scientific and Industrial Research Organization (CSIRO), a research organization of the Australian government, filled an infringement suit against Cisco Systems for one patent essential to the 802.11 standard in the Eastern District of Texas. The patent in suit, which solved the "multipath" problem in wireless networking, had been involved in litigation since 2005 against many of the major ICT companies deploying the WiFi standard in their products. In 2013 the parties agreed to try the case solely on damages having agreed on liability and scope of accused products.

In 2014 after a 4-day bench trial, the court issued its ruling on the royalty as follows:[65]

- 802.11 SEP – Royalty Rate: $0.83 per unit (pro-rated average)

The judge in the case rejected the expert damages models of both parties and crafted his own royalty rate determination using the end product as a royalty base. The decision was appealed to the CAFC and several large tech firms submitted amicus briefs in support of Cisco, including Apple, Intel, Dell, and Hewlett-Packard. The CAFC vacated the district court's decision and remanded for revision of the damages award.[66]

Table 1 provides comparative information across the four district court cases.

COMPARATIVE ANALYSIS

In this section the evidence, arguments, and outcomes of the *Microsoft*, *Ericsson*, *Innovatio* and *CSIRO* cases are presented and comparatively analyzed in relation to market positions and value logics of the actors, the application of economic principles to the FRAND valuation context, the ex ante and ex post techno-economic evaluation of the SEPs, and the royalty base and valuation methods put forward by the market actors and evaluated by the courts.

Positioning of Actors

The four initial cases offer different perspectives on the nature of business models deployed to extract value from technology standards. Figure 3 shows the positions of the actors involved in the disputes in relation to their business models for the standards involved (i.e. H.264 and 802.11).[70]

The Motorola and Microsoft dispute is between two practicing entities where Motorola is looking to generate revenue from its proportionally stronger patent position within the H.264 and 802.11 standards in addition to a grant-back license to Microsoft's relevant patent portfolio. Ericsson is operating similarly, but with a greater focus on licensing revenue in relation to the particular lack of SEP strength of the implementing firms in this particular case. In negotiation circumstances with other SEP holding firms, Ericsson, as a producer of products employing the 802.11 standard, would also strategically look to obtain grant-back licenses from end product manufacturers. CSIRO as a governmental research lab

Table 1. Comparative overview of initial FRAND cases

Plaintive	Defendants	Trial	Standard	No. of SEPs	Value Base*	FRAND Rate
Microsoft[67]	Motorola Mobility (Google)	Bench	H.264 802.11	16 11	MPEG-LA patent pool Mixed base	\$0.00555 \$0.03471
Ericsson	D-Link, Netgear. Belkin, Dell, HP, Acer, Toshiba, Intel	Jury	802.11	3	Previous industry licenses with 3rd parties	\$0.15[68]
Innovatio	Cisco, Motorola Solutions, SonicWALL, Netgear, HP	Bench	802.11	19	Chipset profits	\$0.0956
CSIRO	Cisco Systems	Bench	802.11	1	End products	\$0.83[69]

* The term value base is used to denote the source from which the FRAND royalty was calculated

[a]Based on an average of Via patent pool rates, prior non-FRAND 3rd-party component-level license agreement, and a prior consultancy valuation report

Source: Official trial documents

Figure 3. Plot of actors based on value logic[71]

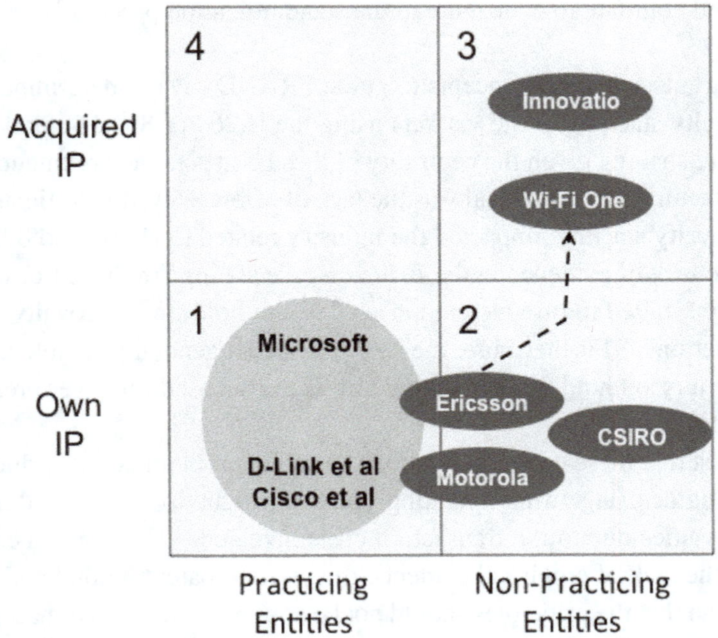

is a pure NPE that focuses on R&D and uses licensing as a primary means for technology transfer in a similar fashion as universities operate. Innovatio is a classic example of a patent assertion entity that has acquired an 802.11 SEP portfolio from a WiFi chip manufacturer with the goal of monetizing the portfolio against users of the infringing products.[72] Wi-Fi One is an offensive aggregation firm that has acquired a portfolio of approximately 200 patents, including the three patents in dispute in the *Ericsson* case. In the context of the H.264 and 802.11 standards, Microsoft, as well as the other computer, router, and chipset manufacturers, is focused primarily on achieving freedom-to-operate to sell their physical and virtual products.[73] Though all four SEP holders seeking royalties in the cases employ different business models, they all seek to license the end product manufacturer for either strategic or revenue purposes or both.

Application of FRAND Valuation Principles

Three of the four cases presented testimony on the concepts of patent holdup and royalty stacking in relation to FRAND commitments and the development of standards. However, only *Microsoft* and *Innovatio* elaborated on the use of the concepts in their overall methodology to determine the FRAND rate.[74] Both *Microsoft* and *Innovatio*, which applied the FRAND valuation guidelines developed in *Microsoft*, used the following three main concepts to evaluate proper methods and determine FRAND rates:

1. FRAND royalties should represent the value of the patented technology separate from the value of its inclusion in the standard, where the later is considered the holdup value.[75]
2. FRAND royalties should apportion value to the patented technology in relation to all other FRAND royalty obligations facing the licensee for other SEP holders on the standard.[76]

3. FRAND royalties should provide a return to the SEP holders that incentivizes them to develop new technologies and continue to contribute to the standardization process.[77]

Despite the importance of these concepts to guide FRAND royalty determinations, no evidence of patent holdup or royalty stacking in the markets using the H.264 or 802.11 standards was presented at any of the trials. More striking given the centrality of these issues to the argumentation of the valuation models of the manufacturing firms at trial was the lack of effort to even investigate at all to what extent patent holdup and royalty stacking impacted the industry related to H.264 and 802.11 standards.[78] The CAFC confirmed the lack of evidence in the *Ericsson* case stating "in this case, we agree with the district court that D- Link failed to provide evidence of patent hold-up and royalty stacking sufficient to warrant a jury instruction".[79] Furthermore, the CAFC issued a general principle that "the district court need not instruct the jury on hold-up or stacking unless the accused infringer presents actual evidence of hold-up or stacking".[80]

This would suggest that these concepts are not a systemic problem possibly due to the fact that most SEP holders are manufacturing firms operating under a production logic or that many SEP holders simple don't seek royalties due to the transaction costs involved.[81] However, according to the District Court in *Microsoft*, the lack of empirical evidence of systemic patent holdup and royalty stacking was not interpreted to mean that the principles should not be applied in the determination of FRAND royalty rates for a given SEP portfolio in the context of a given standard.[82]

The *CSIRO* court did not apply a specific FRAND methodology as the SEP in suit was not judged to be under a FRAND commitment for 802.11 revisions beyond 802.11(a). This was based on the refusal of CSIRO to sign the IEEE letter of assurance for later revisions, and the implicit refusal of industry actors to accept CSIRO's voluntary offer of FRAND licensing in the years prior to litigation. The CAFC judged that the district court erred in not taking into consideration the standard essential status of the CSIRO patent in its damages determination.[83]

Ex Ante Evaluation

Table 2 summarizes the finding of the techno-economic comparison among the four cases from an ex ante perspective in the sections below.

Both the *Microsoft* and *Innovatio* courts agreed that a FRAND royalty rate should take into account the importance of the SEPs to the performance of the standard in relation to alternative technologies that could have been adopted by the SSO as a means of determining the proper value of an SEP portfolio (i.e.

Table 2. Comparison of ex ante valuation results

Plaintive	Standard	Ex Ante Value
Microsoft	H.264 802.11	Low Low
Ericsson	802.11	N/A
Innovatio	802.11	Mod-High
CSIRO	802.11	High

ex ante valuation).[84] Thus the use of ex ante evaluations of alternative technologies was accepted as an appropriate model for apportioning value to SEPs separate from the added value they receive from being adopted in the standard as discussed in principle 1 in section 4.2 above. *Innovatio*, however, limited the definition of an alternative to include only those technical contributions put before the SSO. Furthermore, the court limited the reduction of value for alternatives that are not in the public domain (i.e. other patent contributions), reasoning that reducing the relative value of patented contributions to zero would disincentivize participation in the standardization process as put forward in principle 3 above.[85]

In *Microsoft*, the 802.11 SEP portfolio, for example, was deemed of low significance, while the 802.11 SEP portfolio in *Innovatio* was deemed of moderate to high significance based on the results of the respective ex ante evaluations[86]. Thereafter, the qualitative results of the ex ante valuation were quantified in *Microsoft* through the decision to use the Via patent pool rates and an application of a very low (i.e. 1%) royalty stacking adjustment factor[87]. In *Innovatio*, the moderate to high evaluation was modeled by placing the SEP portfolio in the upper 10% category of patents, which account for 84% of the value of electronics patents according to Schankerman (1998).[88] Thus the qualitative ex ante evaluation had a significant impact on the final FRAND royalty determination in both cases.

In *CSIRO* the court made an ex ante evaluation through the application of Georgia-Pacific factors 9-10 and found that the CSIRO SEP warranted an upward adjustment based on the prior art and due to the fact that the technology was included in several revisions of the standard even without a letter of assurance from CSIRO.[89]

In *Ericsson*, a market method based on prior comparable license information was used for the royalty determination and thus an ex ante evaluation was theoretically unneeded as market prices should already contain this information intrinsically in the price.

Royalty Base and FRAND Valuation Methods

Table 3 summarizes the findings of the royalty base and valuation method comparison among the four cases categorized by business model/value logic as well as the resulting FRAND rate in relation to the original offer. Further analysis of the royalty base determinations and valuation logics is presented in the following sections below.

Table 3. Comparison of valuation methods presented and accepted by the different courts

			Production Logic			Technology Logic		FRAND Rate	
Case	**Standard**	**Royalty Base**	**Patent Pool**	**Comp License**	**Profit Margin**	**Product License**	**Feature Factor**	**SEP Holder**	**Trial Result**
Microsoft	H.264 802.11	Product Mixed	IMP IMP	- IMP	- -	SEP SEP	- IMP	$4.50 $4.50	$0.00555 $0.03471
Ericsson	802.11	Product	-	IMP	-	SEP	SEP	$0.15	$0.15
Innovatio	802.11	Component	IMP	IMP	IMP	SEP	SEP	$4.72	$0.0956
CSIRO	802.11	Product		IMP	SEP	COURT90		$1.5491	$0.8392

IMP = Implementing firm, SEP = SEP licensor – this signifies who put forward the valuation method.[93] **Bolded** entries signify those valuation methods accepted by the court

Royalty Base Determination

The fundamental concept of a royalty base was used in different contexts across the four initial cases.[94] In all cases, the royalty base was determined on the credibility of the valuation evidence presented at trial, which resulted in different determinations of the royalty base in each case as opposed to a specific principle defining the appropriate level to set a FRAND royalty base.[95]

In *Microsoft*, regarding the H.264 standard, the use of patent pool rates implies a product-level royalty base as the pool seeks licenses with end product manufacturers though the product price is not explicit in the determination of the FRAND rate.[96] However, the *Microsoft* court applied a mixed set of royalty bases in its determination of 802.11 royalties by using both patent pool rates and a feature factor method that implicates the product level and a chip license that implicates the component level. In *Ericsson*, the royalty base was implicitly linked to the product level through the use of prior product licenses as comparables though the jury also heard testimony suggesting the component level as the appropriate royalty base. In both *Ericsson* and *CSIRO* the end product price was not explicitly used in the FRAND royalty calculation as the prior agreements intrinsically provided an apportionment of the market value of the SEPs.[97] Only in the *Innovatio* case was the royalty base an explicit determination by the court from which a royalty rate was calculated. However, the choice of the component level in this case could be seen as default decision based on Innovatio's inability to provide credible testimony for a product level royalty base.[98]

This shows a relative openness by the courts to listen and apply different valuation theories in the FRAND context. The CAFC reiterated "apportionment" as the governing rule stating that "as with all patents, the royalty rate for SEPs must be apportioned to the value of the patented invention", and reasoning that "logically, an economist could do this in various ways".[99]

FRAND Valuation Logics

Due to the implicit and ambiguous use of the concept of the royalty base discussed above, this section will analyze the FRAND royalty determination from the perspective of the value logics behind the valuation methods. Table 3 identifies the different valuation methods addressed in the initial cases categorized by the production and technology logics introduced in section 2.1. These different methods for the determination of FRAND royalty rates are briefly discussed below:

Patent Pool

The use of patent pool rates as the value base is a market method based on a production value logic as the goal of the patent pool is to indirectly facilitate the creation of a market for physical products not capture returns from the underlying technology directly[100]. It was presented in *Microsoft* and *Innovatio* but only accepted by the court in *Microsoft*.

Component License

The use of prior component license agreements as the value base is a market method based on the production value logic where the value of technology is found within the price of the physical component in the value chain as opposed to the value to end users. This was presented in all four cases but only accepted in *Microsoft* in relation to the 802.11 SEP portfolio.

Profit Margin/Premium

The use of profit margins/premium is an income method for apportioning the value of technology in a product/component based on the profit that it generates on the market. This method applies a production logic because it limits the apportionment of value to technology to the residual difference between the price and the production cost as opposed to user value[101]. It was presented in both the *CSIRO* and *Innovatio* cases but accepted only in the *Innovatio* case, where it was applied to a component level royalty base.

Product License

The use of prior product license agreements as the value base is a market method based on the technology value logic where the market value of the technology is determined based on arms-length negotiations between technology owners and product manufacturers, where the value of the technology is based on the value proposition of the technology as determined by the licensee employing the technology. Prior product license agreements were presented in all four cases but were only accepted in the *Ericsson* case based on actual prior agreement and in *CSIRO* based on their historical Voluntary Licensing Program.

Feature Factor

The feature factor method is a means of apportioning the user value of the end product to the technology (i.e. a technology value logic). The determination of the portion of the end product's value that is attributable to the technology is an income method that uses the end product price as the royalty base. It was presently explicitly at trial in *Microsoft* and *Innovatio*, where it was accepted by the former and rejected by the later. *Ericsson* applied the principle implicitly in their apportionment logic regarding computers and routers.[102]

Theoretically, valuations based on a production logic will be lower than a technology logic as implementing firms use standards to develop markets where they can sell their products while technology firms look for a return on investment for the technology in the standard itself. However, the cases show mixed results, which are discussed briefly below.

- For the majority of cases, practicing entities introduced valuation models with a production logic and non-practicing entities put forward models with a technology logic as expected.[103]
- When prior product licenses (i.e. a technology logic) were applied by the court in *Ericsson* and *CSIRO*, they produced the highest of the royalty rates determined by the four courts as expected.[104]
- All four cases made royalty determinations based on different combinations of royalty base and valuation methods even in regard to the same 802.11 standard[105]. Furthermore, while *Innovatio* followed the FRAND valuation guidelines in *Microsoft*, it rejected the use of the Via 802.11 patent pool and the ARM component license applied in *Microsoft* as not indicative of a FRAND rate. This indicates that there is still a fundamental ambiguity in relation to which royalty base and valuation methods are applicable to the FRAND context.

- The royalty proposals of the four SEP holders differed greatly even though they all put forward similar valuation methods and royalty bases. In particular, the rate put forward by the SEP holder in *Ericsson* is approximately an order of magnitude less than that in *Microsoft* and *Innovatio* and is capped in regard to the price of the end product[106]. This suggests a different licensing strategy and set of expectations on the part of the different SEP holders possibly based on historical norms or their different business models shown in figure 3.
- A comparison of the results of the feature factor method in *Microsoft* 802.11 and the profit margin method in *Innovatio* show that the particular assumptions used in the calculations of the methods can overcome the hypothetical norms associated with royalty base and value logic[107]. In this case, the specific feature factor method used in Microsoft produced the lowest FRAND rate of all accepted methods in all cases, producing a FRAND rate of $0.008 even though it employed a technology logic and product-level royalty base. Conversely, the profit margin method in *Innovatio* generated a rate of $0.0956, which is over one order of magnitude greater even though it employed a production logic and a component-level royalty base. Thus it is important to compare methods at the level of assumptions and implementation to be able to obtain a detailed comparison[108].

KEY FINDINGS AND IMPLICATION ON ECONOMIC PERFORMANCE AND EFFICIENCY

Below is a discussion of four key conceptual areas supported by the evidence of the four initial cases that together form a framework for further research regarding the value of SEPs, in particular, and an enhanced theory of patent value in the knowledge economy in general.

Transition From a Production Logic to a Technology Logic

The fact that both practicing and non-practicing entities are employing a technology logic to extract value from their patent portfolios, including SEP portfolios, has a normative impact on patent value and business strategy. In particular, a growing division of innovative labor in the telecommunication industry is generating new business models that engage in a more dynamic use of patent portfolios through market transactions.[109] Therefore it is not surprising that the transition to the greater use of IP and knowledge-based business models has created new types of disputes, such as the FRAND cases investigated in this paper, as firms look to maximize their economic performance in different parts of the same value chain.

The four SEP holders under study exemplified the increasing use of SEPs to generate revenue beyond the traditional production of physical products by both practicing and non-practicing entities. Specifically, the four cases illustrated a diverse set of contexts and business models as described in section 2.1, including a mixed use of value logics as follows:

1. Practicing entities who deploy the standard in their own products but also look to monetize their superior SEP portfolio over other producing firms (e.g. Motorola and Ericsson).
2. Non-practicing entities that acquire patents from other market actors (including practicing entities) with the sole purpose of patent assertion to monetize the assets (e.g. Innovatio).
3. Non-practicing entities whose main function is only to perform R&D and rely completely on license-based business models to transfer their technology to the market (e.g. CSIRO)

Given that Innovatio acquired its SEP portfolio from Broadcom, a practicing entity operating primarily under a production logic, the line between firms employing a production or technology value logic is becoming increasingly difficult to draw clearly. The more firms that look to leverage their SEP portfolios beyond traditional product manufacturing, the more acceptable knowledge-based business models will become, which, in turn, will impact the value of SEPs. Thus, the results of the *Microsoft* and *Innovatio* rulings lower the expected value of SEPs, while the *Ericsson* and *CSIRO* results raise the expected value and incentivize greater division of innovative labor and new business models to leverage intellectual property. Further empirical and theoretical research on the impact of the new division of innovative labor on the industrial value chain is required to better ascertain how different knowledge-based modes of firm action affect economic performance and efficiency.

The Context of Standards as a Determinant of SEP Value

In addition to the emergence of new roles and value logics deployed by firms discussed above, there is evidence that the contextual nature of the standard itself may have the greatest impact on the value of SEPs. The in-depth valuation efforts undertaken by the experts and courts regarding the H.264 and 802.11 standards revealed a surprisingly negligible amount of existing, credible FRAND licensing agreements for the courts to draw upon as comparable market transactions.[110] Given that 802.11 was first launched in 1997 and H.264 in 2003, the lack of applicable agreements suggests that these standards were primarily developed under a relatively pre-competitive, production logic, where the main goal was to facilitate the creation of large, global markets as opposed to generating revenue from the technology in the standard. In other words, the standards were designed as means, not as ends in themselves.[111]

Thus the profile of how a standard was developed, in particular, the primary value logic of the stakeholder firms, can be observed by the licensing and litigation history involving the standard. Widespread infringement would suggest a technology logic while extensive cross-licensing or the formation of a successful patent pool would suggest a production logic. Both H.264 and 802.11 seem to have been formed under a production logic, which makes the technology logic difficult to demonstrate (i.e. only Ericsson and CSIRO were able to show applicable prior license agreements).[112] However, other standards, such as cellular, which have a well-defined technology logic culture based on both practicing and non-practicing entities with strong technology and SEP portfolios, may likely find more success in the courts employing valuation models with a technology logic as a long history of applicable prior license agreement will be readily available. There is a need for further research on the techno-economic typology of different standards to confirm this proposition, however, this would imply that one-size fits all business norms, valuation methods, and policy measures will likely not be economically efficient.

Conflicting Norms on Appropriate Royalty Base and Valuation Models

The changing landscape of IP strategies and business models was also apparent in the different value logics that underpinned the choice of valuation methods put forward by the different actors across the four cases. The Ericsson and CSIRO 802.11 SEPs were given the highest valuation of the four portfolios valued by the four courts. One common denominator in these cases was the ability to successfully show proof of the market value of its portfolio through existing licensing agreements and licensing programs. Many valuation methods accepted by the courts were based on valuation methods using a production logic based on a modified version of case law dating back to 1970.[113] Furthermore, *Microsoft* and *In-*

novatio disagreed on the royalty base and the use of several identical market comparable methods put forward to both courts. Evidence from the US courts recent adjudication of FRAND royalty disputes suggests the courts are accepting of various damages theories as long as they comport with the general rule of apportionment, limiting the principles of the smallest saleable patent-practicing unit and the entire market value as secondary considerations based on the specific context of the case.[114] While the damage awards by the district courts in both *Ericsson* and *CSIRO* were vacated and remanded, the CAFC has reiterated its support for the use of prior license agreements as a basis for the determination of SEP damages given that the rule of apportionment is applied in the analysis.[115] Recent district court decisions in *St. Lawrence* and *Core Wireless have* confirmed that the use of adjusted prior license agreements is admissible.[116] This could be seen as contradicting the decisions by the district courts in *Microsoft* and *Innovatio* that found prior agreements as inadmissible and strengthening Georgia Pacific Factors 1-2 that emphasize the probative nature of prior license agreements.

While the US courts have a long history of determining reasonable royalties, it could be argued that these determinations have been primarily made within a production logic (Heiden and Andreasson, 2016). For instance, legal norms for setting the royalty base, such as the rule implicating the smallest saleable patent-practicing unit (SSPPU) or entire market value rule (EMVR), are designed from the perspective of an industrial value chain and may require further adaption to the new roles of IP and knowledge-based business models that often license different parts of the value chain. Therefore, it is not surprising, that in a changing paradigm from a production to a technology logic, the current business norms are mismatched with traditional legal norms, which leads to argumentation aligned to the existing legal norms instead of the current business reality.

For example, the wireless functionality delivered by the 802.11 standard is fundamentally important to many modern devices and its value to the customer is by definition greater than the production cost of a WiFi chip. However, it will be very difficult for any subset of patents, no matter how important to the 802.11 standard, to claim the total demand of a WiFi-enabled device, such as a computer, that incorporates the wireless technology.[117] In this regard evidence from the court cases suggests that market-based methods will likely be more successful for firms employing a technology logic in their business model than income based methods that are forced to calculate a specific royalty.[118] Thus, any apportionment of value to technology outside of the traditional material value chain will first need to be proven through market transactions that reify the value as part of the business norms.[119] Therefore, the cases show that SEP holders will likely need to show evidence of the market value of their SEPs through arms-length market transactions to prove the value through a technology logic.[120] More research is needed to explain how the current legal norms address knowledge-based business models and its subsequent impact on economic performance and efficiency.

The selection of different royalty bases and valuation methods in the cases studied suggests that the courts have shown an openness to a wide variety of arguments regarding the apportionment of value In the face of conflicting legal and business norms, while still upholding a relatively high threshold of credibility to allow a valuation method into evidence. One could argue that ambiguity is bad as it creates uncertainty and confusion in the market, however, given the historical record of success of FRAND-enabled standards, this ambiguity could have provided the normative space required for incomplete contracts, such as FRAND, to facilitate complex transactions among divergent actors and business models, such as those found in the telecommunication value chain. More research is required to better understand the systemic effect both from a historical perspective and in relation to the current and emerging context of technology convergence and connectivity (e.g. in relation to IoT).

Relevance of Patent Holdup and Royalty Stacking

Given the theoretical importance of the concepts of patent holdup and royalty stacking to the determination of SEP value, the dearth of evidence put forward by the market actors in the four initial cases suggests that they are not a significant empirical market issue in the context of the H.264 and 802.11 standards. In particular, the lack of evidence in the *Ericsson* case was echoed by the CAFC in its appellate ruling as grounds for not instructing a jury to consider patent holdup or royalty stacking unless sufficient evidence has been put forward "in relation to both the RAND commitment at issue and the specific technology referenced therein."[121] This further implies that FRAND contracts in their incomplete, historical interpretation have been successful in regulating patent holdup and royalty stacking issues in technology transactions among market actors at least in these standards. One proposition suggested in section 5.2 is that both standards were developed and implemented primarily by practicing entities applying a production logic, which implies that most SEPs should be exhausted within the components and products of the various actors. The simple fact that Broadcom was able to sell a portion of its SEP portfolio to a 3rd-party (i.e. Innovatio) indicates that not all SEPs are cross-licensed (i.e. exhausted) among WiFi chipset manufacturers at the component level. This creates opportunities for offensive aggregation and privateering in the 802.11 standard towards end producers and users as witnessed in *Innovatio*. However, the strong rhetoric about ill-effects of patent holdup and royalty stacking generated by practicing entities will need to be reexamined If it is found that those same practicing entities are providing non-practicing entities with SEPs to hold-up other actors in the value chain.[122] The practical nature of licensing in technology standards, such as 802.11, requires further investigation to better understand the industry norms in order to ascertain the potential impact of business models with a technology logic on value chains with a predominant production logic.

The lack of systemic holdup and stacking doesn't mean that a theoretical royalty stacking factor should not be employed in calculating a FRAND rate as a means to properly apportion the SEPs in question in relation to the overall set of technical contributions that define the standard. The challenge will be to determine the FRAND rate when some actors deploy technology through physical products and some through licensing in the same value chain to the same customer. However, one important insight stemming from the lack of evidence of systemic patent holdup and royalty stacking in the studied cases is that policy measures designed to eliminate these problems are possibly unnecessary and could potentially alter the balance, creating patent holdout that could reduce the quality of standards and reduce overall social welfare (Heiden and Petit, 2017). The recent district court decision in *Core Wireless* where enhanced damages were added to the judgment based on "holdout" behavior by the SEP infringer is an initial indicator.[123] More research is required that models the impact of policy interventions from both a dynamic and static efficiency perspective.

CONCLUSION

The pendulum of patent strength has swung back and forth over the long history of the patent system. Where once weak patents were seen as a disincentive to invest in innovation and needed strengthening to stop free riding, now patents are considered by some as holding-up innovation and restricting the development of markets and therefore should be weakened.[124] For instance, Shapiro (2001) states that it is fair to ask whether the pendulum has swung too far in the direction of strong patent rights. The challenge, however, is to answer that question empirically and implement policy changes that maximize economic efficiency instead of simply swing the pendulum too far in the other direction.

The landmark *Microsoft* ruling together with the *Ericsson, Innovatio, and CSIRO* cases provide a good indication of the current normative space for interpretation of the value of SEPs. While these verdicts may be seen as contradicting the proposition that the US court system is unable to adjudicate reasonable royalties in the context of multi-component products, the question remains whether the methodologies used in these court decisions provide proper guidance to business actors and subsequent judicial proceedings to create a new set of sustainable norms governing the use of FRAND agreements in practice. The appeals to the *Microsoft, Ericsson, and CSIRO* decisions and supporting amicus briefs by major market actors indicate that stakeholders from industry and civil society disagree on whether the courts have gone too far or not far enough.

This paper contradicts what has become a mainstream belief echoed by Rysman and Simcoe (2011) that FRAND commitments are not a workable solution to SSOs' intellectual property problem in that they fail to clearly define a reasonable price. On the contrary, given the historical success of FRAND-enabled standards, such as H.264 and 802.11, it can be argued that incomplete contracts are designed to manage situations that are inherently complex and ambiguous and that attempts to reduce the complexity down to an artificially complete set of parameters may only create greater problems. Thus the US court system has shown an initial capacity to handle SEP disputes without the need for sweeping policy and legislative reform.

This paper also addressed the relevance of the transition from a production to a technology logic fueled by a new division of innovative labor and showed how these logics were used by the market actors in the cases to both capture value from the market and communicate the value of their SEPs through different valuation methods. Practicing and non-practicing entities were shown to employ different value logics as expected, but the results also showed a mixed use of value logics depending on the commercial nature of the actor.

In addition to the increased use of a technology logic by market actors as a determinant of SEP value, this paper also put forward a proposition regarding the path dependency of the commercial context of the standard itself. In other words, the fact that a standard was developed within a production logic (i.e. primarily by implementing firms) may have a strong normative impact on the value of SEPs as actors looking to use standards to build product markets don't value SEPs in the same way as upstream R&D firms with strong IP positions. The evidence of few comparable FRAND licensing agreements from a technology logic in 802.11 and a very successful patent pool in H.264 support this proposition, but more research is needed.

Regarding the ability to manage conflicting legal and business norms, the evidence suggests that the US courts have shown a strong openness to different valuation theories, which have resulted in the determination of a diverse set of royalty bases and valuation methods. While the results from the four cases suggest that a set of valuation principles regarding (1) patent holdup, (2) royalty stacking, and (3) incentivizing patent holders to participate in standards have become generally accepted as a means to govern FRAND royalty determinations, the CAFC has pushed back on the use of patent holdup and royalty stacking theory for jury instructions when insufficient empirical evidence is provided in court.

This re-emphasizes the need to differentiate theoretical from empirical patent holdup and royalty stacking for courts, but more importantly, for policy makers. In particular, the lack of evidence of systemic patent holdup and royalty stacking should provide caution to policy makers when considering new rules and regulations as potential solutions could enhance static efficiency at the expense of dynamic efficiency, which could reduce aggregate social welfare.

REFERENCES

Bekkers, R., & Updegrove, A. (2012). *A study of IPR policies and practices of a representative group of Standards Setting Organizations worldwide, US National Academies of Science, Board of Science, Technology, and Economic Policy*. STEP.

BCG. (2015). *The Mobile Revolution – How Mobile Technologies Drive a Trillion Dollar Impact*. The Boston Consultancy Group.

Bekkers, R., Verspagen, B., & Smits, J. (2002). Intellectual property rights and standardization: The case of GSM. *Telecommunications Policy*, *26*(3-4), 171–188. doi:10.1016/S0308-5961(02)00007-1

Berman, B. (2013). *Public IP companies (PIPCOs) – a business model whose time has come*. Intellectual Asset Management Magazine, March/April.

Bessen, J., & Meurer, M. (2009). *Patent Failure: How Judges, Bureaucrats, and Lawyers Put Innovators at Risk*. Princeton University Press. doi:10.1515/9781400828692

Blind, K. (2004). The economics of standards. Theory, evidence, policy. Cheltenham, UK: Academic Press.

Boldrin, M., & Levine, D. (2013, Winter). The Case Against Patents. *The Journal of Economic Perspectives*, *27*(1), 3–22. doi:10.1257/jep.27.1.3

Contreras, J. L. (2013). Fixing FRAND: A Pseudo-Pool Approach to Standards-Based Patent Licensing, *79. Antitrust Law Journal*, 47.

Contreras, J. L. (2015). *CSIRO v. Cisco: The Convergence of RAND and Non-RAND Royalties for Standards-Essential Patents*. Available at: https://papers.ssrn.com/sol3/papers.cfm?abstract_id=2749653

Contreras, J. L., & Gilbert, R. J. (2015). (Forthcoming). A Unified Framework for RAND and Other Reasonable Royalties. *Berkeley Technology Law Journal*.

Denicolo, V., Geradin, D., Layne-Farrar, A., & Padilla, J. (2008). Revisiting Injunctive Relief – Interpreting eBay in High-Tech Industries with Non-Practicing Patent Holders. *Journal of Competition Law & Economics*, *4*(3), 571–608. doi:10.1093/joclec/nhn028

Dorsey, E., & McGuire, M. R. (2013). How the Google Consent Order Alters the Process and Outcomes of FRAND Bargaining. *George Mason Law Review*, *20*, 979.

Elhauge, E. (2008). Do Patent Holdup and Royalty Stacking Lead to Systematically Excessive Royalties? *Journal of Competition Law & Economics*, *4*(3), 3. doi:10.1093/joclec/nhn027

European Commission. (2014). *Patents and Standards: A modern framework for IPR-based standardization*. European Commission – Directorate-General for Enterprise and Industry.

Ewing, T. (2012). Indirect Exploitation of Intellectual Property Rights by Corporations and Investors: IP Privateering and Modern Letters of Marque and Reprisal. *Hastings Science and Technology Law Journal, 4*.

Farrell, J., Hayes, J., Shapiro, C., & Sullivan, T. (2007). Standard Setting, Patents, and Hold-up. *Antitrust Law Journal*, *74*, 3.

Galetovic, A., & Gupta, K. (2016). *Royalty Stacking and Standard Essential Patents: Theory and Evidence from the World Mobile Wireless Industry*. Academic Press.

Galetovic, A., & Haber, S. (2017). The Fallacies of Patent-Holdup Theory. *Journal of Competition Law & Economics*, *13*(1), 1–44. doi:10.1093/joclec/nhx006

Galetovic, A., Haber, S., & Levine, R. (2015). *An Empirical Examination of Patent Hold-Up*. Nat'l Bureau of Econ. Research, Working Paper No. 21090.

Galetovic, A., Haber, S., & Zaretzki, L. (2016). *A New Dataset on Mobile Phone Patent License Royalties: September 2016 Update*. Hoover IP2 Working Paper 16011.

Geradin, D., & Rato, M. (2007). Can Standard-Setting Lead to Exploitative Abuse: A Dissonant View on Patent Hold-Up, Royalty Stacking and the Meaning of Frand. *European Competition Journal, 3*.

Golden, J. (2013). Patent Privateers: Private Enforcement's Historical Survivors. *Harvard Journal of Law & Technology*, 26.

Granstrand, O. (1999). *The Economics and Management of Intellectual Property*. Edward Elgar.

Grindley, P., & Teece, D. (1997, Winter). Managing Intellectual Capital: Licensing and Cross-Licensing in Semiconductors and Electronics. *California Management Review*, *39*(2), 2. doi:10.2307/41165885

Heiden, B. (2016). The viability of FRAND: How the seminal landmark Microsoft ruling could impact the value of standard essential patents and the future of telecom standards. *Telecommunications Policy*, *40*(9), 870–887. doi:10.1016/j.telpol.2016.03.004

Heiden, B., & Andreasson, J. (2016). Reevaluating Patent Damages in the Knowledge Economy: The Intellectual Value Chain and the Royalty Base for Standard-Essential Patents. *Criterion J. on Innovation*, *1*, 229.

Heiden, B., & Petit, N. (2017). *Patent Trespass and the Royalty Gap: Exploring the Nature and Impact of Patent Holdout*. Available at https://papers.ssrn.com/sol3/papers.cfm?abstract_id=2981577

Iversen, E. (1999) Standardization and Intellectual Property Rights: ETSI's controversial search for new IPR-procedures. *IEEE Conference on Standardisation and Innovation in Information Technology*, Aachen, Germany.

Jaffe, A., & Lerner, J. (2007). *Innovation and Its Discontents: How Our Broken Patent System is Endangering Innovation and Progress, and What to Do About It*. Princeton University Press.

Katz, M. L., & Shapiro, C. (1985). Network Externalities, Competition, and Compatibility. *The American Economic Review*, *75*(3), 424–440.

Klein, B., Crawford, R. G., & Alchian, A. A. (1978). Vertical Integration, Appropriable Rents, and Competitive Contracting Processes. *The Journal of Law & Economics*, *21*(2), 297–326. doi:10.1086/466922

Krechmer, K. (2006). Open Standards Requirements. *International Journal of IT Standards and Standardization Research*, *4*(1), 1. doi:10.4018/jitsr.2006010103

Layne-Farrar, A., Padilla, A. J., & Schmalensee, R. (2007). Pricing Patents for Licensing in Standard-Setting Organizations: Making Sense of FRAND Commitments. *Antitrust Law Journal*, *74*, 3.

Layne-Farrar, A. (2014). *Patent Holdup and Royalty Stacking Theory and Evidence: Where do we Stand After 15 Years of History*. Available at http://www.oecd.org/officialdocuments/publicdisplaydocument pdf/?cote=DAF/COMP/WD%282014%29%2084&doclanguage=en

Layne-Farrar, A., & Wong-Ervin, K. W. (2015). An Analysis of the Federal Circuit's Decision in Ericsson v. D-Link. *CPI Antitrust Chronicle*, (1).

Lemley, M. A. (2002). Intellectual Property Rights and Standard-setting organizations. *California Law Review*, *90*(6), 1889–1990. doi:10.2307/3481437

Lemley, M. A. (2007). Ten Things to Do about Patent Holdup of Standards (and One Not To). *Boston College Law Review. Boston College. Law School*, 48.

Lemley, M. A., & Shapiro, C. (2007). Patent Holdup and Royalty Stacking. *Texas Law Review*, *85*, 1991–2050.

Lemley, M. A., & Shapiro, C. (2013). A Simple Approach to Setting Reasonable Royalties for Standard-Essential Patents. *Berkeley Tech. L.J.*, *28*.

Lemley, M., & Melamed, D. (2013). Missing the Forest for the Trolls. *Columbia Law Review*, *113*, 8.

Lerner, J., & Tirole, J. (2015). Standard-essential patents. *Journal of Political Economy*, *123*(3), 547–586. doi:10.1086/680995

Mallinson, K. (2015). Cumulative mobile-SEP royalty payments no more than around 5% of mobile handset revenues. *IP Finance*.

Maurer, N., & Haber, S. (2017). *Patent Trolls or Patent Elves? Evidence from Publicly-Traded Patent Assertion Entities*. Hoover IP2 Working Paper 17003.

Maskus, K., & Merrill, S. A. (Eds.). (2013). *Patent Challenges for Standard-Setting in the Global Economy: Lessons from Information and Communication Technology*. National Academy of Sciences.

Nazier, D. (2014). *Infamous Wi-Fi Patent Troll Settles For Peanuts*. Electronic Frontier Foundation. Retrieved from www.eff.org

Ostrom, E. (1990). *Governing the Commons – The Evolution of Institutions for Collective Action*. Cambridge University Press. doi:10.1017/CBO9780511807763

Padilla, J., Geradin, D., & Layne-Farrar, A. (2007). *Royalty Stacking In High Tech Industries: Separating Myth From Reality*. Working Paper, CEMFI.

Petrusson, U. (2005). *Intellectual Property & Entrepreneurship: Creating Wealth in an Intellectual Value Chain*. Gothenburg: Center for Intellectual Property.

Petrusson, U., & Heiden, B. (2009). *Assets, Property and Capital in a Globalized Intellectual Value Chain. In From Assets to Profits*. John Wiley & Sons.

Rysman, M., & Simcoe, T. (2011). A NAASTy Alternative to RAND Pricing Commitments. *Telecommunications Policy*, *35*(11), 11. doi:10.1016/j.telpol.2011.07.003

Schmalensee, R. (2009). Standard-Setting, Innovation Specialists and Competition Policy. *The Journal of Industrial Economics*, *57*(3), 526–552. doi:10.1111/j.1467-6451.2009.00388.x

Schankerman, M. (1998). How Valuable is Patent Protection? Estimates by Technology Field. *The Rand Journal of Economics*, *29*(1), 1. doi:10.2307/2555817

Scott Morton, F., & Shapiro, C. (2016). Patent Assertions: Are We Any Closer to Aligning Reward to Contribution? *Innovation Policy and the Economy*, *16*(1), 89–133. doi:10.1086/684987

Shapiro, C. (2001). Navigating the Patent Thicket: Cross Licenses, Patent Pools, and Standard-Setting. In A. B. Jaffe et al. (Eds.), *Innovation Policy and the Economy I* (pp. 119–150). Cambridge, MA: MIT Press. doi:10.2139/ssrn.273550

Shapiro, C., & Varian, H. (1999). The Art of Standards Wars. *California Management Review*, *41*(2), 2. doi:10.2307/41165984

Sidak, G. J. (2008). Holdup, Royalty Stacking, and the Presumption of Injunctive Relief for Patent Infringement: A Reply to Lemley and Shapiro. *Minnesota Law Journal, 92*.

Sidak, G. J. (2013). The Meaning of FRAND, Part I: Royalties. *Jnl of Competition Law & Economics, 9*(4).

Sidak, G. J. (2015). The Antitrust Division's Devaluation of Standard-Essential Patents. *104 Georgetown L.J. Online (Bergheim)*, *48*, 61.

Sidak, G. J. (2016). Apportionment, FRAND Royalites, and Comparable Licenses after Ericsson v. D-Link. *U. Ill. L. Rev.*, 1809.

Siebrasse, N., & Cotter, T. F. (2015). A New Framework for Determining Reasonable Royalties in Patent Litigation. *Florida Law Review*.

Swanson, D., & Baumol, W. (2005). Reasonable and Nondiscriminatiory (RAND) Royalties, Standards Selection, and Control of Market Power. *Antitrust Law Journal*, *73*, 1.

Wright, J. D., & Ginsburg, D. H. (2014). Patent Assertion Entities and Antitrust: A Competition Cure for a Litigation Disease. *79 ANTITRUST L.J. 501, 509.*

ENDNOTES

1 In 2014 the mobile value chain generated approximately $3.3 trillion in revenue globally (BCG, 2015)

2 Cisco predicts that 50 billion devices will be connected to the Internet by 2020 (http://share.cisco.com/internet-of-things.html).

3 See Blind (2004) for a thorough examination of the economics of standards.

4 The IEEE defines that an *Essential Patent Claim* shall mean any patent claim the use of which was necessary to create a compliant implementation of either mandatory or optional portions of the normative clauses of the [Proposed] IEEE Standard when, at the time of the [Proposed] IEEE Standard's approval, there was no commercially and technically feasible non-infringing alternative. (source: IEEE-SA Standards Board By-laws).

5 For this paper, FRAND is used as an equivalent to RAND in meaning. See Bekkers and Updegrove (2012) and Lemley (2002) for a thorough evaluation of IPR policies in SSOs.

6 Note that firms can be both technology owners and implementers with different degrees of power in the technology and product markets based on the strength of their patent portfolios and product offerings, respectively.

7 The 2014 EC Report on Patents and Standards defines patent hold-up as "a situation where implementers are locked into a certain technology by adopting a standard, so that a patent holder can negotiate a significant higher license fee (ex-post) than it could have asked for its technology before such a lock-in occurred (ex-ante)".

8 See Layne-Farrar, Padilla, and Schmalensee (2007) and Swanson and Baumol (2005) for economic modeling of technology valuation ex ante of its inclusion in the standard.

9 See Sidak (2016), Galetovic and Gupta (2016), Galetovic, Haber, and Zaretzki (2016), Galetovic, Haber and Levine (2015), Mallinson (2015), Layne-Farrar (2014), and Padilla, Geradin, and Layne-Farrar (2007). In addition, the recent studies by National Academy of Science (2013) and European Commission (2014) do not cite empirical evidence of a negative economic impact. This doesn't mean that specific cases don't exist only that evidence of systemic market failure through patent holdup is lacking.

10 See Lemley and Shapiro (2007), who provide a short case study section of empirical evidence of royalty stacking in the 3G and 802.11 standards that in actuality only provides evidence of patent counts and theoretical speculates on royalty stacking problems that have not been substantiated in recent history as WiFi and mobile devices markets have grown exponentially since 2007. See Scott-Morton and Shapiro (2016), who hypothesize on patent holdup problems in the emerging IoT area.
In contrast, see Ostrom (1990), who documents how theoretical problems over shared resources do not necessary lead to empirical problems.

11 Typically cited cases include Rambus Inc. v. Infineon Technologies AG, 318 F. 3d 1081 - Court of Appeals, Federal Circuit 2003 and Qualcomm Inc. v. Broadcom Corp., 2007 U.S. Dist. LEXIS 57136 (S.D. Cal. Aug. 6, 2007).

12 Apple was the most highly litigated firm by NPEs (totally 191) in the period 2009-2013. However, this must be put in perspective of the operating size of the company. For example, Apple suffered 42 NPE litigations in 2013 with a rough transaction cost estimate of $120 million using the AIPLA estimate of patent litigation costs of $2.8 million through final deposition. Compared to Apple's

revenue of $171 billion and net income of $37 billion in 2013, there is no reason for Apple to see this as more than a minor cost of doing technology-based business unless of course, these cases have real merit.

13 In fact, there is preliminary evidence that SEP holders are experiencing a "royalty gap", as recent regulatory, judicial, and IPR policy changes have created an incentive for SEP implementers to delay or avoid FRAND royalty payments – see Heiden and Petit (2017).

14 In fact the H.264 and 802.11 standards involved in the initial FRAND court cases addressed in this paper are examples of extremely successful standards with widespread adoption.

15 In 2010, over 90% of the world's population was covered by a mobile cellular network (http://www.itu.int/net/pressoffice/stats/2010/06/#.U8UPEf3JBuY).

16 The ITU reports a near ten-fold growth in mobile broadband from 268 million subscriptions in 2007 to 2,315 million in 2014.

17 These include *Ericsson v. D-Link Systems et al.*, *Innovatio IP Ventures*, and *CSIRO v. Cisco Systems*. Further district court rulings determining FRAND royalty rates have followed in 2014, including Realtek Semiconductor Corp. v. LSI Corp, where a jury awarded 0.19% of the Realtek chipset price for the two SEPs in case and $3.8 million to Realtek for LSI's breach of (FRAND) contract. Two additional cases, GPNE Corp. v. Apple Inc. and Golden Bridge Technology Inc. v. Apple Inc. concluded in findings of non-infringement.

18 See Layne-Farrar and Wong-Ervin (2014) for an overview of all the initial SEP/FRAND disputes in US district court.

19 Industry actors have been used primarily as sources of insight and clarification of the telecommunication context regarding market practice and historical events. These actors have included IP and licensing executives from firms operating across the different business models in the telecommunication value chain.

20 The distinction between practicing entities (PE) and non-practicing entities (NPEs) are often used in business and legal discourse, where the latter are defined as actors that do not manufacture or sell end products or services in relation to their IP position. The strategic implication of NPEs is that they are not vulnerable to counter-assertions as they have no infringing products or services. The distinction is important but the terms implementing and licensing entities have been used to avoid the pejorative nature of the PE/NPE labels. The four logics are depicted as distinct, but there is of course overlap and the potential for the same entity to operate in multiple logics.

21 See Grindley and Teece (1997).

22 This could also be seen as a platform battle between iOS and Android.

23 Ibid.

24 Ericsson and Qualcomm are integrated firms offering IP licensing as well as different product offerings. Qualcomm's licensing revenue was approximately 29% of total revenue in 2014, while Ericsson's licensing revenues only represented around 4%. Interdigital and Nokia Technologies, in telecom, and ARM and Rambus, in semiconductors, are examples of innovative specialists focused solely on IP/technology licensing.

25 See intellectualventures.com and acaciaresearch.com.

26 Patent privateering can occur for several reasons, including reduced exposure to counter-claims, increased licensing revenue, and increased royalty burden on competitors. See Lemley and Melamed (2013), Golden (2013), Ewing (2011).

27 While these firms are not practicing entities themselves their members are practicing entities, which is why they have been categorized as such.

28 Berman (2013) report 27 publicly traded intellectual property companies (PIPCOs) of diverse sizes on the US, UK, and Canadian stock exchanges.

29 PatentFreedom tracks over 750 NPEs and reports an increase in NPE litigation from 235 to 3134 patent law suits during the period 2004 to 2013 (www.patentfreedom.com). For critique of the patent system see Boldrin and Levine (2012), Jaffe & Lerner (2011), and Bessen & Meurer (2008).

30 See Maurer and Haber (2017) for an empirical study showing little evidence that patent assertion entities pose a substantial economic burden on operating companies.

31 See Bekkers and Updegrove (2012). Also, see Krechmer (2006) for a typology of ten requirements that enable open standards.

32 For example, the MPEG-LA patent pool for the ITU H.264/AVC standard consists of 30 licensors, over 2400 essential patents, and 1248 licensees. Source: www.mpegla.com (retrieved on July 2013).

33 Examples include Bluetooth (industry consortium), IEEE (professional association), and ITU-T (International NGO).

34 Standards can also be developed and governed by individual firms (e.g. Microsoft Windows) and competing standards can emerge leading to standards wars – see Shapiro and Varian (1999).

35 See, for example, the ITU/ISO/IEC Common Patent Policy.

36 See Katz and Shapiro (1985) regarding the effects of network externalities and Farrell and Klemperer (2003) for a discussion on coordination and lock-in.

37 See Bekkers et al. (2002) for an early example of the power of SEPs to affect the structure of the nascent GSM market in the late 1980s and early 1990s.

38 See Contreras (2015) for a brief history of FRAND.

39 See Iversen (1999) for a historical account of ETSI's struggle to develop an effective and acceptable IPR policy in its formative years from 1989-2004.

40 See Miller (2007) for a transaction cost perspective on FRAND as a governance mechanism that allocates an access right to the adopter community. See also Remarks of Commissioner Joshua D. Wright, Federal Trade Commission, *SSOs, FRAND, and Antitrust: Lessons from the Economics of Incomplete Contracts,* at 19–20 (Sept. 12, 2013) (Wright Remarks) available at http://www.ftc.gov/sites/default/files/documents/public_statements/ssos-frand-and-antitrust-lessons-economics-incomplete-contracts/130912cpip.pdf.

41 See 802.11 (IEEE). H.264 (ITU). LTE (3GPP) for examples of open, FRAND-enabled technology standards.

42 See Klein, Crawford, and Alchian (1978) for an early discussion of opportunism (and holdup) as a case of appropriable quasi-rents to contracted specific assets, where opportunism can take place in either direction (i.e. holdup or holdout). See Heiden and Petit (2017) for a discussion on the misuse of "holdup" and "holdout" in the patent context from the perspective of mainstream economic theory.

43 Here, systemic patent holdup refers to the widespread condition of holdup by SEP holders such that market formation is blocked or severely restricted (i.e. market failure). The term systemic is used to differentiate from contractual/transactional holdup, which is limited to specific actors and doesn't greatly impact economic efficiency. See Heiden and Petit (2017) for a discussion on different degrees of economic impact from transactional to systematic to systemic. See Reply Submission on the Public Interest of Federal Trade Commissioners Maureen K. Ohlhausen and Joshua D. Wright,

In the Matter of – Certain 3G Mobile Handsets and Components Thereof, International Trade Commission (Inv. No. 337-TA-613, 2015), Section 2 entitled, "Empirical Evidence Suggest no Systemic Problem With Holdup." See Wright and Ginsburg (2014) for a discussion on contractual opportunism and antitrust implications, and Balkenborg, Kaplan, and Miller (2010) differentiating holdup situations that merely redistribute income (i.e. transactional) from those that block investment and create inefficiencies (i.e. systemic). The concepts of ex post vs. ex ante opportunism is another way to view the difference between transactional and systemic holdup, where the latter has a much greater impact on investment and market formation.

44 For example, neither the 2014 EC report on Patents and Standards or the 2013 National Academy of Science report on Patent Challenges for Standard-Setting in the Global Economy – Lessons from Information and Communication Technology does not mention empirical evidence of systemic problem related to patent hold-up beyond an increase FRAND-related court cases. However, the NAS report specifically states the committee found "no empirical evidence showing that royalty stacking currently suppresses the adoption or use of standard-compliant products."

45 See Layne-Farrar and Wong-Ervin (2015) for several reasons for this. It should be noted that patent holdup cannot exist without injunctive relief as implementing firms, which have already received the standard specifications, can simply choose not to except an offer from an SEP licensor.

46 See Dorsey and McGuire (2013). which describes holdup as fundamentally a problem of rent appropriation; that is, it is an issue of determining – as between patent holders, licensees, and, of course, end-consumers – who gets which benefits and in what amounts.

47 See Geradin et al. (2008) for an early critique of patent holdup remedies.

48 *Microsoft Corp. v. Motorola, Inc.*, No. 10-cv-1823 (W.D. Wash.).

49 The ITU developed the standard together with ISO and IEC.

50 *Microsoft Corp. v. Motorola, Inc.*, 864 F. Supp. 2d 1023 (Dist. Court, WD Washington 2012).

51 Microsoft Corp. v. Motorola, Inc., 696 F. 3d 872 (Court of Appeals, 9th Circuit 2012)

52 See *Microsoft Corp. v. Motorola, Inc. et al*; Case No 14-35393 (9th Cir, July 30, 2015).

53 See https://www.ft.com/content/7ecab5c6-67be-11e5-97d0-1456a776a4f5

54 Ericsson Inc. et al. v. D-Link Corp. et al., No. 6:10-cv-00473, (E.D. Texas).

55 Ibid.

56 See *Ericsson v. D-Link Sys. Inc., et al.*, 773 F.3d (Fed. Cir. 2014). The logic is still deemed to be valid as it is not yet obvious if new jury instructions will impact a FRAND royalty determination that was based primarily on a market method that could be considered to include the FRAND criteria in the base.

57 In re Innovatio IP Ventures LLC, No. 11-cv-09308 (N.D. Illinois).

58 It is interesting to note that Broadcom testified at the trial to the problem of patent ambush in the industry, while they were the one who sold the SEP portfolio to Innovatio in the first place – a practice that is known as patent privateering.

59 See Nazer (2014) at https://www.eff.org/deeplinks/2014/02/infamous-wi-fi-patent-troll-settles-peanuts.

60 http://blogs.cisco.com/news/innovatio-case-victory-for-cisco-customers-makes-the-case-for-patent-reform.

61 Ibid

62 In re Innovatio IP Ventures, LLC Patent Litig., MDL 2303, 2013 WL 3874042 (N.D. Ill. July 26, 2013).

63 See supra note 60.

64 *CSIRO v. Cisco Systems*, 6:11-cv-343 (E.D. Texas, 2014).

65 It should be noted that the court determined that CSIRO's SEP in suit was not obligated to a FRAND commitment other than for the 802.11(a) standard, which was less than 1% of the scope of infringed products. Nonetheless, the valuation methods and reasoning of the court exposes the SEP logic and FRAND considerations and is thus a useful case for comparative purposes.

66 *CSIRO v. Cisco Systems*, 809 F.3d 1295 (Fed. Cir. 2015).

67 Note that Microsoft sued for breach of contract, so Motorola/Google is the actor seeking FRAND royalties for their SEP portfolio.

68 The FRAND rate has been vacated and remanded with instructions from CAFC.

69 The FRAND rate has been vacated and remanded with instructions from CAFC. The royalty rate, for the most part, is not explicitly based on a FRAND commitment due to the historical circumstances of CSIRO relations with the 802.11 standard and the wireless industry.

70 It must be noted that the actors could operate with different logics in different business areas. For example, Ericsson operates within box 1 in relation to patents regarding mobile infrastructure.

71 While Motorola and Ericsson are integrated firms with both products and SEPs, they both acted primarily as licensing entities in relation to the opposing firms in the specific court cases studied based on the relative strength of their SEP portfolios. Some of the implementing entities have small SEP portfolios as well, such as Microsoft in the H.264 standard. Additionally, Ericsson has assigned the SEPs in their case to Wi-Fi One, LLC, which is a firm operating in box 3 in the figure.

72 This is also an example of patent privateering where a manufacturing actor operating in box 1 sells its patents to an NPE to assert against others in the industry.

73 Note that the chip company, Broadcom, was the actor that sold Innovatio SEPs in what was described in section 2.1 as a patent privateering model, as opposed to a freedom-to-operate model. Contrast this with the fact that Broadcom filed for *inter partes* review of Ericsson's patents, which clearly shows the complexity of interests and incentives in the telecommunication value chain.

74 In *Ericsson*, testimony regarding patent holdup and royalty stacking was presented to the jury, which made its own interpretation of its impact on the determination of the final FRAND rate. The defendants in the case cited the lack of clear FRAND instructions to the jury in their post-trial motion, which was denied by the *Ericsson* court but accepted in their appeal by the Federal Circuit.

75 See *Microsoft*, comment 71,74 and *Innovatio*, pp. 14-17.

76 See *Microsoft*, comment 72 and *Innovatio*, pp. 17-19.

77 See *Microsoft*, comment 73 and *Innovatio*, pp. 19-21.

78 For example see *Innovatio*, p. 36, which states that the manufacturers' expert never identified an actual royalty stack; he never even attempted to determine the actual amount of royalties Defendants currently pay for 802.11 patents.

79 See *Ericsson* 773 F.3d, p. 54.

80 Ibid.

81 These hypotheses were put forward by interviewees that work specifically with licensing related to the WIFi standard, but they were not tested in this paper.

82 See *Microsoft*, comment 460 for a similar line of reasoning.

83 See *CSIRO* 809 F.3d.

84 See *Microsoft* comments 75-76 and *Innovatio*, pp. 36-40. This process was unnecessary in *Ericsson* due to the use of prior product license agreements as comparables as this apportionment is deemed inherent in market prices.

85 See *Innovatio*, pp. 36-40.

86 See *Microsoft*, comments 352-406 and *Innovatio*, pp. 40-58.

87 See *Microsoft*, comments 577, 595.

88 See *Innovatio*, pp.84-85. Note that Schankerman (1998) investigated patents (in general, not SEPs) applied in France from 1969-82 and simulated patent value from patent renewal information, which lacks value information for patents held to full term.

89 See *CSIRO* pp. 27-28.

90 Having rejected the damages models of both parties, the court crafted is own model based on the hypothetical use of an existing of CSIRO's Voluntary Licensing Program and adjusted through the application of the Georgia-Pacific factors. The CAFC vacated the damages award citing that the district court erred by not accounting for the standard-essential status of CSIRO's patent and by leaving out a relevant prior licensing agreement.

91 This is a prorated sum based on two per-unit royalties of $2.07 and $1.53.

92 This is a prorated sum based on two per-unit royalties of $1.70 and $0.80.

93 Motorola and Ericsson utilize standards in their own products and thus operate in both a production and technology logic (see Figure 3). However, they are categorized in the table as SEP as they were the SEP licensors seeking FRAND royalties.

94 A royalty base typically represents a saleable object in the value chain at the component or product level. However, the impact of the royalty base can change based on the context of its use. For instance, the royalty base can correspond to (1) a price level at which a royalty rate is applied as a percentage (e.g. a 4% royalty rate applied to an end product price of an Xbox of $300 as with the 802.11 feature factor method in *Microsoft*) or (2) a volume of units sold where a fixed royalty rate is applied (e.g. $0.15 royalty applied to the number of computers or WiFi routers sold as with the 802.11 product license method in *Ericsson*).

95 This in spite of the clarification of the use of the Entire Market Value Rule and the smallest saleable patent-practicing unit by the CAFC in LaserDynamics, Inc. v. Quanta Computer, Inc., 694 F. 3d 51 - Court of Appeals, Federal Circuit 2012.

96 See MPEG LA v. AUDIOVOX ELECS., 33 Misc. 3d 802 - NY: Supreme Court 2011, which upheld MPEG-LA's claim for royalties from the supplier of products to end users.

97 The use of prior market agreements was affirmed by the CAFC in both Ericsson and CSIRO as appropriate methods of applying the apportionment rule and not in violation of the secondary principle of the smallest saleable patent-practicing unit (SSPPU).

98 For example, the *Innovatio* court stated "Innovatio's application of its approach did not credibly apportion the value of the end-products down to the patented features. In light of that failure of proof, the court has no choice based on the record but to calculate a royalty based on the Wi-Fi chip". (*Innovatio* pp. 26-27).

99 See *Ericsson* 773 F.3d, pp. 39,50.

100 It could be argued that both market formation and revenue creation are addressed in a patent pool, though companies focused primarily on earning licensing income on standards often do not join pools exemplified by the fact that none of the SEP holders are members of patent pools in the standards related to the cases studied.

101 Using a profit approach, the value of technology would go to zero in a competitive, static efficient market providing no compensation to a patent holder even though the technology is employed by the manufacturer to generate user value.

102 For example, arguments were put forward in all cases that the value of 802.11 SEPs should be apportioned to the value of WiFi in a computer in relation to its other functions, while WiFi routers have 802.11 Wifi communication as their main function, and thus could be said to require little further apportionment.

103 The exceptions were the feature factor method introduced by Microsoft for the 802.11 standard and profit premium method put forward by CSIRO. The CSIRO valuation could be said to be technology logic focused due to the use of the end product as the royalty base even though profits instead of sales were applied.

104 In *Microsoft* and *Innovatio*, the court rejected the prior product licenses for similar reasons, including lack of apportionment of relevant patents, lack of evidence of FRAND negotiations, and negotiations under duress of litigation. In CSIRO, the court rejected the use of license agreements negotiated much further in time than the hypothetical negotiation date. However, the judge used a general license offer under CSIRO's Voluntary Licensing Program as the royalty base for its decision.

105 *Microsoft* relied highly on patent pool rates, *Ericsson* on prior product license agreements, and *Innovatio* on an income method using component profit margins.

106 The court FRAND royalty determination in *Microsoft* for H.264 and 802.11 was approximately three and two orders of magnitude less, respectively, and approximately 50 times less in *Innovatio*, indicating that the market offers of these firms were much higher than FRAND. However, without the threat of injunctive relief, there can be no systemic patent holdup as implementing firms can simply refuse the terms. The fact that Microsoft could petition the court and win damages for breach of contract is an indication that SEP concerns are transactional, not systemic, in nature and can be resolved through application of contract and patent law is courts without the need for policy intervention.

107 , For example, the SEP portfolio size is similar in this comparison but the strength of the Innovatio patents were deemed much greater than the Motorola portfolio.

108 This level of investigation is an interesting topic for further research but beyond the scope of this paper.

109 In addition to Ericsson, companies such as Qualcomm and Nokia have a revenue generating focus to their patent portfolios (i.e. Box 2 logic) and the sale of SEPs by Broadcom to Innovatio is indicative of a trend for practicing entities to engage offensive aggregators (i.e. box 3 logic) to generate income.

110 The only exception was a number of 802.11 licensing agreements by Ericsson and licensing attempts by CSIRO. The court in *CSIRO* cited the fact that infringement was rampant in the wireless industry thus also suggesting a lack of licensing norms (see *CSIRO* p. 22).

111 For example, CSIRO did not receive any licensing revenue from its SEP prior to 2008 based on litigation even though it made FRAND licensing offers to the industry (see *CSIRO* p.22). One reason for the success of the MPEG H.264 AVC patent pool could be that the H:264 standard was primarily a pre-competitive standard to support downstream value creation by the major actors in its development. For example, the large, initial technical development of the H.264 standard was transferred royalty free by Telenor.

112 For example, the H.264 standard has an extremely well-functioning patent pool, MPEG-LA H.264 AVC.

113 See *Georgia-Pacific Corp. v. U.S. Plywood Corp.*, 318 F. Supp. 1116, 1120 (S.D.N.Y. 1970).

114 See *Core Wireless Licensing v. LG Electronic et al.*, Case No. 2:14-cv-911-JRG (E.D. Tex, 2016) in support of CAFC rulings in *Ericsson* and *CSIRO*.

115 See *Ericsson* 773 F.3d, pp. 37-44 and *CSIRO* 809 F.3d., pp. 10-15.

116 See *Core Wireless v. Apple*, Case No. 15-cv-05008 NC (N.D. Cal, 2016) and *Saint Lawrence Communications LLC v. ZTE Corporation et al*, No. 2: 2015cv00349 - Document 441 (E.D. Tex. 2017).

117 This was however proposed in the CSIRO's expert valuation. While the court rejected the CSIRO damages model it more emphatically rejected the idea of using the chipset as the royalty base proposed by Cisco's damages model, stating "the inventive aspect of the '069 Patent is carried out in the PHY layer of the wireless chip, the chip itself is not the invention."

118 See Sidak (2013) that defines the end product as the smallest saleable unit in the context of SEPs.

119 See *Ericsson*, p. 46 that upholds Ericsson's right to license fully compliant products based on their IEEE Letter of Assurance, citing that other companies have adopted the same licensing norms.

120 This is supported by the fact that none of the SEP holders in the cases formally implicated the EMVR, while at the same time, actively attempting to the set the royalty base on the product level.

121 This opinion was echoed by the recent district court decision in the Daubert hearing in *Core Wireless v. Apple*, Case No. 15-cv-05008 NC (N.D. Cal, 2016).

122 For example, Broadcom testified in court in *Innovatio* and stated in a 2011 letter to the FTC that patent holdup was a significant problem in the industry even though they provided Innovatio with the SEPs in suit being alleged of patent holdup. (see *Innovatio*, pp.15-16).

123 *Core Wireless Licensing v. LG Electronic et al.*, Case No. 2:14-cv-912-JRG (E.D. Tex, 2016).

124 See Granstrand (1999).

Section 3
Implementing Standards

Chapter 9
An Investigation Into the Critical Success Factors of Implementing Information Technology Service Management Frameworks

Ahad Zare Ravasan
Mehralborz Institute of Higher Education, Iran

Mohammad Mehrabioun Mohammadi
Allameh Tabataba'i University, Iran

Homa Hamidi
Mehralborz Institute of Higher Education, Iran

ABSTRACT

During the last decade, many organizations have adopted information technology service management (ITSM) frameworks such as information technology infrastructure library (ITIL). However, such efforts are challenging and costly and not all have been successful. Keeping this in mind, this chapter seeks to identify critical success factors (CSFs) that impact successful implementation of ITSM frameworks with a focus on ITIL. Hence, based on a mixed method, at first, with the semi-structured interviews and literature review, 22 success factors were identified. The data of 122 valid questionnaires were analyzed using robust exploratory factor analysis (EFA) to validate and categorize identified factors. Finally, 22 CSFs are categorized under five "organizational," "human resources," "project management," "managerial," and "process" main groups. The findings of this study have provided a very useful reference for IT standardization scholars and practitioners to identify the important issues of ITSM frameworks implementation projects in research and practice.

DOI: 10.4018/978-1-5225-5320-5.ch009

INTRODUCTION

Today, information technology (IT) has become the backbone of businesses to the point where it would be impossible for them to survive and compete in the market without employing IT facilities. Due to the increasing role of IT in enterprises, its function has shifted from a technology provider and supporter to a strategic partner (Salle & Rosenthal, 2005). The traditional function of IT management (i.e. hardware and software installation, network management, applications management, etc.) now includes business-oriented service support, in which IT services are planned and managed according to their contributions to the required business processes (McNaughton, Ray, & Lewis, 2010).

IT service management (ITSM) -as a concept to support this radical transformation- is a strategy whereby information systems (ISs) are offered under contract to customers and performance is managed as a service (C. Pollard & A. Cater-Steel, 2009). ITSM provides a framework to structure IT operations that enable organizations to deliver quality IT services to meet business needs and adhere to service level agreements (SLAs) (Mesquida, Mas, Amengual, & Calvo-Manzano, 2011). Various ITSM frameworks have been developed to provide guidelines and best practices to help managers improve IT operations. However, the Information Technology Infrastructure Library (ITIL) has become more popular due to the drivers such as: the pressure to reduce cost or do more for less cost, the push for end-to-end service management, introduction of Service Level Agreements (SLAs) for measuring user experience, and requirement of IT to comply with legislations (McNaughton, et al., 2010; C. Pollard & A. Cater-Steel, 2009). Also, according to the results of an online survey conducted by Forrester, Inc. on 92 global IT decision-makers, the ITIL v3 was the most applied methodology for organizational setup or reorganization efforts (as asserted by 38% of respondents) following by Business Process Reengineering (BPR) and ITIL V2 (34% and 33% respectively) (Hubbert, 2010).

There is evidence of the growing global interest to ITIL. For example, in a recent global study, Axios Systems (2008) surveyed 255 IT professionals from global organizations at a series of service management events (conferences, seminars and workshops) across the UK, Australia and America. The results revealed that 64% of IT professionals believe following ITIL is a key to improve IT reputation. The study also depicted that 87% of the organizations followed ITIL guidelines. Further, it can be inferred from the statistics of the study done by Axios-Systems(2008) 33% of organizations intended to adopt ITIL within a year, and 36% were considering its adoption.

The key benefits expected from implementing an ITSM framework such as ITIL are: alignment of IT services with current and probable future business needs, more customer-focused and improved quality of IT services, better internal communication and communication with suppliers and customers, increased IT predictability and efficiency, and a reduction in the long-term costs of service provision (Addy, 2007; Cater-Steel & Toleman, 2009; McNaughton, et al., 2010; Peak, Guynes, & Kroon, 2005; C. Pollard & A. Cater-Steel, 2009; Van Bon et al., 2007; Yamakawa, Noriega, Linares, & Ramírez, 2012; Zare Ravasan, Mansouri, Mohammadi, & Rouhani, 2014). ITIL with introducing a consistent set of processes will highlight potential weaknesses in the previous operations and encourages pro-active improvements. Shortening resolution times, better management control, more reliable services, implementation of permanent solutions to formally acknowledged problems are some of the many ways ITIL will improve the quality of IT services via its proposed guidelines and best practices. As an example, service desk at JPMorgan Chase was able to achieve 75 percent first-call resolution and 93 percent customer satisfaction ratings. 80 percent of the calls coming into the service desk were answered within 20 seconds. The net result of improvement from ITIL implementation was the elimination of 500,000 calls to the service desk by tracking common problems and providing fixes at the root of the problem (Kumbakara, 2008). In another case, the IT providers at Caterpillar reduced response time to resolve web incidents after the firm implemented ITIL procedures. Proctor and Gamble saved $125 million by implementing ITIL-based service management processes (Galup, Dattero, Quan, & Conger, 2009).

It is noticeable that although many organizations worldwide are successfully taking up ITSM, not all are experiencing positive outcomes and many of them are confused about how to implement ITIL successfully. This is consistent with findings reported by Cater-Steel and Tan (2006) that only 56% of 108 Australian companies surveyed felt that ITIL implementations had met or exceeded their expectations. Introducing ITIL in an organization is a complex endeavor, depending on the characteristics and the ambitions of the target organization. ITIL will affect almost every member of staff; it may lead to rearranging organizational structures, work practices and worker roles. New competence is needed, specifically in ITIL, and more generally in process thinking. ITIL requires a devoted focus on IT services and customers' needs, which more technically-oriented organizations may find challenging. In addition, an integrated IT service management system must be implemented in order to support and control the various processes and tasks (Jon Iden & Langeland, 2010). A full adoption of ITIL may take years, and requires the dedication of managers as well as personnel. Studies have demonstrated that most companies start with establishing a service desk followed by the implementation of the incident, change, and service level management processes (Cater-Steel & Tan, 2005; J Iden, Steindal, & Stokke, 2007; Zare Ravasan, et al., 2014).

There are factors might influence on the ITSM implementation projects success and benefits acquisition such as top management support, personnel training, careful software selection, use of consultants, interdepartmental communication, organizational culture, and customer-focused metrics (C. Pollard & A. Cater-Steel, 2009). Processes are another key success factor of IT service management and it must be taken into account that adequate techniques and tools must be applied (Lucio-Nieto, Colomo-Palacios, Soto-Acosta, Popa, & Amescua-Seco, 2012; Zare Ravasan, et al., 2014). These factors are also identified as Critical Success Factors (CSFs) in the literature (Carol Pollard & Aileen Cater-Steel, 2009). Most of the research conducted on identifying ITSM frameworks implementation projects is devoted to developed rather than developing countries. Despite the existing challenges in ITSM frameworks implementation projects, little exhaustive study has been done so far to investigate the main CSFs of such frameworks in less developed countries (Al-Karaghouli, Al Azri, & Al Salti; Nfuka & Rusu, 2011; Othman, Chan, & Foo, 2011). Although, it is assumed that organizations in different countries face similar difficulties and challenges in implementing general IT applications, but studies indicate that large IT initiatives in developing countries confront specific difficulties that differ from those found in industrialized ones (Amid, Moalagh, & Ravasan, 2012; Z. Huang & Palvia, 2001; Soja, 2008).

Regarding the fact that many firms in developing countries are now adopting ITSM frameworks, the subject of studying ITSM CSFs in these countries could be of more interest among scholars. Iran is one of these developing countries that faced a significant increase in implementing such ITSM frameworks, confronting with crucial challenges and failures. Despite the existing challenges in its ITSM frameworks implementation projects, there is no comprehensive study which investigated main success factors of these systems in the context of less developed countries. Hence, the purpose of this study is to explore the most crucial ITSM frameworks Implementation (especially ITIL) CSFs in Iranian industries as a case of a developing country. Understanding these factors will deepen the perception of ITSM related standards in practice and help avoid execution mistakes, thereby increasing the success rate of such standards implementation in less developed countries as a different cultural and organizational context.

The paper is organized as follows: the following section reviews the literature on the concept of ITSM, and the most related work in the research area. The next section addresses ITSM frameworks implementation projects' CSFs identified through literature review and semi-structured interviews. Then, by using questionnaire data, a classification model is formulated for these CSFs in Iranian industries. Based on this model, some recommendations can be made to impede failure or avoid potential inconveniences that might occur in such projects. In the final section, the implications of the proposed model and revealed factors are discussed.

LITERATURE REVIEW

This section encompasses a brief introduction of ITSM and related frameworks, and then the most related research in the field are introduced.

ITSM and Related Frameworks

According to ITIL version 3, also known as ITIL V3, a service is a means of delivering value to customers by facilitating outcomes the customers want to achieve without the ownership of specific costs or risks. Similarly service management is defined as a set of specialized organizational capabilities for providing value to customers in the form of services (Van Bon, et al., 2007). IT Service Management is concerned with the delivery and support of IT services that are appropriate to the business requirements of an organization. ITSM provides specific processes, frameworks, methodologies and guidance to manage planning, implementation and assessment of IT service processes to optimize tactical and strategic IT operations-related activities (Mesquida, et al., 2011; Zare Ravasan, et al., 2014). Some of the most internationally accepted ITSM standards and frameworks are briefly introduced below:

- **ISO/IEC 15504 IT – Process Assessment:** Also known as SPICE (Software Process Improvement and Capability Determination) is a set of technical standards documents for the computer software development process and related business management functions. It can be used by any organization to determine the current and potential capability of its own processes as well as to define areas and priorities for process improvement to perform a process assessment conformant with the assessment model defined in ISO/IEC 15504-2 (ISO/IEC, 2003)
- **ITIL V3:** ITIL is a series of documents that contains a set of best practices to aid the implementation of a framework for ITSM. ITIL gives a detailed description of most of the important processes in an IT organization and includes checklists for tasks, procedures and responsibilities which can be used as a basis for tailoring to the needs of individual organizations. ITIL V3 approaches service management from the service lifecycle including service strategy, service design, service transition, service operation, and continual service improvement (Van Bon, et al., 2007).
- **ISO/IEC 20000: ISO 20000:** (formerly known as BS15000), is the world's first standard for IT service management. The standard specifies a set of inter-related management processes, and is based heavily upon the ITIL framework. These ISO standards emphasize the definition, description and design of processes (BS15000-Associates-Group, 2012; Van Bon, et al., 2007).
- **Capability Maturity Model Integration (CMMI):** This process improvement method was developed by the Software Engineering Institute (SEI) of Carnegie Mellon University. CMMI pro-

vides both a staged and a continuous model. In the continuous representation, improvement is measured using capability levels. Maturity is measured for a particular process across an organization. In the staged representation, improvement is measured using maturity levels, for a set of processes across an organization (Van Bon, et al., 2007).

- **HP ITSM Reference Model**: In 1996, HP consultants around the world were brought together to discuss the creation of an IT process model. The result is a model that combines the best that ITIL has to offer with the best that industry experience has to offer. The reference model is structured around five groups: business-IT alignment, service design and management, service delivery assurance, service development and deployment, service operations (HP, 2003).
- **Control Objectives for Information and related Technologies (COBIT):** COBIT 4.0 is a high-level process model that organizes a broad range of IT activities in 34 processes. COBIT consists of a selection of popular management tools and techniques described in an IT management context. COBIT 5 builds and expands upon COBIT 4.1 by integrating other major frameworks, standards and resources, including ISACA's Val IT and Risk IT, ITIL and related standards from the ISO. (ISACA, 2012).
- **Microsoft Operations Framework (MOF):** MOF is Microsoft's vision for IT service management. MOF is composed of three core models – process, team and risk – that include the principles and practices that business people and IT practitioners need in order to manage IT operations effectively on the Microsoft platform. MOF provides comprehensive technical guidance for achieving mission-critical production system reliability, availability, supportability, and manageability on Microsoft's products and technologies (Microsoft-Inc, 2012; Salle & Rosenthal, 2005).
- **IBM's Systems Management Solution Lifecycle:** This roadmap is consistent with the ITIL framework and leads an enterprise through the four phases of implementing ITSM, as design the solution approach, design the solution, implement the solution, and deliver service. The lifecycle encompasses methods, knowledge, data and analytics (IBM-Global-Services, 2004).

Related Research

With the ubiquity of information technologies and the impact of information systems on organizations' productivity, planning, implementing, and presenting IT services have become very decisive and its successful implementation is brought up as one of the of research concerns in this field. In the earliest studies related to this topic, Potgieter, Botha, and Lew (2005) investigated the impact of ITIL framework on customer satisfaction. According to their research, the use of the ITIL framework will not only raise customer satisfaction, but also simultaneously improves operational performance of the organization. Based in this argument, there is a positive correlation between the use of the ITIL framework and customer satisfaction. In the same year, using a case study, Hochstein, Tamm, and Brenner (2005) studied six successful organizations in terms of the implementation of ITSM frameworks. According to their findings, six critical success factors including demonstrating "quick wins", strive for continuous improvement, market campaigns in order to create acceptance and understanding, management support, training, and formation of virtual teams, were recognized.

Bringing up 3 questions 1)what are ITIL framework impacts on organizations? 2)do the organizations that use the ITIL framework follow common patterns?, and 3)what are the challenges to successfully implementing ITIL framework?, Cater-Steel, et al., (2006) evaluated the effectiveness of the ITIL framework implementation in organizations and used case study methodology to accomplish their research objectives. They investigated five successful organizations that had implemented the ITIL framework. The Results suggest that using the ITIL framework exerts a positive effect on the organization. Such an effect is derived from the organization's concentration on IT services management, better control of measurement and system changes, better organizational infrastructures, faster identification of problems associated with the IT services, and better communication of IT department with others.

Another result of Cater-Steel, et al., (2006) study indicates that in establishing the ITIL framework, organizations encounter several challenges. The most important challenges include: lack of management support, cultural issues in term of employees' resistance, delays in choosing an appropriate tool, and management problems for resources such as time, staff and money. In another research in 2006, the same authors investigated different frameworks in relation to IT service management. Their findings suggest that combining COBIT, CMMI and ISO 9001 frameworks with ITIL framework not only promotes the success rate of its implementation, but also helps organizations to establish a strong IT governance (Cater-Steel, Tan, & Toleman, 2006).

Studying an Australian government agency, Tan, et al., (2007) classified ITIL implementation CSFs in four perspectives: management (senior management commitment, management of achieved benefits, appropriate communication among vendor, consultant and project manager), customer (change management, training and customers satisfaction), process (process methodology and process performance) and tools (set of tools related to IT services management).

Using the case study method, Spremic, Zmirak, and Kraljevic (2008) considered implementation of ITIL Framework in a financial organization and compared key performance indicators in the areas of processes and services, before and after the implementation of this framework, to evaluate the effectiveness of its implementation. Based on the results of their research, not only implementing this framework leads to direct benefits of the organization but it also brings about many other intangible benefits to the organization. Similarly, they have suggested that board of directors' support, training and in-depth feasibility and functional analysis of required solution is essentials for successful implementation of ITIL framework. According to studies of Pollard and Cater-Steel (2009), key factors in successful implementation of ITIL in the U.S. and Australian organizations include top management support, training and staff awareness, virtual project team, selecting an ITSM tool set, using proper consultants, interdepartmental communication and collaboration, process as a priority, service-oriented culture and use of customer focused metrics.

Carrying out research by a case study in Mexico to identify the factors influencing the successful implementation of IT services management within organizations, Lucio-Nieto, et al. (2012) concluded that establishing a good service management office in organizations is likely to increase the probability of successful implementation of IT services management. Such a mechanism could tie work methods to tools and communication processes; consequently, it increases the effectiveness of decisions related to IT services management in organizations. Huang, Wu, and Chen (2012), stating that IT services management is an appropriate strategy towards integrating IT with business goals, studied critical factors in implementing IT services management in organizations benefiting from opinions of IT experts in Asia using Delphi method. Based on the researchers' findings, successful implementation of IT services management within an organization requires a balance among change management, continuous management of processes, technology management, strategic planning and project management.

Recently, Alreemy, Chang, Walters, and Wills (2016) proposed IT governance frameworks CSFs into five main groups of strategic alignment, environmental effect, organizational effect, performance management, and resource management. Ndlovu and Kyobe (2016) explored the barriers to successful migration to CoBIT 5 (Control Objectives for Information and related Technologies) and identified the key factors that influence effective migration. They suggested that there are four distinct aspects of structure, managerial task/ process, people, and technology need to be reconciled.

To gain a better understanding of research conducted in the field of ITSM, relevant research and their findings have been summarized in Table 1.

According to the Table 1, to identify the CSFs in ITIL implementation, few quantitative methods have been employed so far. The reason is that IT services management is relatively a novel field and researchers have resorted to the use of qualitative methods to get a better understanding of the phenomenon

Table 1. Relevant research on ITSM and ITIL implementation

Id	Key Findings	Research Method	Source
1	Identifying four main advantages of IT services management: improvement in quality of IT services, efficiency and optimization of processes and transparency, comparability through process documentation, and process monitoring	Qualitative (case study)	(Hochstein, et al., 2005)
2	Investigating the effect of the ITIL implementation on customer satisfaction and service quality and finding direct correlation between customer satisfaction, service quality and the use of ITIL	Quantitative (SERVQUAL)	(Potgieter, et al., 2005)
3	Identifying critical success factors in the implementation of ITIL projects that include top management support, training and staff awareness, virtual project team, selecting an ITSM "Toolset", use of consultants, interdepartmental communication and collaboration, process as priority, service-oriented culture and use of customer focused metrics.	Qualitative (case study)	(Carol Pollard & Aileen Cater-Steel, 2009)
4	Investigating ITSM frameworks such as ITIL, COBIT, CMMI, and ISO 9001, and possible motivations and challenges for their adoption.	Mixed method (survey method and case study)	(Cater-Steel, Tan, et al., 2006)
5	Identification of factors that challenge the use of ITIL: lack of management support, cultural issues in term of employees' resistance, delays in choosing an appropriate tool, and poor resource (time, human and money) management	Qualitative (case study and structured interviews)	(Cater-Steel, M, & Tan, 2006)
6	Selecting and evaluating key performance indicators before and after the ITIL adoption.	Qualitative (case study)	(Spremic, et al., 2008)
7	Proposing six ITIL adoption benefits as improvements in customer satisfaction, improvements in internal processes, standardization of processes, improvements in service quality, increasing in efficiency, and improvements in return on investment.	Qualitative (case study)	(Marrone, Kiessling, & Kolbe, 2010)
8	Stressing the importance of Service Management Office in increasing the likelihood of success in implementing the ITIL framework	Qualitative (grounded theory and unstructured interviews)	(Lucio-Nieto, et al., 2012)
9	Identification of ITIL CSFs including top management support, interdepartmental communication and collaboration, proper consultants, timely and carefully selection of ITSM "Toolset", service-oriented culture, process as priority and customer focused metrics	Qualitative (case study)	(Jon Iden & Langeland, 2010)
10	Successful implementation of ITIL framework is needed to create a balance among change management, continuous management of processes, technology management, strategic planning and project management	Qualitative (Delphi method)	(S.-J. Huang, et al., 2012)
11	Proposing IT governance frameworks CSFs into five main groups of strategic alignment, environmental effect, organizational effect, performance management, and resource management.	Quantitative (survey)	(Alreemy, et al., 2016)
12	Exploring the barriers to successful migration to CoBIT 5 as structure, managerial task/ process, people, and technology need to be reconciled.	Quantitative (survey)	(Ndlovu & Kyobe, 2016)

under study. In this research, a qualitative approach based on the interview with experts and literature review is followed. Once the factors have been gathered and CSF pool has been created (see Table 2), in the next step, quantitative approach is followed. Consequently, the research method has somehow taken a mixed nature.

Research Methodology

The research steps, i.e., identifying CSFs, conducting semi-structured interviews, instrument development, data collection, data analysis using robust EFA, are shown in Figure 1.

Conducting Semi-Structured Interviews

Although, authors have conducted an in-depth review on current literature to explore the more cited ITIL implementation CSFs, there are factors that might have not been well documented in available research

Table 2. ITSM frameworks Implementation CSFs identified through literature review

(Hochstein, et al., 2005)	(A. Cater-Steel, et al., 2006)	(Tan, et al., 2007)	(Spremic, et al., 2008)	(Carol Pollard & Aileen Cater-Steel, 2009)	(Tan, Cater-Steel, & Toleman, 2009)	(Iden & Langeland, 2010)	(Lucio-Nieto, et al., 2012)	(Huang, et al., 2012)	(Ahmad & Shamsudin, 2013)	CSFs Title
		●			●					Proper vendors
		●		●						Customer focused metrics
		●		●						Proper consultants
		●		●	●	●	●			Process as a priority
●	●			●						Staff awareness
●	●	●		●		●				Staff training
		●						●		Projects' support
●	●	●		●		●		●		Top management support
					●				●	Project governance
					●				●	Change-friendly culture
			●						●	Continuous monitoring and evaluation of processes
●					●					IT and business alignment

Figure 1. The research steps

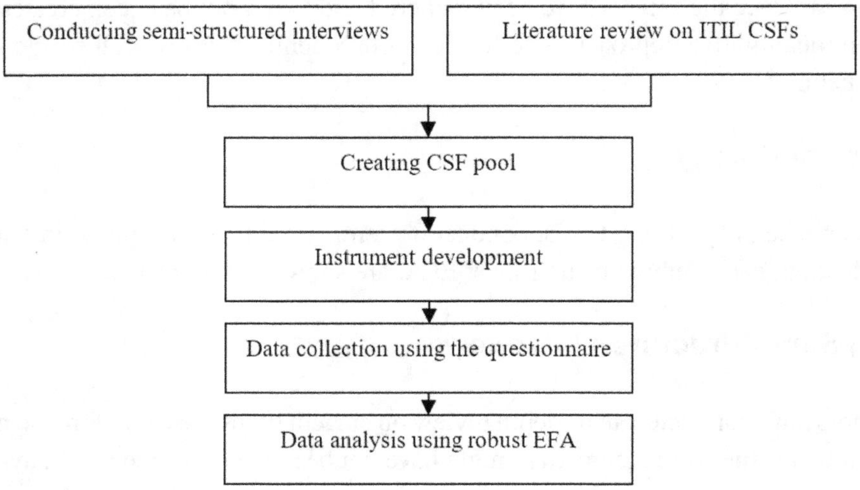

works. Also, there are some existing factors in special contexts, cultures, economies, and etc. So, for conducting a holistic analysis on ITIL implementation CSFs, a hybrid approach using both interviews and literature review is used in the paper. For interviews, 13 ITIL implementation project managers are interviewed. The quotes of such interviews are summarized in the Appendix. As a result, ten new factors are identified and incorporated into initial 12 factors. These new factors seem to be prevalent in Iranian industries that have not been well introduced in extant literature.

Instrument Development

A 22 items questionnaire has been developed by integrating factors identified through interviews and literature review. The survey instrument asked the respondents to rate the impact of 22 identified CSFs using a 9 point scale with items ranged from 1 (strongly low) to 9 (strongly high). Face or content validity of the questionnaire is conducted through the literature review and experts judgment. To ensure this, at first six ITIL implementation project managers of high academic levels and more than 5 year experience reviewed the questionnaire. They had some comments on the length and the clarity of each question. Their suggestions were incorporated into the final version of the questionnaire. The content validity of the instrument was thereby addressed. Also, for evaluating the reliability of the questionnaire, test-retest method was used. For conducting test- retest method, authors asked 15 project managers in a 14-day interval to participate in the study. The resulted Cronbach alpha estimated to be 0.83 (greater than 0.7) that implies good reliability of the instrument (Nunnally, 1978).

Data Collection and Sample Size

The target of this study was ITIL implementation project team members in Iranian industries which have at least 3 years of experience. In order to gather data from the companies, the following procedure was performed: first, the eligible companies and related team members were identified and their contact information were gathered; second, authors asked them to participate in the study; third, the questionnaire were sent to them, and finally, they filled and replied the questionnaires. Totally, 185 questionnaires were sent, 122 questionnaires were gathered and 112 usable questionnaires were used for the analysis (response rate: 0.66). Sample size of 112 seems to be adequate for conducting robust EFA (recommended ratio of 5:1). However some authors argued that such ratios might be an oversimplification (MacCallum, Widaman, Preacher, & Hong, 2001).

Results

EFA is a frequently used method to discover patterns of multidimensional constructs that are subsequently used for the development of measurement scales. Its major objective is to reduce a number of observed variables to fewer factors in order to enhance interpretability and detect hidden structures in the data. Here, robust EFA (Treiblmaier & Filzmoser, 2010) was employed to perform the analysis. Robust EFA has many advantages over classical factor analyses in helping a researcher to choose multitude of options, such as various types of data transformation, the choice of the factor extraction method, factor rotation, and the number of factors to be chosen. Also, it makes less restrictive assumptions about data distribution than classical factor analysis and reduces the influence of outliers, which finally leads to more valid results.

Prior to factor analysis, a test was conducted to verify the adequacy of the data for FA. The Kaiser-Meyer-Olkin (KMO) measure of sampling adequacy is popular measure for assessing the extent to which the indicators of a construct belong to one another. Large values for the KMO indicate that a factor analysis of the variables is a good idea. The KMO is 0.814 which is above the "Mediocre" threshold of 0.5 (Kaiser, 1974). We also performed a Bartlett sphericity test, which was statistically significant ($p < 0.05$), indicating the eligibility of the data. Then, we used a Shapiro–Wilk test to determine whether our sample had a normal distribution. We found that none of our variables was normally distributed. Thus, principal component analysis (PCA) was our choice for the factor extraction method as proposed in robust EFA. Another option should be selected for conducting robust EFA is factor rotation method. There are various rotation methods available for use in several widely used statistical software packages. Here, since it was reasonable to expect that the dimensions of ITIL implementation CSFs would correlate with one another. We used Oblimin rotation, which is proposed in robust EFA (Treiblmaier & Filzmoser, 2010). Also, other methods such as Varimax rotation, usually results in a highly dominant first factor and a rather weak discrimination of the other factors according to their loadings that is not desired here. Finally, the number of factors to be extracted from the data should be determined. Here, Eigen values greater than one, and an absolute value of factor loading greater than 0.60 has been selected. It has been suggest that a factor loading higher than 0.60 on the ''parent factor'' and a loading of less than 0.40 on a ''foreign factor'' indicate convergent and discriminant validity of the constructs (Chin, Gopal, & Salisbury, 1997). As a result, 2 out of 22 factors were dropped from the initial pool and remained 20 factors were grouped into 5 components. The results can be seen in Table 3.

Data Interpretation

To indicate the meaning of the components, they have been given short labels indicating their content. Since the results of this stage were open to several interpretations, we decided to use experts' opinions here. So, we invited 3 project managers and finalized components' name by discussion and general agreement. In this way, "Organizational", "Human Resources", "Project Management", "Managerial",

Table 3. The results of robust EFA

	1	2	3	4	5
Change-friendly culture	0.85				
Project governance	0.79				
Achieving agreement among different stakeholders	0.64				
IT and business alignment	0.66				
Staff awareness		0.82			
Staff training		0.77			
Involving employees in the project		0.70			
Proper vendors			0.75		
Proper consultants			0.77		
Good relationships with project consultant			0.75		
Proper project management			0.74		
Good project team			0.80		
Top management support				0.80	
Management belief in project outcomes				0.78	
Optimized budget allocation				0.75	
Continuous performance assessment					0.85
Continuous monitoring and evaluation of processes					0.76
Process as a priority					0.82
Customer focused metrics					0.75
An integrated process and service based approach					0.77
% of variance	20.166	19.066	16.636	11.855	10.595
Cumulative %		39.232	55.868	67.723	78.317

Extraction method used is Principle Component Analysis and the rotation method used is Oblimin.
[a]Only loading of 0.4 or above are shown.

and finally "Processes" are the names which have been assigned to the extracted components. The final results are shown in Table 4.

DISCUSSION

This study was conducted to identify the CSFs in implementing IT services management frameworks. This goal was achieved by following a hybrid literature review and semi-structured interviews with field experts. Using robust exploratory factor analysis, five main categories of organizational, human resources, project management, managerial, and processes were identified, which will be discussed more in this section.

Organizational category is the most important CSF group in the implementation of ITIL framework. Adopting service-based strategies requires some sorts of changes in the organizational structure, systems, procedures, and processes which imposes a degree of risks into an organization. Besides, IT service

Table 4. Extracted components and their related factors

Component ID	Component Name	Factor ID	Factor Name
C1	Organizational	OR01	Change-friendly culture
		OR02	Project governance
		OR03	Achieving agreement among different stakeholders
		OR04	IT and business alignment
C2	Human Resources	HR01	Staff awareness
		HR02	Staff training
		HR03	Involving employees in the project
C3	Project Management	PM01	Proper vendors
		PM02	Proper consultants
		PM03	Good relationships with project consultant
		PM04	Proper project management
		PM05	Good project team
C4	Managerial	MG01	Top management support
		MG02	Management belief in project outcomes
		MG03	Optimized budget allocation
C5	Processes	PR01	Continuous performance assessment
		PR02	Continuous monitoring and evaluation of processes
		PR03	Process as a priority
		PR04	Customer focused metrics
		PR05	An integrated process and service based approach

management is a combination of IT technical and business issues which calls for proper IT- business alignment initiatives. Therefore, it is necessary to make a wide variety of changes at different organizational levels while implementing ITIL framework. Employees might resist change due to some reasons such as misunderstanding about the need for change, fear of the unknown Lack of competence, low trust, poor communication, and so on (de Sousa Pereira & da Silva, 2010).

As it has been stated in the previous research, project governance is the key factors in successful implementation of the ITIL framework. Such a factor, clarifies the objectives, reduces implementation risks and allows managers and executives to see the exact costs and outcomes. Project governance emphasize on doing right projects. Especially project governance is tied with decision making about delivering value to the projects stakeholders. Appropriate project governance helps organizations save resources and money by ensuring that all conducted project activities are aligned with organization's strategic plans. Project governance leads to a framework in which project decisions are made in compliance with business objectives.

The multifaceted nature of the IT projects is such that its implementation requires the involvement of a variety of people with diverse perspectives insisting on their own interests. Such circumstances can emerge conflicts within an organization. For this reason, one of the major factors of organizational category is reaching to an agreement among different interests. The last organizational factor is the business and IT alignment. Business-IT alignment is a dynamic state in which a business organization is able to use

IT effectively to achieve business objectives, typically improved financial performance or marketplace competitiveness. Business-IT alignment integrates the information technology to the strategy, mission, and goals of the organization (Hochstein, et al., 2005; Tan, et al., 2007).

Another category in the successful implementation of ITIL framework is human resources. Three factors are involved in this area: employees' knowledge, employees' training, and employees' involvement in the project. Employees' awareness is one of the most important issues in successful implementation of ITIL framework in organizations because employees' knowledge can reveal worth of ITIL framework as well as the necessity of its implementation in an organization. This factor, not only coordinates different operational and managerial organizational levels with the project requirements, but also reduces employees' resistance (Cater-Steel, Toleman, et al., 2006; Hochstein, et al., 2005). Employees' awareness could be achieved through staff training. Therefore, staff training should be continuously followed throughout project phases (Jon Iden & Langeland, 2010; Carol Pollard & Aileen Cater-Steel, 2009).

ITIL is presented as a series of successful experiences and best practices. Therefore, if an organization wishes to be successful in ITIL implementation, it should be able to learn from these successful experiences that, in turn, it needs organizational learning capabilities. Given the fact that the ITIL implementation projects require considerable changes in organizations, especially in the area of IT processes, they usually face employees' resistance. Identification of opposing patterns of employees' behavior in the project and following the proper course of action may also increase the probability of the project success.

Project management is another category in the successful implementation of ITIL. The first factor in this category is strong vendor for the selected solution because ITSM solutions usually are acquired from market vendors. Acquiring such off-the-shelf ITSM solutions reduces organizations' costs, comparing to in-house developed solutions. However, there are some risks such as vendor luck-in, improper vendor selection or not well developed contracts (Abdullah & Verner, 2012; Chou & Chou, 2009; Rouhani & Zare Ravasan, 2014) in such efforts. Therefore, it is necessary to evaluate the associated risks as well as the impacts on successful implementation. Strong vendor may simultaneously increase the operational efficiency while reduce associated risks. Employing a strong consulting team together with a strong vendor is another factor influencing the success of organizations in implementation of the ITIL framework (Tan, et al., 2007). It is worth mentioning that strong consulting team can effectively assist the organization with the transfer of related knowledge (Carol Pollard & Aileen Cater-Steel, 2009; Tan, et al., 2007). Project management in ITIL framework implementation is of high importance. Many changes brought about by ITIL framework implementation should be defined as improvement actions or projects. Also, management of continuous improvement, as one of the pillars of systems' life cycle, requires definition of improvement projects. It critically demands the proper project management. If project governance dealt with doing right projects, proper project management emphasizes on doing projects right. A proper project management leads to more clarified budgets estimation, optimized fund using and following project activities efficiently. Through proper project management, stakeholders can assure that all project targets have been achieved due to time and budget constraints. Another important issue in proper management is stakeholder awareness about project's outcomes. Stakeholders and high level managers expect to see quick-wins from their investment which calls for rapid tangible outputs.

The fourth CSF category in the successful implementation of ITIL framework in organizations is managerial group. This category involves: top management support, management belief in the project and proper allocation of financial resources. Top management support of ITIL implementation projects is one of the most cited CSFs in the literature (Carol Pollard & Aileen Cater-Steel, 2009). In case, implementation of the ITIL frameworks is tied to and aligned with strategies of the organization, it is more likely to get such support (S.-J. Huang, et al., 2012; Jon Iden & Langeland, 2010; Carol Pollard & Aileen Cater-Steel, 2009). Also, based on the results, it can be implied that management belief in the project outcomes plays an effective role in the successful implementation of ITIL framework. It is obvious that if the top management does not believe in the outcomes of project, he/she will not provide the project with necessary support and will be hesitant about the allocation of required financial and human resources. Then, it could be assumed that such beliefs would enhance top management support. To secure the management belief in the project, it is necessary to develop an appropriate culture among organization's managers to increase their awareness level through training (S.-J. Huang, et al., 2012; Thakor, 1999).

The last category affecting the successful implementation of ITIL framework in organizations is process that includes factors of continuous performance assessment, continuous monitoring and assessment of operational processes, giving priority to processes over tools, use of customer- focused metrics and an integrated process and service-based approach. Continuous and appropriate performance assessment is in line with continuous improvement of services and processes in ITIL framework. Operational efficiency and customer satisfaction are viewed as the main goals of IT service management. Therefore, one of the important factors in the process category is following proper customer-focused measures pertaining to continuous monitoring of processes. The results of the evaluation of customers' opinions help not only the project team with obtaining information on end-users' satisfaction but also help managers assess the effectiveness of project outcomes (Carol Pollard & Aileen Cater-Steel, 2009; Tan, et al., 2007). They may also raise the bargaining power of the project team against senior managers. Finally, process perspective and its prevalence in the organization is another CSF impacting on the successful implementation of IT services management. In Iranian organizations, due to the lack of a process oriented view and predominance of functional structures in organizations, execution of process-centric frameworks such as ITIL has typically encountered major challenges (Zare Ravasan & Mansouri, 2014, 2016).

CONCLUSION

Nowadays, many organizations adopt ITSM best practices to enhance their service related processes. Despite all the benefits such framework brings for organizations, the failure rate and implementation challenges of such efforts are still of real concern. This has encouraged some researchers to seek for the cause of its implementation failure as well as identification of CSFs in its successful implementation. This study was conducted with this in mind and as a response to this question: "What are critical success factors in implementation of the ITIL framework?" To answer this question, the study enjoyed a mixed approach, including survey and semi-structured interviews. After reviewing the subject literature and semi-structured interviews and the extraction of CSFs, in the next step, categories and concepts entailing factors were extracted using robust exploratory factor analysis. Findings of the analysis clarify that five "organizational", "human resources", "project management", "managerial", and "processes" categories exert a significant effect on the successful implementation of ITIL framework.

The research results provide some implications for IT standardization research and practice community. For research community, the results of the study imply that there are some CSFs which might be more prevalent in less developed countries due to their lower IT maturity level or different cultural context. Meanwhile, understanding these factors will deepen the perception of ITSM related standards in practice and help avoid execution mistakes, thereby increasing the success rate of such standards' implementation in less developed countries as a different context which could be considered as the main practice implication of the research.

This study suffers from some limitations. At first, this study is by no means comprehensive enough to address all issues related to ITIL framework CSFs. As another limitation of this study, it could be difficult to make generalizations based on the contents of the work, and that relatively few articles could be found for most countries, especially for developing countries. As a potential for future work, the proposed CSFs could be used in other countries to test its applicability. Also, researchers may follow qualitative research methods such as case studies to investigate such factors in similar or other settings. Case study is a useful method in studying such a subject. Moreover, future works could focus on more specific areas such as project management, organizational structure or the impact of organizational culture on ITSM frameworks implementation so that more detailed and in-depth information or deep-rooted failure-success reasons could be identified. Moreover, future research could move beyond listing CSFs and explore the interrelationships between them. For example, it is of vital importance to investigate whether mismatches between factors such as management, processes, human resources, structures and technology serve as the reason for problems. It would also be valuable to relate CSFs to project implementation or IT service lifecycle phases.

REFERENCES

Abdullah, L. M., & Verner, J. M. (2012). Analysis and application of an outsourcing risk framework. *Journal of Systems and Software*, *85*(8), 1930–1952. doi:10.1016/j.jss.2012.02.040

Addy, R. (2007). *Effective IT Service Management: To ITIL and Beyond!* Berlin: Springer.

Al-Karaghouli, W., Al Azri, A., & Al Salti, Z. (n.d.). *E-Government and EIS Change Management and Critical Success Factors: An Omani Success Story*. Academic Press.

Alreemy, Z., Chang, V., Walters, R., & Wills, G. (2016). Critical success factors (CSFs) for information technology governance (ITG). *International Journal of Information Management*, *36*(6), 907–916. doi:10.1016/j.ijinfomgt.2016.05.017

Amid, A., Moalagh, M., & Ravasan, A. Z. (2012). Identification and classification of ERP critical failure factors in Iranian industries. *Information Systems*, *37*(3), 227–237. doi:10.1016/j.is.2011.10.010

Axios-Systems. (2008). *ITIL Adoption Surges Despite Confusion*. Author.

BS15000-Associates-Group. (2012). *The ISO 20000 (BS15000 / BS 15000)*. ITSM Standard.

Cater-Steel, A., M, T., & Tan, W.-G. (2006). *Transforming IT Service Management – the ITIL Impact*. Paper presented at the 17th Australasian conference on information systems, Adelaide, Australia.

Cater-Steel, A., & Tan, W.-G. (2005). *Implementation of IT Infrastructure Library (ITIL) in Australia: Progress and success factors*. Paper presented at the IT Governance International Conference, Auckland, New Zealand.

Cater-Steel, A., Tan, W.-G., & Toleman, M. (2006). *Challenge of adopting multiple process improvement frameworks*. Paper presented at the 14th European Conference on Information Systems, Goteborg, Sweden.

Cater-Steel, A , & Toleman, M. (2009). Education for IT service management standards. *International Journal of IT Standards and Standardization Research, 5*(2), 27–41. doi:10.4018/jitsr.2007070103

Cater-Steel, A., Toleman, M., & Tan, W.-G. (2006). Transforming IT service management-the ITIL impact. *Proceedings of the 17th Australasian Conference on Information Systems (ACIS 2006).*

Chin, W. W., Gopal, A., & Salisbury, W. D. (1997). Advancing the theory of adaptive structuration: The development of a scale to measure faithfulness of appropriation. *Information Systems Research, 8*(4), 342–367. doi:10.1287/isre.8.4.342

Chou, D. C., & Chou, A. Y. (2009). Information systems outsourcing life cycle and risks analysis. *Computer Standards & Interfaces, 31*(5), 1036–1043. doi:10.1016/j.csi.2008.09.032

de Sousa Pereira, R. F., & da Silva, M. M. (2010). *A maturity model for implementing itil v3.* Paper presented at the Services (SERVICES-1), 2010 6th World Congress on.

Galup, S. D., Dattero, R., Quan, J. J., & Conger, S. (2009). An overview of IT service management. *Communications of the ACM, 52*(5), 124–127. doi:10.1145/1506409.1506439

Hochstein, A., Tamm, G., & Brenner, W. (2005). *Service-oriented IT-Management: Benefit, Cost and Success Factors.* Paper presented at the 13th European Conference on Information Systems, Regensburg.

HP. (2003). *HP IT Service Management (ITSM); Transforming IT organizations into service providers.* Retrieved August 02 2012 from http://www.hp.com/hpinfo/newsroom/press_kits/2007/businesstechnology/wp_it_transformation.pdf

Huang, S.-J., Wu, M.-S., & Chen, L.-W. (2012). Critical success factors in aligning IT and business objectives: A Delphi study. *Total Quality Management & Business Excellence*, 1-22.

Huang, Z., & Palvia, P. (2001). ERP implementation issues in advanced and developing countries. *Business Process Management Journal, 7*(3), 276–284. doi:10.1108/14637150110392773

Hubbert, E. (2010). *Market Overview: IT Service Management Support Tools. An Overview Of The Modern Service Desk And Other Technologies.* Forrester Research, Inc.

IBM-Global-Services. (2004). *IBM and the IT Infrastructure Library. How IBM supports ITIL and provides ITIL-based capabilities and solutions.* Author.

Iden, J., & Langeland, L. (2010). Setting the Stage for a Successful ITIL Adoption: A Delphi Study of IT Experts in the Norwegian Armed Forces. *Information Systems Management, 27*(2), 103-112. doi: 10.1080/10580531003708378

Iden, J., Steindal, F., & Stokke, B. (2007). The Implementation of IT Infrastructure Library (ITIL) in Norway: Progress, success factors and benefits. In L. Habib (Ed.), *Norsk konferanse for organisasjoners bruk av informasjonsteknologi. NOKOBIT 2007* (pp. 43–56). Trondheim, Norway: Tapir Akademisk Forlag.

ISACA. (2012). *Control Objectives for Information and related Technologies.* COBIT.

ISO/IEC. (2003). *ISO/IEC 15504- 2:2004 Software Engineering – Process Assessment – Part 2: Performing an Assessment.* Author.

Kaiser, H. F. (1974). An index of factorial simplicity. *Psychometrika, 39*(1), 31–36. doi:10.1007/BF02291575

Kumbakara, N. (2008). Managed IT services: The role of IT standards. *Information Management & Computer Security, 16*(4), 336–359. doi:10.1108/09685220810908778

Lucio-Nieto, T., Colomo-Palacios, R., Soto-Acosta, P., Popa, S., & Amescua-Seco, A. (2012). Implementing an IT service information management framework: The case of COTEMAR. *International Journal of Information Management, 32*(6), 589–594. doi:10.1016/j.ijinfomgt.2012.08.004

MacCallum, R. C., Widaman, K. F., Preacher, K. J., & Hong, S. (2001). Sample size in factor analysis: The role of model error. *Multivariate Behavioral Research, 36*(4), 611–637. doi:10.1207/S15327906MBR3604_06 PMID:26822184

Marrone, M., Kiessling, M., & Kolbe, L. M. (2010). *Are we really innovating? An exploratory study on Innovation Management and Service Management.* Paper presented at the Management of Innovation and Technology (ICMIT), 2010 IEEE International Conference on.

McNaughton, B., Ray, P., & Lewis, L. (2010). Designing an evaluation framework for IT service management. *Information & Management, 47*(4), 219–225. doi:10.1016/j.im.2010.02.003

Mesquida, A. L., Mas, A., Amengual, E., & Calvo-Manzano, J. A. (2011). IT Service Management Process Improvement based on ISO/IEC 15504: A systematic review. *Information and Software Technology, 54*(3), 239–247. doi:10.1016/j.infsof.2011.11.002

Microsoft-Inc. (2012). *Microsoft Operations Framework.* MOF.

Ndlovu, S. L., & Kyobe, M. E. (2016). *Challenges of CoBIT 5 IT Governance Framework Migration.* Paper presented at the CONF-IRM.

Nfuka, E. N., & Rusu, L. (2011). Critical Success Factors Framework For Implementing Effective IT Governance In Public Sector Organizations In A Developing Country. *AMCIS 2011 Proceedings.*

Nunnally, J. (1978). *Psychometric theory.* McGraw-Hill.

Othman, M. F. I., Chan, T., & Foo, E. (2011). *IT governance adoption in Malaysia: a preliminary investigation.* Academic Press.

Peak, D., Guynes, C. S., & Kroon, V. (2005). Information technology alignment planning—A case study. *Information & Management, 42*(5), 635–649. doi:10.1016/j.im.2004.02.009

Pollard, C., & Cater-Steel, A. (2009). Justifications, Strategies, and Critical Success Factors in Successful ITIL Implementations in U.S. and Australian Companies: An Exploratory Study. *Information Systems Management, 26*(2), 164–175. doi:10.1080/10580530902797540

Pollard, C., & Cater-Steel, A. (2009). Justifications, strategies, and critical success factors in successful ITIL implementations in US and Australian companies: An exploratory study. *Information Systems Management, 26*(2), 164–175. doi:10.1080/10580530902797540

Potgieter, B., Botha, J., & Lew, C. (2005). *Evidence that use of the ITIL framework is effective.* Paper presented at the 18th annual conference of the national advisory committee on computing qualifications, Tauranga.

Rouhani, S., & Zare Ravasan, A. (2014). A Fuzzy TOPSIS based Approach for ITSM Software Selection. *International Journal of IT/Business Alignment and Governance*, *5*(2), 1–26. doi:10.4018/ijitbag.2014070101

Salle, M., & Rosenthal, S. (2005). Formulating and Implementing an HP IT program strategy using CobiT and HP ITSM. *Proceedings of the 38th Hawaii International Conference on System Sciences*. doi:10.1109/HICSS.2005.276

Soja, P. (2008). Difficulties in enterprise system implementation in emerging economies: Insights from an exploratory study in Poland. *Information Technology for Development*, *14*(1), 31–51. doi:10.1002/itdj.20086

Spremic, M., Zmirak, Z., & Kraljevic, K. (2008). *IT and business process performance management: Case study of ITIL implementation in finance service industry*. Paper presented at the Information Technology Interfaces, 2008. ITI 2008. 30th International Conference on.

Tan, W.-G., Cater-Steel, A., & Toleman, M. (2007). *Implementing centralised IT service management: drawing lessons from the public sector*. Paper presented at the 18th Australasian Conference on Information Systems, Toowoomba, Australia.

Thakor, A. V. (1999). Information technology and financial services consolidation. *Journal of Banking &amp. Finance*, *23*(2–4), 697–700. doi:10.1016/s0378-4266(98)00104-6

Treiblmaier, H., & Filzmoser, P. (2010). Exploratory factor analysis revisited: How robust methods support the detection of hidden multivariate data structures in IS research. *Information & Management*, *47*(4), 197–207. doi:10.1016/j.im.2010.02.002

Van Bon, J., de Jong, A., Kolthof, A., Pieper, M., Tjassing, R., van der Veen, A., & Verheijen, T. (2007). *Foundations of IT Service Management: based on ITIL*. Van Haren Publishing. Retrieved from www.vanharen.net

Yamakawa, P., Noriega, C. O., Linares, A. N., & Ramírez, W. V. (2012). Improving ITIL compliance using change management practices: A finance sector case study. *Business Process Management Journal*, *18*(6), 1020–1035. doi:10.1108/14637151211283393

Zare Ravasan, A., & Mansouri, T. (2014). A FCM-Based Dynamic Modeling of ERP Implementation Critical Failure Factors. *International Journal of Enterprise Information Systems*, *10*(1), 32–52. doi:10.4018/ijeis.2014010103

Zare Ravasan, A., & Mansouri, T. (2016). A dynamic ERP critical failure factors modelling with FCM throughout project lifecycle phases. *Production Planning and Control*, *27*(2), 65–82. doi:10.1080/09537287.2015.1064551

Zare Ravasan, A., Mansouri, T., Mohammadi, M. M., & Rouhani, S. (2014). Supporting the module sequencing decision in ITIL solution implementation: An application of the fuzzy TOPSIS approach. *International Journal of Information Technologies and Systems Approach*, *7*(2), 41–60. doi:10.4018/ijitsa.2014070103

APPENDIX

Table 5. The interviews' quotes

Factor	The Interview Quotes
Project time management	When it comes to improvement in IT services, managers expect quick- wins. Hence time management comes to the fore. Time is the managers' top priority in decision-making (5*).
	Many organizations use the ITIL to increase their service quality and decrease undesirable outcomes. In this case, if organizations have an improper time management efforts; they will look for some alternative methods to improve their services, which may adversely affect the project outcomes (12).
Continuous performance assessment	IT service management requires an interaction between different organizational units. That's why the way salaries and rewards are compensated is a matter of concern (9).
	It's necessary to monitor the performance of everyone involved in a project as well as the processes closely and constantly (8).
Proper project management	For successful IT service management, a skilled manager is a must. Such a manger can clarify the costs of a project appropriately, make good use of the resources, and enhance the performance properly (7).
	An important issue in successful IT service management is setting an exact budget, because when a project is carrying out, many different financial issues might happen and it's necessary to predict them in advance (1).
Optimized budget allocation	In an IT service management project, the IT manager must be able to predict a near exact and optimized budget despite all economic fluctuations in order to meet the expectations of the executive team and let the project be carried out at an acceptable rate (11).
Achieving agreement among different stakeholders	One of the most important things to have a successful IT service management project is to consider various viewpoints of different parties. It can lead to a successful execution of the ITIL. Each unit has different expectations from the ITIL according to its own level of understanding and unique requirements. This issue can result in a slow progress in execution of the plans. Therefore, reconciliation between different viewpoints is inevitable (4).
Good relationships with project consultant	A good relation between the client and project consultant in a way that client don't view the consultant as a person outside the project, is critical. Such a good relationship can help the parties adjust themselves to project needs and requirements (8).
	Communication with the project consultant is a matter of paramount importance to have successful ITIL projects, because if the executive body cannot establish a good relation with the project consultant, he/she will no longer view the project as his/her first priority. It's necessary for the executive team to engender a feeling in the consultant that the plans are being carried out according to his/her advises and that there is no obstacle for his contribution in the project (2).
	It's worth mentioning that one of the issues that can affect the success of IT service management is to bring to the fore the role of the executive team. On many cases, the achievements of a project are presented in a way that all those achievements are exclusively contributed to the consultant and that the executive body doesn't have a big role in the success of the project. This issue will diminish their sense of belonging to the project and therefore they will make less effort in the project. That's why the relation between the executive body and the consultant is very important (13).
Involving employees in the project	Successful IT service management demands cooperation between different departments of an organization because if one department fails to cooperate in what it has been assigned to, then the project will be affected badly. All employees in related organizational departments should be involved in the project to get required project support (6).
Good project team	A critical success factor in the ITIL projects is to create a sense of harmony among different members of a project. Since these projects may encounter unexpected problems, having a proper empowered team, is very important (7).
Management belief in project outcomes	The achievements of the ITIL are both tangible and intangible and managers usually emphasize on the intangible ones. That's why when the results of the projects are offered, the intangible dimensions of them must be emphasized to draw the attention of managers (1).
	My successful experiments show that whenever the senior manager has a good understanding of the ITIL and executive processes, employees' resistance to change will decrease considerably (7).
An integrated process and service based approach	It is the organization's processes that can propose a value to the customer and the organization's services would be a tool for a better relation between these processes and human resources. That's why when an organization is managing IT services, it should measure the effects of them on organizational process and avoid considering these two as separate issues (10).
	Identifying the business of an organization and its IT capabilities are very important. The business of an organization boils down to its processes, which demands an efficient relationship between the business of an organization and IT services. A successful ITIL project requires a successful relationship between business-IT service strategies (3).

* Numbers denoted interviewee ID

Chapter 10
On Challenges for Implementing ISO Standards in Software:
Can Both Open and Closed Standards Be Implemented in Open Source Software?

Björn Lundell
University of Skövde, Sweden

Jonas Gamalielsson
University of Skövde, Sweden

Andrew Katz
University of Skövde, Sweden & Moorcrofts LLP, UK

ABSTRACT

Over the years, the importance of open standards has been acknowledged in EU and national policies. Formal (e.g., ISO) standards are often referred to in software development and procurement. Use of formal (ISO) standards and to what extent ISO standards can be implemented in open source software is considered, with particular reference to patent licensing. It is shown that not all formal standards are open standards and that FRAND commitments may impose major challenges for use of such standards. Further policies and procedures set by standards setting organisations (SSOs) regarding the notification of standards-essential patents (SEPs) present challenges for organisations wishing to implement standards in software. This chapter elaborates implications and suggests ways of addressing the challenges identified. Use of formal standards may create barriers for implementation in open source software and inhibit an open and inclusive business-friendly ecosystem.

DOI: 10.4018/978-1-5225-5320-5.ch010

INTRODUCTION

'Openness' including open standards and open source software is increasingly prevalent, but presents a number of challenges requiring effective policy and strategic initiatives. The European Commission (EC, 2013a, 2013b) and countries, such as the Netherlands (NOC, 2007), Portugal (Ballard, 2012), and the U.K. (UK, 2012a, 2015), have acknowledged the importance of open standards and have implemented initiatives accordingly.

Open standards have been discussed by researchers (e.g. Bird, 1998) and policy makers in the EU and different member countries (EU, 2004; SOU, 2009; EC, 2016; Permanent Representatives Committee, 2016) for a long time. Some member countries mandate use of open standards, based on definitions which require that standards are provided on royalty-free conditions, as part of national policy (e.g. NOC, 2007; UK, 2012a). Such policies aim to promote use of standards which have certain open properties and can thereby be used as a basis for implementation in software under different (proprietary and open source) software licenses. For example, the U.K. Government has a national policy which promotes and mandates use of specific open standards (UK, 2012a, 2012b, 2014, 2015). In Sweden, the minister responsible for municipalities has expressed support for the definition of 'open standard' set out in the European Interoperability Framework version 1.0 (Odell, 2009) and national framework agreements for public sector procurement of software in Sweden refer to open standards (EU, 2004; SOU, 2009; Kammarkollegiet, 2013, 2014a, 2014b, 2016) in relation to the standards which can be referenced in procurement. Further, in a response to the view that ICT standardisation needs an IPR policy which is "based on FRAND licensing terms" (EC, 2016) within the EU there are also seven EU members which express concerns and instead emphasise the importance of Open Standards "relying on Royalty Free intellectual property models in regard to software" (Permanent Representatives Committee, 2016) as a strategy for removing "barriers to innovation, particularly for SMEs" (Permanent Representatives Committee, 2016).

At the same time, there is confusion related to use of the term 'standard' and research shows that practitioners may regard products and applications (e.g. Microsoft Word) as standards (e.g. Lundell, 2011; Lundell et al., 2016). The study by Lundell et al. (2016) which is commissioned by the Swedish Competition Authority "found that many IT-projects in the Swedish public sector refer to closed standards which cannot be implemented in open source software" (Lundell et al., 2017). Further, the Director General for the Swedish Competition Authority expresses concerns for closed standards, as follows: "From a competition perspective it is often problematic when public sector organisations conduct IT procurement and express requirements for closed standards" (Lundell et al., 2016 (our translation)).

It has also been shown that there is confusion amongst policy makers between the two concepts of open standard and open source software (e.g. Egyedi & Enserink, 2013). Previous research results also show that many standardisation organisations neglect implementation issues and conclude that standards development and implementation activities "cannot be meaningfully separated" (Egyedi, 2007, p. 612). In particular, implementation of standards for representation of data over long life-cycles, beyond the life-cycle for any specific software, is of particular importance for long-term maintenance of data (Lundell, 2012). For these reasons, this study considers standards for representation of data and the potential for implementation of such standards in software, with a specific focus on the extent to which different standards can be implemented in open source software (i.e. software provided under a license which is recognised by Open Source Initiative (OSI, 2017)).

Previous research shows various positive effects from use of open standards (e.g. Friedrich, 2011; Ghosh, 2005; Krechmer, 2005; Lundell, 2012; Simcoe, 2006) and its potential for promotion of innovation has been stressed in recent research (e.g. Lundell, 2012; Lundell et al., 2016; Lundell et al., 2017). Further, reports from the European Commission (EC, 2013a) and the U.K. Government (UK, 2012a, 2012b, 2015) show considerable potential for innovation from the use of open standards, which can also reduce certain risks, for example to enable interoperability and prevent different kinds of lock-in effects with associated unwanted dependencies on suppliers and proprietary technologies. Friedrich (2011) states that the "prime example for how Open Standards can boost innovation are the internet and the world wide web". Open standards facilitate collaboration in development of software which can be provided under different types of licenses, including open source software. Such open collaboration represents an early exemplar of open innovation (Lundell & van der Linden, 2013) and open standards and open source software are used by most innovative organisations. For example, on 5 May 2014 Rachael King reported in the Wall Street Journal that a Samsung representative stated during an open source business conference: "Today, you can't build a product without using Open Source"[1].

It is noted that "standards are subject to legal rights which impact upon, not only their development, but also their implementation" (Fitzgerald & Pappalardo, 2009, p. 467). Specifically, writing software to implement the technical specifications embodied in standards also requires addressing a number of legal issues since "technical standards may incorporate patented technologies, while the specification documents of standards are protected by copyright" (Fitzgerald & Pappalardo, 2009, p. 467). Legal experts have argued that some commonly used (F)RAND licences are incompatible with open source licensing owing to the inability of the licensee to sub-licence to downstream recipients (EC, 2012) and the European Commission acknowledges that such licensing conditions for standards "create barriers for Open Source projects to implement the technical specification" (EC, 2013b). Further, in a public response to an open consultation concerning establishment of a national open standards policy representatives for the World-Wide-Web Consortium argued: "If a standard is covered by a patent or is in a FRAND system potentially covered by a patent, an open source developer risks his/her economic survival by implementing it because the patent owner can always go back and ask for past royalties." (Dardailler et al., 2012)

There has been a long tradition of using and referencing formal standards when developing and procuring software. For many years it has been permitted to explicitly reference formal standards (as opposed to informal standards) in public sector procurement (Lundell, 2011). However, there is limited knowledge concerning the relationship between formal and open standards, despite inclusion of requirements for open standards in policies in several countries.

The *overarching goal* of our study is to clarify and characterise use of formal and open standards in national policy and implications for implementation in software. Based on this, the study addresses *three specific objectives*. First, we review and report on conditions for use of ISO standards which are to be implemented in software. Second, we report on insights concerning open and formal standards, and elaborate on conditions for use of formal standards in scenarios when national policy imposes requirements for use of open standards. Third, we establish under what conditions open standards can be implemented in open source software, and contrast this with conditions for implementation of formal standards in open source software with a view to suggest ways for resolving potential inhibitors.

Specifically, the chapter makes *three novel contributions*. First, we elaborate conditions for use of ISO standards and highlight inhibitors for their implementation in software. Second, we elaborate conditions for use of open standards and formal standards, and present a conceptual model which can be used as an analytical device for analysis of specific standards. Third, we elaborate conditions for implementation of specific open and formal standards in open source software and thereby illuminate why certain formal standards are not open standards, and based on this elaboration suggest a number of ways of resolving certain inhibitors to implementation.

ON FORMAL AND OPEN STANDARDS

Definitions of 'standards' and the potential business benefit from use of standards in various contexts have been issues for ongoing discussion in the IT-field since the 1990s (e.g. Bird, 1998). Standards have a function of creating norms and can thereby "establish requirements that, though not expressed in formal legal instruments, are in practice mandatory and must be implemented by participants in certain fields of technical or business activity" (Fitzgerald & Pappalardo, 2009, p. 473). Formal standards are provided by organisations recognised as formal standardisation organisations (SSOs) (de Vries, 2006), which include ISO (International Organization for Standardization), ITU (International Telecommunication Union), ETSI (European Telecommunication Standards Institute), and national standardisation organisations (e.g. British Standards Institute). Industry consortia and other bodies (e.g. W3C (World Wide Web Consortium) which do not have the status of formal standardisation organisations also create standards which are adopted *de facto* by industry. Some of these informal standards are submitted to formal SSOs and each such standard may become a formal standard if the standard is adopted by a formal SSO. Broadly, formal standards[2] may, in accordance with relevant legislation[3], be specified as part of a procurement process, whereas informal standards do not automatically have this status[4] (although there exist circumstances in which informal standards may be specifically referenced).

Over the past decades, many consortia and industry fora have become involved in ICT standards setting (Jakobs & Mora, 2008). During this period there has also been a shift of control concerning content in technical standards from government to industry, something which in turn has led to difficulties in distinguishing IT-standards from proprietary technology controlled by individual companies. As stated by Krechmer (2001): "At the end of the nineteenth century, governments controlled the technical standards domain. In the past 100 years, the voluntary consensus standards process has developed and expanded to the point that we have difficulty distinguishing between a standard and a vendor's proprietary technology" (Krechmer, 2001, p. 100).

It has been argued that openness (as the term is used in relation to standards) "describes the fairness of the standardization process to all possible interest groups" (Krechmer, 2001). Further, Bird (1998, p. 76) states that "accessibility of the standard and the control of the standard" are two key principles for any definition of an open standard. First, concerning the principle for accessibility of a standard, it is argued that "any standard must be available to be implemented in product without encumbrance, no royalties, no excessive charges to gain access to the document" (Bird, 1998, p. 76). Second, concerning the principle for control of a standard, it is argued that the "standard must be evolved through a known and predictable process that is open to input and influence by all interested parties" (Bird, 1998, p. 76). However, we contend that these are not the only criteria determining openness. A further crucial criterion concerns its licensing conditions.

Besides copyright for "legal protection to access the standard documentation, there could also be some industrial property rights (that is, patents) on the technical solutions included and described in the standard itself. Therefore whoever acquires such documentation could still be prevented from adopting and implementing the standard, unless by paying another royalty to the possible patent holders." (Aliprandi, 2011, p. 12) In some cases, patentees are willing to license their patents impacting on a particular standard under so called (F)RAND ('(fair,) reasonable and non-discriminatory') licensing terms, something which would imply that all would be allowed licences on supposedly fair royalty terms (Lea & Hall, 2004).

As stated by Contreras (2015): "FRAND commitments are made voluntarily by participants in standards-development activities, among other things, to induce others to adopt their patented technology in a standard". Licensing of patented technology may involve significant revenues for companies. For example, Jakobs (2017) reports that: "Ericsson, Microsoft, and Nokia, with a combined annual income of over $1 billion from the licensing of patents, are examples of firms benefitting from the licensing of IP." (p. 85) Different SSOs have established their "own terms for a FRAND commitment, which could be phrased as an offer to negotiate a license on fair and reasonable terms – it is not a commitment to negotiate a contract at a set rate." (Pentheroudakis & Baron, 2017, p. 33) However, it should be noted that 'fairness' in specific licensing terms for IT-standards are context-dependent and often cannot be assessed a-priori since "the license terms are usually kept secret" (Lea & Hall, 2004, p. 83). For this reason, when an IT-standard is provided under such unknown (F)RAND licensing terms it seems clear that an organisation which plans to use such a standard for implementation in software will face significant challenges.

Based on a comprehensive analysis of cases, a study which analysed interpretation of FRAND commitments and the definition of FRAND royalties found that "Most cases before courts and competition authorities concerning SEPs are related to patent infringement damages, injunctions or antitrust." (Pentheroudakis & Baron, 2017, p. 10). Further, Lerner & Tirole (2015) found that declarations of SEPs that companies do to SSOs contributes to disputes between companies and it is claimed that many of "the critical assertions in these disputes relate to the commitments that firms have made to standard-setting bodies during the standard-setting process." (Lerner & Tirole, 2015, p. 548)

There are several factors which may contribute to significant uncertainty related to implications of FRAND commitments for different parties (e.g. Contreras, 2015, 2017; EC, 2012; ECSIP, 2014; Pentheroudakis & Baron, 2017). For example, when SEPs are transferred to new owners this contributes to increased difficulty in clarifying conditions for use of standards in that some new owners may be difficult or impossible to identify and reach. This may be due to the fact that declarations of SEPs to a patent database maintained by an SSO (e.g. the ISO patent database) are rarely updated, something which contributes to uncertainty concerning FRAND commitments. Concerns related to transfer of SEPs have been analysed in a commissioned study published by the EC (ECSIP, 2014), which found that: "SEP transfers contribute to the lack of transparency about SEP ownership, because their occurrence and the identity of new owners are not public information. A possible means to address this problem is the notification of SEP transfers by recordation." (ECSIP, 2014, p. 189) Further, in a situation where a patent owner decides to sell, or in some other way, transfer, a SEP that has previously been declared in a patent database maintained by one (or several) SSOs this introduces uncertainty concerning the applicability of the commitment made by the initial declarant to their successors in title with respect to the relevant SEPs, something which has been elaborated as follows by Contreras (2017): "When a SEP holder makes a commitment to license SEPs to manufacturers of standardized products, it is not always clear whether that commitment applies only to the SEP holder making the commitment, or whether it binds subsequent holders of the SEP." (Contreras, 2017, p. 38, 39)

It should be noted that not all informal standards bodies work on a Royalty-Free (RF) basis and many have adopted a RAND-based IPR-policy (e.g. OMA[5] and SMPTE[6]). Further, some standards bodies (e.g. OASIS and GS1) allow for both Royalty-Free and RAND-conditions. For example, OASIS allows for different IPR-models and when a Technical Committee (TC) is established the TC selects to operate under one of four different IPR-modes[7] (including RF and RAND terms). Similarly, the GS1 Intellectual Property (IP) Policy[8] states that "GS1 seeks to develop standards that can be practised on a royalty-free basis to the greatest extent possible", even if they also allow for RAND-based licensing commitments from companies.

Standardisation organisations have recently shown an increased interest in exploring the potential for strategic adoption of open source development practices and utilising open source projects for development and provision of standards. For example, several SSOs have organised workshops in order to scrutinise the potential adoption of open source in the traditional SSO contexts in order to improve standardisation processes. These workshops have led to various experiences and suggestions being discussed (e.g. Clark, 2016; Lundell et al., 2016b). From investigation of a specific example, it should be noted that it has been found that open source work practices may contribute to improved standardisation processes (Lundell & Gamalielsson, 2017).

RESEARCH APPROACH

For addressing the *first objective*, we reviewed conditions for use of ISO standards with a view to specifically considering implementation of standards in software. We considered patent policy and other information provided by ISO, such as information related to the ISO patent database. For addressing the *second objective*, we reviewed conditions for use of formal standards in scenarios when national policy imposes requirements for use of open standards. In so doing, we present a conceptual model aimed to clarify the dimensions openness and formality of standards. For addressing the *third objective*, we identified and analysed a relevant set of specific standards for populating and characterising the conceptual model to thereby establish insights and report on conditions (and potential inhibitors) for implementation of standards in open source software. To elaborate on the dimensions openness and formality of standards we undertook a review of conditions for use of standards provided by different formal and informal organisations (including ISO, IETF, W3C, etc.) in order to identify a relevant set of standards for analysis. To this end, our review considered information provided by standardisation organisations (ISO, IETF, W3C, etc.) concerning patent disclosures related to different standards and the extent to which specific standards have been recognised as open standards according to national policy. Our goal was to cover a representative set of standards which have either been recognised as open standards according to national policy (NOC, 2007; Standaardisatie, 2014; Standardisation Forum, 2011), or not so recognised, and within that set, to include standards provided by formal standards organisations on the one hand, and informal standards organisations on the other. Further, within the subset of formal standards, we focus on those for which declarations of patents have been made to the relevant SSO. Further, the study focused on standards for representation of graphics as such constitute a relevant group of (similar) standards for representation of data. An investigation of this type of standards is important for a number of reasons, including the fact that these constitute a basis for maintenance of valuable data for many organisations, sometimes over very long life-cycles. As an outcome of this, the analysis includes three specific examples of formal (ISO) standards (PNG[9], JPEG 2000[10], and TIFF/EP[11]). The analysis also includes a standard (PNG) which in addition to recognition by ISO is also provided as a W3C standard and one specific informal (W3C) standard (SVG[12]) which is not recognised by ISO. By inclusion of these four standards for analysis we cover examples of formal standards (PNG, JPEG 2000, and TIFF/EP), open standards (SVG and PNG), and standards which are both formal and open (PNG).

Specifically, we reviewed patent disclosures for all four selected standards provided by ISO and W3C. For the two standards provided by W3C (SVG and PNG) we investigated specific statements concerning patent disclosures provided at the W3C website[13]. Further, for the three ISO standards we reviewed the content in the ISO patent database[14] related to specific formal standards and collect data from all organisations that have declared patents related to specific standards. Data collection was undertaken by sending letters[15] (with reminders sent more than one month after initial requests) to organisations that have declared IPR related to the specific standard documents using contact information provided for each organisation. Responses led to additional requests for clarifications in several cases and in one case a conference call involving the researchers and legal representatives from the organisation controlling IPR related to specific standard documents.

RESULTS

On Conditions for Use of ISO Standards

ISO (International Organization for Standardization) is a formal standards body that has "published over 19 500 International Standards that can be purchased from the ISO store or from our members" (ISO, 2015a). A standard is defined by ISO as follows: "A standard is a document that provides requirements, specifications, guidelines or characteristics that can be used consistently to ensure that materials, products, processes and services are fit for their purpose." (ISO, 2015a)

The three formal standardisation organisations ISO, IEC and ITU have adopted a common patent policy which is applicable for ISO deliverables, IEC deliverables, ITU-T Recommendations, and ITU-R Recommendations. From their common patent policy, it follows that "a patent embodied fully or partly in a Recommendation | Deliverable must be accessible to everybody without undue constraints" (ISO, 2007, 2012, 2015d). In other words, the patent policy and associated guidelines clarify that a patent which impacts on use of a standard must be accessible to everybody. However, the standardisation organisations are not in any way involved in such arrangements, which is clarified as follows: "The detailed arrangements arising from patents (licensing, royalties, etc.) are left to the parties concerned, as these arrangements might differ from case to case." (ISO, 2007, 2012, 2015d) It follows that an organisation wishing to use a specific standard from any of these formal standardisation organisations must identify and obtain all necessary rights which are required for all patents impacting on the standard.

Patent databases are provided by the formal standardisation organisations as a means for clarifying which organisations control patents related to specific standards. The content in each database is based on information provided in patent declarations provided by patent holders and submitted to the standardisation organisation (ISO, 2012). For example, the ISO patent database contains 2854 declarations (24 October 2015) from organisations that control patents related to specific standards (ISO, 2015b). It should be noted that several of these declarations cover several patents impacting on a specific standard and that patent declarations made for one (or several) normative references are not visible when searching for patent declarations made for a specific standard number in the ISO patent database[16]. Further, most organisations have not disclosed the granted patent number(s) (or application number(s) if pending).

The patent databases are populated from data provided by organisations which make declarations to ISO. The current patent statement and licensing declaration form (which is uniform across ISO, ITU-T, ITU-R and IEC) currently allows the declarant to select one of three options. The options are numbered 1, 2 and 3, with the first being most favourable to a potential licensee, and the third being least favourable. The options are:

- **Option 1 ('Free of Charge'):** The Patent Holder is prepared to grant a Free of Charge license to an unrestricted number of applicants on a worldwide non-discriminatory basis and under other reasonable terms and conditions to make, use and sell implementations of the [standard]. [There are, further, two reciprocity options not discussed in this section of the chapter].
- **Option 2 ('RAND'):** The Patent Holder is prepared to grant a license to an unrestricted number of applicants on a worldwide non-discriminatory basis and under other reasonable terms and conditions to make, use and sell implementations of the [standard]. [There is a further reciprocity option not discussed in this section of the chapter].
- **Option 3 ('Unwilling to Grant'):** The Patent Holder is unwilling to grant licenses in accordance with the provisions of either 1 or 2 above. [In this case, the ISO strongly desires the declarant to notify ISO of relevant patents].

The ISO patent policy only impinges on parties which are involved in the standards-setting process. For this reason, it is possible that relevant patents held by other non-involved organisations may exist which are not reported (and hence not contained in the ISO database).

On Conditions for Use of Open and Formal Standards

This sub-section elaborates on use of open standards in national policy and reports on potential inhibitors for use of formal standards. In so doing we present a conceptual model aimed to clarify the dimensions openness and formality of standards.

The national policy in the Netherlands adopted the same definition of an open standard as the European Interoperability Framework (EU, 2004). According to the definition adopted in the Netherlands, a "standard is fully 'open' if:

1. The standard is adopted and will be maintained by a not-for-profit organisation, and its ongoing development occurs on the basis of an open decision-making procedure available to all interested parties (consensus or majority decision etc.).
2. The standard has been published and the standard specification document is available either freely or at a nominal charge. It must be permissible to all to copy, distribute and use it for no fee or at a nominal fee.
3. The intellectual property – i.e. patents possibly present – of (parts of) the standard is made irrevocably available on a royalty-free basis.
4. There are no constraints on the re-use of the standard." (NOVFS, 2011)

An important principle underlying the idea of an open standard is that it ensures that data can be interpreted independently of the software which generated it (Lundell, 2012). Further, a central characteristic of "open standards is that there are no restrictions regarding their use by ICT users and providers. Open standards are the opposite of closed standards, which do have restrictions" (NOVFS, 2011). Hence, of particular importance with open standards is that a standard which conforms to this definition can be implemented in software that is provided under different proprietary and open source software licenses. In addition to including a definition of open standard (e.g. EU, 2004) in national policies, some countries also provide specific guidance to their agencies. For example, based on the same definition (EU, 2004) the "Swedish National Procurement Services has published a list of open standards (see Kammarkollegiet, 2016), which all can be referenced as mandatory requirements in public procurement" (Lundell et al., 2017, p. 82). Further, all these "standards can be implemented and distributed under different licenses" for proprietary software and under all licenses approved by the Open Source Initiative for open source software (Lundell et al., 2017).

If a standard is open, both in the sense that the process leading to its adoption is open to all, and that the ability to implement it is not encumbered by difficult or expensive access to the standards documentation itself, as well as challenges raised by obtaining licences to patents which are required to implement the standard without infringing, then the largest number of actors, from small companies through to multinational organisations, will be able to be involved in implementation of the standard.

A number of initiatives from government (for example, in the UK the G-Cloud/Digital Marketplace project is intended to attract the maximum number of potential suppliers for cloud-based services to the public sector) (UK, 2012c) are based on the premise that increasing the number of actors, and levelling the playing field to ensure that SMEs are not excluded from the procurement process because of barriers that are not explicit, but implicit in the complexity of the process, will increase competition, and therefore lower the cost to purchasing organisations. Likewise, where the standards are able to be implemented by the widest possible range of organisations, this can be expected to have a similar pro-competitive effect. Naturally, these arguments do not apply solely in the public sector procurement process, and a recognition that open standards have the potential to lower costs for all purchasers, from individuals to private organisations to the public sector, will be welcomed by many organisations.

Even though organisations that control patents which they believe impact on specific standards are encouraged to declare details of such it is important to note that declarations are voluntary and that not all organisations that control patents impacting on specific standards are involved in a process with the standardisation organisations. In addition, there may also be several other organisations that control patents which impact on a specific standard that cannot be found in the database, perhaps because these organisations have no interest in standardisation and therefore have not declared that they control patents to the standardisation organisation. Further, to maintain an up-to-date content in the patent database the guideline stresses the importance of providing "contact information that will remain valid over time" (ISO, 2012, 2015d) and that contact information therefore "should be generic" (ISO, 2012, 2015d). However, timeliness and the validity of the content in a patent database is considered as a challenge for a standardisation organisation: "The ITU Telecommunication Standardization Bureau (TSB), the ITU Radiocommunication Bureau (BR) and the offices of the CEOs of ISO and IEC are not in a position to give authoritative or comprehensive information about evidence, validity or scope of patents or similar rights" (ISO, 2007, 2012, 2015d).

Figure 1 presents a conceptual model aimed to provide support for analysis of specific standards. The model consists of two orthogonal dimensions. One dimension in the model concerns openness of standards (open vs. closed standard), whereas the other dimension is formality of standards (formal vs. informal standard). For example, the ISO standard PNG (ISO/IEC 15948:2004) is considered to be an open standard in the Netherlands (by the Forum Standaardisate) according to the outcome of specific assessments (NOVFS, 2011) which require that a specific standard conforms to their definition of an open standard (EU, 2004). Hence, based on the outcome of their assessment, some ISO standards (e.g. PNG) belong in the upper right quadrant, whereas other ISO standards (e.g. JPEG 2000) belong in the upper left part of the conceptual model (see Figure 1). On the other hand, some industry consortia use closed processes for development and maintenance of standards (e.g. MXF-standards from SMPTE) which clearly do not fulfil the definition of an open standard (EU, 2004). Such standards belong in the lower left quadrant in Figure 1, and are therefore unsuitable for use in the public sector according to national policy in EU countries that have adopted policy for use of open standards. Hence, it follows that not all informal standards are open standards.

Note that both axes are continuous: a standard may be more or less open depending on any one of the applicable criteria. All other things being equal, a standard, the documentation for which is available free of charge, will be regarded as more open than a standard which costs EUR 5,000 to obtain, and there is clearly a gradation in between. Likewise, the degree of formality varies from standards bodies. For example, previous research indicate that informal processes utilised by consortium SSOs (e.g. OASIS and IETF) are feasible for small companies (e.g. Gamalielsson et al., 2015, p. 41) as contributing to such typically involves remote participation via the web compared to participation in formal standardisation which typically involves additional need for travel.

Figure 1. Openness vs. Formality of standards

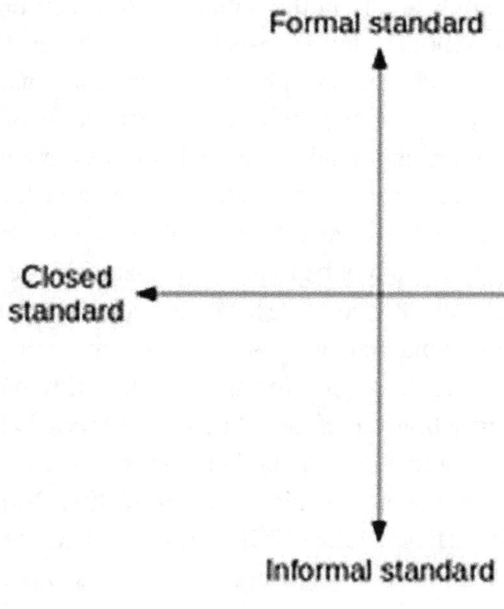

Conditions for Implementing Open and Formal Standards in Software

In this sub-section we present a review of four specific standards with respect to the definition of an open standard and populate the conceptual model with these standards. Of the four specific standards analysed, one informal (consortium) standard (SVG) is provided as a W3C standard (but not as an ISO standard). Further, one formal (ISO) standard (PNG) is included[17] in the explicit list of open standards published in the Netherlands (Standaardisatie, 2014), whereas two other formal (ISO) standards (JPEG 2000 and TIFF/EP) are not included in the explicit list and are consequently considered to be closed standards.

The SVG (Scalable Vector Graphics) standard ("Scalable Vector Graphics (SVG) 1.1 (Second Edition)", W3C Recommendation 16 August 2011[18]) is an informal (W3C) standard for representation of graphics included in the list of open standards in the Netherlands (Standaardisatie, 2014). SVG has been developed and maintained by the W3C SVG Working Group since 1998 and reached standard status (as "W3C recommendation") in 2001 (version 1.0). The current version (1.1) was released in 2011, and is currently widely deployed.

The outcome of the assessment made in the Netherlands implies that the informal (W3C) standard for the SVG standard is considered as an open standard and therefore belongs in the upper right quadrant of the conceptual model (see Figure 1). Considering the third criterion in the definition of an open standard (EU, 2004) and based on the outcome from our own analysis of patent disclosures provided by W3C it seems reasonable that the informal (W3C) standard has been included in the list of open standards (Standaardisatie, 2014).

The information provided by W3C concerning patents disclose that: "the SVG Working Group participants and the W3C are not aware of any royalty-bearing patents that are essential to implement the deliverables of the SVG Working Group, which includes all versions of the SVG specification and the SVG Mobile Profiles." Further, the same web page clarifies that one patent has been disclosed by one company and also acknowledges that the company does not believe it currently has any essential claims that fall within the specification of the recommendation as currently understood and interpreted by the company for implementors of SVG. Further, from information provided by W3C it is clear that several widely deployed (proprietary and open source licensed) software projects have implemented support for SVG[19], including several projects provided under the GPL license (e.g. Blender, Inkscape, and Scribus[20]). For these reasons it may be unsurprising that SVG is recommended for use in national policy concerning open standards, such as in the Netherlands.

The PNG (Portable Network Graphics) standard (ISO/IEC 15948:2004) is a formal (ISO) standard for representation of graphics included in the list of open standards in the Netherlands (Standaardisatie, 2014). Before accepted as an ISO standard, PNG was initially published in 1996 by the Internet Engineering Task Force (as RFC 2083) and soon after as a W3C standard (also in 1996).

The outcome of the assessment made in the Netherlands implies that the formal (ISO) standard for the PNG standard is considered as an open standard and therefore belongs in the upper right quadrant of the conceptual model (see Figure 1). Further, it should be noted that PNG is also provided as a W3C standard[21] (as "W3C recommendation" since 1 October 1996, i.e. before PNG was adopted as an ISO standard) and therefore also belongs in the lower right quadrant of the conceptual model (see Figure 1). Hence, based on adoption by both ISO and W3C, PNG belongs in both the upper right quadrant and the lower right quadrant of Figure 1. Considering the third criterion in the definition of an open standard (EU, 2004) and based on the outcome from our own analysis of the content in the ISO patent database it seems reasonable that the formal (ISO) standard has been included in the list of open standards (Standaardisatie, 2014). We note that no organisation is listed in the ISO patent database for the PNG standard (ISO/IEC 15948:2004).

The information provided by W3C concerning patents disclose that[22]: "the International Standards Organisation (ISO), the PNG Development Group and the W3C are not aware of any royalty-bearing patents that are essential to implement the Portable Network Graphics specification." Like for SVG, it is clear that several widely deployed (proprietary and open source licensed) software projects have implemented support for PNG[23], including several projects provided under the GPL license (e.g. GIMP, Inkscape, and TuxPaint[24]). For these reasons it may be unsurprising that PNG is recommended for use in national policy concerning open standards, such as in the Netherlands.

The JPEG 2000 standard[25] is a formal (ISO) standard for the representation of graphics which has not been included in the list of open standards in the Netherlands (Standaardisatie, 2014). The JPEG 2000 standard is developed by the Joint Photographic Experts Group[26] (i.e. a joint committee between ISO/IEC JTC1 and ITU-T). The standard consists of 14 parts (parts 1-6 and 8-15 since part 7 has been abandoned) and a number of technical corrigendum have been published for several parts. For example, part 1 of the ISO standard for JPEG 2000 (ISO/IEC 15444-1:2004) with all technical corrections and amendments consists of 12 standard documents in total and the other 12 parts consist of 33 standard documents in total which implies that the entire ISO standard for JPEG 2000 currently consists of 45 standard documents[27]. The total price for buying the 12 standard documents related to part 1 of the ISO standard is 310 CHF, whereas the total price for buying all parts (i.e. the 45 standard documents) of the ISO standard from ISO will be 2718 CHF. With this, the total cost for all these standard documents may be perceived as an inhibitor for adoption, in particular for small companies wishing to bid for contracts involving implementation of this standard. Further, in this case, the cost may seem to exceed what may be considered as reasonable for a "nominal charge" according to the second criterion in the definition of an open standard (EU, 2004).

Since the ISO standard JPEG 2000 is not included in the list of open standards (Standaardisatie, 2014) it is clear that the formal (ISO) standard for the JPEG 2000 standard is considered as a closed standard and therefore belongs in the upper left quadrant of the conceptual model (see Figure 1). Considering the third criterion in the adopted definition of an open standard (EU, 2004) and based on the outcome from our own analysis of the content in the ISO patent database it is clear that the formal (ISO) standard should not be considered an open standard and it is therefore not surprising that it is has not been included in the list of open standards (Standaardisatie, 2014). We note that nine organisations are listed in the ISO patent database for part 1 of the JPEG 2000 standard (ISO/IEC 15444-1:2004) and that, in total, 16 organisations are also listed in the ISO patent database for all parts of the standard. Further, no information is provided in the database for most of these organisations concerning which patents these organisations control and some organisations even declare that they are willing to provide patent licenses under conditions which are not compatible with open source software.

To investigate the situation further we undertook a specific analysis aimed to clarify under which conditions the ISO standard can be used. This involved contacting each organisation listed in the ISO patent database with a set of questions sent in a letter[28] (sent via air-mail since email addresses were not available for most organisations in the database). In total, we sent questions to 16 organisations and after reminders (sent more than one month later) we have received some responses from three organisations. In total, letters from five organisations have so far been returned[29] since the contact information provided in the ISO patent database was incorrect or outdated (with the message "recipient unknown").

Amongst received responses, one organisation responded that they were unwilling to grant a license for their patents that would allow implementation in software to be provided under the GPLv3 license (i.e. a license which is recognised by both Open Source Initiative (OSI, 2017) and Free Software Foundation (FSF, 2017)). Another organisation explicitly stated that they decline to respond, whereas another declined to provide us information concerning which patents they control (the response was "we have at least 3 patents" on the specific standard). It should be noted that an annex in part 1 of the ISO standard (ISO/IEC 15444-1:2004) explicitly lists several organisations which have mentioned that they are willing to provide their patents (for part 1) free-of-charge but also that the patent database contains information with additional organisations since new patents have been declared after the publication of the ISO standard. Further, in one case the organisation that had declared patents declined to clarify conditions for use of the standard and instead referred our request to ISO. However, from a dialogue with ISO representatives we note that ISO as an organisation does not engage in clarifying conditions for use of standards for which patents have been declared. Hence, in this case specific questions concerning conditions for implementation of the standard were left unanswered (as one organisation controlling patent(s) for JPEG 2000 referred to ISO, and vice versa). In addition, it should also be noted that the ITU-T patent database contains information about a different set of organisations which have declared patents.

The TIFF/EP standard[30] is another of the formal (ISO) standards for representation of graphics which has not been included in the list of open standards in the Netherlands (Standaardisatie, 2014). The TIFF/EP (Tag Image File Format / Electronic Photography) standard is developed and maintained by the International Organization of Standardization (ISO). We note that 19 organisations are listed in the ISO patent database for the TIFF/EP standard. From this it follows that TIFF/EP is a closed standard according to the third criterion in the definition (EU, 2004) and that it therefore belongs in the upper left quadrant of the conceptual model (see Figure 1).

To investigate the situation further we undertook a specific analysis aimed to clarify under which conditions the ISO standard for TIFF/EP can be used. This involved contacting each organisation listed in the ISO patent database with a set of questions sent in a letter (sent via air-mail since email addresses were not available for most organisations in the database). In total, we sent questions to 19 organisations and after reminders (sent more than one month later) we did not receive any responses. In total, letters from four organisations were returned[31] since the contact information provided in the ISO patent database was incorrect or outdated (with the message "recipient unknown").

ANALYSIS

First, as one objective for our study involves analysing the conditions for use of formal (ISO) standards where national policy imposes requirements for use of open standards, it seems clear that the conceptual model presented can also support an analysis of the situation with respect to informal standards (e.g. those developed and maintained by various consortia, such as the W3C). However, it should be noted that several standards are recognised and maintained by more than one organisation. For example, the PNG standard is maintained both by ISO and by the W3C which implies that this specific standard could be seen as both a formal standard and an informal standard. Further, previous research shows that there are "many misconceptions and significant unawareness" concerning differences between standards for representation of data and their implementation in software amongst decision makers in public sector organisations (Lundell & Gamalielsson, 2013). By way of example, the Dutch Parliament requested in

2010 the Court of Audit to undertake an assessment of the benefits of open standards and open source software in the government IT which resulted in a criticised report that was published in March 2011 (Egyedi & Enserink, 2013). According to a critical review by Egyedi and Enserink (2013), the assessment was criticised for omitting to address effects of open standards on the market.

Second, our results show that the third criterion (about IPR) in the definition of an open standard adopted in the Netherlands brings with it several complexities for any organisation wishing to use a specific standard. Since different organisations developing and maintaining standards have different policies concerning patents it is important to realise that not all formal standards are open standards. Lea and Hall (2004) bring clarity to this complex issue as follows: "At the time patent policies first emerged, it was clear that few patentees would be willing to license on a 'royalty-free' (RF) basis in the sense of absolutely free: therefore, in some quarters, royalty-free came to mean 'for a lump sum up front' and, subsequently, the alternative concept of '[fair,] reasonable and non-discriminatory' ([F]RAND) licensing was developed, whereby all-comers would be allowed licenses but on royalty terms supposedly both fair across the industry as a whole and as between each of the licensees. This latter often cannot be proven, since the license terms are usually kept secret)." (Lea & Hall, 2004, p. 83)

Third, our results clearly illuminate differences between the investigated formal standards. Our results show clear differences between formal standards which are also 'open' compared to those which are also closed standards (e.g. the ISO standard for JPEG 2000). Our results show that a key issue impacting on the conditions under which formal standards can (or cannot) be implemented in open source software concern the third criterion in the definition of an open standard that has been adopted in the Netherlands (EU, 2004). From this, it is clear that while some formal (ISO) standards (e.g. the ISO standard for PNG) can (and have been) implemented in a variety of different open source software applications, it is also evident that for other formal (ISO) standards (e.g. the ISO standard for JPEG 2000) it is not possible to obtain licenses for the standard to permit implementation in open source software. For the investigated standards, our results clearly illuminate that amongst organisations which control patents impacting on the standard there is no interest in providing licences for those patents which would allow implementation in software to be provided under the GPLv3 licence. Further, given that the vast majority of organisations that have declared that they control patents impacting on the investigated standards either cannot be reached or decline to respond it is clear that any organisation considering an implementation of the standard under any open source software license would face significant risks. Hence, implementing a standard without first having obtained all necessary rights for use of the standard cannot be recommended to anyone.

Fourth, our results show that for several formal (ISO) standards a number of organisations have voluntarily declared that they control patents which are believed to impact on conditions for use of the specific ISO standard and also that investigations for clarifying the conditions are not easy to do in practice since several organisations cannot even be reached. As most organisations that have declared patents in the ISO database related to JPEG 2000 and TIFF/EP cannot be reached, it is implied that there is considerable uncertainty concerning conditions for use of such standards. Further, the complexity related to RAND conditions has been elaborated in previous research (e.g. Fomin et al., 2008). For example, in their study they report that: "Another expert elaborates on why (F)RAND issues, mentioned above, are important. He notes that IPR is the secret (in many senses of the word) tool of vendors to manipulate the standards process. The patent process can be easily manipulated to exclude competition, and flagrant abuses of it can be used to preclude challenge by smaller companies who cannot afford the fight even when they are right" (Fomin et al., 2008). Further, the same study argues that "without meaningful policies on

IPR, operating on a global basis, standardization will be manipulated" using such business strategies (Fomin et al., 2008).

Fifth, adoption of requirements for use of standards in national policy may have a number of different effects for different stakeholder groups. Our results show that some formal (ISO) standards are provided under conditions that are unclear even after significant efforts for clarifying conditions for use of such standards. Such uncertainty may inhibit competition and impose challenges related to requirements expressed in public sector procurement. In the European context, legislation and directives for public procurement (Directives 2004/17/EC and 2004/18/EC) aim to achieve procurement practices that stimulate a fair and competitive market based on the important principles of transparency, non-discrimination and equal treatment (Lundell, 2011). Specifically, references to a technical specification "shall not refer to a specific make or source, or to a particular process, or to trade marks, patents, types or a specific origin or production with the effect of favouring or eliminating certain undertakings or certain products." (Directive 2004/17/EC (Article 34) and Directive 2004/18/EC (Article 23)). For this reason, results from an analysis of current practices in Swedish public sector organisations show significant lock-in and illuminate inclusion of requirements for specific standards which refer to specific products, trademarks and imply dependencies to access of specific patent licenses (Lundell, 2011; Wessman, 2013). Further, results from an analysis published by the Swedish competition authorities (Wessman, 2013) show that only a minority of decisions impacting on procurement consider any strategy for avoiding lock-in effects.

Finally, based on our results it is evident that some formal (ISO) standards analysed in this study cannot be implemented and deployed as open source software because of lack of clarity concerning IPRs. Further, when conditions for use of specific ISO standards cannot be clarified and all necessary patent licenses therefore cannot be obtained (as experienced in this study), it follows that such standards cannot be used for implementation in software under any software license without significant risks.

IMPLICATIONS FOR USE OF ISO STANDARDS AND IMPACT ON POLICY

Based on the analysis of our results, from having investigated the process of obtaining information on the conditions for use of specific standards from W3C and ISO, this section elaborates on the implications of our findings for practice. First we elaborate on the implications for small companies[32] wishing to implement ISO standards in software, and second, we suggest how the ISO declaration form can be improved to allow a further option for holders of standards-essential patents (SEPs), to avoid some of the problems identified.

Implications for Companies Wishing to Implement Standards in Software

A small company wishing to implement a standard in software will typically have one or more of the following characteristics:

1. It will not have or control patents of its own;
2. It will not be party to any patent pool arrangement[33];
3. It will not have access to an in-house legal department or have in-depth knowledge of the patent landscape;
4. It will not be actively involved in any standards setting process; and

5. It will wish to implement the standard using open source software components, which may include software licensed under the GPL family.

In seeking to implement an ISO standard in software, a prudent company may take the following steps, each of which raises a potential barrier to such a company implementing a standard, the implementation of which is dependent on patents:

1. The company must acquire a copy of the standard (the cost of which may itself be prohibitive, especially where the standard contains multiple parts and normative references to other standards copies of which themselves may need obtaining in a similar way. By way of example, the results of our analysis show that the cost to acquire the JPEG 2000 standard documentation, even without considering the cost of documentation relating to normative references, would be 2718 CHF per copy);
2. It must review the ISO database to determine whether the standard (including other standards it relies on, such as normative references) is subject to declared patents;
3. It must attempt to determine what patents are included in each declaration covering the standard (including other standards it relies on), OR if that is not possible, it must determine whether it is going to engage in a negotiation for licences as another way of reducing potential risk;
4. It must determine whether its own implementation of the standard covers the patents in question, and if so, it must contact, and engage in negotiations with, the declarant organisation (or, to the extent that it is possible to do so, decide to exclude provision of its software from the specific market(s) in which relevant patents are registered).

Steps 3 and 4 must be repeated for every declaration contained in the database which is listed against the standard (including those listed against relevant and normative references referred to in the standard) which the company wishes to implement. Clearly, a failure to reach an agreement with the declarant organisation in even one case where the relevant patent impinges on the standard will potentially render the implementing company open to a patent infringement claim.

The above steps present a number of challenges in terms of cost and risk to the implementing company, which we call ‘process barriers’. We elaborate further on some of these issues.

To obtain a licence, the implementing company first needs to know what the relevant patents are. Several SSOs (including ISO) have databases populated by organisations which have taken part in the standards setting process, and which have declared that they hold patents which are essential in implementing the standards. They do not necessarily declare (and it is not required as part of the ISO declaration process to declare) which specific patents they hold.

From our experience during the data-gathering process, it is by no means straightforward to obtain the information necessary to enable identification of the relevant patents. We even had problems in many cases eliciting any form of response from organisations listed in the database which claim they are in control of SEPs. We found that contact information provided in the database was no longer accurate for some of the declarations, leading to bounced emails and undelivered airmail letters.

The SSOs are clear that all patent licence negotiations are strictly between the company seeking a licence, and the organisation making the declaration, with the SSO taking no part in that process. However, where it is impossible even to engage the vast majority of declaring organisations, this process is clearly not fit for purpose.

As identified in this study, in the case of the ISO standard TIFF/EP, none of the 19 organisations identified in the ISO database responded to our questions. In the case of JPEG 2000, we were able to establish contact with three organisations from the 16 identified in the database, all of which we attempted to contact.

From the information provided in the dialogues where we were able to engage with the declaring organisations, we were able to establish that one referred us to ISO (which, as we have established, is not prepared to become part of licensing discussions between parties), one was prepared to license, in theory, on RAND terms (but did not provide a draft license), and the third was not prepared to grant a licence which was compatible with the GPLv3 licence. Even if we had received licences from the three respondents that we were able to engage in a dialogue, in the absence of licences from the other declarants (and assuming that those declarants genuinely held SEPs), we would still have been unable to implement the standard without infringing.

Were an enquiring organisation to proceed to implement the standard in software without a complete set of appropriate licences to SEPs, it would be at risk of patent claims from the holders of the SEPs (other than those from whom it had received licences). Those claims could lead to monetary damages (which may, under the US patent regime, have been tripled, given that disclosure on the database would put the enquirer on notice that patents potentially existed) and an injunction restraining any further use or distribution of the software. It is no guarantee of safety that a company is operating solely in one jurisdiction where no SEPs exist. Since software development practices currently tend to involve US based repositories such as GitHub, and since, in any event, exploitation and marketing of the software is likely to occur on the internet and hence worldwide, claims may arise from any jurisdiction where SEPs are registered.

SSOs do not, in general, guarantee completeness of the database or that it is up to date. For example, ISO (2015c) states that "ISO does not verify the veracity or accuracy of the information nor the relevance of the identified patents/patent applications to ISO Standards."

Even assuming that the database is comprehensive, it will not be clear to the implementing company whether:

1. The (part of the) standard it wants to implement is covered by a patent held by the disclosing company or not;
2. The patent(s) in question are still in force, or have expired;
3. The patent(s) in question in fact impinge on the implementation of the standard at all.

Assuming that implementation of the standard in question will infringe a patent which is disclosed, the implementing company will have to enter into negotiations with the declarant organisation, on the basis, as selected by that organisation of either Option 1 (Free of Charge) or Option 2 (RAND) (as such terms are set out in section 4.1).

This process is expected to occur outside the context of ISO. It also assumes that the company wishing to implement is, in practice, able to find, and successfully engage with, the declarant organisation.

If, on the other hand, the company wishing to implement is a larger organisation, by reason of its greater resources, it may, in practice, have a number of other options open to it. It may be able to:

1. Determine that the means by which it wishes to implement the standard does not, in fact, infringe any patent, (or modify the means of implementation to avoid the patent) and so no licence is required;

2. Determine that it already has access to a licence to the relevant patent, either through a patent pool arrangement, or through a pre-existing licence or cross-licence;
3. Negotiate with the patent holder outside the scope of the free-of-charge or RAND option, possibly by cross licensing its own patents, or by joining a relevant patent pool;
4. Seek to invalidate the patent; or
5. Assess the risk that there may be a successful patent claim, and determine it is prepared to accept that risk.

Options a, d and e (above) are also potentially accessible to the smaller company, but are likely to be at a disproportionate cost.

In either case, the implementing company has the option to implement the standard and risk a claim from a patent holder (option e above). If that occurs, the larger company will have the advantage of:

1. Using its own patent portfolio to offer a cross-licensing deal;
2. Using its superior resources to fight the claim, either by invalidating the patent or demonstrating that its implementation does not infringe the underlying patent.

An organisation will experience minimal process barriers where it is possible for that organisation to implement the standard in software without concern for patent infringement, and be permitted to use software which is covered by any of the entire range of licenses approved by OSI (2017) and FSF (2017). It is the case that companies, even where they have no desire to supply software under an open source licence, will nonetheless frequently use open source code as a component in their product as a consequence of near-universal modern software development practices. Hardly any new system is built without using the high quality, easily accessible code which is often available as part of open source development projects (Simeonova, 2015; Milinkovich, 2015). Barriers to entry are likely to impact smaller companies disproportionately[34].

For a company wishing to implement a standard in software it is critical that it can obtain all necessary licenses to do so. In seeking to implement a standard, such a company should initially review the relevant SSO's database to determine whether its implementation is likely to infringe any patents which have been declared by a declarant organisation.

Since details of the specific patents declared by the declarants are not necessarily disclosed in the database (although the declarants are encouraged by the SSO to do so), further investigation may be necessary to even determine what patents may possibly be infringed. For example, previous research shows that it is not uncommon for companies to declare that they own essential patents without specifying any details about these patents (Bekkers & West, 2009, p. 83).

If a company is comfortable that its implementation will not require a licence under a particular patent – perhaps because

- It is not implementing the part of the standard which is covered by the patent, or
- The patent has expired, or
- It does not wish to implement or provide the software product in the jurisdiction where the patent is registered, or
- On close reading of the patent, it is not essential in implementing the standard

then it will, naturally, not need to approach the relevant declarant. In all other cases, to minimise its risk of infringement in implementing the standard, it will need to ensure that relevant licences are in place and engage with the patent holder accordingly.

If any patents are declared under Option 3, this presents a serious risk to a company wishing to implement the standards (in practice, this is unlikely to occur, as the ISO's wider patent policy excludes from the standardisation process essential patents which are unavailable under RAND, or Free of Charge terms). Where patents are declared under Options 2 or 1, the company wishing to implement will need to engage with each relevant declarant organisation to negotiate a patent licence (on RAND or Free of Charge terms).

It is important to recognise that when an organisation declares to ISO that it is willing to provide its patents related to the specific standard at hand according to Option 1 (or indeed Option 2) this does not imply that the organisation *has* provided a license for these patents or that a license is automatically available on defined terms. A declaration under Option 1 implies that the organisation has declared that it is committed to provide its patents on Free of Charge terms and that it is willing to engage in a negotiation with the organisation which wishes to use the standard on that basis.

For many companies it is important to be able to implement the standard in open source software that is to be provided under common open source licences, including licences from the GPL-family. As GPL-licensed open source software is the most common type of open source software (under various accepted measures), and previous research from the embedded systems domain has showed that this is particularly essential for consultant companies, it is important for many companies to be able to use and implement standards in software that is provided under the GPL and its family of licences (Lundell et al., 2011).

Towards an Improved Standardisation System for Stakeholders

The standards ecosystem involves a number of different stakeholder groups. These include: legislators; SSOs; organisations which are involved in the standards setting process (and which may hold patents which impinge on the standards); and companies which wish to implement the standards.

For companies wishing to implement standards it would clearly be of significant benefit if the friction inherent in this process could be reduced by minimising the number of decision points that occur in the process. As our results have shown, one set of decision points arises in relation to the implementation of ISO standards at the stage where the implementing company undertakes an analysis of the patents which may possibly be infringed by the implementation of the standard in software, based on the data disclosed in the ISO database (or possibly elsewhere), and entering into negotiations with the parties it identifies as (possibly) having relevant patents to obtain appropriate licences. Any initiative which simplifies this process and brings clarity will be welcomed by the implementing company by both reducing the time and effort involved in the process, and providing improved data on which the company can assess its commercial risk.

One way forward would be to introduce a mechanism whereby a relevant licence may, if selected by the patent holder as part of the patent declaration process applicable to a specific standard, automatically be available on suitable terms covering patents which may possibly impinge on an implementation of the standard. In such a case, the implementing company would no longer need to:

- Assess whether the declared patent did cover its implementation of the standard;
- Consider not trading in a jurisdiction where a relevant patent was registered, if it perceives negotiating the licence to be too complex (to the extent that it is possible to do so, given the global reach of the internet, and its almost universal role in software distribution);
- Assess the risk of implementing the standard without a suitable licence;
- Enter into negotiations with the declaring body;
- Reject methods of implementation (for example, the use of GPL code) which may be incompatible with any licence granted;

all of which potentially represent a significant investment and time and, possibly (for small companies in particular), the purchase of external professional advice.

To minimise friction, and maximise the ability for all companies, including SMEs, to implement standards under transparent conditions in a way which enables them to more accurately assess and manage risk (thus encouraging competition and innovation by allowing more actors with diverse interests to enter the market), we propose that an additional option for patent-holders declaring their approach to patent licensing with respect to a specific standard may be offered to patent-holders as follows (we call this 'Option Zero'):

1. By selecting this box, the declarant agrees to license, perpetually, and irrevocably except as specifically set out below, on a worldwide, royalty-free, non-exclusive basis to all third parties ('licensee') seeking to make, use and sell software implementations of the above document solely to the extent that such software implements the standard;
2. The licence set out in (1) above arises automatically in favour of all licensees without the need to execute any document;
3. The licence set out in (1) above may only be revoked against a specific licensee where the licensee is in breach of the licence set out in (1) above, and, having received notice specifying such breach fails to cure it within 30 days.

The most significant characteristics of Option Zero, in contrast with Option 1, are:

- An Option Zero licence is automatically available. There is no need to negotiate it;
- An Option Zero licence is explicitly perpetual and irrevocable, except in specified circumstances. As well as protecting the licensees' investment in development of a compatible solution, this is intended to aid compatibility with various definitions of 'open standard' (including version 1 of the European Interoperability Framework); and
- An Option Zero licence is explicitly designed to be compatible both with a proprietary software development model, and an open source development model.

This last point is somewhat problematic as regards certain open source licences. It has been noted (Mitchell & Mason, 2010), that there may be a fundamental mismatch between the requirements of some open source software licences (notably those of the GPL family), and parallel patent licences, unless the terms of those licences are extremely liberal.

In brief, as an analysis of this point is beyond the scope of this chapter, there is a spectrum of opinion as to the interaction between the GPL and typical patent licensing structures. On the one hand, the least problematic view is that so long as the implementer of a standard using software licensed under a GPL family licence can distribute that software in such a way that the recipient has no fewer rights that the implementer has, then there is no problem (in other words, if the implementer has the benefit of a specific patent licence, any recipient of the implementer's code will also need to have the benefit of a patent licence on the same terms, either directly from the licensor, or as a sub-licence from the implementer). On the other hand, a more problematic view is that the implementer of a standard containing third party GPL code has to cause any recipient of that code to receive a licence (including a patent licence) which would enable the recipient to exercise ***all*** rights permitted under the GPL, including the right to modify the code so that it no longer implements the standard, but nonetheless still has the benefit of a relevant patent licence. Clearly, such a broadly drafted patent licence would render the economic value of the patent close to zero, so it would be unlikely to be acceptable.

The approach we have taken falls midway between these two extremes and is guided by the scope of the patent licence in GPLv3 (which does not expect a distributor of GPL software to provide a blanket patent licence to recipients covering all possible modifications of the code transferred). If the ISO were to adopt something similar to Option Zero, we would expect this to follow consultation with relevant stakeholders in the worlds of both proprietary and open source software, and, in particular, seek assurances from organisations like the Free Software Foundation (custodian of the GPL family of licences) that an approach like this would be compatible with both the spirit and the legal terms of the relevant licence (noting that such organisations' comments are at best only persuasive: legal interpretation of the licence is ultimately the responsibility of the courts of any jurisdiction in which the licence is litigated).

Many companies involved in the standards setting process are also companies which are heavily involved in the development of open source code, and will be familiar with the concept of licensing their patents on a royalty-free basis (e.g. HP, 2015) in a way which is typically compatible with open source licences. Many licences (GPLv3 and LGPLv3, Mozilla 2.0 and Apache 2.0) contain widely accepted and understood patent licensing clauses, which typically limit the licence granted to the scope of the claims implemented by the version of the software as distributed by the patent licensor. In other words, and by way of example, if a recipient received some word-processing software from the licensor, the recipient cannot expect the patent licence to cover the recipient's use of it should the recipient choose to modify the word processing software to act as a control system for a nuclear power plant. Any open source licence will grant the recipient under its copyright provisions the freedom to make those changes, the patent sections of the same licence may not. By analogy, we believe it is within the scope of acceptable open source practice to limit the scope of the patent licence granted to implementations of the standard. The recipient may, under the open source licence, modify the software so that its functionality no longer meets the requirements of the standard, but, should Option Zero apply, it will not receive the benefit of the patent licence in so doing.

By adopting a mechanism which is similar to that which organisations, including those with large patent portfolios, are already employing to license their patents under various open source licences, we believe that the introduction of similar licence terms as an option in the ISO patent declaration would not be an unusual step. We stress that Option Zero is, as the name implies, only an option which the patent-holder may select, and there is no compulsion in selecting Option Zero. For the many organisations which already license software under licences such as GPLv3 or the Apache licence which contain a similar patent licensing clause, we submit that this option is entirely consistent with, and readily reconcilable with, their current patent licensing policies.

Option Zero does raise a number of interesting issues, which could be considered as sub-options. The patent sections of open source licences frequently contain provisions which are intended to foster reciprocity by ensuring that any actor taking advantage of such a patent licence loses the licence if they themselves assert patent rights against a third party in certain circumstances. We would suggest that suitable wording to implement this would be to allow the declarant to terminate the licence against a specific licensee where:

the licensee asserts any patent claim against a third party, where that third party is implementing the standard and the patent claim covers the implementation of that standard (where 'implementation' includes distribution and importation of software which implements the standard)[35].

In other words, where someone takes the benefit of an Option Zero licence, that licence will terminate if that person starts issuing claims for infringement of its own patents where the claim covers the alleged infringer's implementation of the relevant standard.

Note that the currently-existing Options 1 and 2 also contain sub-options which allow (in the case of Option 1) the declarant to specify that it is only prepared to grant a licence to entities which are prepared to grant a similar Free of Charge licence themselves. Similarly, in the case of Option 2, there is an equivalent sub-option which allows the declarant to specify that it is only prepared to grant a licence to those entities which are prepared to grant a RAND licence (which may not be Free of Charge) themselves.

Another way of achieving a similar aim, which may be neater conceptually, is to state that the Option Zero reciprocity sub-option requires that any licensee taking the declarant's licence is itself deemed to have granted a licence to its own SEPs under Option Zero.

Notwithstanding these suggestions, we have identified two further issues which may impose risk for potential standards-implementing entities.

The first issue involves reliance on the information contained in the patent disclosure database itself. The second issue involves patent ambush, where an organisation initiates a patent claim relating to an implementation of a standard, where that patent has not been disclosed at all. We acknowledge that for the ISO standards investigated in this research, the question of undisclosed patents did not become relevant (because, as our results show, we were unable even to establish contact with many declarant organisations). However, this is an issue which is commonly understood in this field (see, for example, Baird (2007); EC (2007, 2009); Updegrove (2009)), and in the interests of complete disclosure we consider it appropriate to refer to this issue and the extent to which it impacts on our proposals.

In relation to the first issue, we have found that the content in the ISO database is not always accurate or up to date, and, at the very least, ISO should (1) take steps to check the content of its database from time to time, and/or (2) consider implementing sanctions on declarants which fail to keep their entries up to date. To improve the situation, one possibility would be to implement a rule that if a declarant fails to notify ISO on an annual basis that its information remains the same, or if it has changed, the nature of the changes, then that declarant is deemed to have selected Option Zero. Note that once an option has been selected, ISO's rules state that it may only be reclassified to a lower number (e.g. a declarant may make a submission that it is selecting Option 1 in relation to a specific disclosure, only where it has already made an Option 3 or 2 disclosure, or none at all). To retain consistency with this rule, an Option Zero selection would be perpetual (at least until all potentially relevant patents have expired). Note that an inaccuracy in an Option Zero declaration (given that it contains the licence itself, and not an invitation to negotiate a licence), is less impactful on licensees and potential licensees than any of the other options.[36]

From the declarants' perspective, an Option Zero may be attractive (compared to Option 1 – given that both Options Zero and 1 bear no fee or royalty), for reasons of reducing the time and effort involved in responding to, and negotiating individual licences. Further, Option Zero may also have the positive effect of creating an ad-hoc patent pool since declarants are themselves likely to be implementing the standard, and will therefore require their own licences from the other declarants. However, if Option Zero is selected, there may also be possible disadvantages for the declarant:

- The declarant loses the opportunity to negotiate individual terms with the implementer;
- The declarant will not automatically have a list of implementing companies.

These issues also arise when comparing Option Zero with Option 2. In this case, there is the additional disadvantage, from the declarant's perspective, that it will be unable to charge a fee to the licensee.

In relation to the second problematic issue, we have identified concerns relating to patent ambush. Patent ambush occurs when an organisation initiates patent litigation in relation to implementations of a standard, where is has not disclosed those patents during the standards-setting process, either because the entity in question chooses not to disclose the patent in breach of the relevant SSO's rules, or because the entity in question is not part of the standards setting process itself in the first place. The only solutions to this issue which present themselves to us require either the non-trivial implementation of legislation, which, because of the international nature of standardisation would necessarily require a treaty, or some form of international concerted effort requiring states to coordinate approaches in competition/antitrust law. One form of international legislation may provide that entities which hold a patent which impinges on a proposed standard, drafts of which standard have been made publicly available through ISO or another recognised international SSO, are deemed, unless they respond to the standards setting organisation and select a specific option in the patent declaration, to have granted an Option Zero licence in respect of that patent. Alternatively, it may be that competition or anti-trust law provides a solution, possibly by mandating compulsory licensing in this context (but see Chronopoulos (2009), which also covers a number of other possible solutions, analysis of which is beyond the scope of this chapter). We are under no illusion as to the significant challenges raised in implementing these suggestions.

CONCLUSION

Our study highlights that open standards are important for a number of different reasons, both in private and public sectors, something which is apparent in some countries based on recognition and adoption in national policy.

The study shows that while some formal (ISO) standards (e.g. the ISO standard for PNG) are also considered open standards according to the definition adopted in several national policies (including the one adopted in the Netherlands), it is also clear that other formal (ISO) standards (e.g. the ISO standard for JPEG 2000) are not considered open standards by the same policies. Further, from our analysis of the unclear situation related to patents which several organisations have declared related to the ISO standard for JPEG 2000 it is clear from this example that some ISO standards cannot be considered open standards.

Findings from the study contribute to reducing confusion concerning the use of standards in software. In particular, the conceptual model aimed to clarify the dimensions 'openness' and 'formality' of standards may provide an important means for analysing specific standards. Further, it seems apparent that any decision maker involved in specification of requirements for development and procurement of software in which standards are implemented needs to understand and account for conditions under which formal and open standards can be used.

Our study has identified a number of issues for any company wishing to implement an ISO standard. These issues include the ability to determine which organisations hold patents which may impinge on the standard, the complexity of the process under which such a company is expected to engage with the patent-declaring organisation. We found that information contained in ISO's database was, for two of the ISO standards investigated, so inaccurate as to cause attempts even to contact the relevant patent-declaring organisation to fail. In considering these issues, we suggest a number of ways in which they can be effectively addressed.

Our research has shown that there are significant risks associated with use of formal (ISO) standards for which conditions for use are unclear. For any organisation it is essential to clarify conditions for use of any standard before undertaking efforts to implement such in software, irrespectively under which license the software is to be provided. In situations when most organisations that have declared patents related to a specific standard do not respond and cannot be reached (as in the case of JPEG 2000 and TIFF/EP), it seems clear that the content of the patent database (as currently structured) provides limited support for any organisation wishing to clarify conditions for implementation of the standard in software. Our research has elaborated on the implications of the implementation of standards in software for which the patent landscape is unclear. Further, our contribution proposes a mechanism for addressing challenges related to unclear patents by suggesting the addition of an Option Zero to the ISO patent policy and associated patent declaration form.

In practice, as shown through this study, it is evident that FRAND commitments may involve a number of significant challenges for any organisation wishing to obtain all necessary rights to allow for implementation of file formats standards in software. In particular, the study shows that for some ISO standards it may even be impossible to clarify conditions and obtain all necessary rights that allow for implementation of file format standards (provided under ISO IPR rules) in software. There are a number of implications from this which in turn inhibit implementations in open source software (it being a fundamental principle of open source licensing that software under such a licence *must* be capable of distribution to a third party). In light of the observations from practice concerning file formats, with identified challenges concerning FRAND commitment due to unclear conditions for use of file format standards that may be critical for many organisations and the broader society, the situation may call for serious consideration amongst all stakeholders involved that need to maintain their own data in these formats over long life-cycles. From an organisational and broader societal perspective, to be dependent upon file formats which cannot be implemented in open source software may constitute a serious risk. As such situations may inhibit fundamental requirements concerning longevity of data and mission-critical digital assets that need to be maintained over very long life-cycles, envisaged proposals for standardisation (such as 'option zero') may be one way forward to improve the situation, at least for some types of file formats and standards that are either business critical or otherwise important for broader societal reasons. Increased transparency and predictability concerning declared SEPs may improve the situation, especially since transfers of SEPs may take place after[37] the publication of a standard. In order to achieve sustainable IT solutions, that allow for software provided under open source licenses, it is critical that file formats allow for implementation in open source software, which can be transferred to third party without any additional restriction[38]. We conjecture that a future standardisation landscape, with numerous declarations and transfer of SEPs in increasingly complex networks that involve many individuals and organisations, may call for file formats provided under option zero, as outlined for ISO standards in this chapter. Through implementation of option zero amongst SSOs, this will ensure that fundamental data which needs to be maintained over long life-cycles through software solutions that implement file formats can effectively address societal needs.

In summary, the study shows that some formal standards create barriers for implementation in open source software, a finding which confirms previous observations expressed in reports from the European Commission. Thereby, the results from this study provide an important contribution to a more comprehensive understanding concerning conditions (and inhibitors) for implementation of standards in open source software. From these results it follows that inclusion of explicit requirements for use of some formal standards in development and procurement of software may significantly inhibit an open and inclusive business friendly ecosystem. In contrast, a relatively simple change to the process by which patent holders declare patents to the ISO and other SSOs may have the effect of promoting such an open and inclusive ecosystem, something which is of particular importance for small companies that are essential players in an innovative and international society.

REFERENCES

Aliprandi, S. (2011). Interoperability And Open Standards: The Key To True Openness And Innovation. *International Free and Open Source Software Law Review*, *3*(1), 5–24. doi:10.5033/ifosslr.v3i1.53

Baird, S. (2007). The Government at the Standards Bazaar. *Stanford Law & Policy Review*, *18*(1), 35–100.

Ballard, M. (2012). Portugal's prescribed open standards - full list. *Computer Weekly*. Retrieved from http://www.computerweekly.com/blogs/public-sector/2012/11/-tablemytable-border-collapsec.html

Bekkers, R., & West, J. (2009). The limits to IPR standardization policies as evidenced by strategic patenting in UMTS. *Telecommunications Policy*, *33*(1-2), 80–97. doi:10.1016/j.telpol.2008.11.003

Bird, G. B. (1998). The Business Benefit of Standards. *StandardView*, *5*(2), 76–80. doi:10.1145/301688.301691

Chronopoulos, A. (2009). Patenting Standards - A case for US Antitrust Law or a Call for Recognizing Immanent Public Policy Limitations to the Exploitation Rights Conferred by the Patent Act? *International Review of Intellectual Property and Competition Law*, *40*(7), 782–816.

Clark, J. (2016). *Convergence, Collaboration and Smart Shopping in Open Standards and Open Source*. In Open Source and Standards in 5G, ITU / NGMN Joint Workshop on OS, San Diego, CA. Retrieved from http://www.itu.int/en/ITU-T/Workshops-and-Seminars/itu-ngmn/Documents/Abstracts_and_Presentations/Jamie_Clark_v2.pdf

Contreras, J. L. (2015). A Brief History of FRAND: Analyzing Current Debates in Standard Setting and Antitrust Through a Historical Lens. *Antitrust Law Journal*, *80*(1), 39–120.

Contreras, J. L. (2017). *Technical Standards, Standards-Setting Organizations and Intellectual Property: A Survey of the Literature (with an Emphasis on Empirical Approaches).* Utah Law Faculty Scholarship 11. Retrieved from http://dc.law.utah.edu/scholarship/11

Dardailler, D., Archer, P., & Wenning, R. (2012). *W3C Response to UK Cabinet Office Open Standards Consultation.* Retrieved from http://www.w3.org/2012/04/openstandards.html

de Vries, H. J. (2006). IT Standards Typology. In K. Jakobs (Ed.), *Advanced Topics in Information Technology Standards and Standardization Research* (Vol. 1, pp. 1–26). Hershey, PA: Idea Group Publishing. doi:10.4018/978-1-59140-938-0.ch001

EC. (2007). *Antitrust: Commission confirms sending a Statement of Objections to Rambus.* MEMO/07/330, European Commission.

EC. (2009). *Antitrust: Commission accepts commitments from Rambus lowering memory chip royalty rates.* IP/09/1897, European Commission.

EC. (2012). *Implementing FRAND standards in Open Source: Business as usual or mission impossible?* Report summary from a workshop organised by the European Commission (EC) and the European Patent Office.

EC. (2013a). *European Commission: Press Release, Digital Agenda: Open standards would save public sector €1 billion a year.* Retrieved from http://europa.eu/rapid/press-release_IP-13-602_en.htm

EC. (2013b). *Commission staff working document: Guide for the procurement of standards-based ICT — Elements of Good Practice.* SWD(2013) 224 final, European Commission. Retrieved from http://ec.europa.eu/information_society/newsroom/cf/dae/document.cfm?doc_id=2326

EC. (2016). *ICT Standardisation Priorities for the Digital Single Market.* Communication from the Commission to the European Parliament, the Council, the European Economic and Social Committee and the Committee of the regions, COM(2016) 176 final, European Commission.

ECSIP. (2014). Patents and Standards: A modern framework for IPR-based standardization. Final Report: A study prepared for the European Commission Directorate-General for Enterprise and Industry, European Competitiveness and Sustainable Industrial Policy Consortium, Ref. Ares(2014)917720 – 25/03/2014.

Egyedi, T. (2007). Standard-compliant, but incompatible?! *Computer Standards & Interfaces, 29*(6), 605–613. doi:10.1016/j.csi.2007.04.001

Egyedi, T. & Enserink, B. (2013). Measuring the Benefits of Open Standards: A Contribution to Dutch Politics. *Standards Today, 12*(1).

EU. (2004). *European Interoperability Framework for pan-European eGovernment Services, Version 1.0.* European Commission. Retrieved from http://ec.europa.eu/idabc/servlets/

Fitzgerald, A., & Pappalardo, K. (2009). Moving Towards Open Standards. *SCRIPted, 6*(2), 467–483.

Fomin, V. V., Pedersen, M. K., & de Vries, H. J. (2008). Open Standards and Government Policy: Results of a Delphi Survey. *Communications of the Association for Information Systems, 22*(1), 25.

Friedrich, J. (2011). Making Innovation Happen: The Role of Standards and Openness in an Innovation-Friendy Ecosystem, In *Proceedings of 7th International Conference on Standardization and Innovation in Information Technology (SIIT)*. IEEE Computer Society. doi:10.1109/SIIT.2011.6083609

FSF. (2017). *Free Software Foundation*. Retrieved from www.fsf.org

Gamalielsson, J., Lundell, B., Feist, J., Gustavsson, T., & Landqvist, F. (2015). On organisational influences in software standards and their open source implementations. *Information and Software Technology*, *67*(Nov), 30–43. doi:10.1016/j.infsof.2015.06.006

Ghosh, R. A. (2005). *An Economic Basis for Open Standards*. FLOSSPOLS. Retrieved from http://www.flosspols.org/deliverables/FLOSSPOLS-D04-openstandards-v6.pdf

HP. (2015). *HP Response to Consultation on Patents and Standards*. Hewlett-Packard.

ISO. (2007). *ISO/IEC/ITU common patent policy*. Retrieved from http://isotc.iso.org/livelink/livelink/fetch/ 2000/2122/3770791/Common_Policy.htm

ISO. (2012). *Guidelines for Implementation of the Common Patent Policy for ITU-T/ITU-R/ISO/IEC, Revision 1*. Retrieved from http://isotc.iso.org/livelink/livelink/fetch/2000/2122/3770791/Common_Guidelines_01_March_07.pdf

ISO. (2015a). *Standards: What is a standard?* Retrieved from http://www.iso.org/iso/home/standards.htm

ISO. (2015b). *Patent declarations submitted to ISO*. Retrieved from http://isotc.iso.org/livelink/livelink/13622347/Patents_database.xls?func=doc.Fetch&nodeId=13622347

ISO. (2015c). *ISO Standards and Patents*. Retrieved from http://www.iso.org/iso/standards_development/patents

ISO. (2015d). *Guidelines for Implementation of the Common Patent Policy for ITU-T/ITU-R/ISO/IEC, Revision 2*. Retrieved from http://www.itu.int/dms_pub/itu-t/oth/04/04/T04040000010004PDFE.pdf

Jakobs, K. (2017). Two dimensions of success in ICT standardization – A review. *ICT Express*, *3*(2), 85–89. doi:10.1016/j.icte.2017.05.008

Jakobs, K. & Mora, M. (2008). Co-ordinating Rule Setters – Co-operation in ICT Standards Setting. *CONF-IRM 2008 Proceedings*, Paper 17.

Kammarkollegiet (2013). Programvaror och tjänster 2013, Statens Inköpscentral, Kammarkollegiet, 5 February, Dnr. 96-40-2013.

Kammarkollegiet (2014a). Programvaror och tjänster 2014 - Kontorsstöd, Dnr 96-33-2014, 2014-09-11.

Kammarkollegiet (2014b). Programvaror och tjänster 2014 - Grundläggande it, Dnr 96-34-2014, 2014-09-11.

Kammarkollegiet. (2016). *Open IT-standards, National Procurement Services (Kammarkollegiet), 7 March, Dnr 96-38-2014*. Retrieved from http://www.avropa.se/globalassets/open-it-standards.pdf

Krechmer, K. (2001). The Need for Openness in Standards. *IEEE Computer*, *34*(6), 100–101.

Krechmer, K. (2005). The Meaning of Open Standards, In *Proceedings of the 38th Hawaii International Conference on System Sciences*. IEEE Computer Society. doi:10.1109/HICSS.2005.605

Lea, G., & Hall, P. (2004). Standards and intellectual property rights: An economic and legal perspective. *Information Economics and Policy*, *16*(1), 67–89. doi:10.1016/j.infoecopol.2003.09.005

Lerner, J., & Tirole, J. (2015). Standard-Essential Patents. *Journal of Political Economy*, *123*(3), 547–586. doi:10.1086/680995

Lundell, B. (2011). e-Governance in public sector ICT-procurement: what is shaping practice in Sweden? *European Journal of ePractice, 12*(6), 66-78.

Lundell, B. (2012). Why do we need Open Standards? In M. Orviska & K. Jakobs (Eds.), *Proceedings 17th EURAS Annual Standardisation Conference 'Standards and Innovation'* (pp. 227-240). The EURAS Board Series.

Lundell, B., Marr, D., Opie, E., Piana, C., & van Rooijen, A. (2016b). *Panel discussion @ Workshop on Open source and standardization: Legal interactions*. ETSI, Sophia Antipolis, France.

Lundell, B., & Gamalielsson, J. (2013). Usage of Open Standards and Open Source in Swedish schools: On Promotion of Openness and Transparency, In E. Petrinja et al. (Eds.), Open Source Software: Quality Verification (pp. 207-221). IFIP Advances in Information and Communication Technology.

Lundell, B., & Gamalielsson, J. (2017). On the potential for improved standardisation through use of open source work practices in different standardisation organisations: How can open source-projects contribute to development of IT-standards? In *Digitalisation: Challenge and Opportunity for Standardisation: Proceedings of the 22nd EURAS Annual Standardisation Conference*. EURAS Contributions to Standardisation Research.

Lundell, B., Gamalielsson, J., & Tengblad, S. (2016). IT-standarder, inlåsning och konkurrens: En analys av policy och praktik inom svensk förvaltning. *Uppdragsforskningsrapport, 2016*(2). (in Swedish)

Lundell, B., Gamalielsson, J., Tengblad, S., Hooshyar Yousefi, B., Fischer, T., Johansson, G., ... Lönroth, E. (2017). Addressing lock-in, interoperability, and long-term maintenance challenges through Open Source: How can companies strategically use Open Source? In *The 13th International Conference on Open Source Systems (OSS 2017), IFIP AICT 496*. Springer.

Lundell, B., Lings, B., & Syberfeldt, A. (2011). Practitioner perceptions of Open Source software in the embedded systems area. *Journal of Systems and Software*, *84*(9), 1540–1549. doi:10.1016/j.jss.2011.03.020

Lundell, B., & van der Linden, F. (2013). Open Source Software as Open Innovation: Experiences from the Medical Domain. In J. S. Z. Eriksson et al. (Eds.), *Managing open innovation technologies* (pp. 3–16). Berlin: Springer. doi:10.1007/978-3-642-31650-0_1

Milinkovich, M. (2015). *How the Eclipse community works*. Keynote presentation @ the 11th International Conference on Open Source Systems, Florence, Italy.

Mitchell, I. G., & Mason, S. (2010). Compatibility Of The Licensing Of Embedded Patents With Open Source Licensing Terms. *International Free and Open Source Software Law Review*, *3*(1), 25–58. doi:10.5033/ifosslr.v3i1.57

NOC. (2007). *The Netherlands in Open Connection: An action plan for the use of Open Standards and Open Source Software in the public and semi-public sector*. Ministry of Economic Affairs. Retrieved from https://www.ictu.nl/archief/noiv.nl/files/2009/12/Action_plan_english.pdf

NOVFS. (2011). *Open Document Standards for the Government Guide*. Programma Nederland Open in Verbinding & Forum Standaardisatie. Retrieved from https://www.forumstandaardisatie.nl/fileadmin/os/documenten/Handreiking_ODF_Engelse_versie.pdf

Odell, M. (2009). European Public Sector Award. Public speech by Swedish minister, Maastricht.

OSI. (2017). *Open Source Initiative*. Retrieved from opensource.org

Pentheroudakis, C., & Baron, J. A. (2017). *Licensing Terms for Standard Essential Patents: A Comprehensive Analysis of Cases*. JRC Science for policy report, European Commission.

Permanent Representatives Committee. (2016). Draft Council conclusions on the "Digital Single Market Technologies and Public Services Modernisation" package. Statement by the United Kingdom, Estonia, Belgium, Slovenia, Poland, Latvia and Malta, 8735/16 ADD 1, 26 May, Brussels.

Simcoe, T. S. (2006). Open Standards and intellectual property rights. In H. Chesbrough et al. (Eds.), *Open Innovation researching a new paradigm*. Oxford, UK: Oxford University Press.

Simeonova, D. (2015). *Digital Solutions: Can we 'Open Source' the Future? Commission en direct #21*. European Commission.

SOU. (2009). *Strategi för myndigheternas arbete med e-förvaltning*. Statens Offentliga Utredningar: SOU 2009:86, e-Delegationen, Finansdepartementet, Regeringskansliet, Stockholm. (in Swedish)

Standaardisatie (2014). *Lijst met open standaarden, Forum Standaardisatie*. Retrieved from https://lijsten.forumstandaardisatie.nl/lijsten/open-standaarden/

Standardisation Forum. (2011). *A guide for government organisations Governance of Open Standards*. The Standardisation Forum, The Hague.

UK. (2012a). *Open Standards Principles: For software interoperability, data and document formats in government IT specifications*. HM Government, UK. Retrieved from https://www.gov.uk/government/uploads/ system/uploads/attachment_data/file/183962/Open-Standards-Principles-FINAL.pdf

UK. (2012b). *Open Standards in Government IT: A Review of the Evidence*. HM Government, UK. Retrieved from https://www.gov.uk/government/uploads/system/uploads/attachment_data/file/78891/Review-of-the-Evidence_CIPPM.pdf

UK. (2012c). *CloudStore opens for business, Cabinet Office and The Rt Hon Lord Maude of Horsham*. Retrieved from https://www.gov.uk/government/news/cloudstore-opens-for-business

UK. (2014). *Standards Hub*. CabinettOffice, UK Government. Retrieved from http://standards.data.gov.uk/

UK. (2015). *Open Standards Principles, Updated 7 September 2015*. HM Government, UK. Retrieved from https://www.gov.uk/government/uploads/system/uploads/attachment_data/file/459075/OpenStandardsPrinciples2015.pdf

Updegrove, A. (2009). *The EC Settlement: Rambus, Writs and the Rule of Law*. Retrieved from http://www.consortiuminfo.org/standardsblog/article.php?story=20090622043038212

Wessman, R. (2013). *Upphandling av IT - inlåsningseffekter och möjligheter, Uppdragsforskningsrapport 2013:2, ISSN-nr 1401-8438*. Swedish Competition Authorities.

KEY TERMS AND DEFINITIONS

Formal/Informal Standard: Formal standards are provided by organisations which are recognised as formal standardisation organisations, whereas information standards are provided by organisations which are recognised as informal (or consortia) standardisation organisations. For example, an ISO standard that is provided by ISO (International Organization for Standardisation) is considered as a formal standard, whereas a W3C standard that is provided by W3C (the World-Wide-Web consortium) is considered as an informal (or consortia) standard.

FRAND: An acronym for "fair reasonable and non-discriminatory" terms on a similar (or identical) basis to RAND: it is debated whether there is a material difference between RAND and FRAND.

Open Source Software (OSS): Software made available under a software license which has been approved by the Open Source Initiative (see: opensource.org).

Open/Closed Standard: Open standards are standards which are provided on royalty-free terms and such standards can be implemented in open source software without restriction, whereas closed standards are provided on RAND (or other) terms that cannot be implemented in open source software. For example, the W3C standard SVG is considered an open standard, whereas the ISO standard JPEG2000 is considered a closed standard.

RAND: An acronym for "reasonable and non-discriminatory terms." It is used by standardisation organisations to describe a form of licensing commitment. A patent owner that has agreed with a standardisation organisation to provide their patents which impinge on a standard on RAND terms is bound by an offer from any prospective licensee to negotiate a license on reasonable and non-discriminatory terms.

Royalty: A payment made by the licensee to the licensor for the right to use an asset. For example, a payment made by a user of a standard to the owner of a standard-essential-patent. It may be one-off, periodic, or based on another metric (for example, per licensed unit sold).

Royalty-Free: Refers to the right to use an asset (e.g., copyright or patent) without the need for payment. For example, a licensor of a standard-essential patent which provide the patent on royalty-free terms allows a user of the standard to use the patent without payment.

Standard Essential Patent: Refers to a patent which impinges on the conditions for use of a standard. A patent which impinges on a standard is a standard essential patent in case it is necessarily infringed when the technical specification of the standard cannot be implemented without infringing on the patent (i.e. if the technical specification of the standard cannot be implemented without infringing the standard essential patent).

ENDNOTES

1 http://blogs.wsj.com/cio/2014/05/05/open-source-eating-software-world-samsung/

2 It should be noted that active participation in international meetings organised in standardisation projects (e.g. ISO and SMPTE) may constitute major inhibitors for a small company, as experienced by representatives from small companies in previous collaborative research involving authors of this manuscript (e.g. Gamalielsson et al., 2015).

3 See Directive 2014/24/EU, http://eur-lex.europa.eu/legal-content/EN/TXT/PDF/?uri=CELEX:32014L0024&from=EN; Regulation (EU) No 1025/2012, http://eur-lex.europa.eu/LexUriServ/LexUriServ.do?uri=OJ:L:2012:316:0012:0033:EN:PDF

4 For example, the Regulation (EU) No 1025/2012 (EC, 2012) clarifies that consortia standards must not "limit the possibilities for implementers" to use the standard. Further, this EU regulation states that some consortia developing 'informal standards' do not fulfil requirements according to this regulation: "Some ICT technical specifications are not developed in accordance with the founding principles" (i.e. "coherence, transparency, openness, consensus, voluntary application, independence from special interests and efficiency"). From this it follows that 'informal standards' from such consortia cannot be referenced in public sector procurements according to this regulation. However, a legal review of this regulation is beyond the scope of this chapter.

5 Open Mobile Alliance (OMA), http://openmobilealliance.org/, clarifies that "No license to any patent, trademark, copyright or other proprietary right is granted under this Agreement or through any disclosure hereunder except as expressly stated in this Agreement", see http://openmobilealliance.org/about-oma/policies-and-terms-of-use/use-agreement/

6 Society of Motion Picture & Television Engineers (SMPTE), https://www.smpte.org/, clarifies that documents "may include technology that is subject to Essential Claims" only "if all known patent holders are prepared to agree to terms that are RAND for all Essential Claims" (i.e. a claim which "is necessarily infringed by implementing the Normative Text of that Engineering Document and is 'necessarily infringed' only when there is no commercially-reasonable non-infringing alternative for implementing the Engineering Document"), see https://www.smpte.org/sites/default/files/SMPTE_IP_Policy_2013-08.pdf.

7 The OASIS IPR-rules clarifies that when a TC is established it can chose to operate under one (of four different) IPR Modes. At time of initial data collection and analysis, we note that for the 89 currently active TC:s in OASIS we find that: 52 have selected to operate under the mode "RF on Limited Terms"; 18 have selected to operate under the mode "RF on RAND Terms"; 18 have selected to operate under the mode "Non-Assertion"; and 1 has selected to operate under the mode "RAND" (for an overview of the 4 models, see further https://www.oasis-open.org/policies-guidelines/ipr#tcformation). In addition, there are also additional "Legacy IPR rules" for other OASIS TC:s (e.g. see https://www.oasis-open.org/committees/tc_home.php?wg_abbrev=wss; https://www.oasis-open.org/committees/wss/ipr.php).

8 The OASIS IPR-rules clarifies that when a TC is established it can chose to operate under one (of four different) IPR Modes. At time of initial data collection and analysis, we note that for the 89 currently active TC:s in OASIS we find that: 52 have selected to operate under the mode "RF on Limited Terms"; 18 have selected to operate under the mode "RF on RAND Terms"; 18 have selected to operate under the mode "Non-Assertion"; and 1 has selected to operate under the mode "RAND" (for an overview of the 4 models, see further https://www.oasis-open.org/policies-guide-

lines/ipr#tcformation). In addition, there are also additional "Legacy IPR rules" for other OASIS TC:s (e.g. see https://www.oasis-open.org/committees/tc_home.php?wg_abbrev=wss; https://www.oasis-open.org/committees/wss/ipr.php).

9 ISO/IEC 15948, "Information technology -- Computer graphics and image processing -- Portable Network Graphics (PNG)", ISO/IEC JTC1/SC24.

10 ISO/IEC 15444, "Information technology -- JPEG 2000 image coding system", ISO/IEC JTC1/SC29.

11 ISO 12234-2, "Electronic still-picture imaging -- Removable memory -- Part 2: TIFF/EP image data format", ISO/TC 42.

12 W3C Recommendation, "Scalable Vector Graphics (SVG) 1.1 (Second Edition)", 16 August 2011.

13 www.w3.org

14 http://www.iso.org/iso/standards_development/patents

15 Initial letters (air-mail) were sent in May 2014. Further, requests were also sent via email to all organisations that included email addresses as part of their contact details in the patent database.

16 For example, the ISO standard PDF/A-2 (ISO 19005-2:2011) contains several normative references to other standards (maintained by ISO and other SDOs) which thereby constitute inherent parts of the ISO 19005-2:2011 standard. One of these normative references is Part 2 of the ISO/IEC standard JPEG 2000 for which several patent declarations can be identified in the ISO patent database. However, the same declarations cannot be found when searching for declarations made for ISO 19005-2:2011 in the ISO patent database (and it is instead necessary to manually search the ISO patent database for all normative references at all levels). Hence, it follows that a simple search in the ISO patent database gives a misleading indication of the scope of how patents declared for the investigated standards impact on other standards.

17 It should be noted that PNG is also provided as a W3C standard, but it is only included as an ISO standard in the explicit list of open standards in the Netherlands.

18 http://www.w3.org/TR/2011/REC-SVG11-20110816/

19 http://www.w3.org/Graphics/SVG/WG/wiki/Implementations

20 http://www.blender.org/; https://inkscape.org; http://wiki.scribus.net

21 http://www.w3.org/TR/2003/REC-PNG-20031110/

22 http://www.w3.org/Graphics/PNG/Disclosures

23 http://www.w3.org/TR/PNG/; http://www.libpng.org/pub/png/png-sitemap.html

24 http://www.gimp.org/; https://inkscape.org; http://www.tuxpaint.org/

25 For the rest of this chapter, we write "JPEG 2000" when referring to the complete ISO standard ISO/IEC 15444.

26 For an overview of the JPEG 2000 standard as presented by the JPEG committee,see http://jpeg.org/jpeg2000/index.html

27 In addition, there are a number of draft specifications and several outdated standard documents, which also may be relevant for implementation in software in different usage scenarios (e.g. for scenarios with long life-cycles)

28 For the six organisations that provided email addresses as part of their contact details in the patent database related to JPEG 2000 we observed that the email addresses to four organisations were invalid (most likely outdated) which implied that we used air mail addresses instead.

29 We consider data collection to be finished since more than 3 years have elapsed since the initial request. Further, the most recently returned letter was received in March 2015 (i.e. more than eight months after the last reminder).

30 For the rest of this chapter, we write “TIFF/EP” when referring to the complete ISO standard ISO 12234-2:2001.

31 We consider data collection to be finished since more than 3 years have elapsed since the initial request. Further, the most recently returned letter was received in March 2015 (i.e. more than eight months after the last reminder).

32 In this section, we use the terminology ‘companies’ and ‘implementing companies’, but this term should be understood to embrace other organisations, such as open source projects, which may also want to implement software which implements the standard, and many of the issues investigated will also impact larger companies.

33 Organisations exist, such as the Open Invention Network (OIN), which act as the nexus of cross-licensing arrangements between their member organisations, typically in relation to a specific application or vertical market (in the case of the OIN, this covers the various components that they define as the ‘Linux System’). Essentially, this enables the members to pool their patent portfolios. The pool may also assist members when third parties claim that the relevant technology infringes their own patents. The OIN, in addition to holding licences from members, also holds patents that it licenses to members.

34 It should be noted that decisions concerning whether or not small companies will participate in international standardisation is also affected by (perceived and experienced) barriers to entry, as indicated by experiences from previous research published by authors of this chapter: “From our own experiences of participation in IETF and W3C standardisation it is highly feasible for small companies to contribute due to low barriers for participation” (Gamalielsson et al., 2015, p. 41).

35 This mechanism, and to a lesser degree, the specific wording, implements a mechanism which will be familiar to many organisations of all sizes which are involved in the licensing of open source software, given that a similar mechanism appears in many common open source licences.

36 One would hope that such a declaration would remain, under all relevant jurisdictions, both irrevocable and also pass with the underlying patent, should the patent be assigned to a third party. The legal analysis of these points will vary from jurisdiction to jurisdiction and is beyond the scope of this chapter. In most common-law jurisdictions (Ireland being an exception that immediately comes to mind), the ISO rules could be amended to render potential licensees under Option Zero third party beneficiaries. There are other possible mechanisms to facilitate transfer of the commitment alongside the patent (deed poll, promissory estoppel, for example), but an analysis of them is beyond the scope of this chapter.

37 For example, concerning transfer of SEPs it has been reported that “A large majority of the transfers (83.5%) took place after standard release.” (ECSIP, 2014, p. 190)

38 The first principle of the Open Source Definition states: “The license shall not restrict any party from selling or giving away the software...The license shall not require a royalty or other fee...”. For any licence to qualify as an Open Source Licence, it must adhere to this principle. https://opensource.org/osd-annotated

Chapter 11
Metastructuring for Standards:
How Organizations Respond to the Multiplicity of Standards

Ronny Gey
Friedrich Schiller University Jena, Germany

Andrea Fried
Linköping University, Sweden

ABSTRACT

This chapter focusses on the appearance and implementation of process standards in software development organizations. The authors are interested in the way organizations handle the plurality of process standards. Organizations respond by metastructuring to the increasing demand for standardizing their development processes. Standards metastructuring summarizes all organizational mechanisms for facilitating the ongoing adaption of global standards to the organizational context. Based on an in-depth single case study of a software developing organization in the automotive technology sector, the authors found four areas of metastructuring, four roles for standard mediation, and four types of metastructuring activities. With the case study, they encourage further research that proves standards in use and how organizations respond to the challenges of standardization.

INTRODUCTION

The multiplicity and plurality of standards is one of the challenging issues for modern working organizations. In the light of the recent economic developments and the imposed dependency of economic decisions on standards it is important to elaborate on these challenges, changes and risks which confront organizations within an audit society of organized uncertainty (Power, 1997). In this context, we will contribute to the question of how organizations respond to and cope with the consequences evolving from the multiplicity and plurality of process standards. We chose an exploratory study on the micro level within a single software development organization.

DOI: 10.4018/978-1-5225-5320-5.ch011

Research on standards has many facets. On the one hand, we distinguish research on standards focusing on the design mode and, on the other hand, the use mode of standards (following Orlikowski, 1992). This distinction emphasizes the occurrence of social construction, both, before and after a standard is enacted in an organization. It refers to processes of setting and following standards. The *design* mode describes the process of standardization on the standard setting bodies' and/ or the related stakeholder's side. Hereby, researchers have inquired how standards emerge, how standard setting bodies are organized and how these bodies determine the content of standards, as well as, how they convince potential stakeholders to certify their organizations (Blind and Mangelsdorf, 2016; Brunswicker et al., 2015; DongBack, 2013; Egyedi, 2008; Furusten, 2000; Marimon et al., 2009; Tamm Hallström, 2004, Wiegmann et al., 2017).

For this paper, the *use* mode of standards is of special interest. The use mode literature on process standards is chiefly shaped by the research on the ISO 9000s and ISO 14000s process quality norms, culminating in the following two topics in the early 2000s. Firstly, there is an extensive amount of literature about the motivation and barriers of the implementation of standards (Boiral, 2003; Niazi et al., 2005; Zeng et al., 2007). This literature focuses on the influence of the implementation of process standards on product quality, on the organizational performance and the obstacles and limitations while putting a standard into effect. Secondly, there is a discussion around process standards that reflects on the requirements for a 'good' organization. This can be measured through customer satisfaction, defined responsibilities, the reduction of production and management mistakes, quality assurance, documentation of all processes, decisions and related audits (Lawrence and Phillips, 1998). Finally, little research has been done on the social construction of standards, in particular, on standards as a form of regulation or as a code of corporate governance and on the functioning of these code regimes (Jakobs, 2006; Power, 1997; Seidl, 2007; Wieland, 2005). Thereby, code regimes are specific types of standards which regulate corporate behavior based on a 'comply-or-explain rule'. Seidl (2007) investigates code regimes in view of parameters that influence the effectiveness of a code regime. Important for our understanding of standards is here the conclusion that the de-facto content of a standard is determined to a significant extent by standard followers and not by standardization bodies.

Besides these studies, we can resume that the social construction of standards by organizations plays a minor role on the use mode of standards in the existing research. Thus, we deem it as important to elaborate on this subject since we do not sufficiently know how organizations themselves adopt multiple standards in reaction to a number of institutional pressures. With this contribution we intent to provide empirical insights on a micro level addressing organizational issues of standard multiplicity and plurality.

The contribution of this article is threefold. Firstly, we elaborate on the institutional answers how organizations deal internally with the plurality and multiplicity of standards. In contrast to standardization *by* organizations and standardization *as* organization, we are dealing with the standardization of organizations. Standardization of organizations relates to the question of "how standards are adopted, diffused, implemented, avoided, and altered in the course of their implementations" (Brunsson et al., 2012, p. 614).

Secondly, we focus on the empirical phenomenon of standard-use mediators and their metastructuring activities (Orlikowski et al., 1995). In consequence of the development of a so-called "standard bible" by the software organizations, standard-use mediators occur in organizations as a response to a multi-facet standardization pressure from the institutional environment of organizations. We believe that the way organizations meta-structure multiple process standards has an enormous influence e.g. on the innovating activities of an organization (Fried, 2010; Fried et al., 2013).

Finally, while referring to metastructuring we base our argumentation on structuration theory (Giddens, 1984). Local organizational structures depend on the institutional environment which is propagated by professionals and associations to promote collective goods as standards. Likewise, mutual interferences between the local organizational actions and structures and the institutional environment can be observed. We suppose that, in order to function, process standards are structured and swayed by internal mediators. The structuration of standards is influenced by the mediators' interpretation of their working environment, their access to resources and the normative rules they are confronted with (Barley, 1986; DeSanctis and Poole, 1994; Weick, 1990).

For exploring the phenomenon of standard-use mediators and their metastructuring activities in detail, the chapter is structured as follows. The subsequent section will provide the theoretical background on metastructuring and standard-use mediation. Starting from the theory on structuration by Giddens (1984), we will present metastructuring as a phenomenon which Orlikowski et al. (1995) first introduced in the area of technology research. Section 3 describes the research design of our case study. We chose an exploratory approach within a single software development organization. In particular, process standards will be introduced as the empirical object of interest. The findings of our case study *A-Suppliers* are presented in section four. The empirical investigations will highlight the use mode perspective on standards. We intend to attract the readers' attention to how A-Suppliers has decided to address plurality and multiplicity of process standards. In the discussion, we provide insights into metastructuring as a powerful mechanism in dealing with multiple and plural process standard requirements in organizations. Standard-use mediators are explored and investigated regarding the metastructuring *process*, metastructuring *areas* as well as the different *activity types* of metastructuring connected to these areas. The conclusion section provides supplementary notes on future research and limitations of standard metastructuring.

METASTRUCTURING OF STANDARDS IN THE CONTEXT OF USE

In general, standards can be understood as rules. Rules sanction social conduct and constitute meaning (Giddens, 1984). Standards focus on something "to be reached, be it in terms of quality or quantity, extent or intensity, precision or value, technique or morality"(Ortmann, 2010, p. 204). Process standards are the center of interest in our empirical investigations. They represent "how organizational processes should be designed and controlled" (Brunsson and Jacobsson, 2000b, p. 4).

From our point of view, standards are solely those rules (a) which are purposefully set, which are (b) strongly formalized and written down (for instance by standardization bodies) and, finally, which (c) cover and adapt generalized, comparable occasions among organizations (e.g. processes, technical or social requirements, performance). Thus, standards are only those rules "which tend to be [...] developed and adopted through explicit procedures that historians can trace" (Timmermans and Epstein, 2010, p. 71). Moreover, standards are rules which are often connected to a certification by an organization-external party (certification body). Thereby, certification is an evaluation process of the compliance with predefined requirements. It results in a document (certificate) that is trusted to be evidence of requirement compliance, assuming a comparability among certified organizations. Alternatively for certification the terms *appraisal* (e.g. for the CMMI[1] standard) or *assessment* (e.g. for the ISO/IEC 15504 standard) are used for the external evaluation by a certification body and its assessors.

By choosing a structuration theory viewpoint, we also need to address the relation between the design and the use of standards. This relation is to be seen as "recursive, i.e. that the formulation of rules (possibly) constitutes their application and vice versa" (Ortmann, 2010, p. 205). People are knowledgeable agents who can construct alternative meanings of once designed standards. They are able to redefine and modify standards in an emancipatory way and shape the meaning, purpose and ways of standard usage. Standards do not exist without human activity; they are enacted by actions (Fried et al. 2013).

In the scope of this paper, we report on a specific type of structuring activity around process standards within organizations. We focus on a process Orlikowski et al. (1995) called *metastructuring*. Metastructuring was primarily used in the context of technology research. Yates et al. (1999) observed that "a few members of the project group explicitly engaged in activities that facilitated both early and ongoing use of the new technology by the rest of the project group" (Yates et al., 1999; for technology structuring see also Okamura et al., 1994 and Orlikowski et al., 1995). These activities included (1) guidelines for the use of a new (conference) technology, (2) provision of qualification and follow-ups in using the technology as well as (3) modification of the technology according to contextual, project-specific requirements. Orlikowski et al. (1995) summarized metastructuring "as an organizational mechanism for facilitating the ongoing adaptation of technologies, their use, and organizational contexts to each other and to changing conditions" (Orlikowski et al., 1995, p. 441).

Four types of metastructuring were classified in technology research: establishment, reinforcement, adjustment and episodic change (Orlikowski et al., 1995). During *establishment,* the mediators come into play and enact their roles in the organization. Mediators are a group of individuals who intervene significantly to modify and to adapt new material artifacts to the organization in an appropriate manner. They pre-shape the material artifact before the actual users apply it. Therefore this activity is called metastructuring. Moreover, in addition to other actors like promoters or trainers, mediators do not only intervene reactively in the initial stages of establishment. Mediators act pro-actively. They are institutionally recognized and play an ongoing role. Internal technology consultants or the employees of the IT department can be considered as mediators. Of course, they set up the infrastructure and implement the new material artifact. They also need to promote and to support it, write user-guides and especially convince managers or key personnel in the organization to adopt the new technology. When it comes to *reinforcement*, mediators maintain their operations and reinforce its vision. An ongoing training on the new material artifact is offered. Moreover, they ease the use of the system, collect user feedback and, as a consequence, try to intensify the usage. During *adjustment,* they react on the given feedback and usage tendencies. This includes smaller and less radical modifications of the artifact in use. The most dramatic changes are considered during the *episodic change* period. Significant improvements and corrections to the functionality or the technical core of the system are accomplished.

What started in the context of technology use gained further influence especially in socio-technical studies on electronic communication or collaboration. Henriksen et al. (2002) explored web-based group ware, Bansler and Havn (2004) a virtual workspace and Janneck and Finck (2006) researched online communities. Clear and MacDonell (2011) analyzed email data of global virtual teams. Metastructuring is also very prominent in studies on medical technology. Novak et al. (2012; 2012a), Davidson and Chiasson (2005) and Davidson and Chismar (2007) all studied the implementation of medical IT systems. The study by Rodon et al. (2011) portrayed the post-implementation phase of inter-organizational information systems and especially focused on managerial interventions and users' appropriation. They reported that metastructuring might have an important impact on the structuring processes of other users while applying the different types of metastructuring activities. Overall, these authors characterized the process as very complex and rather complicated to understand. Thus, they call for additional research on different technologies (Bansler and Havn, 2004) and raise the question of other potential sources of mediation (Rodon et al., 2011).

In the metastructuring discourse, the scope of technology is restricted to material artifacts of various hardware and software configurations (Orlikowski, 1992). It is emphasized that there is a distinction between the material nature of technologies and the human activities designing and using these artifacts. Similarly, for standards as artifacts and their rule-like character, the need to be enacted in order to be effective in organizational life is evident. On this basis, we ask whether the implementation of other artifacts than technologies - like standards - also trigger such metastructuring activities in organizations. Standards are powerful rules in a globalized world. The way organizations deal with facing a plurality and multiplicity of standards is an urgent matter that needs clarification. Accordingly, our interest is to observe whether similar powerful mechanisms can be found within the 'world of standards' (Brunsson and Jacobsson, 2000c) and which influence metastructuring can have on the effectiveness of software development processes.

CASE STUDY DESIGN

We use a case study approach to investigate contemporary phenomena in their real-life context (Yin, 2003). As typical for qualitative studies, the personal realm of experience of each interviewee lies in the center of our analysis. We chose interviews as the main data gathering technique. The result is a single exploratory case study. We analyzed the interview data with a qualitative content analysis to explore qualified hypotheses (Krippendorff, 2004; Kohlbacher, 2006). It enables the researcher to include textual information and to identify its properties systematically.

Eight interviews were conducted in October 2010. Each of them lasted between 80 and 100 minutes. Later on, two additional interviews were necessary to clarify some data and to gain some more fine-grained information. With an emphasis on the Capability Maturity Model Integration (CMMI) as the process standard all interviews were part of a larger, multi-case study undertaking (Fried et al., 2013) exploring the relationship between standardization and innovation in software development organizations. CMMI is a standard for software process improvement (Humphrey, 1992). Since 2002, it has been the successor to CMM. When we speak about CMMI in this study, we refer to either CMMI (since 2002) or to CMM (before 2002). The interviews involved strategic planners, managers responsible for the implementation of process management standards, team leaders and software engineers of one working context (see Table 1). They were semi-structured to allow flexibility and to ensure that we focus on interesting phenomena. Further, a narrative interview style was chosen to stimulate reports about a variety of expressions and experiences in terms of process standards.

The subjects of the interviewees' narrations and requests by the interviewer were centered on the history of the organization, the history, implementation and usage of process management standards, the procedure of certification audits as well as the individual and organizational experiences with process management standards. The interviewees were asked to reflect on changes in the course of time.

The recorded interviews were transcribed and afterwards coded and analyzed with *Atlas.ti* (Muhr, 1991), a software for qualitative data analysis. The anonymized publication of our findings was requested by the organization. Where necessary, sensitive data was paraphrased and marked by square brackets.

Table 1. Data of Interviewees

Name	Position	Experience	
		CMMI	A-Suppliers
A1	Process Coach (PC)	10 years	>15 years
A2	Head of QM	5 years	>15 years
A3	Former CIO, now Board Member	5 years	>15 years
A4	SW Developer, Project Leader	5 years	>15 years
A5	Process Coach (PC)	5 years	12 years
A6	SW Developer, SW Architect	5 years	8 years
A7/A9/A10	Head of a Process Group (PG) unit	12 years	5 years
A8	Former Head of Process Group (PG)	5 years	13 years

The analysis of the case is based on a code-name schema. The initial code-names and their mutual relations are depicted in Figure 1.

Among the team members, we agreed on the initial code-name schema in advance. However, it was constantly extended during the analysis. In the end we coded additional relational information and thus extended the code-name schema during the analysis. We ended up with 31 different code names.

Moreover, to achieve an adequate level of validity we used multiple sources of evidence. Most of the data was gathered from the transcriptions of the interviews. Additionally, we participated in a guided tour through the production area as well as the software development office where the interviews were subsequently conducted. After each interview and during the lunch and coffee breaks we prepared protocols of the past interviews. We focused not only on context information but also on the emotional setting of the interview situation. The next section provides our findings on metastructuring at A-Suppliers.

Figure 1. Code-names for Content Analysis

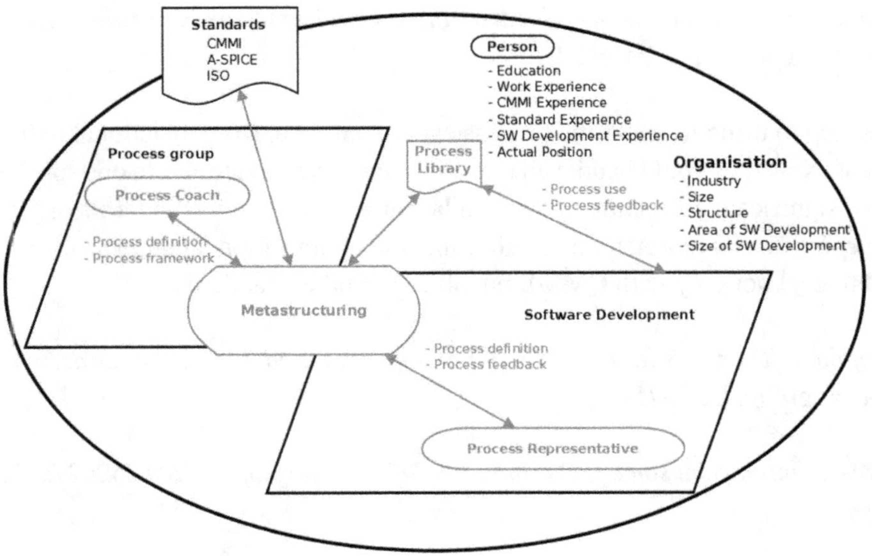

FINDINGS ON METASTRUCTURING AT A-SUPPLIERS

Findings

The investigated organization *A-Suppliers* acts in the automotive technology sector and produces electronic control units which includes systems for ABS, ESP, exhaust and engine controls. They employ about 5,000 system developers, of which about 3,000 are directly involved in software development. The employees who worked in the department at that time mostly had an electrical engineering background. As an automotive supplier they have to consider several industry-related standards (e.g. ISO 26262, A-Spice, ISO 16949). Automotive SPICE (A-SPICE) is thereby the most important standard for the automotive industry. A-Suppliers has long-term experience in process management.

In 1999, the head of software development came up with the idea of continuous process improvement: "*[...] he always had this state of the art development with a process improvement infrastructure in mind.*" *(A1:069)*. The first trigger for CMMI came from inside the company but was also motivated by external customers starting to demand CMMI certification: *"[the customer] finally required it from [A-Suppliers]. So the pressure from outside of [A-Suppliers] came in, from the customer." (A1:069).*

In the software development unit, they set up a pilot project to accomplish a CMMI certification. We regard it as first *establishment* step since they implemented a new standard, CMMI, and they established their own roles as process coaches (PC):

And the project leader [...] had to build up a project team and he took some of the specialists out of various departments. And I was one of them, especially for the whole quality management stuff and subcontract management at that time. So it was three of us, so-called coaches [...] trying to establish some processes. (A1:17)

After more than three years, in 2002, they successfully got CMMI level 3 certification. Directly after their certification, they turned the pilot project into an own department:

There have been some people assigned but it was more or less a project. And then [...] it was established as a department. So there was a budget allocated, there have been head count allocated, there was work space and everything allocated. (A8:021)

Taking this step to bring together the process experts under one organizational unit was considered highly important. Now they could handle all relevant standardization topics in one specialized organizational unit. As a significant correction of the standard-use at A-Suppliers in terms of CMMI, we assess this step as an *episodic change*. After a certain time, the organization again expanded its competences since they had to deal not only with CMMI, but also with other standards:

CMMI is one of them. There is still A-SPICE and all the other models and standards, which are handled within the process group. (A1:045)

And also we have to implement some ISO topics, for Safety management, ISO 26262 and 16949. (A7:39)

As a result, the process experts could now focus entirely on standardization and process-related tasks - a significant and *episodic change*. They gained more expertise in the area while collaborating closely in the new department. Therefore, the time necessary to achieve a new certification was reduced by more than a year:

And this makes it much easier, because if you do something you do it for all projects. It makes it much easier [...] So we reached CMM level 3 in 2002 and level 3 CMMI in 2004. (A1:101)

Whereas formerly only standards from the domain of software development were part of its work items, in 2006 the process group (PG) took over the responsibility for all the process standards, organization-wide (*episodic change*). Again, we regard it as a reaction to the increased attention and importance of standardization at A-Suppliers and their environment. Since CMMI continuously gained influence in the automotive industry, the suppliers consequently had to suit these needs. Since then, the PG has consequently acted as a standard-use mediator for the whole organization. In fact, the engineers were faced with several standards which made their work more complex and difficult to handle. In some cases, it led to a decreased productivity: "*[...] if you use the wrong interpretation for the requirements or for the standard and then you end up with a lower productivity.*" *(A2:067)*. Thus, reducing complexity by mediating between user and standards was the main idea of a distinct PG:

[...] we are reducing complexity for the engineer. Because otherwise, we would challenge our experts for hardware or software to interpret the standard in the right way and this is absolutely not necessary. (A2:087)

Instead, the process experts now deal with the multiplicity of standards. It necessarily means that all standards "*[...] are inputs to the process development. So now you have to find a process design which jointly or comparably fulfills these requirements.*" *(A2:079)*. Interviewee [A9] described this process in detail as follows:

So, we have a process library where all processes for our product development can be found. And we mapped all artifacts, all work products with CMMI or A-Spice. This means, we have the models and the best practices in the database [....] We have the models in the process library and we have the processes of the different products in the library. And then, we link the artifacts or work products with the requirements of the models or standards. With that, we know about our evidences, what we cover and where we still have gaps which need to be closed. With this way, we ensure that if you follow the process library then you fulfill the standards or models. (A9:61)

The PCs interpret each single standard; there are several experts for each. The interpretation of standards can be seen as *establishment* as well as *adjustment*. While interpreting a new standard for the first time, they establish the standard in the company, whereas minor *adjustments* in the interpretation happen on a regular basis. Furthermore, they compare the standards among each other to find overlapping parts and keep the process descriptions as lean and therefore as user-friendly as possible - a sign for *reinforcement*. Creation (*establishment*) and maintenance (*reinforcement*) of the standard script are an immense effort, which is mutually accomplished by the PG in collaboration with the process representatives (PR).

The process library is based on a web-based technology, where all the mappings between the different standards and the product processes only need to be done once. Subsequently, it can be consistently displayed. A-Suppliers even reviews the standard conformity of the process library by internal or external assessments:

[...] after we mapped A-Spice we had an A-Spice assessment by an external A-Spice principal. He reviewed how far the mapping is correct. [...] And we do it for CMMI as soon as we renew the mappings for our appraisal. [...] So, either we review or we get assessed externally. (A9:73)

For ISO 26262, we had a [technical assessment organization] doing the assessment, also for our process library. [...] and for ISO 16949 we actually had our quality surgeon in the organization. He did an ISO 16949 audit for the mapping. (A9:81)

Once the mapping is done, the different standards are not explicitly specified to the user. It reduces complexity since now they only have to interact with the process library instead of all the different standards itself:

[...] this is a documentation of all the processes and you will never find a position inside all these processes, where CMMI or A-Spice is mentioned in reference to. So we hide all these frameworks, all the requirements from our engineers. They only have to follow our internal processes, process descriptions. (A2:083)

Regarding the standard script or the production process, all organizationally relevant information is collected in the process library. As a result, the process library is a highly valuable and critical tool in the organization:

If you look in our process system, you see the processes and the tasks. You see the roles and linked to the roles are the requirements, the capabilities and also the trainings to achieve the know-how. (A2:211)

The interviewees confirmed that the process library dramatically eases the training on-the-job period for new employees or for the achievement of further job qualifications since: "*they start to work and they can be sure if they follow the process library it will be okay.*" *(A7:315).* The process coaches now accomplish a part of the corporate training, the one regarding all process issues (reinforcement). All interviewees demonstrated an exclusively positive attitude towards the process library (PL). Some actually expressed being proud of what they achieved with this tool:

We built up this tool and now we are in the version 3.3 of the process library [...] So it took nearly 3 years to set up the whole system because it is, you can imagine, it is a rather complex and consistent system. So an auditor, whatever kind he is, has no chance to detect gaps or weaknesses. He can do internal checking and so on. So at least the instruction side is complete. And the auditors give up [...] (A2:295)

In this system, the so-called process representatives that work in the relevant operational departments define the operational processes. This happens when the technology is established, as well as when *adjust-*

ments to the relevant processes are required. The PR is linked to the PG and cooperates in all respective process tasks - an integrated and participative approach between the PG and the operational departments:

We have some process representatives [in the operational departments]. And there, we make the process definition. So the process representative is responsible for the process definition. He gets support from my side or from other people from the PG. And the developers were included in discussions in certain groups. Not all, but certain from different departments to get different opinions and different needs. So normally, we had teams, the project leader was the process representative. And there was a PG coach in the process team. And there were different representatives for the different departments in the project team for this certain process group. (A5:113)

The user participation continues once the processes are defined and implemented. Literally everybody in the organization can provide feedback on the processes, either reporting it to the PRs: "*We get indirect feedback via the process representatives.*" *(A2:115)* or by using the process library: "*There is a feedback button for each process. And they can give feedback, change requests.*" *(A2:095)*. In this way, an organized feedback system for the users of the processes was established in order to interact about the process landscape. The PR has the responsibility for handling the feedback requests and, as a main user of the process itself, he also actively produces feedback (*reinforcement*).

Due to the organization-wide interaction with it, the process library is the central technology in the standard mediation process at A-Suppliers. Design and maintenance receives a high attention, especially from the PG. Reflecting the size of the organization and the amount of standards which need to be served, they decided to dedicate a single full-time position to the editorial maintenance of the PL, the script editor (SE):

It is absolutely necessary, it won't work otherwise. So, the PRs describe, they are responsible for the content. But, you need somebody who is doing the formal review of the content. If everybody writes what he wants, everybody has an own style of writing. And it has to be used, it has to be understood. (A9:141)

Above all, this continuous development required support and motivation from the very top of the organization:

And PQ is the successor organization of the original PG. So it is also sustainable in the organization. But without [A3] as strong mentor on the higher management level, this wouldn't have had happened. And we had a second mentor in the board here in [A-Suppliers], which is [the CMMI board member]. He was also convinced that it will help. So also [A3] had a backup on [the executive management level]. (A8:041)

With the support of the higher management, the implementation of a PG at A-Suppliers became a broad change management success. A further significant proportion of it can be ascribed to the slow implementation of CMMI and the whole process landscape at A-Suppliers:

And it depends on how you communicate it and how you work with the people. If you just send an email: 'Here is the process library, you have to use it.' It will not work. You will have to do shit work [continu-

ous motivation, explanation and convincing]. And to communicate it, change management is the best way to describe it. (A7:339)

As highlighted before, the role of the PRs and the established feedback procedures stand for the participative and integrated approach. This is a crucial factor in the standardization process at A-Suppliers highlighted by interviewee A2, the head of the quality management department:

Because we want to have the experts for the products also writing proper processes for them. And we, we supply the model competence let me say. We do some coaching for the process representatives, to bring in the model competence for ISO 26262 and so on. So this is the corporation model and I think, if you want, you can call it decentralized process organization. This is distributed responsibility. (A2:115)

The PG is now a fully established actor in the organization of A-Suppliers. As such, they have concrete tasks, roles dedicated to these tasks and well-defined processes for their actions. Further, they recently showed signs of confidence towards their own role and its importance to the company. They reported on a study about "*process modeling for the process library of the future*" *(A9:209)*. In this way, they emphasized their pro-active role in the company towards processes:

[...] we as process group, we feel ourselves appointed to look ahead. We don't want to stand still. With the new product generation, we are also looking at process development and how we can achieve it in the future. (A9:213)

As the interviewee mentions, the A-Suppliers' PG intends to predict how the standard and process handling within the organization can be enhanced. Further, they are likewise engaged directly or indirectly on the side of the standard-setting bodies in the creation and maintenance of the standards they use. They not only translate and use the different standards while creating and maintaining their process library; they also give feedback and help with designing the standards they use as members of the standardization bodies. Whether it be the ISO standards, A-Spice or CMMI, for each an own standardization and/or certification body promotes and supervises the standard implementation. The PG of A-Suppliers participates in these activities:

[...] we are sitting in the standard committees of A-Spice and ISO 26262. We also have people working directly with the SEI. We have these people. These are central positions at A-Suppliers. And we do communicate directly with them. So, when someone sits in the committee of A-Spice and they discuss the next version [...] no matter which subject, then it always ends up on my desk. Then, I do check whether I have comments from my point of view. Does it fit or not for my requirements. (A9:249)

Summary of Findings

We present the summary of our findings along three different aspects of the metastructuring approach. First, we outline the process of metastructuring and its evolution over time. Then, we describe the areas in which metastructuring activities appear. Finally, we highlight different types of metastructuring activities that we found in the A-Suppliers case.

The Process of Metastructuring

A-Suppliers developed step-by-step an entire department as standard-use mediators (see Figure 2). They reacted to the increasing amount of process standards they are confronted with (CMMI, A-SPICE, ISO26262, ISO16949). Since 2000, A-Suppliers has been dealing with an increased standard complexity due to the advent of standards like CMMI or A-Spice. First, they gathered initial experiences within a pilot project to learn about CMMI implementation requirements (cf. Figure 2 - step 1). Then, the number of required standards increased and the software-related standard script is developed to integrate different standard requirements (cf. Figure 2 - step 2).

After this step, the appearance of metastructuring results in a new department, the process group (PG) (cf. Figure 2 - step 3), which then handles requirements of process standards organization-wide. The project group's task is the fulfillment of the standard requirements for the whole organization. The group manages the process library which was developed to ease maintenance and enhancement of the standard script. Within the PG, the relevant positions executing these tasks are the process coaches (PC). They are supported by the PRs. In the end, a new department and its roles as well as new artifacts, the standard script and the process library, emerge.

Figure 2. Process of Metastructuring at A-Suppliers

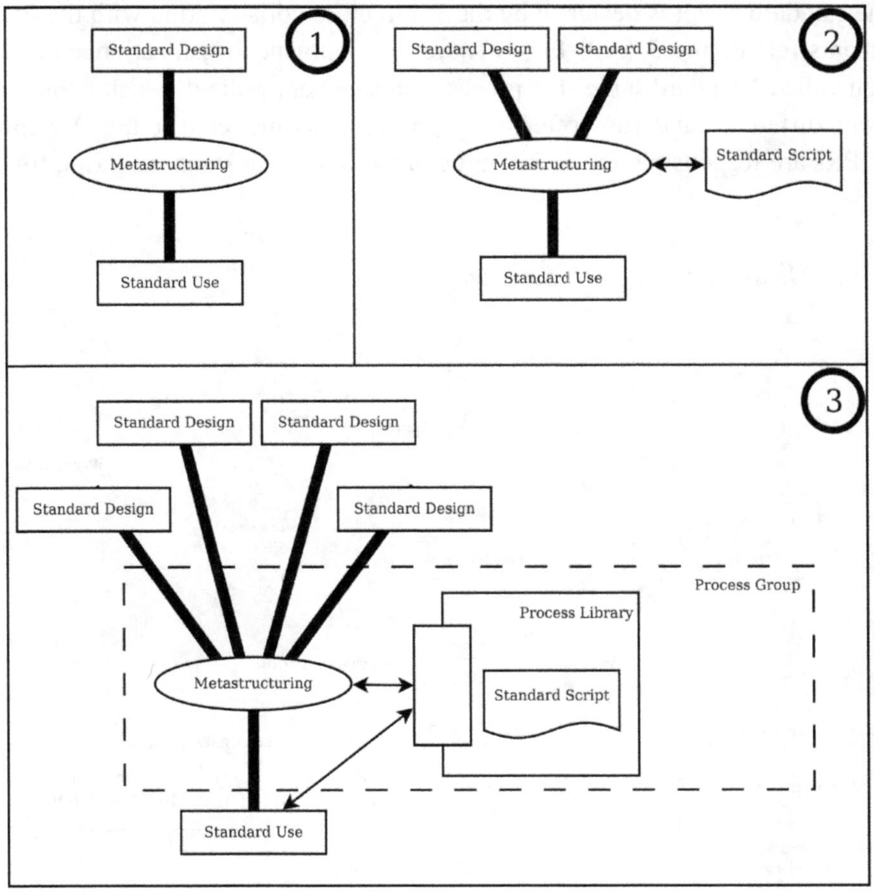

The Areas of Metastructuring

At A-Suppliers there are four areas of metastructuring: (A) managing the standard script, (B) managing the process library, (C) handling internal/corporate standards and (D) coping with external/international standards (Figure 3).

To start with (D), the PG takes external standards, interprets and applies them as input for the design of the corporate standard script embodied in the process library (cf. Figure 3 (D)). This is mainly the task of the PCs. Little involvement of the PRs is needed. Feedback for the creators of standards (e.g. standardization bodies) is sometimes provided. For most of the applied standards, A-Suppliers is represented with at least one person in the standardization body. This could also be verified by the authors by researching the internet regarding the relevant committees. However, influence on the design process of CMMI is reduced to feedback via change requests (Software Engineering Institute, 2012).

As a second metastructuring area, the PG has to take into consideration corporate standards (cf. Figure 3 (C)): directives towards process, products or even corporate identity. The corporate standards change the design of the standard script but the process itself was not described explicitly throughout the interviews. A-Suppliers has full control over design and usage of their own corporate standards. The interviewees described this process as integrated and participative and we assume that this metastructuring area is a full and open structuring process. This task is primarily accomplished by the PCs of the process group, although the nature of an internal standard allows a broader participation of the PRs in the structuring of these standards.

Thirdly, the standard script is designed by the PG in close collaboration with the PRs of the operational departments (cf. Figure 3 (A)). In (software development) organizations, standard scripts are internally often called 'standard bible' or 'process management policy' which contain an aggregated version of several different standards – often overlapping, sometimes conflicting. A-Suppliers' tasks are separated; the PRs are responsible for defining the processes in the standard script, for the training on

Figure 3. Areas of Metastructuring at A-Suppliers

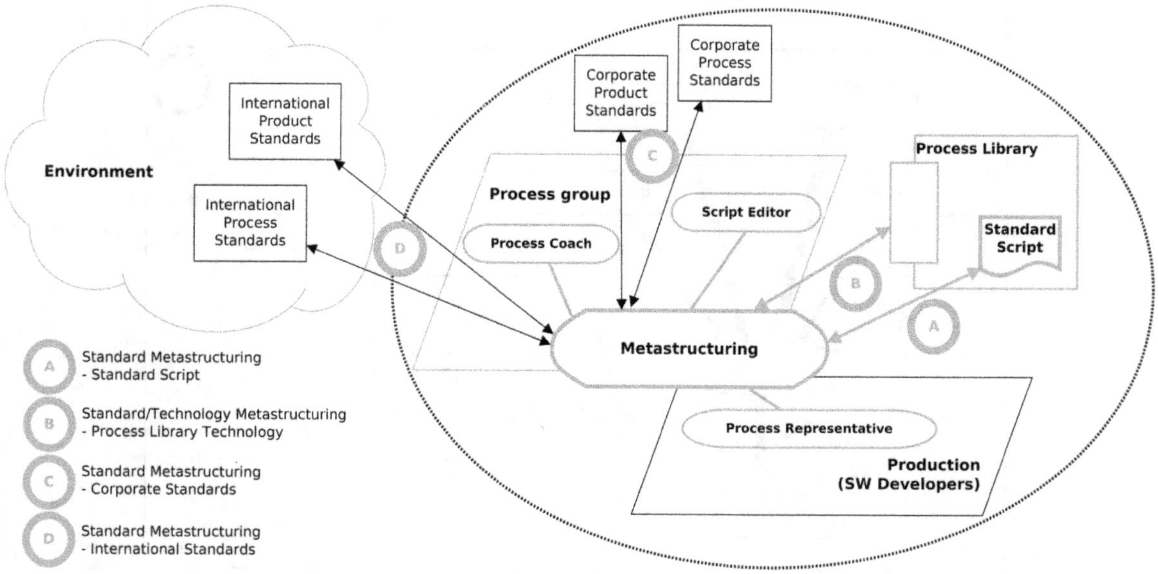

these processes and also for feedback and improvement of the process while the PC defines the broader process framework for the whole organization and further consults the PRs in their tasks. The standard script is then used by the software engineers (cf. Figure 3 (A)). Different feedback loops, either by using the process tool or via the PRs, were reported. Thus, the creation and use of a standard script embodied into the process library can be regarded as a vital structuring process. The standard script is the result of a translation process of several external and internal standards into a single internal one (cf. Figure 3 (C) and (D) as input for (A)).

Finally, we found another metastructuring area in the design, usage and maintenance of the (web-based) technology on which the process library is based (cf. Figure 3 (B)). A service provider developed the underlying technology of the process library as a customer development. As such, A-Suppliers had and still have full control over the implemented functionality. On the one hand, they considered standard specific requirements for the design of the technology. On the other hand, the technological restrictions influenced the representation of the standard script within the process library.

The Types of Metastructuring Activities

At A-Suppliers, the PG accomplishes the metastructuring activities to cope with the multiplicity and plurality of standards. Nevertheless, the process of metastructuring itself is not fully separated from the activities of the software engineers. A-Suppliers set up an integrated and participative environment where structuring in respect of the different standards is executed by different roles, both in the PG as well as in the software development department. Whereas a process coach (PC) works in the PG, the process representatives (PR) are based directly in the operational department, acting as standard-use mediators.

In Table 2 we summarized the different mediation activities related to four different roles - *process representative, process coach, script editor and process library consultant* – categorized along the four types of metastructuring activities: *establishment*, *reinforcement*, *adjustment* and *episodic change* (Orlikowski et al., 1995). At A-Suppliers, a significant part of the responsibility over standard structuring still remains with the operational departments. Furthermore, the *script editor* can be seen as a proof for this approach: the recently installed role fulfills the tasks of a standard script mediator. She/he streamlines the process descriptions of the different authors. Literally everybody at A-Suppliers can describe, maintain or give feedback on the processes. Hence, the process library is considered as a mass media within A-Suppliers which needs to be read and understood by every employee. The script editor can be regarded as a consequence of the integrative approach towards processes. Everybody participates but the uncontrolled growth needs to be harmonized professionally afterwards. The script editor is another example for the step-by-step development of an entire mediator department that reflects the complexity of the standard environment of A-Suppliers.

DISCUSSION

Our findings show that the powerful mechanism of metastructuring, originally described in technology research, exists for software development and process standards in the same way. A-Suppliers is an example for the contextualization of standards in use and provides insights in dealing with multiple and plural standard requirements in organizations. We term this specific type of metastructuring *standard metastructuring* since it structures users of standards "by influencing their interpretations and interac-

Table 2. Metastructuring Activities of the Different Mediatior Roles along the Phases of the Metastructuring Process

	Process Coach (PC)	Process Representative (PR)	Script Editor (SE)	Process Library Consultant (PLC)
Establishment	• set up standard pilot project (e.g. CMMI) • defining their own role as PC • set up process infrastructure • interpretation of each standard • creation of the standard script • creation of the process library	• creation of the standard script • defining the operational processes		creation of the process library
Reinforcement	• lean and user- friendly process descriptions • maintenance of the standard script • maintenance of the process library • training for standard- related tasks	• maintenance of the standard script • collect, produce and treat user feedback • process training for owned processes	• lean and user- friendly process descriptions • maintenance of the standard script	maintenance of the process library
Adjustment	• changing the interpretation of each standard • smaller changes to the process library	changing the operational processes		smaller changes to the process library
Episodic Change	• expanding responsibilities (more standards & sub-units to serve) • significant changes to the process library			significant changes to the process library

tions, by changing their institutional context of use and by modifying the [standard] itself" (Orlikowski et al., 1995, p. 425). We verified that standard metastructuring is an explicit, to a large degree conscious and ongoing set of activities and a powerful mechanism in dynamic environments of software development organizations.

Roles and Organizational Sub-Units in Standard Metastructuring

Looking at the process, areas and activity types, metastructuring of the standard script is considered as a very complex task for which standard mediators as experts have to be appointed and trained. The process group (PG), process consultants (PCs), process representatives (PRs) and the script editor (SE) act as standard mediators, or what Adler (2005, p. 417) calls shepherds. On this, we agree with Brunsson (2000a) and Tamm Hallström (2000) that "many of the actors in the area of quality are so-called intermediaries; they pass the standard on to users" (Tamm Hallström, 2000, p. 91). However, some of these intermediaries can be users as well: the PRs. The standard mediators now solve tasks which were formerly in the hands of each software engineer: assessment of new or changed standards, interpretation of standards, definition of processes, training on processes, preparation and execution of standard appraisals or handling the feedback for each standard. The role of the standard mediator is comparable to the role of the technology mediator since they serve as (1) an organizational mechanism (2) to facilitate the adaptation of technologies/the interpretation of standards (3) and the organizational context (4) ongoing and to changing conditions (Orlikowski, 1995).

When Orlikowski et al. (1995) first presented their theory on metastructuring; they made general propositions about the specific characteristics of mediators towards the mediation process. Therefore, she called for further research on characteristics and differentiations of mediators. With our study, we show first evidence regarding the differentiation of specific metastructuring roles. In our case, the organization differentiated a rather user-centered role, the PR, a standard-driven role, the PC, a standard and technology-driven role, the SE and a technology-driven role, the PLC. The differentiation of roles depends on the scope and complexity of the mediation task. We analyzed two bifurcations: firstly, when more and more standards come into play complexity increases and the organization differentiates between user- and standard-centered roles; secondly, when the organization starts to develop a standard script based on a technology infrastructure, enhancement and maintenance of the script and technology become crucial for the enactment of standards. The organization appoints specific roles for these tasks, for instance a SE or specific technology consultants, the PLCs. The role of the script editor is designed to streamline the process descriptions of the different authors for the standard script. Several distinct employees right up to everybody can participate in describing, maintaining or at least in giving feedback on the standard script. However, it needs to be read and understood by every employee and the uncontrolled growth needs to be harmonized professionally afterwards.

In standard metastructuring, the user-centered role of the process representative (PR) needs further explanations. As stated by Fleck (1994), the configuration of technology is highly important for a successful implementation and use of the technology and thus demands substantial inclusion of the user. This counts for process standards likewise. Processes and their contexts are very much organization-specific and demand a sensible handling in situations of organizational change. This leads to the conclusion that user-centered metastructuring in the process standard context is of high importance and highly relevant for organizations. This renders the role of PRs as central for a standard metastructuring that respects the use-mode of standards.

While looking at standard mediators and the differentiation in roles and organizational sub-units, researchers of technology-use can also draw conclusions from the alignment research stream. Here, "the question of stability and change in role relationships" is investigated and therefore "the use of a technology might alter or confirm an existing social order" (Leonardi and Barley, 2010, p. 25). Alignment to new standards leads to the creation of new roles and new organizational sub-units. First there were experienced users who enacted an implicit structuring process. The transition from implicit to explicit structuring was accomplished when these users were selected to exercise the new role as standard-mediators. This argumentation is supported by the findings of Yates et al. (1999) on implicit and explicit structuring of mediators for an electronic communication media.

The Process of Standard Metastructuring

Technology mediators "create policies, procedures, guidelines, templates, access mechanisms, applications, and physical configurations which alter the technology itself" (Orlikowski et al., 1995, p. 437). In the context of standard metastructuring, the same mediating activities exist. During *Establishment,* standard mediators set up the process infrastructure and define the operational processes. They define their own role and they start interpreting each applicable standard. At this point, the mediators could also create the standard script and the process library. However, these two tasks could also be accomplished at a later stage of metastructuring as an *Episodic Change*. In *Reinforcement,* mediators improve process descriptions and maintain the standard script and the process library. They set up and undertake standard and process training and they collect and treat user feedback. The user feedback and smaller changes to the interpretation of standards and the process library lead to changes of these items during the *adjustment* phase. During the *episodic change,* mediators make significant changes to the standard script or the process library and they expand their responsibilities. They incorporate either new standards to structure or new organizational sub-units which would then start using their standard metastructuring service.

Differences between standard and technology metastructuring tasks originate from the very differences between the artifacts to structure and the practices that appear around these artifacts: between standards and information systems as the somewhat classical artifacts of technology structuring. In organizational practice, organizations often have to consider several standards (Djelic and den Hond, 2014). For various reasons, some of these standards and its requirements overlap each other, yet they are to be considered likewise in the standard script. In technology structuring, there is no such thing as a standard script. Further, organizations avoid enacting several technologies for the same task. Hence, chances are rare to integrate several technologies into a single, central one. Consequently, the differences between standard and technology metastructuring concern the tasks around the creation and maintenance of the standard script and, during e*pisodic change,* the task of expanding responsibilities to incorporate more standards in the metastructuring activities.

Orlikowski et al. (1995, p. 437ff) described the process of metastructuring after the *establishment* phase as periodic changes from *reinforcement* and *adjustment* to *episodic change.* In standard metastructuring, we find the same steps of periodic changes. Orlikowski (1996) and Leonardi (2007) described these steps as “a phased sequence of emergent and cumulative changes” (Leonardi and Barley, 2010, p. 28). Standard metastructuring develops from implicit structuring, to explicit structuring with ongoing and further differentiating roles and own organizational structures like differentiated groups or departments.

Standard Script and Process Library

For the process of standard metastructuring, the artifacts of standard script and process library play a significant role. With the standard script, we emphasize that standards have to be encoded in the organizational practices to be effective. Following Barley and Tolbert (1997) and Gioia and Poole (1984) on scripts we can describe standard scripts as ’realized standards’ that are and have to be encoded and enacted in the organization’s stock of knowledge and practices. In the case of multiple standards, organizations enact a single standard script that fits all the standards that are to be followed by the organization. The more standards the organization follows, the more complex the enactment of the standard script becomes. In that situation, organizations revert to sophisticated tools that organize the different standards requirements and its references in the standard script. We call these tools a process library. Once a process library is implemented, the organizational members access the documentation of the standard script only through the process library interface. For them, the process library represents the standard script and the both are irrevocably entangled.

When looking deeper into the practices that emerge around the process library and the standard script, researchers could draw further insights from research on *boundary objects* (Carlile, 2002; Levina and Vaast, 2005; Star and Griesemer, 1989; Star, 2010). We consider the process library or the standard script as a boundary object that resides between social worlds (e.g. standard world, engineering world, world of HR), where the tool has different meanings (standard world: to integrate different standards and to design the standard script; engineering world: to create and derive engineering processes, world of HR: to plan and execute professional training) (Star and Griesemer, 1989). The standard mediators could be analyzed as boundary-spanners-in-practice (Levina and Vaast, 2005) whose task is to process and transform knowledge (Carlile, 2002).

Standard Metastructuring and Standardization Research

Brunsson (2000a), Tamm Hallström (2000) as well as Timmermans and Epstein (2010) argued about the relation between *designer* and *user* of a standard. They concluded that there is a sharp distinction between both: "Standards are presumed to be in the public interest, but the public to whom standards apply is usually not directly represented in standard creation" (Timmermans and Epstein, 2010, p. 77). Whereas this might be true for specific types of standards, we have to differentiate in the case of process or product standards that apply to organizations. In our study, the organization engages in the process of standard setting via participation in standard setting bodies. Further, the organizational users of the standard engage in the internal standard setting process via feedback options and via different roles that structure the internal use of standard by setting the organizational standard script. "The trick in standardization appears to be to find a balance between flexibility and rigidity and to trust users with the right amount of agency to keep a standard sufficiently uniform for the task at hand" (Timmermans and Epstein, 2010, p. 81). This raises the question of how much "distance in time and space" (Brunsson, 2000a, p. 27) we actually speak when it comes to design and use of a standard, especially when dealing with procedural standards (Timmermans and Epstein, 2010, p. 72). In the area of technology structuring, Orlikowski (1992, p. 407) speaks about the interpretive flexibility of technology. This flexibility refers to the engagement of users of a technology in its constitution. With an increasing distance in time and space, "the likelihood that the technology will be interpreted and used with little flexibility [...]" (Orlikowski, 1992, p. 421) also increases. This phenomenon occurs for technology as well as for standards.

With our research on the micro level of standardization, we start to counter the fact that "studies of standardization have neglected the crucial role of technical experts on the receiving end of standards" (Sandholtz, 2012, p. 670). Sandholtz (2012) described two different cases of decoupling (malignant and benign) while implementing the ISO 9000 standard. Further, Fried et al. (2013) analyzed organizations that temporarily loosely couple or decouple their innovating activities from the reigning standard script to avoid disturbing their innovating activities. As highlighted in this research, organizations could refer to standard metastructuring to avoid disturbances at the receiving end of standards. We regard process consultants, process representatives, script editors and process library consultants as the "group of internal experts to standardize their own work processes in a manner consistent with external certification demands" (Sandholtz, 2012, p. 676).

However, whether a standard is experienced as coercive or enabling by its users depends not only on decoupling approaches or standard metastructuring. Various contextual factors should be taken into account: management commitment (Adler et al., 2005a; Boiral, 2003), staff participation (Adler et al., 2005a; Allison and Merali, 2005; Boiral, 2003; Lavellée and Robillard, 2012; Sandholtz, 2012), culture (Braa and Hedberg, 2002; Müller et al., 2010; Ngwenyama and Nielsen, 2003) or the appropriation of standards (Gey, 2011). For the latter, standardization research could benefit tremendously from contemporary technology studies on appropriation (Leonardi et al., 2016, Quinones et al., 2013, You et al., 2015).

CONCLUSION

In this contribution, we examined how software development organizations deal with the multiplicity and plurality of standards. We used a social constructivist viewpoint of technology use in organizations and focused on the use-mode of standards (Orlikowski, 1992). We found that metastructuring is an important mechanism for organizations to handle complexity arising from the use of multiple process standards. Relating to Orlikowski et al. (1995) on technologies, we called it *standard* metastructuring since it has a considerable influence on how users perceive, interpret and apply these standards. We used the concept of technology structuring and applied it to standards as the artifacts to structure. The results show that the structuring of standards has a lot in common with the structuring of technologies, even though they are not technologies themselves. Additionally, because of the increased number of standards to handle, the organization developed a process library. This software solution consists of all standards embodied in the standard script as well as options to access information, to define processes and to give feedback on processes.

We understand this exploratory study as a starting point for a more systematic examination on standard metastructuring and the process of metastructuring. We tried to adopt the metastructuring concept from the technology world for the world of standards. Therefore, it was necessary to discuss the differences between standards and technology as well as the resulting differences for the metastructuring concept. We further enhanced the metastructuring concept by a proposition towards different types of metastructuring activities. We assume that the more complex the standard landscape an organization is confronted with, the more likely is the differentiation in several mediator roles.

The scope of this study is limited and calls for further research. For instance, more in-depth research is needed regarding the specific activities of standard metastructuring. We suggest further investigation of different types of metastructuring and their coercive and enabling character. This might involve ambiguities of the different standards while asking how these ambiguities are resolved during the metastructuring process. Empirical studies could also show to what extent standard-use mediation varies with the type of standard and individuals involved.

Most prominently, there is the question of how much "*distance in time and space*" (Brunsson, 2000a, p. 27) should be taken into consideration when it comes to the design and use of a standard. Prospective studies could explore the mentioned distance in time, space and other categories, e.g. social distance. Researchers in the area of standardization should always keep in mind that "*differences between presentation and practice, between formal structures and actual operations, and between what people say and what they do*" (Brunsson and Jacobsson, 2000b, p. 130) can be encountered.

In summary, a well-managed process of standard metastructuring represents a relevant factor for organizations introducing and using new process standards as a response to changing market requirements or to an increased focus on process quality and process maturity of organizations. However, the process does not automatically result in a successful outcome, which calls for a comprehensive and careful implementation and continuous adaption to a changing environment.

REFERENCES

Adler, P. (2005). The evolving object of software development. *Organization, 12*(3), 401–435. doi:10.1177/1350508405051277

Adler, P. S., McGarry, F. E., Irion-Talbot, W. B., & Binney, D. J. (2005a). Enabling process discipline: Lessons from the journey to CMM Level 5. *MIS Quarterly Executive, 4*(1), 215–227.

Allison, I., & Merali, Y. (2007). Software process improvement as emergent change: A structurational analysis. *Information and Software Technology, 49*(6), 668–681. doi:10.1016/j.infsof.2007.02.003

Bansler, J., & Havn, E. (2004). Technology-use mediation. Making sense of electronic communication in an organizational context. *Scandinavian Journal of Information Systems, 16*, 57–84.

Barley, S. (1985). Technology as an occasion for structuring: Evidence from observations of ct scanners and the social order of radiology departments. *Administrative Science Quarterly, 31*(1), 78–108. doi:10.2307/2392767 PMID:10281188

Barley, S., & Tolbert, P. (1997). Institutionalization and structuration: Studying the links between action and institution. *Organization Studies, 18*(1), 93–117. doi:10.1177/017084069701800106

Blind, K., & Mangelsdorf, A. (2016). Motives to standardize: Empirical evidence from Germany. *Technovation, 48-49*(1), 13–24. doi:10.1016/j.technovation.2016.01.001

Boiral, O. (2003). ISO 9000: Outside the iron cage. *Organization Science, 14*(6), 720–737. doi:10.1287/orsc.14.6.720.24873

Braa, J., & Hedberg, C. (2002). The Struggle for District-Based Health Information Systems in South Africa. *The Information Society, 18*(2), 113–127. doi:10.1080/01972240290075048

Brunsson, N. (2000a). Organizations, Markets, and Standardization. In N. Brunsson & B. Jacobsson (Eds.), *A World of Standards* (pp. 21–40). Oxford, UK: Oxford University Press.

Brunsson, N. (2000b). Standardization and uniformity. In N. Brunsson & B. Jacobsson (Eds.), *A World of Standards* (pp. 138–151). Oxford, UK: Oxford University Press.

Brunsson, N., & Jacobsson, B. (2000c). The contemporary expansion of standardization. In N. Brunsson & B. Jacobsson (Eds.), *A World of Standards* (pp. 1–19). Oxford, UK: Oxford University Press.

Brunsson, N., & Jacobsson, B. (2000c). *A World of Standards*. Oxford, UK: Oxford University Press.

Brunsson, N., & Jacobsson, B. (2000d). Following standards. In N. Brunsson & B. Jacobsson (Eds.), *A World of Standards* (pp. 127–138). Oxford, UK: Oxford University Press.

Brunsson, N., Rasche, A., & Seidl, D. (2012). The dynamics of standardization: Three perspectives on standards in organization studies. *Organization Studies, 33*(5-6), 613–632. doi:10.1177/0170840612450120

Brunswicker, S.; Rodriguez, J. A. & Wareham, J. D. (2015). Governing Standardization through Formal and Informal Standard Development Organizations. *Academy of Management Proceedings 2015*.

Carlile, P. R. (2002). A Pragmatic View of Knowledge and Boundaries: Boundary Objects in New Product Development. *Organization Science, 13*(4), 442–455. doi:10.1287/orsc.13.4.442.2953

Clear, T., & MacDonell, S. G. (2011). Understanding technology use in global virtual teams: Research methodologies and methods. *Information and Software Technology, 53*(9), 994–1011. doi:10.1016/j.infsof.2011.01.011

Davidson, E., & Chiasson, M. (2005). Contextual influences on technology use mediation: A comparative analysis of electronic medical record systems. *European Journal of Information Systems, 14*(1), 6–18. doi:10.1057/palgrave.ejis.3000518

Davidson, E., & Chismar, W. (2007). The interaction of institutionally triggered and technology-triggered social structure change: An investigation of computerized physician order entry. *Management Information Systems Quarterly, 31*(4), 739–758.

DeSanctis, G., & Poole, M. S. (1994). Capturing the complexity in advanced technology use: Adaptive structuration theory. *Organization Science, 5*(2), 121–147. doi:10.1287/orsc.5.2.121

Djelic, M. L., & den Hond, F. (2014). Introduction: Multiplicity and plurality in the world of standards. *Business and Politics, 16*(1), 67–77. doi:10.1515/bap-2013-0034

Egyedi, T. M. (2008). An implementation perspective on sources of incompatibility and standards' dynamics. In T. M. Egyedi & K. Blind (Eds.), *The Dynamics of Standards* (pp. 28–43). Cheltenham, UK: Edward Elgar.

Fleck, J. (1994). Learning by trying: The implementation of configurational technology. *Research Policy, 23*(6), 637–652. doi:10.1016/0048-7333(94)90014-0

Fried, A. (2010). Performance measurement systems and their relation to strategic learning: A case study in a software-developing organization. *Critical Perspectives on Accounting, 21*(2), 118–133. doi:10.1016/j.cpa.2009.08.007

Fried, A., Gey, R., Pretorius, A., & Günther, L. (2013). Decoupling from standards - process management and technical innovation in software development organisations. *International Journal of Innovation Management, 17*(4), 1–34. doi:10.1142/S1363919613500126

Furusten, S. (2000). The knowledge base of standards. In N. Brunsson & B. Jacobsson (Eds.), *A World of Standards* (pp. 71–85). Oxford, UK: Oxford University Press.

Gey, R. (2011). Appropriation of Software Process Improvement Standards: An Empirical Study in Software Development Organizations. *Proceedings of the 8th International Conference on Intellectual Capital, Knowledge Management & Organisational Learning*, 182–189.

Giddens, A. (1984). *The constitution of society: Outline of the theory of structuration*. Cambridge, UK: Polity Press.

Gioia, D., & Poole, P. (1984). Scripts in organizational behavior. *Academy of Management Review*, *9*(3), 449–459.

Henriksen, D., Nicolajsen, H., & Pors, J. (2002). Towards variation or uniformity? Comparing technology-use mediations of web-based groupware. *Proceedings of the European Conference on Information Systems 2002*, 1174–1184.

Humphrey, W. (1992). Introduction to software process improvement. Technical report, Software Engineering Institute, Carnegie-Mellon University.

Jakobs, K. (2006). Shaping user-side innovation through standardisation: The example of ICT. *Technological Forecasting and Social Change*, *73*(1), 27–40. doi:10.1016/j.techfore.2005.06.007

Janneck, M., & Finck, M. (2006). Appropriation and mediation of technology use in stable self-organised online communities. *Web Based Communities, Proceedings of IADIS International Conference*, 149–156.

Kohlbacher, F. (2006). The use of qualitative content analysis in case study research. *Forum Qualitative Social Research*, *7*(1).

Krippendorff, K. (2004). *Content analysis: An introduction to its methodology*. Thousand Oaks, CA: Sage Publications.

Lavallée, M., & Robillard, P. N. (2012). The impacts of software process improvement on developers: a systematic review. *Proceedings of the 2012 International Conference on Software Engineering*, 113-122. doi:10.1109/ICSE.2012.6227201

Lawrence, T., & Phillips, N. (1998). Commentary: Separating play and critique. *Journal of Management Inquiry*, *7*(2), 154–160. doi:10.1177/105649269872010

Leonardi, P. (2007). Activating the informational capabilities of information technology for organizational change. *Organization Science*, *18*(5), 813–831. doi:10.1287/orsc.1070.0284

Leonardi, P., & Barley, S. (2010). What's under construction here? Social action, materiality, and power in constructivist studies of technology and organizing. *The Academy of Management Annals*, *4*(1), 1–51. doi:10.1080/19416521003654160

Leonardi, P. M., Bailey, D. E., Diniz, E. H., Sholler, D., & Nardi, B. (2016). Multiplex Appropriation in Complex Systems Implementation: The Case of Brazil's Correspondent Banking System. *Management Information Systems Quarterly*, *40*(2), 461–473. doi:10.25300/MISQ/2016/40.2.10

Levina, N., & Vaast, E. (2005). The Emergence of Boundary Spanning Competence in Practice: Implications for Implementation and Use of Information Systems. *Management Information Systems Quarterly*, *29*(2), 335–363.

Marimon, F., Heras, I., & Casadesús, M. (2009). ISO 9000 and ISO 14000 standards: A projection model for the decline phase. *Total Quality Management*, *20*(1), 1–21. doi:10.1080/14783360802614257

Muhr, T. (1991). Atlas/ti - a prototype for the support of text interpretation. *Qualitative Sociology*, *14*(4), 349–371. doi:10.1007/BF00989645

Müller, S. D., Mathiassen, L., & Balshøj, H. H. (2010). Software Process Improvement as organizational change: A metaphorical analysis of the literature. *Journal of Systems and Software*, *83*(11), 2128–2146. doi:10.1016/j.jss.2010.06.017

Ngwenyama, O., & Nielsen, P. A. (2003). Competing values in software process improvement: An assumption analysis of CMM from an organizational culture perspective. *IEEE Transactions on Engineering Management*, *50*(1), 100–112. doi:10.1109/TEM.2002.808267

Niazi, M., Wilson, D., & Zowghi, D. (2005). A maturity model for the implementation of software process improvement: An empirical study. *Journal of Systems and Software*, *74*(2), 155–172. doi:10.1016/j.jss.2003.10.017

Novak, L., Brooks, J., Gadd, C., Anders, S., & Lorenzi, N. (2012a). Mediating the intersections of organizational routines during the introduction of a health IT system. *European Journal of Information Systems*, *21*(5), 552–569. doi:10.1057/ejis.2012.2 PMID:24357898

Novak, L. L., Anders, S., Gadd, C. S., & Lorenzi, N. M. (2012). Mediation of adoption and use: A key strategy for mitigating unintended consequences of health IT implementation. *Journal of the American Medical Informatics Association*, *19*(6), 1043–1049. doi:10.1136/amiajnl-2011-000575 PMID:22634157

Okamura, K., Orlikowski, W., Fujimoto, M., & Yates, J. (1994). Helping CSCW applications succeed: the role of mediators in the context of use. *Proceedings of the 1994 ACM conference on computer supported cooperative work*, 55-65. doi:10.1145/192844.192871

Orlikowski, W. (1992). The duality of technology: Rethinking the concept of technology in organizations. *Organization Science*, *3*(3), 398–427. doi:10.1287/orsc.3.3.398

Orlikowski, W. (1996). Improvising organizational transformation over time: A situated change perspective. *Information Systems Research*, *7*(1), 63–92. doi:10.1287/isre.7.1.63

Orlikowski, W., Yates, J., Okamura, K., & Fujimoto, M. (1995). Shaping electronic communication: The metastructuring of technology in the context of use. *Organization Science*, *6*(4), 423–444. doi:10.1287/orsc.6.4.423

Ortmann, G. (2010). On drifting rules and standards. *Scandinavian Journal of Management*, *26*(2), 204–214. doi:10.1016/j.scaman.2010.02.004

Power, M. (1997). *The audit society: rituals of verification*. Oxford, UK: Oxford University Press.

Quinones, P.-A., Teasley, S. D., & Lonn, S. (2013). Appropriation by unanticipated users: looking beyond design intent and expected use. *Proceedings of the 2013 conference on computer supported cooperative work*, 1515-1526. doi:10.1145/2441776.2441949

Rodon, J., Sese, F., & Christiaanse, E. (2011). Exploring users' appropriation and post-implementation managerial intervention in the context of industry IOIS. *Information Systems Journal*, *21*(3), 223–248. doi:10.1111/j.1365-2575.2009.00339.x

Sandholtz, K. W. (2012). Making standards stick: A theory of coupled vs. decoupled compliance. *Organization Studies*, *33*(5-6), 655–679. doi:10.1177/0170840612443623

Seidl, D. (2007). Standard setting and following in corporate governance: An observation-theoretical study of the effectiveness of governance codes. *Organization*, *14*(5), 705–727. doi:10.1177/1350508407080316

Seo, D. B.DongBack. (2013). Analysis of Various Structures of Standards Setting Organizations (SSOs) that Impact Tension among Members. *International Journal of IT Standards and Standardization Research*, *11*(2), 46–60. doi:10.4018/jitsr.2013070104

Software Engineering Institute (SEI). (2012). *Change Requests*. Retrieved January 15, 2017 from http://www.sei.cmu.edu/cmmi/solutions/crs/index.cfm

Star, S. L. (2010). This is not a boundary object: Reflections on the origin of a concept. *Science, Technology & Human Values*, *35*(5), 601–617. doi:10.1177/0162243910377624

Star, S. L., & Griesemer, J. R. (1989). Institutional Ecology, 'Translations' and Boundary Objects: Amateurs and Professionals in Berkeley's Museum of Vertebrate Zoology. *Social Studies of Science*, *19*(3), 387–420. doi:10.1177/030631289019003001

Tamm Hallström, K. (2000). Organizing the process of standardization. In N. Brunsson & B. Jacobsson (Eds.), *A World of Standards* (pp. 85–100). Oxford, UK: Oxford University Press.

Tamm Hallström, K. (2004). *Organizing international standardization. ISO and the IASC in quest of authority*. Cheltenham, UK: Edward Elgar.

Timmermans, S., & Epstein, S. (2010). A World of Standards but not a Standard World: Toward a Sociology of Standards and Standardization. *Annual Review of Sociology*, *36*(1), 69–89. doi:10.1146/annurev.soc.012809.102629

Weick, K. (1990). Technology as equivoque: sensemaking in new technologies. In P. S. Goodman & L. S. Sproull (Eds.), *Technology and Organizations* (pp. 1–44). San Francisco, CA: Jossey-Bass.

Weick, K. E. (1976). Educational organizations as loosely coupled systems. *Administrative Science Quarterly*, *21*(1), 1–19. doi:10.2307/2391875

Wiegmann, P. M., de Vries, H. J., & Blind, K. (2017). Multi-mode standardisation: A critical review and a research agenda. *Research Policy*, *46*(8), 1370–1386. doi:10.1016/j.respol.2017.06.002

Wieland, J. (2005). Corporate governance, values management and standards: A European perspective. *Business & Society*, *44*(1), 74–93. doi:10.1177/0007650305274852

Yates, J., Orlikowski, W., & Okamura, K. (1999). Explicit and implicit structuring of genres in electronic communication: Reinforcement and change of social interaction. *Organization Science*, *10*(1), 83–103. doi:10.1287/orsc.10.1.83

Yin, R. K. (2003). *Case Study Research*. Thousand Oaks, CA: Sage Publications.

You, S., Robert, L. P. Jr, & Rieh, S. Y. (2015). The Appropriation Paradox: Benefits and Burdens of Appropriating Collaboration Technologies. *Proceedings of the 33rd Annual ACM Conference Extended Abstracts on Human Factors in Computing Systems*, 1741-1746. doi:10.1145/2702613.2732919

Zeng, S., Shi, J., & Lou, G. (2007). A synergetic model for implementing an integrated management system: An empirical study in China. *Journal of Cleaner Production*, *15*(18), 1760–1767. doi:10.1016/j.jclepro.2006.03.007

ENDNOTE

1 For a collection of reference models for process improvement, see Humphrey (1992).

Section 4
1000 Flowers

Chapter 12
Impact of Industry Conditions on Innovation:
Pre-Existing Standards and Regulations

J. Roland Ortt
Delft University of Technology, The Netherlands

Tineke Mirjam Egyedi
Delft Institute for Research on Standardization, The Netherlands

ABSTRACT

This chapter underscores the importance of timing by focusing on the effect of pre-existing standards and regulations on the innovation and diffusion of new high-tech product innovations. The effect is assessed in terms of the time interval between the invention of a technological principle and the introduction of the first marketable product (development phase), and the successive time interval up to the start of large-scale industrial production and diffusion (adaptation phase). Fifty heterogeneous cases of new high-tech product innovations from 1850 onward are analysed. Results indicate that pre-existing standards and regulations significantly shorten the adaptation phase, an effect not found for the development phase. The shortening effect on the adaptation phase is particularly evident for more radical innovations and for innovations that are more interrelated with a larger technological system. This accelerating effect on the diffusion of innovations is highly relevant for innovation managers and policy makers alike.

INTRODUCTION

The question whether standards and regulations hamper or enable innovation has created controversy among scientists, while the policy and managerial importance of resolving it goes undisputed. A key issue explaining the controversy is the lack of specificity about the nature of the relationship. With the current chapter we aim to contribute towards its further clarification. We focus on *pre-existing* standards and regulations. Pre-existing standards and regulations refer to the set of standards, formal guidelines, rules, laws and conventions in force prior to the invention of a new product i.e., prior to the demonstration of the first rudimentary version of the product. These standards and regulations apply

DOI: 10.4018/978-1-5225-5320-5.ch012

industry-wide and have not specifically been set to address the new product . That is, they form part of the industry conditions within which a new product is developed and later on diffuses. An example of such standards and regulations are safety and quality requirements prevalent in an industry to which a new product has to comply.

The focus is on radically new high-tech products. The contraceptive pill and Nylon, for example, have in common that, at the time of their introduction, they were radically new to the market and technologically state-of-the-art in their respective disciplines. Radically new to the market means that a product's functionality is new to the market; technologically state-of-the-art means that a product's price-performance ratio is much better than that of contemporary products or that it is based on new technical principles (e.g., contraception via hormones). Following the typology proposed by Garcia and Calantone (2002), we therefore refer to them as radically new.

In some cases, industry conditions, notably pre-existing standards and regulations, facilitate the innovation and diffusion of new products. An example is SMS in its time. SMS is a short text message service sent via the control channel of the network. The control-channel was originally introduced as part of the ISDN standard but later became part of subsequent cellular mobile telephony standards during which period the idea of using the control-channel for SMS emerged (Brusoni and Corrocher, 2006). The availability of this standard and of regulation about the structure of the telecommunication network eased both the innovation and subsequent diffusion of the SMS-service (Brusoni and Corrocher, 2006; Lacohée, Wakeford and Pearson, 2003; Taylor and Vincent, 2005).

However, in other cases industry-wide standards and regulations may hamper the innovation and market diffusion of radically new high-tech products. Kay (2002) and Constant (1980), for example, illustrate this with respect to the innovative jet-engine developments in both Germany and Britain at the start of the Second World War. In both countries the entire military air force system was organized around and restricted to facilitate propeller-powered airplanes. The available budget for developing a jet-engine was limited as were the possibilities for the military to use it once developed.

In the afore-mentioned scientific controversy some scholars perceive standards and regulations as limiting variety and, more specifically, as restricting innovation (Temple, 2005) and product variety (Wölker, 1996). In contrast, others emphasize the positive economic effects of standards, such as allowing the build-up of critical mass and enabling economies of scale (e.g., Swann, 2000; Blind, 2004). A possible cause for these contradictory findings is lack of specificity about the phase of the technology life-cycle concerned. Standards and regulations may already be in place at the time of the invention. However, they may also be developed and effectuated at later stages. Whereas the adverse effects of standardizing and regulating too early or too late are well-noted (Sherif, 2003; Blind, 2004; Ho and Sullivan, 2016), no systematic enquiry has been made into the way the timing thereof affects innovation (West 2003; Temple, 2005; Egyedi and Sherif, 2010, Ho, 2017; Koch, 2017). In the following the authors refer to the effect of standards and regulations on the length of the innovation and diffusion process as the 'timing effect'.

While from a scientific perspective it is highly relevant to resolve the controversy, improved insight into the timing effect of standards and regulations may also help design institutional and policy frameworks that better catalyse innovation and diffusion processes. Such frameworks are needed to support more effective innovation policies at the industry and company level.

In this chapter, the objective is to clarify a few issues that cloud the debate and start by addressing part of the timing effect. In particular, the chapter will explore the effect of pre-existing standards and regulations on the time required for different phases in the innovation trajectory of radically new high-tech products. This effect is investigated for a heterogeneous set of cases from different historical periods and industries.

The chapter is structured as follows. First, the authors address theoretical considerations, such as the controversy about the impact of standards and regulations on innovation. Next, the methodology and research findings will be presented. The chapter closec with a conclusion and discussion section and a section on future research.

THEORY

The Effect of Standards and Regulations on Innovation and Diffusion

Past studies are not unanimous about the effect of standards and regulations on innovation and diffusion, as overviews of the literature show (Swann, 2000, 2010; Blind, 2004; SOU, 2007). Some emphasize the terse relationship between the two, and in particular the inherent variety reducing effect of standards. Standards can be seen as freezing or fixing technology development (Hanseth, Monteiro and Hatling, 1996; de Vries, Verhagen and Beerepoot, 2008). Blind presents a list of authors who view standards as constraining innovation (2004, Table 4.1, pp. 22-26). They argue that standards limit product diversity and consumer choice (*e.g.,* Wölker, 1996). Where de facto standards are concerned, this may lead to market concentration and monopolies (Blind, 2004, p. 22). Monopolies impede the transition from an old to a new technology (vendor lock-in) and reduce incentives to innovate. Where committee standards are concerned, Senden and Wöckel (1997) speak of *inflexible* standardization that "cement[s] the state of technology" and "cut[s] down on other variations in the interest of standardization"; and Salop and Sheffman (1983) point to the possible occurrence of 'regulatory capture', whereby minimum quality standards (*e.g.,* for safety) are defined in a way that highly raises the costs of innovation for rivals. These negative effects are due to flaws in committee standardization: an ill-advised choice of what is to be standardized (inflexible standardization), a 'politicized' standards process (regulatory capture), or of wrong timing and inadequate standards implementation (Blind, 2004, Table 4.2). While difficult to operationalize, some studies use the term 'effective standards' (Swann, 2000, 2010; SOU, 2007) to distinguish the effects of *flawed* standards from the effects of standards in general. These and other authors highlight that standards and standardization can also enable technology innovation (Swann, 2000, 2010; Blind, 2004; SOU, 2007). As summarized by Swann (2010, p. 9):

standardization helps to build focus, cohesion and critical mass in the formative stages of a market (e.g., Krechmer 1996; Swann and Watts, 2002). Standardization of measurements allows innovative producers to demonstrate to the satisfaction of the customer that products are as innovative as they claim to be (e.g., Tassey, 1982; Swann, 1999). Standardization codifies and diffuses state of the art technology and best practice (e.g., Krechmer 2000, 2005; Blind and Grupp, 2000). Open standards are desirable to enable a competitive process of innovation-led growth (e.g., Krechmer, 1998; Swann, 1990).

Explanations for the innovation-enabling nature of standards are, firstly, that standards form an infrastructure and shared basis for technology development. As such, standards reduce transaction costs, facilitate trade, build critical mass and allow economies of scale. Secondly, standards embed information that spurs innovation. Standards codify and share state-of-the-art scientific and engineering knowledge which is put to use by innovators.

This chapter explores the impact of pre-existing committee standards and pre-existing regulations on innovation and diffusion building on Swann (2000) and others (DTI, 2005; King, 2006; Swann and Lambert, 2010). To address the question 'Do standards enable or constrain innovation', Swann analyses the response of companies to two questions in the Community Innovation Survey (CIS3; period 1998-2000; DTI, 2005, p. 80).

- Question 8.1 asks the respondent to comment on a range of factors that may inhibit the enterprise's ability to innovate[1], including the 'Impact of regulations or standards'. The respondent is asked to grade the importance of these constraints.
- Question 12.1 asks respondents to rank the different sources of knowledge or information used in innovation activities. The sources include 'Technical standards, Health and safety standards and regulations, and Environmental standards and regulations'.

Swann demonstrates that the two issues are positively correlated. Respondents who say that standards are a source of information for innovation activities also indicate that standards and regulations constrain their innovation activities, and vice versa (DTI, 2005; Swann, 2010, p. 10). Swann and Lambert (2010) therefore rephrase the research question as 'Why do standards enable *and* constrain innovation?'

King (2006) specifies why this may be the case. Using additional CIS data on the 'effects of a firm's innovation activity' (King, 2006, p. 72-73), he concludes that innovation activities of companies may well be standards- or regulation-oriented:

It is clear that the pattern for the constraint imposed by regulations paralleled that (...) for the use of standards as a source of information. This suggests that those who make use of standards to help them achieve a particular goal feel to some extent constrained by them. Moreover, Swann argued that the 'informing' and 'constraining' roles of standards were complementary, and these results support his claim. Standards documents provide guidance and stipulations concerning best practice for ensuring rigorous quality control, and specifications to enable compatibility and minimum levels of performance. Hence, standards inevitably constrain a firm's activities if they wish to receive the benefits that standardization brings. Finally, regulations may force firms to innovate and adapt their practices in order to comply with regulation, and in such circumstances firms are necessarily constrained by regulation. In essence, taking account of standards and regulations is part of the routine that successful firms follow. (King, 2006, p. 76)

In sum, according to this set of studies, the question whether standards and regulations enable or constrain innovation contains a false antithesis (Swann and Lambert, 2010, p. 371). Standards and regulations can at the same time be a constraining as well as an enabling force for innovation.

To explore this 'false antithesis', the effect of industry-wide standards and regulations in force on technology innovation and diffusion over time will be investigated.

Innovation and Diffusion Processes

The timescale of the innovation and diffusion of new high-tech products can vary widely. Some of these products show a life cycle that unfolds within a few years while for other products this may take more than a century. The literature distinguishes different types of life cycles. The idea that different types of innovations (incremental and radical) can be distinguished in the innovation and diffusion of technological systems is widely recognized (Abernathy and Utterback, 1978; Tushman and Anderson, 1986; Tushman and Rosenkopf, 1992; Tushman *et al.*, 1997) and is referred to as the punctuated equilibrium. Based on these conceptual notions, Ortt and Schoormans (2004) propose a model in which three phases are distinguished in the innovation and diffusion of radically new products: the development, adaptation and stabilization phase (see Figure 1).

The development phase comprises the period from the invention of a technological principle up to the first application of that principle in a new high-tech product in the market. The invention is defined to be the first time the working of the main components of a product are demonstrated in practice. In many cases, the invention is merely the demonstration of a crude working principle that is far from a marketable product or useful process. The second phase, referred to as the adaptation phase, begins after the first application of the product in practice and ends when the application of the product takes off. After the first application, instead of a smooth S-curve, in practice an erratic process of diffusion may occur (Clark, 1985). The diffusion is often characterized by the periodic introduction, decline and re-introduction of multiple products in multiple small-scale applications, as is illustrated with many telecommunication products and services (Carey and Moss, 1985). In standard innovation and diffusion models the adaptation phase is often omitted whereas Ortt (2010) shows that in 80% of the radically new high-tech products this phase can be distinguished. The third phase, referred to as the stabilization phase, begins when the diffusion of the new high-tech product takes off. It ends when the type of product is substituted by another type of product. In this phase, the diffusion of a product most closely resembles an S-curve. Whereas similar, but industry-specific phases are distinguished by Agarwal and Bayus (2002) and Tushman and Anderson (1986), Ortt and Schoormans' (2004) description of the innovation and diffusion process is applicable to widely different industries.

The first phase is called development phase because it refers to the development of the invention into an initial product. The second phase is called adaptation phase because in a kind of trial and error process product variants are developed that are adapted to meet the needs of specific customer segments. The third phase is called stabilization phase because therein the type of product and customer segments have

Figure 1. Phases in the process of innovation and diffusion

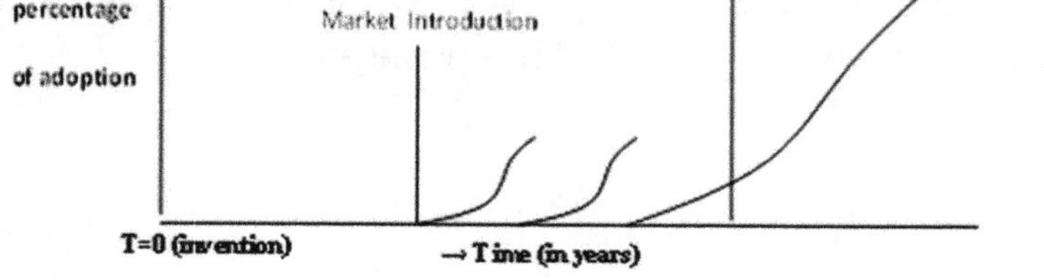

more or less crystalized and attention focuses on optimizing the product, its production, distribution and marketing. Innovation may occur during all three phases.

This innovation model and the three phases it distinguishes diverges considerably from the mainstream model of innovation and diffusion. The latter model sees innovation as a project after which the innovation (usually a new product or system) is introduced in the market and diffuses according to the classic S-curve . However, the mainstream model completely ignores the notion that the development phase is more than just a project. In the development phase some networks of companies may continue with the scientific development and improvement of technological principles while others work in parallel on developing similar products. The mainstream model also ignores the adaptation phase, a phase in which a new socio-technical system may need to be built around the new product, a product that may have to compete with existing as well as other new products. Moreover, this is also typically the phase in which innovators experiment with developing different product variants for specific customer segments.

Note that while Ortt and Schoormans' innovation model is generic, the specific pattern of product innovation and diffusion can differ per case. In specific cases the development or adaptation phase may even be insignificant. Medicines, for example, often require a long time to develop, but once developed their diffusion may take off immediately. In such cases, the innovation process does not include an adaptation phase (Ortt, 2010). Moreover, their development is not just a project. It is often plagued by considerable delays because of lack of funding which may ultimately block further development and is referred to as the 'Valley of death' (Markham et al., 2010). Analogously, some inventions can almost directly be developed into a marketable product. An example is dynamite, which was introduced within a year after it's invention (Brown, 1998). In this case, the duration of the development phase was almost insignificant and the entire process looked like the mainstream model. In other cases a long adaptation phase is visible. In the case of the microwave oven it spanned about 25 years, a period in which the oven was introduced in multiple market niches before large-scale diffusion started. Finally, some inventions never make it to the stabilization phase. The mechanical television, for example, was available as an experimental product but became outdated before large-scale diffusion could take place. The underlying technology was completely substituted by its electronic successor (Van Den Ende et al., 1997/1998).

That is, the mainstream model of innovation and diffusion (summarized as an innovation project that precedes an S-shaped diffusion curve) is a special case of Ortt and Schoormans' model of innovation and diffusion (distinguishing three phases). The latter more generic model is better suited for a heterogeneous set of cases. This chapter focuses on the first two phases of the model also referred to as the pre-diffusion phases, i.e. the development and adaptation phases. The chapter explores the effect of pre-existing standards and regulations on the length of these two phases.

The Effect of Pre-Existing Standards and Regulations on the Innovation and Diffusion Process

In the following, the chapter aims to contribute to theory on the relation between standards and innovation by, first, including the influence of timing of standards and regulation on the technology life-cycle. In the work of Swann and others no distinction is made between 'pre-existing standards and regulations', which refer to the set of standards, rules, laws and conventions available in the industry *prior to* the innovation of a radically new product, and those standards and rules that come into place after the invention has been made. Regarding the latter, Sherif (2006) distinguishes anticipatory, enabling and responsive standards during the technology life-cycle. Their timing roughly corresponds to early, mid-term and late standardization in the technology trajectory. The focus is on a period that precedes his classification, i.e. on pre-existing standards and regulations. These do not specifically target an invention or a new product but constitute part of the industry preconditions. Second, the work in this chapter differs from other studies by focusing on a yet unaddressed area of innovation: the effect of standards and regulation on the speed of innovation – in particular on the length of the pre-diffusion phases. Third and last, several studies use the publicly accessible ICS data set, and thus depend on the data gathering and methodological choices of third parties. Having a data set based on own study allows refinement of the research questions accordingly. The data includes a large set of case studies from different industries. Different lines of reasoning may explain why pre-existing standards and regulations might enable or hamper innovation and, accordingly, lengthen or shorten the development phase *i.e.,* the phase from invention to first introduction. On the one hand, these standards and regulations may contain requirements that have to be met and which therefore may lengthen the development phase. On the other hand, they may also limit the options to be explored during the development phase and thereby shorten it. The net result of these two effects is as yet unclear.

In a similar vein, different lines of reasoning can be followed to argue why pre-existing standards and regulations might shorten or lengthen the adaptation phase *i.e.,* the phase from an innovation's first introduction up to a product's initial industrial production and large-scale diffusion. On the one hand, they might facilitate the process of mutual adaptation of product variants and customer segments by, for example, easing interoperability with complementary products and services if these standards and regulations are informative and contain guidance on the design, construction process, application requirements and/or use of the radically new high-tech product. Such information could prevent the emergence of competing alternative designs and facilitate the new product's usability for specific applications. Hence, pre-existing standards and regulations could shorten the adaptation phase. On the other hand, standards and regulations could also hamper this adaptation process if incompatible with the new product and its intended use. Hence, pre-existing standards and regulations could also lengthen the adaptation phase.

In sum, there are grounds to argue that pre-existing standards and regulations can both stimulate and hamper innovation, and therefore shorten as well as lengthen the development and adaptation phase, respectively. Pre-existing standards and regulations that can be built upon (and provide initial clarity about the design space, including design requirements) and are well-aligned with the new product and its application context are most likely to have a stimulating effect and shorten the development and adaptation phase.

- **Hypothesis 1:** Pre-existing standards and regulations that can be built upon in developing the new product, will shorten the development and adaptation phase.

Unlike standalone products such as a whisk or a vase, certain new products will need to interwork with existing products or within an existing technological system. For example, a new machine. Such a machine will have to take into account pre-existing standards (*e.g.,* on nuts and bolts) and regulations (*e.g.,* on safety). The more system elements the new product has to interoperate or interact with, the higher its degree of 'technical interrelatedness', as it is called for lack of a better term. For 'highly interrelated' new products pre-existing standards and regulations are likely to be more relevant than for standalone products. Therefore, it is expected that for such products the effect of standards and regulations on the length of the adaptation and development phase is likely to be more prominent. That is, the degree to which a high-tech product is to operate within a wider technological system (captured by the term 'interrelatedness') is seen as a variable that moderates the effect of pre-existing standards and regulations on the length of the development and adaptation phase. Furthermore, it is expected that for product innovations that are more radical at the time of introduction, pre-existing standards and regulations will have a more prominent effect. Radically new product innovations diverge considerably from already existing products in the market and usually embody new technologies. The more radically new the technology in an innovation the more important the role will be of prevailing standards and regulations.

- **Hypothesis 2:** The more an innovation is to be part of a technological system (i.e., interrelated with system setting), the more it will reinforce the shortening effect of pre-existing standards and regulations on the length of the development and adaptation phase.
- **Hypothesis 3:** The more radically new the technology in the product innovation, the more it will reinforce the shortening effect of pre-existing standards and regulations on the length of the development and adaptation phase.

All three hypotheses are shown in Figure 2. Hypothesis 1 indicates a direct effect of pre-existing standards and regulations on the length of the development and adaptation phase. Hypothesis 2 and 3 indicate that the strength of the previous effect depends on two moderating variables, i.e. the high-tech product's interrelatedness with the wider system of which it is to be part and the radicalness of the technology involved.

METHOD

Data Gathering

To explore the impact of pre-existing standards and regulations on the innovation and diffusion process, we selected fifty cases of radically new product innovations. Because a more heterogeneous set of cases

Figure 2. Model of how pre-existing standards and regulations in combination with two moderating variables determine the length of the development and adaptation phases

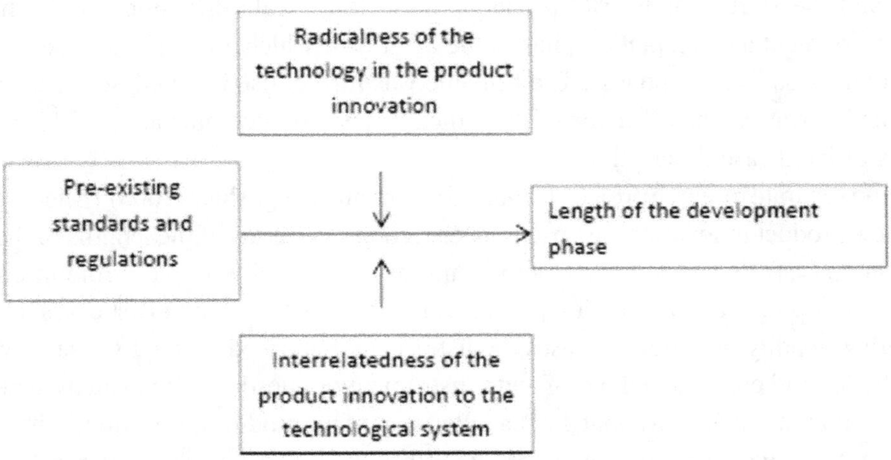

promised to deliver richer insights, cases were chosen that differ with respect to industry sector and time of introduction. The selected industries were (1) 'Materials', including chemicals, materials and metals; (2) 'Pharmaceuticals', including medicine and healthcare equipment; (3) 'Telecommunications', including telecommunication products and services, and (4) 'Electronics', including electronic components and products. These industries differ strongly in the way new products are developed and introduced. For example, the (relative) length of their development and adaptation phases differs significantly (Ortt et al., 2009). Regarding the time of introduction, to avoid anachronism, cases were selected based on timeslots from the mid-nineteenth century onwards. This period can reasonably be argued to be a starting point for many industries. Telegraphy, one of the first forms of electronic telecommunication, was applied from then on (Huurdeman, 2003). In the same period, semi-synthetic materials such as nitrocellulose started to be systematically investigated. These materials are the precursors of modern chemical industry (Friedel, 1983). Similar arguments apply to pharmaceuticals and electronics.

For each industry radically new high-tech product innovations are identified. Following the typology from Garcia and Calantone (2002), these innovations are identified by assessing whether the innovations met at least one out of three criteria at the time of their invention. The first criterion was that the innovation had a functionality that was new to the market (*e.g.*, telegraphy, telephony, X-ray and Prozac), the second criterion was a price-performance ratio that was much better than contemporary products (*e.g.*, Nylon, Kevlar and DVD) and the third criterion was that the innovation was based on a new technical principle (*e.g.*, the contraceptive pill was the first contraceptive to use hormones, and the transistor which was the first solid-state electronic amplifier).

To compare and contrast processes of innovation and diffusion in different industries it is decided to select a significant number of cases per industry. After balancing the time required to complete the analyses of cases (several days per case) and the wish to base findings on as many cases as possible, it is decided to investigate about a dozen cases per industry. Within each industry prominent cases from each time-period are selected (See Table 1 for the selection of cases).

For each case an extensive literature search is completed using Google Scholar. Systematic searches were completed for sources using several case-related keywords. To start with, a wider search was done by combining different names for the same product with the keyword 'history'. More specific searches followed that combined the different product names with the keywords 'invention', 'introduction', 'diffusion', 'innovation' and 'development'. From these search results, a selection was used for further study. Per case multiple sources are analysed and cross-checked to assess the approximate year of invention, introduction and the start of industrial production and large-scale diffusion. These three hallmarks define the development and adaptation phases, the lengths of which are hypothesized to be affected by pre-existing regulations and standards. The timeline was further used to analyse the actors and factors relevant for the innovation and diffusion of the radically new product innovations. This resulted in a set of chronically ordered case descriptions.

In a prior and separate research project, (reported in Ortt and Delgoshaie, 2008) 18 heterogeneous cases of radically new product innovations are used (heterogeneous in terms of the lengths of the development and adaptation phases) to explore which actors and factors typically played a role in determining the length of the development and adaptation phase. During the analysis of the first cases, the list of actors and factors grew steadily but after ten cases the list became saturated. After 18 cases it was decided to consolidate the list and construct a kind of audit instrument or questionnaire to analyse new cases. This questionnaire contained questions about the characteristics of the product innovation, competing products, the characteristics of the companies involved, and other market factors and actors relevant during the

Table 1. Cases of radically new high-tech products.

↓ Time of Introduction	Industry →				
	Chemicals, Metals & Materials	Medicines	Telecommunication Equipment	Electronic Equipment	
1839-1900	- Nitroglycerine - Rayon - Celluloid - Gun cotton - Dynamite	- Aspirin - X-ray	- Telegraphy - Telephony	- Phonograph	10
1901-1925	- Cellophane - Monel - Bakelite	- Salversan - Insulin	- Radio telephony - Radio broadcast	- Magnetic recording - Vacuum tube	9
1926-1950	- Nylon - Teflon - PET - PVC - DDT	- Penicillin	- Electronic television	- Microwave oven - Fluorescent light	9
1951-1975	- Kevlar - Dyneema - Nitinol	- Paracetamol - Methylphenidate - Polio vaccine - Minoxidil	- GNSS (GPS)	- Transistor - Video Cassette Recorder - LED - Microprocessor - LCD - Plasma display	14
1976-	- Twaron	- Viagra - SSRI (Prozac)	- Bluetooth - SMS - Cellular mobile telephony	- Computer mouse - DVD	8
Total	17	11	9	13	50

development and adaptation phase. Standards and regulations emerged as relevant market factors. For all questions covering relevant market actors and factors, alternative answers were construed to enable coding and further analysis of the cases. The questions that were used to codify the information that is used in this article, can be found in Appendix 1.

The questionnaire was used for the 50 cases reported in the current chapter. To codify the information, for each case two experts were asked to read the entire case description and score the questionnaire items. The first expert typically had expertise in the case-related industry area. Since research in this chapter covers several industries, different industry experts were invited to join. The second expert was held constant for all cases to ensure a consistent assessment across cases. The industry expert typically analysed multiple cases and to ensure a proper coding of the information in the questionnaire, questions were explained and discussed prior to their use. So experts that coded the information were trained to prevent possible misunderstandings. While in most cases, the expert evaluations coincided, in some cases discussion was needed to reach agreement.

To summarize, the method was designed to meet two goals. The first goal was to codify information from multiple cases to enable statistical analyses. To attain that goal a questionnaire was construed and couples of experts were requested to complete this questionnaire for each case. The second goal was to design process that limited or even compensated evaluation errors typically made by experts. Experts are often found to be unable to validly forecast or assess technological developments (Brandes, 2009; Galbraith et al, 2007; Schnaars, 1989; Tichy, 2004). In order to limit or compensate evaluation errors the following procedures are adopted:

- Experts were not asked to predict future technological and market developments but were asked to assess historical cases.
- Two different experts were invited to assess each case. Galbraith et al (2006) show how experts can add to the validity when different types of experts are combined. Similar advices to increase the quality of assessments are provided by (Tichy, 2004).
- To enable good assessments, for each case information was systematically searched and literal information from multiple sources was ordered to reflect the developments in the particular case over time. This information was passed on to both experts.
- Experts were requested to answer a questionnaire to codify the available information. Galbraith et al (2006) show how experts can add to the validity when their assessments are based on factors determined before. In the questionnaire, the experts for example assessed to what degree standards and regulations were available in the market at the time of the invention of the radically new product innovation.
- The coded information was then entered in a database and analysed statistically.

An important choice in the procedure is that experts are not requested to assess whether the standards or regulation actually had an effect on the process of innovation and diffusion. Whether the availability of standards and regulations might have had an effect on the process of innovation and diffusion was assessed by looking at correlations after the coded information was analysed statistically.

Analyses

Two analyses were carried out. The first one focused on the relationship between "pre-existing standards and regulations", on the one hand, and "the length of the development and adaptation phase", on the other. This relationship was analysed using both parametric and non-parametric correlations. Parametric correlations (Pearson's r) are often used for five-point scales. A parametric correlation assesses an effect assuming that the variables are interval-scaled and the relationship is linear. Non-parametric correlations (Kendall's tau and Spearman's rho) assume that the variables are ordinal-scaled (which is in fact the case) and are able to detect non-linear relationships.

The second analysis further scrutinized the strength of the above correlations by taking into account the possible effect of the two moderating variables: "Interrelatedness" and "Radicalness" of the technology. Two groups of about equal size were formed for each moderating variable, that is, product innovations that were more and less interrelated (i.e. embedded in a technological system) and product innovations that were more and less radically new. For all groups the first analysis were repeated and the correlations between the availability of "pre-existing standards and regulations" and "the length of the development and adaptation phase" are assessed.

RESULTS

The Impact of Pre-Existing Standards and Regulations

For the 50 cases in the sample, the average length of the development and adaptation phase is about 12.5 and 7.5 years, respectively. That is, the time required for radically new product innovations to start diffusion on a large scale after its invention is roughly two decades in total. The information about the process of innovation and diffusion for the cases can be found in Appendix 2. Table 2 depicts the relationship between the length of the development or adaptation phase and the availability of pre-existing standards and regulations.

The results in Table 2 show that pre-existing standards and regulations are not related to - and therefore have no clear effect on - the length of the development phase. This finding is consistent for all three types of correlations. In contrast, a significant relationship is found between 'pre-existing standards and regulations' and 'the length of the adaptation phase'. That is, hypothesis 1 is confirmed for the adaptation phase but not for the development phase.

The Moderating Effects of Interrelatedness and Radicalness

Before further investigating the relationship between 'pre-existing standards and regulations' and 'the length of the adaptation phase', the two moderating variables 'interrelatedness of the high-tech product to the technical system' and 'radicalness of the technology' were analysed. For both variables, cases were divided in groups with relatively low and high values. Table 3 summarizes how the cases are spread according to their values on both moderating variables.

Cases that score relatively low on interrelatedness mostly come from two industries: (1) Chemicals, metals & materials and (2) Medicines. Cases that are relatively high on interrelatedness are mostly from the telecommunication and electronic industries. For the moderating variable 'radicalness' such a distinction per industry is not evident. All industries contain innovations that are relatively more and less radical.

The next step was to look at the relationship between 'pre-existing standards and regulations', on the one hand, and 'the length of the adaptation phase', on the other, for the different values of the moderating variables (see Table 4).

The results in the second row of Table 4 are consistent: for product innovations in a technological system with relatively low interrelatedness no effect of pre-existing standards and regulations is found; whereas for product innovations in a technological system with a relatively high interrelatedness a very large and significant effect is found. This means that hypothesis 2 is confirmed: The more interrelated

Table 2. Correlation between pre-existing standards and regulations and the length of phases

	Development phase			Adaptation phase		
	Pearson Correlation	Spearman's Rho	Kendall's Tau	Pearson Correlation	Spearman's Rho	Kendall's Tau
Pre-existing standards and regulations	0.006 p=0.483 N=50	0.147 p=0.154 N=50	0.105 p=0.163 N=50	-0.299 p=0.018* N=50	-0.323 p=0.011* N=50	-0.227 p=0.017* N=50

(*) Significant correlations are indicated with *, significance means here p<0.05

Table 3. Case groupings based on interrelatedness and radicalness of the technological system.

	Relatively Low Score on Radicalness		Relatively High Score on Radicalness		
Relatively low score on interrelated-ness	Kevlar Nylon Bakelite Dyneema Magnetic recording (wire) Dynamite Cellophane PVC PET Rayon SSRI	Monel Teflon Nitroglycerine Celluloid Paracetamol DDT Twaron Polio medicine Minoxidil Salversan	Nitinol Aspirin Penicilin	Viagra Methylphenidate Insuline	27
Relatively high score on interrelatedness	DVD Bluetooth Transistor SMS	LED Fluorescent lamp Guncotton	GNSS Microwave oven Cellular mobile telephony (Electrical) Telegraphy Plasma tv VCR Electronic television X-Ray	Computer mouse Vacuum tube Telephony Microprocessor LCD Radio broadcasting Radio telephony Phonograph	23
	28		22		50

Table 4. Correlation between 'pre-existing standards and regulations' and 'the length of the adaptation phase' for cases with low and high values on the moderating variables

Correlation between Pre-Existing Standards and Regulations and the Length of the Adaptation Phase	Products in Systems With a Relatively Low Score on Interrelatedness	Products in Systems With a Relatively High Score on Interrelatedness
Pearson correlation **Spearman's rho** **Kendall's tau**	-0.076; p=0.353; N=27 -0.024; p=0.453; N=27 -0.016; p=0.457; N=27	-0.512; p=0.006*; N=23 -0.643; p=0.000*; N=23 -0.463; p=0.002*; N=23
	Product innovations with a relatively low score on radicalness	**Product innovations with a relatively high score on radicalness**
Pearson correlation **Spearman's rho** **Kendall's tau**	-0.216; p=0.270; N=28 -0.214; p=0.275; N=28 -0.160; p=0.273; N=28	-0.389; p=0.074 ; N=22 -0.466; p=0.029*; N=22 -0.322; p=0.052 ; N=22

the product innovation in a technological system, the more pre-existing standards and regulations shorten the adaptation phase.

The results in the bottom row show that for product innovations with relatively low scores on radicalness no effects of pre-existing standards and regulations are found; whereas for product innovations with relatively high scores on radicalness a modest effect is found (almost significant effect for the Pearson correlation and Kendall's tau and significant effect using Spearman's rho). This means that hypothesis 3 is confirmed: The more radically new the technology in the product innovation, the more pre-existing standards and regulations shorten the adaptation phase.

Although the results in Table 2 indicated that pre-existing standards and regulations had no significant effect on the length of the development phase, their effect on the development phase was retested after entering the two moderating variables, 'interrelatedness' and 'radicalness'. The inclusion of the moderating variables did not have a significant effect indicating that the length of the development phase remained unaffected by pre-existing standards and regulations.

CONCLUSION AND DISCUSSION

On average, for a wide range of radically new high-tech products, pre-existing standards and regulations are associated with a short time interval from first market introduction to large-scale diffusion and industrial production (referred to as the adaptation phase). A similar effect is not found for the time interval between the invention and the first introduction of a radically new high-tech product (referred to as the development phase).

The direct effect on the length of the adaptation phase is moderated by two variables: the degree of interrelatedness of the innovative product in respect to the setting in which it is to be used and the degree of radicalness of the technology embedded in the product innovation. That is, the effect of pre-existing standards and regulations on the length of the adaptation phase is very strong for product innovations that are to work within a technological system and for product innovations based on more radically technologies.

Methodologically, these results indicate a strong and consistent effect. The direct effect of pre-existing standards and regulations on the length of the adaptation phase is significant. This is remarkable, especially if one realizes that the variable 'interrelatedness' itself, has no direct effect on the length of this phase. Interrelatedness in respect to the setting in which the product is to be used would seem an obvious cause for a long development and adaptation phase, since the physical network requirements, the complexity of the entire technological system, and the importance of compatibility of the innovation with other components of the system, would at first sight call for a longer development and adaptation phase. But again, that effect is not found. The effect of pre-existing standards and regulations is also remarkable taking into account the relatively small sample size and the heterogeneous character of the cases. In such a situation, the statistical power to detect relationships is relatively small and only strong relationships can be detected. The results regarding the moderated effect further confirm the significance of pre-existing standards and regulations.

Scientifically, the starting point was Swann and Lambert's (2010) conclusion that the question whether standards enable or constrain innovation contains a false antithesis. The explanations offered by King (2006) and Swann and Lambert (2010) indicated the enabling *and* constraining influence of pre-existing standards and regulations. But why would this reflect on the length of the adaptation phase rather than the development phase, as one would sooner expect?

The different effects become clearer if one focuses on the way pre-existing standards and regulations can be a point of departure for a new product. In the development phase, compliance to these standards and regulations might require extra effort and thus lengthen the development phase; and they might focus the innovation effort in a particular direction and thereby shorten the development phase. The net effect of these two mechanisms would then be insignificant (no effect detected), which would coincide with the reported finding. However, in the adaptation phase, pre-existing standards and regulations that have been taken into account by the new product will shorten the time required to build complementary products and services; they will ease customer use and implementation. Both elements shorten the adaptation phase. This explanation of the findings implies that pre-existing standards and regulations that do *not* match the requirements of the new product, will lengthen the adaptation phase. Further research is needed to verify this.

The managerial and societal implications of the finding that pre-existing standards and regulations shorten the adaptation phase are significant. On average this phase lasts long – in the sample the average duration of the adaptation phase is 7.5 years - and can be a disastrous period for pioneering companies attempting to commercialize their radically new product. Many companies fail in this phase (Ortt, Shah and Zegveld, 2007), especially pioneering companies. The high failure rate among pioneering companies is referred to as 'the burnout of the pioneers' (Olleros, 1986). For radically new high-tech products this failure rate is significantly higher than for companies that introduce incrementally new products (Pech, 2003; Tellis and Golder, 1996). Societally a long adaptation phase is expensive and creates much uncertainty about and delay in the application of valuable innovations. The effect of pre-existing standards and regulations on the length of the adaptation phase is very large in practice. To illustrate this effect 20% of the cases with the highest availability of pre-existing standards and regulations are selected. For this subgroup of cases the average length of the adaptation phase was 3.1 years compared to 8.5 years for the remainder of the cases. The difference between the means is significant ($p<0.01$). The managerial and societal implication is that pre-existing standards and regulations shorten the adaptation phase on average with 5 years if they match with the innovation! In some separate cases where pre-existing standards and regulations do not match with the innovation, such as for cellular mobile telephony, regulation and standardization are claimed to have delayed the diffusion considerably (Gershon, 2003).

Furthermore, the finding that the benefits from pre-existing standards and regulations are most significant for product innovations that are part of a larger technological system and that are based on relatively more radical technologies, deserves further scrutiny. The results indicate that pre-existing standards and regulations have a strong effect on the length of the adaptation phase in the telecommunication and electronics industry. Additional ex post analysis shows that the need for compatibility largely explains these results. For more 'stand-alone' product innovations, compatibility with complementary products and service is significantly less important than for innovations that are to be part of larger technological systems. This implies that need for and presence of compatibility standards, which are most common in the telecommunication and electronics industry, may play a determining role regarding the length of the adaptation phase. Follow-up research is needed to further explore the different effects of different types of standards on innovation.

FUTURE RESEARCH DIRECTIONS

The findings in this chapter offer several avenues for future research. First of all, further explanation is needed on why pre-existing standards and regulations shorten the adaptation phase. Moreover, also the development phase deserves attention. The results indicate that the development phase is rather long (for the sample the average length of this phase is about 12.5 years). Apparently pre-existing standards and regulations do not shorten this period. Why the development phase takes so much time and how this phase can be shortened are questions to address in further research.

Second, while cross-checking the findings it was found that the shortening effect on the length of the adaptation phase was not only found for 'standards and regulations' together but also for 'standards' separately. This is remarkable given the different nature of - voluntary - standards (compliance is not obligatory) and - mandatory - regulation (compliance is mandatory). A follow-up study is proposed that refines the variables and their interrelations to increase insight into the impact of standards on different phases of the process of innovation and diffusion. Thus, the difference between pre-existing regulations and standards and their distinct effect should be further examined. Where standards are concerned, it is recommended to differentiate between categories of standards, that is, between product (descriptive) and performance standards; measurement standards, minimum/maximum requirements standards, classification standards and compatibility standards. Of interest is whether distinct categories of standards have a different impact on the innovation and diffusion of radically new high-tech products.

Third, the previous provides a stepping stone for research on the key question 'Does the timing of committee-based standardization affect innovation?' Clarifying this would contribute to the debate about the effect of standardizing too soon or too late. It would entail expanding the current research from pre-existing standards to standardization efforts during the innovation trajectory, that is, during the development, adaptation, and stabilization phases. These phases roughly coincide with anticipatory standards (agreed on in the development phase), enabling standards (agreed on during the adaptation phase), and retrospective standards (agreed on during the stabilization phase). This should be cross-checked with the type of innovation involved (Egyedi and Sherif, 2010). Does the effect of timing in standardization depend on the kind of innovation at stake?

ACKNOWLEDGMENT

Mostafa Sherif and three anonymous SIIT reviewers are sincerely thanked for their comments on the work and Peter Swann for helping us with several important sources.

REFERENCES

Abernathy, W. J., & Utterback, J. M. (1978). Patterns of innovation in industry. *Technology Review*, *80*(7), 40–47.

Agarwal, R., & Bayus, B. L. (2002). The market evolution and sales takeoff of product innovations. *Management Science*, *48*(8), 1024–1041. doi:10.1287/mnsc.48.8.1024.167

Blind, K. (2004). *The Economics Of Standards; Theory, Evidence, Policy*. Cheltenham, UK: Edward Elgar Publishing.

Blind, K., & Grupp, H. (2000). *Standards Statistics as New Indicators for the Diffusion of Technology.* Paper presented to the International JA Schumpeter Conference, Manchester, UK.

Brandes, F. (2009). The UK technology foresight programme: An assessment of expert estimates. *Technological Forecasting and Social Change*, *76*(7), 869–879. doi:10.1016/j.techfore.2009.01.005

Brown, G. I. (1998). *The Big Bang; History of Explosives*. Sutton Publishing Limited.

Brusoni, S., & Corrocher, N. (2006). *The development and commercial success of SMS: standard committees as engine of knowledge-based entrepreneurship*. Deliverable 13, Work Package 2A of project Knowledge-based Entrepreneurship: Innovation, Networks and Systems. European Commission Sixth Framework Programme (2002-2006).

Carey, J., & Moss, M. L. (1985). The diffusion of telecommunication technologies. *Telecommunications Policy*, *6*(2), 145–158. doi:10.1016/0308-5961(85)90038-2

Clark, K. B. (1985). The interaction of design hierarchies and market concepts in technological evolution. *Research Policy*, *14*(5), 235–251. doi:10.1016/0048-7333(85)90007-1

Constant, E. W. II. (1980). *The Origins of the Turbojet Revolution*. Baltimore, MD: The Johns Hopkins University Press.

DTI. (2005). *The Empirical Economics Of Standards*. DTI Economics Paper 12. London: Department of Trade and Industry. Retrieved from www.bis.gov.uk/files/file9655.pdf

Egyedi, T. M., & Sherif, M. H. (2010). Standards Dynamics through an Innovation Lens: Next Generation Ethernet Networks. *IEEE Communications Magazine*, *48*(October), 166–171. doi:10.1109/MCOM.2010.5594692

Friedel, R. (1983). *Pioneer Plastic*. Madison, WI: University of Wisconsin Press.

Galbraith, C. S., DeNoble, A. F., Ehrlich, S. B., & Kline, D. M. (2007). Can experts really assess future technology success? A neural network and Bayesian analysis of early stage technology proposals. *The Journal of High Technology Management Research*, *17*(2), 125–137. doi:10.1016/j.hitech.2006.11.002

Galbraith, C. S., Ehrlich, S. B., & DeNoble, A. F. (2006). Predicting Technology Success: Identifying Key Predictors and Assessing Expert Evaluation for Advanced Technologies. *The Journal of Technology Transfer*, *31*(6), 673–684. doi:10.1007/s10961-006-0022-8

Garcia, R., & Calantone, R. (2002). A critical look at technological innovation typology and innovativeness terminology: A literature review. *Journal of Product Innovation Management*, *19*(2), 110–132. doi:10.1016/S0737-6782(01)00132-1

Gershon, R. A. (2003). Cellular Telephony. Encyclopedia of International Media and Communications, 1, 175-187.

Hanseth, O., Monteiro, E., & Hatling, M. (1996). Developing information infrastructure: The tension between standardization and flexibility. *Science, Technology & Human Values*, *21*(4), 407–426. doi:10.1177/016224399602100402

Ho, J. (2017). *A Systems Perspective on Standardisation in Technological Innovation: A Conceptual Framework and a Process Model Supporting Strategic Policy Foresight* (Doctoral thesis). University of Cambridge, UK. 10.17863/CAM.8705

Ho, J.-Y., & Sullivan, E. (2016). The Multi-dimensional Nature of Standardisation in Support of Innovation: A Systematic Analysis of the History of Photovoltaic Technology. *EURAS Proceedings 2016*.

Huurdeman, A. A. (2003). *The Worldwide History of Telecommunications*. Hoboken, NJ: Wiley-Interscience. doi:10.1002/0471722243

Kay, A. L. (2002). *German Jet Engine and Gas Turbine Development 1930-1945*. Shrewsbury: Airlife Publishing Ltd.

King, M. (2006). *Standards and Innovation* (Master of Science Dissertation). School of Social Sciences, University of Manchester.

Koch, C. (2017). Standardization in Energing Technologies: The Case of Additive Manufacturing. *EURAS Proceedings 2017*, T117-133.

Krechmer, K. (1996). Technical standards: Foundations of the future. *StandardView*, *4*(1), 4–8. doi:10.1145/230871.230872

Krechmer, K. (1998). The Principles of Open Standards. *Standards Engineering Society: WorldStandards Day Paper Competition.*

Krechmer, K. (2000). *The Fundamental Nature of Standards: Economics Perspective.* Paper presented at the International JA Schumpeter Conference, Manchester, UK.

Krechmer, K. (2005). The Meaning of Open Standards. *Hawaii International Conference on System Sciences.*

Lacohée, H., Wakeford, N., & Pearson, I. (2003). A social history of the mobile telephone with a view of its future. *BT Technology Journal*, *21*(3), 203–211. doi:10.1023/A:1025187821567

Markham, S. K., Ward, S. J., Aiman-Smith, L., & Kingon, A. I. (2010). The Valley of Death as Context for Role Theory in Product Innovation. *Journal of Product Innovation Management*, *27*(3), 402–417. doi:10.1111/j.1540-5885.2010.00724.x

Olleros, F. (1986). Emerging Industries and the Burnout of Pioneers. *Journal of Product Innovation Management*, *1*(1), 5–18. doi:10.1016/0737-6782(86)90039-1

Ortt, J. R. (2010). Understanding the Pre-diffusion Phases. In J. Tidd (Ed.), *Gaining Momentum Managing the Diffusion of Innovations* (pp. 47–80). London: Imperial College Press. doi:10.1142/9781848163553_0002

Ortt, J. R., & Delgoshaie, N. (2008). Why does it take so long before the diffusion of new high-tech products takes off? In B. Abu-Hijleh, M. Arif, T. Khalil, & Y. Hosni (Eds.), *Creating & Managing a Knowledge Economy* (pp. 2–15). Dubai: The Britisch University in Dubai.

Ortt, J. R., & Schoormans, J. P. L. (2004). The Pattern of Development and Diffusion of Breakthrough Communication Technologies. *European Journal of Innovation Management*, *7*(4), 292–302. doi:10.1108/14601060410565047

Ortt, J. R., Shah, C. M., & Zegveld, M. A. (2007). Strategies to Commercialize Breakthrough Technologies. *International Association for Management of Technology Conference Proceedings.*

Ortt, J. R., & Suprapto, M. (2010). Niches Prior to Large-scale Application of Radically New High-Tech Products. *ERSCP-EMSU Conference proceedings.*

Ortt, J. R., & Suprapto, M. (2011). The Role of Strategic Niches in Creating Large-scale Applications for High-tech Products. *International Association for Management of Technology Conference Proceedings*.

Pech, R. J. (2003). Memetics and innovation: Profit through balanced meme management. *European Journal of Innovation Management*, *6*(2), 111–117. doi:10.1108/14601060310475264

Sahal, D. (1981). *Patterns of Technological Innovation*. Addison-Wesley.

Salop, S. C., & Scheffman, D. T. (1983). Raising Rivals' Costs. *The American Economic Review*, *73*(2), 267–271.

Schnaars, S. P. (1989). *Megamistakes; Forecasting and the Myth of Rapid Technological Change*. New York: The Free Press.

Senden, M. J., & Wöckel, I. (1997). Gelebte Unternehmensphilosophie contra Normierungszwaenge: ISO 9000 im DLR-Wissenschaftsmanagement'. *Wissenschaftsmanagement*, *3*, 269–272.

Sherif, M. H. (2001). A Framework for Standardization in Telecommunications and Information Technology. *IEEE Communications Magazine*, *39*(April), 94–100. doi:10.1109/35.917510

Sherif, M. H. (2003). When is Standardization Slow? *International Journal of IT Standards and Standardization Research*, *1*(1), 19–32. doi:10.4018/jitsr.2003010102

SOU. (2007). Standarder och tillväxt: en kommenterad forskningsöversikt – Standards and growth: a research review. SOU, Statens offentliga utredningar nr. 2007, 83, Stockholm.

Swann, G. M. P. (1990). Standards and the Growth of a Software Network: A Case Study of PC Applications Software. In J. Berg & H. Schumny (Eds.), *An Analysis of the IT Standardisation Process*. Amsterdam: Elsevier Science Publishers. doi:10.1016/B978-0-444-87390-3.50050-5

Swann, G. M. P. (1999). *The Economics of Measurement*. London: Department of Trade and Industry, Standards and Technical Regulations Directorate.

Swann, G. M. P. (2000). *The Economics of Standardization*. London: Department of Trade and Industry, Standards and Technical Regulations Directorate.

Swann G. M. P. (2010). *The Economics of Standardization: An Update*. Report for the UK Department of Business, Innovation and Skills.

Swann, G. M. P., & Lambert, R. J. (2010). Why do Standards Enable *and* Constrain Innovation? *EURAS Proceedings 2010: Service Standardization*, 357-376.

Swann, G. M. P., & Watts, T. P. (2002). Visualisation Needs Vision: The Pre-Paradigmatic Character of Virtual Reality. In S. Woolgar (Ed.), *The Virtual Society? Technology, Cyberbole, Reality*. Oxford, UK: Oxford University Press.

Tassey, G. (1982). Infratechnologies and the Role of Government. *Technological Forecasting and Social Change*, *21*(2), 163–180. doi:10.1016/0040-1625(82)90013-0

Taylor, A. S., & Vincent, J. (2005). An SMS history. In *Mobile World; Past, Present and Future* (pp. 75–91). London: Springer.

Tellis, G. J., & Golder, P. N. (1996, Winter). First to Market, First to Fail? Real Causes of Enduring Market Leadership. *Sloan Management Review*, 65–75.

Temple, P. (2005). *The Empirical Economics of Standards*. DTI economics paper, No. 12, June 2005.

Tichy, G. (2004). The over-optimism among experts in assessment and foresight. *Technological Forecasting and Social Change*, *71*(4), 341–363. doi:10.1016/j.techfore.2004.01.003

Tushman, M. L., & Anderson, P. (1986). Technological Discontinuities and Organizational Environments. *Administrative Science Quarterly*, *31*(3), 439–465. doi:10.2307/2392832

Tushman, M. L., Anderson, P. C., & O'Reilly, C. (1997). Technology Cycles, Innovation Streams, and Ambidextrous Organizations: Organization Renewal Through Innovation Streams and Strategic Change. In M. L. Tushman & P. C. Anderson (Eds.), *Managing Strategic Innovation and Change* (pp. 3–23). Oxford, UK: Oxford University Press.

Tushman, M. L., & Rosenkopf, L. (1992). Organizational determinants of technological change. Towards a sociology of technological evolution. *Research in Organizational Behavior*, *14*, 311–347.

Van Den Ende, J., Ravesteijn, W., & De Wit, D. (1997/1998). Shaping the Early Development of Television. *IEEE Technology and Society Magazine*, *16*(Winter), 13–26. doi:10.1109/44.642518

Vries de, H. J., Verhagen, P., & Beerepoot, M. (2008). Impact of standards on innovation - A case of energy performance standards. *EuroMOT Conference Proceedings*. Retrieved from http://webintec.ceram.fr/euromot2008/conftool/uploads/386/1-EPC_EuroMOT_2008-08-01.pdf

West, J. (2003). The role of standards in the creation and use of information systems. *MISQ special issues Workshop*.

Wölker, Th. (1996). Technische Normen im Verlauf der Geschichte. *Homo Economicus*, *13*, 51–75.

KEY TERMS AND DEFINITIONS

Adaptation Phase: This is a phase in the process of innovation and diffusion. The phase begins after the first application of the product in practice and ends when the application of the product takes off.

Development Phase: This is a phase in the process of innovation and diffusion. The phase lasts from the invention of a technological principle up to the first application of that principle in a new high-tech product in the market.

Interrelatedness: New products often need to interwork with existing products or within an existing technological system. The more system elements the new product has to interoperate or interact with, the higher its degree of "technical interrelatedness."

Invention: The first time the working of the main components of a product are demonstrated in practice. In many cases, the invention is merely the demonstration of a crude working principle that is far from a marketable product or useful process.

Pre-Existing Standards and Regulations: The set of standards, formal guidelines, rules, laws, and conventions in force prior to the invention of a new product (i.e., prior to the demonstration of the first rudimentary version of the product). These standards and regulations apply industry-wide and have not specifically been set to address the new product. That is, they form part of the industry conditions within which a new product is developed and later on diffuses.

Process of Innovation and Diffusion: The timescale and particular pattern of the innovation and diffusion of new high-tech products.

Radically New High-Tech Product: Radically new to the market means that a product's functionality is new to the market; technologically state-of-the-art means that a product's price-performance ratio is much better than that of contemporary products or that it is based on new technical principles (e.g., contraception via hormones).

ENDNOTE

1 SOU (2007, p. 143) questions the use of such composite questions in the CIS (i.e., about regulations and standards), and the lack of clarity of these terms (*e.g.*, nothing is said about the existence of different types of standards or on how they may inhibit innovation). In a similar vein, Swann and Lambert (2010, p. 372) reflect that 'inhibiting innovation' in question 8.1 is probably not read as 'preventing' but rather as 'constraining' options, given their research findings.

APPENDIX 1: OPERATIONALIZATION OF THE VARIABLES

Pre-Existing Standards and Regulations

The two questionnaire items on the pre-existence of case-related standards and regulations are presented in the box below (Table 5). The scores on the two items showed a very high correlation (r=0.76; p=0.000). Moreover, the items formed a very reliable scale (alpha=0.86), a scale which was used to assess whether pre-existing standards and regulations have an effect on the length of the development and adaptation phase.

Interrelatedness of the Product Innovation

Next we formed two subgroups of cases based on the degree of interrelatedness of the product to the technological system of which it is to be part. This variable was compounded by using three questionnaire items (see Table 6), each item with answers ranging from 1-5. The items formed a very reliable scale (alpha=0.92).

Radicalness of the Technology in the Product Innovation

The 'radicalness of the technology in the product innovation' refers to the degree to which the technological principle, the functionality provided by the technology and the concerned product type is considered new to the market at the time of its introduction. We formed two subgroups of cases based on the degree of 'Radicalness of the technology'. This variable was compounded by using three questionnaire items, each item with answers ranging from 1-5. The items formed a reliable scale (alpha=0.79).

Table 5.

Operationalization of Pre-Existing Standards and Regulations
The availability of pre-existing standards on different levels (*e.g.*, OSI for telco system) with regard to diverse aspects of the system that can be used when the technology is applied 1. Very small 2. Small 3. Not large or small 4. Large 5. Very large The availability of pre-existing regulation that can be used when the technology is applied. 1. Very small 2. Small 3. Not large or small 4. Large 5. Very large

Table 6.

Operationalization of the Interrelatedness of the Product Innovation
The technology can be classified as: 1. A singular material (like Kevlar). 2. An alloy or a mixture of materials or substances (like memory metal). 3. A simple assembled component or product (like bolt and nut, ski, can, paperclip). 4. A complex assembled component or product (like a chip or a car). 5. A system consisting of a combination of many products and services (like a telecommunication or railway system). Physical network requirements for large-scale application of the technology. 1. Very small (no network requirements). 2. Small (like compatibility with some peripherals or tools). 3. Not large or small (like dealer/service network with special tools for specific cars). 4. Large (like a hardware software requirements for PC or video). 5. Very large (like large-scale infrastructure for GPS, telecom or transport). Degree to which compatibility of the technology with complementary technologies is important for wide-scale diffusion. 1. Very unimportant (technology can function completely stand-alone). 2. Unimportant (technology should be compatible with some peripherals). 3. Not (un)important 4. Important (technology can function standalone without a network but should be compatible with some peripherals and/or software). 5. Very important (technology should be compatible with a large system consisting of a network, peripherals, other appliances, software, and so on).

Table 7.

Operationalization of the Radicalness of the Technology in the Product Innovation
In comparison to the closest alternative technology available before the first introduction of the new technology, the new technology is 1. Far more simple. 2. More simple. 3. As simple or complex. 4. More complex. 5. Far more complex. The newness (i.e. newness to the world) of the functionality provided by technology *at the time of introduction.* The functionality refers to the function that is performed (like moving/transport, communication, generate or transform energy, and so on). 1. Standard, well-known functionality (clothing: from silk and cotton to nylon). 2. Known functionality with minor extensions (storing data: from gramophone to cd). 3. Known functionality, major extensions (*e.g.,* from classic to digital photography). 4. Known functionality changed fundamentally (*e.g.,* from gramophone to interactive cd). 5. Totally new functionality (telecommunication: from letters to telegraphy to telephony) The newness (i.e. newness to the world) of the product incorporating the technology *at the time of introduction.* The term 'product' can also refer to a component or process. 1. Same product (new product in terms of price/performance but not in form). 2. Renewed product (new product like a *larger* propeller-powered airplane). 3. New product (product is renewed considerably, like the first jet-powered airplane in comparison to airplane with propellers). 4. New product type (new type of product in existing category, like the first helicopter as a new type of airplane). 5. New product category (new product category is formed, like the first airplane).

Length of the Development and Adaptation Phase

Finally, the lengths of the development and adaptation phase were assessed in terms of the time-interval (in years). The development phase refers to the time-interval between invention and first market introduction. The invention is defined as the first time that the technical principle in the radically new product innovation is demonstrated. The first market introduction is defined as the first time the product innovation is put intro practice. The adaptation phase refers to the time-interval between the first market introduction and the start of industrial production and large-scale diffusion. The timing of the three hallmarks ((1) invention, (2) introduction, (3) start of industrial production and large-scale diffusion) is assessed by combining and analysing the information from the selected sources.

APPENDIX 2: PROCESS DATA FOR THE CASES

Table 8.

	Name Case	Invention	Introduction	Start Large-Scale Diffusion and Industrial Production	Length Development Phase	Length Adaptation Phase
1	DVD	1995	1997	1997	2	0
2	GNSS	1958	1964	1978	6	14
3	Bluetooth	1994	2000	2001	6	1
4	Microwave oven	1945	1947	1967	2	20
5	Transistor	1938	1951	1952	13	1
6	Nitinol	1962	1968	1989	6	21
7	Cellular mobile telephony	1962	1983	1983	21	0
8	Kevlar	1965	1971	1983	6	12
9	Nylon	1934	1937	1940	3	3
10	Bakelite	1907	1908	1911	1	3
11	Dyneema	1964	1975	1990	11	15
12	Aspirin	1853	1899	1900	46	1
13	Magnetic recording (wire)	1898	1903	1933	5	30
14	(Electrical) Telegraphy	1832	1839	1846	7	7
15	Penicilin	1928	1941	1942	13	1
16	SMS	1990	1994	1995	4	1
17	Plasma tv	1964	1971	2001	7	30
18	Dynamite	1847	1858	1865	11	7
19	Cellophane	1908	1912	1912	4	0
20	VCR	1951	1956	1971	5	15
21	PVC	1872	1928	1938	56	10
22	Electronic television	1926	1939	1946	13	7
23	X-Ray	1895	1896	1901	1	5
24	Viagra	1991	1996	1998	5	2
25	LED	1951	1964	1968	13	4
26	Computer mouse	1963	1977	1983	14	6
27	Fluorescent lamp	1859	1927	1939	68	12
28	Vacuum tube	1904	1905	1920	1	15
29	Telephony	1875	1877	1878	2	1
30	PET	1941	1946	1953	5	7
31	Microprocessor	1969	1970	1974	1	4
32	Rayon	1855	1891	1906	36	15

continued on following page

Table 8. Continued

	Name Case	Invention	Introduction	Start Large-Scale Diffusion and Industrial Production	Length Development Phase	Length Adaptation Phase
33	LCD	1912	1973	1973	61	0
34	SSRI	1969	1982	1987	13	5
35	Monel	1906	1907	1907	1	0
36	Teflon	1938	1941	1950	3	9
37	Nitroglycerine	1847	1858	1865	11	7
38	Guncotton	1846	1847	1871	1	24
39	Celluloid	1856	1867	1872	11	5
40	Paracetamol	1878	1953	1955	75	2
41	Methylphenidate	1944	1954	1960	10	6
42	DDT	1939	1942	1942	3	0
43	Twaron	1972	1985	1987	13	2
44	Polio medicine	1952	1955	1955	3	0
45	Insuline	1921	1922	1923	1	1
46	Minoxidil	1960	1970	1979	10	9
47	Salversan	1908	1910	1911	2	1
48	Radio broadcasting	1906	1919	1921	13	2
49	Radio telephony	1900	1907	1930	7	23
50	Phonograph	1877	1887	1896	10	9

Chapter 13
ICT Standards Management:
Architectural Challenges and Management Opportunities

Michael B. Spring
University of Pittsburgh, USA

ABSTRACT

This chapter explores opportunities to manage standards and standardization with a particular focus on the information and communication technologies (ICT) sector. It looks at the historical "management" of standards primarily in the United States, highlighting government and industrial approaches and the forces that have shaped the management process. It then turns to the current pressures and forces facing the management of ICT standards and standardization and makes some suggestions for activities that might enhance the management of standards.

INTRODUCTION

This article explores opportunities to manage standards and standardization with a particular focus on the information and communication technologies (ICT) sector. It looks at the historical "management" of standards primarily in the United States highlighting government and industrial approaches and the forces that have shaped the management process. It then turns to the current pressures and forces facing the management of ICT standards and standardization and makes some suggestions for activities that might enhance the management of standards.

AN HISTORICAL PERSPECTIVE: FOCUS ON THE UNITED STATES

It is somewhat misleading to believe that we manage standards and the standardization process. Broadly, standards serve to improve commerce. Nations try to cope with the myriad forces which impact commerce and standards organizations and government agencies work to produce standards and manage standardization are one part of that endeavor. The United States, often held up as a prime example of

DOI: 10.4018/978-1-5225-5320-5.ch013

business and industry led standardization, highlights some of the relationships between government and industry in this process. Over the last fifty years, developments in the ICT arena provide a prime example of how outside forces impact standards and standardization in a period of rapid change. Both reviews can inform a view of how the management process for standards and standardization might be improved.

Standards for communication (languages) and commerce (money) have been with us for more than two millennia. Few would argue that these developments were "managed". In the last millennia, there have been at least two broad periods of standardization. (Some might argue for three or four periods, but this analysis begins with just two.) The first period of managed standardization can be observed to begin with government control of weights and measures and monetary standards written into the constitutions of governments. Which nation was first is not as important as the centuries long tradition of governmental control of instruments deemed essential to commerce. A second period of managed standards begins with the industrial revolution. Scientific and technical standards were called for and to a large degree they became the purview of professional societies and industry. In some countries, governments played a facilitating role. In others, the government was inactive, as was largely the case in the United States. A few examples set the stage.

In order to schedule trains, railroads needed standard time. In Great Britain, standard time for the railroads was adopted in 1847. In the United States, the industry adopted standard time in 1870 (two time zones) and 1883 (four zones). This process was managed by the railroad industry out of a need. The U.S. government did not adopt standard time until 1912. Similarly, the infamous QWERTY keyboard standard arose from the dominance of the early Remington typewriter (No. 2) introduced in 1878 based on the patented design of the 1867 Sholes and Glidden typewriter. The arrangement was intended to prevent jams of hammers striking during the upstroke. The QWERTY keyboard, the IBM PC, the Microsoft Word document format are all examples of the many standards that arose based on various factors that led to market dominance. It would be hard to imagine that they were managed although it might be argued that "fast-tracking" of publicly available specifications is one management response to this situation.

Over the twentieth century, standards were increasingly set by professional societies such as the American Society of Civil Engineers (ACSE) and the American Society of Mechanical Engineers. Indeed, in part due to the prevalence of boiler explosions, the American Society of Mechanical Engineers established standards for boiler construction. ACSE provides standards for the construction of bridges and other civil structures. The development of standards for fire equipment and safety is often linked to the Baltimore fire. The fact is that fires in Chicago, San Francisco and other cities faced similar challenges. In Baltimore, the problem was highlighted by the number of cities that tried to provide help, but were unable to comment their equipment because of incompatible connections. The National Fire Protection Association adopted a national standard for fire hydrant and hose connections. While the standardization of fire hydrants and hose couplings grew, it was not an immediate accomplishment, and even today some cities have equipment that is non-standard. A 2004 study by NIST found:

Most of the major cities in the U.S. do not have standard fire hydrants and fire-hose couplings. In fact, only 18 out of the 48 most populated cities have both small hose and pumper connections on fire hydrants that comply with the NFPA standard. (Seck and Evans, 2004, p.11)

While professional societies provide guidance and assistance, few would suggest a coordinated management plan existed for the adoption of their standards.

In large part, standards were developed when industry saw the need. At the same time, the players did introduce management structures when and where they deemed them appropriate. Perhaps the most significant innovation related to management of the process arose from the efforts of the Pennsylvania Railroad to influence the composition of steel rails for the railroad. At that time the Pennsylvania Railroad was the largest corporation in the US and Charles Dudley, as the head of its chemistry department, was responsible for the quality of the steel, paint, wood, and other products purchased and used. In 1878, Dudley published a report on the durability of different types of steel rails. Dudley was shocked at the reaction of the steel companies. In short, they did not want consumers telling them how to make steel. In search of a solution, Dudley suggested a meeting of consumers and producers to come to consensus about how to produce rails that could be efficiently produced by the producers and that would meet the durability needs of the consumers. The voluntary consensus process was born and Dudley went on to become one of the major forces in the founding of the American Society of Testing and Materials serving as president from 1902-1908. Like so many aspects of the standardization process, the vaunted voluntary consensus process and the mandate of involving consumers and producers was born as part of a solution to a very specific problem. It suggested a particular management process that proved useful when standards crossed industry groups. Interestingly, the British Standards Institution began in 1901 as the Engineering Standards committee and early on was also concerned with standards for steel used for railways.

The role of government in the management of standards, at least in the US, is also born out of a particular need. The articles of confederation and the Constitution gave the federal government of the United States the sole and exclusive right of fixing the standards of weights and measures. Similar to other modern governments, these standards that had a significant impact on commerce were defined as within the scope of the government. Presidents Washington, Jefferson, Adams and Madison all emphasized the importance of having the federal government act on standards, but Congress demurred and ignored the matter, leaving it to the states (Cochrane, 1974, pp 21-23.) What responsibility existed for standards in the government moved from the Department of Treasury to the Department of Navy (maps) and then back to Treasury. With the rise of the industrial revolution and the need for new standards of measurement for electricity, light, chemicals, etc., there was a ground swell of support for a national organization.

And under pressure to produce or to satisfy their own demands for quantitative results, it was the scientists who sought better standards of measurement, better tools, precision instruments, and materials.... Science, better than industry, was aware that only Federal legislation could establish the necessary criteria, criteria that would possess national as well as international validity. Other nations, more advanced in commerce and industry, had long since recognized the need for such legislation and had established national standards laboratories....The meeting of these forces at the end of the 19th century—the growing needs of science and technology, coinciding with a new sense of national the impulse that created the National Bureau of Standards. (Cochrane, 1974, pp 15.)

1901 a bill to form the U.S. National Bureau of Standards was enacted into law and in 1903 the Bureau was transferred from Treasury to the Department of Commerce and Labor. The focus on standards and metrics for the burgeoning engineering and science branches can be clearly seen. The divisions addressed: heat and thermometry, light and optical instruments, resistance and EMF, magnetism and current, inductance and capacity, measuring instruments, and photometry.

In the early 1900s, several important events occurred related to the development and management of standards. What would become the British Standards Institution was founded as the Engineering Standards Committee in London in 1901. In 1906, the International Electromechanical Commission was formed. Also in 1906, the International Radiotelegraph Union was created as a companion to the International Telegraph Union which had been formed in 1865. These organizations would ultimately become the International Telecommunications Union (in 1932), a special agency of the United Nations since 1947. In the U.S. an important event related to the management of standards occurred in 1916 when four organizations that were developing standards (AIEE, now IEEE, ASME, ASCE and AIME, now ASTM) became concerned about the public acceptance of the standards they were developing. They established an impartial body to coordinate standards development, and reduce user confusion about who was developing the standards. The American Engineering Standards Committee (AESC) was founded in 1919 and over time played a role in the establishment of the International Standards Association which would evolve into the International Organization for Standardization. The AESC grew and reorganized as the American Standards Association in 1928 and then the American National Standards Institute in 1969. This suggests that at very least the professional societies developing standards were concerned with establishing a membership organization that could play a role in clarifying the legitimate standards developers for the public. Over the twentieth century, most of the U.S. standards developers, professional society and industry based, followed the practice of using ANSI as a mechanism to provide oversight of their processes. It is worth noting that the US movement to provide standards oversight played a role in the formation of the International Federation of the National Standardizing Associations which would evolve into International Organization for Standardization.

History suggests that, at least in the U.S., there was a general lack of interest in "managing" standards. Standards did emerge from invention, professional concern, industrial need, and other forces. At the same time, through the 18th and 19th century, there was little effort to manage the process strategically, tactically, or in any other fashion. Consensus based on participation of all the involved parties – consumers and producers – grew out of the need to invent a particular solution that would satisfy two industrial giants of the day, steelmakers and railroads. The idea of a voluntary consensus process developed by Dudley was a major strategy for the American Society of Testing and Materials and has become a dominant theme in many standards developing organizations. Stakeholder management has become an area of study for those interested in balanced standards (De Vries, H., Verhuekl, H. & Willemse, H., 2003) The need to provide assurance to the public about their efforts led to the creation of an oversight body that could provide certification of practice. Finally, the need for broad cross industry development of metrics for the sciences was seen as something most appropriate for the central government. This might be seen as an extreme case of a need for or cross industry standardization – i.e. electricity, optical and magnetic measures would impact many scientific and industrial enterprises.

MORE RECENT TRENDS AND FORCES: FOCUS ON COMPUTERS AND NETWORKING

Looking at more recent events, there have been numerous changes in the standards landscape. Some impact a broad array of standards. For example, the growth of free trading blocs has led to an increased use of standards as barriers to trade which has impacted agricultural and ICT products (Yue, C. 2006). The emergence of the European Union has spurred the use of standards to structure trade within the

Union (Buttle, 1997). The growth of the Chinese economy has spurred Chinese efforts in the standards arena which have had impacts on intellectual property and trade (Ernst, 2011).

Other forces have a more narrow impact. Specifically, the demand for new Information and Communication Technology (ICT) standards has been extraordinary. ICT standards tend have a very short life span. They are often lengthy and complex. The organizations developing them are more transient and diverse than those in other sectors. Indeed rather than a half dozen SDOs, there are several hundred groups advocating or developing standards. ICT standards tend, more than standards in other areas to be coupled to intellectual property. Finally, the products being developed based on these standards have a significant and direct impact on many aspects of personal and social life including a significant impact on the privacy and security of information about humans. Below, we examine some of the more interesting developments in the ICT area.

Over the twentieth century, only a few groups were responsible for setting standards in each area of standardization. In the area of information and communications technologies (ICT) there were less than a dozen major players. Today, hundreds of consortia, working groups and organizations supply standards in the ICT arena. Most are aware of the major developers of these standards – ISO, ITU, ECMA, INCITS, ATIS, IETF and W3C. The standards for other of the ubiquitous technologies have emerged in a less controlled fashion from other well-known organizations, e.g. Bluetooth SIG, USB Implementers Forum. The number of standards developers in the ICT arena has exploded and includes, to name just a few, the ZigBee Alliance, 3MF Consortium(3D Printing), Mobile Payment Forum (MPF), Car Connectivity Consortium, SD Card Association Security Industry Association (SIA), Smart TV Alliance, Clinical Data Interchange Standards Consortium (CDISC), EIS Alliance(energy management and smart grid), Genomic Standards Consortium, HDMI Forum, HomePlug Powerline Alliance, Open Automotive Alliance, Open Geospatial Consortium (OGC), Personal Connected Health Alliance. Most of these organizations are engaged outside the historical coordinating structures. Unlike more traditional SDO's, the various fora, consortia, and working groups come and go more quickly. They operate in new ways that are not subject to the traditional approaches recommended for the development of voluntary consensus standards (Baron, J., and Pohlmann, T., 2013; Baron, J., Ménière, Y., and Pohlmann, T., 2014). The growth of consortia and other groups signals a decentralization of standardization without control. It also suggests new forms of negotiation that speed the process of standardization (Polyakova, M., & Rubin, G. 2017). Does this need to be managed? What leverage do we have, and what are the key concerns for management?

Almost a century after the founding of ANSI, on August 19, 2012, the IEEE, the W3C, the IETF, the Internet Architecture Board, and the Internet Society joined to form Open Stand and invited others to join them in establishing a new set of principles to guide standardization. There are many similarities between Open Stand and ANSI. So why a new organization? Open Stand makes a commitment to make standards available free of charge. It is also committed to reducing patent and intellectual property issues that have become especially troublesome in the ICT arena. Is it a shift in power from the center (ANSI) to the SDOs (W3C, IETF, etc.)? Open Stand represent the beginning of a new oversight structure for standardization both within the United States and internationally? Egyedi & Enserink(2013) provide a detailed analysis of the benefits of open standards in the context of Dutch national standards.

The standards developed by the Internet Engineering Task Force (IETF) are different from standards developed by other SDOs. (Oksala, S., Rutkowski, A., Spring, M.B., and O'Donnell, J., 1996) IETF standards tend to emerge from the bottom up rather than the top down. Further, one result of the requirement that IETF candidate standards be working prototypes is that multiple standards sometimes compete for adoption. Does the IETF signal a growth of "wisdom of the crowd" standardization? Will it be standardization by competition? Security, extensibility, and comprehensive application were not the primary goals of the IETF standardization effort. The weak security built into the Internet we know today is often explained as an oversight. Security was not a primary goal when the original protocols were developed. This is particularly interesting given the ARPANet goal of better communication system for the military! Over time, the Internet standards that emerged from the "competitive" standardization approach of the IETF focused on quickly providing capability. The goal was focused standards that maximized capability and sic, minimized complexity. Egyedi has focused much of her research on the implications of competing standards (Egyedi, 2010; Egyedi, 2012).

The impact of Tim Berners-Lee on the direction of standardization undertaken by the W3C has been dramatic. While businesses play the important role in the W3C, as they do in any consortium, the technical staff of the W3C provides a damper on rampant commercial interests and focuses the standardization effort on what they consider the best technical approach. A careful reading of the W3C procedures makes clear that there are multiple checks and balances on standardization efforts that are closely tied to the central technical staff of the W3C.

Members express interest in the form of Member Submissions, and the Team monitors work inside and outside of W3C for signs of interest.... When there is enough interest in a topic ..., the Director announces the development of a proposal for a new Activity or Working Group charter, depending on the breadth of the topic of interest..... When there is support within W3C for investing resources in the topic of interest, the Director approves the new Activity and groups get down to work. (W3C, 2005)

Do the efforts of the W3C signal a return to standards dominated by technical experts? Does the W3C model offer a mechanism for institutionalizing professional guidance of standards development?

The academic community has begun to examine the processes for standards setting. The issues examined, to cite just a few, include the roles of end users (Jakobs, 2002), the productivity of consortia processes (Egyedi, 2012), and the overall organizational structuring of standardization (Brunsson, 2012). More needs to be done in this area, but carefully structured research does exist in a number of areas.

WHAT MIGHT WE WORRY ABOUT MANAGING

Studies by the Office of Technology Assessment (OTA, 1990; OTA, 1992), suggested various roles that might be appropriate for the government in the standards process. In the ICT world, standards may focus on architecture, design, interconnection/interoperability, or implementation/derivation for specific situations. Some of us suggested two decades ago that architectural or design standards would be undersupplied in an environment driven by a business return on investment (Spring, M.B., and Weiss, M.B., 1995). This seems to be true, especially for the standards efforts of the ad hoc organizations developing ICT standards. They tend to be guided by very narrow goals directed at maximizing capability. While few want to see more government control of standardization, there is less aversion to government support of the cost of standardization. It might be the case that government could support industry development of architecture standards that would help to guide particular product efforts. Is there a role for government in helping to support the development of roadmaps where such are possible but don't exist? Is there a role for government in adopting important standards efforts that are orphaned by small consortium efforts? Is there a value in supporting open standards (Egyedi & Enserink, 2013).

With the emergence of the Chinese market and the corresponding growth in standards development by the Chinese, there appears to be growth in trade controlling standards. Will standardization focus increasingly on market control and national interests? Will standards become a new mechanism for controlling trade in a post tariff world of free trade? There is a new role that standards have played in a post tariff world. Is this an appropriate role for standards? Some believe that we are beginning to work are way out of this problem by explicitly addressing the issue in the negotiation of free trade agreements. There is some belief that the TransPacific Partnership may simplify standards issues related to trade rather than complicate them. Do nations need to step in to manage the use of national standards as barriers to trade (Okun-Kozlowicki, 2016)?

The rapid and competitive standardization in the ICT field is occurring with very little centralized control. It would appear that many standards are focused on capabilities with a relative disregard for security or interoperability. While the standards are rich in capability, it is not clear that the players have a game plan when it comes to extensibility, security and privacy. Some organizations, such as the IEEE and NIST continue to think about roadmaps and architectures in their areas (NIST, 2011; IEEE, 2014). Others are more focused on developing capabilities for their specialized technologies with less regard for frameworks. Should the government play a role in assuring that a security framework is an a priori consideration in standardization for Bluetooth, ZigBee, USB, etc.? This is not to suggest that NIST, DHS or DOD should develop the standards, but rather that government might play a role. The government could act as convener or help to provide a structure for less formal organizations to work toward in areas like security and privacy. Further, the national government organizations might commit to financing the development of the reference models that business cannot justify based on a simple return on investment (Spring & Weiss, 1999). Is there a way to maintain the entrepreneurial spirit that leads to rapid development while insuring that the structure put in place accounts at very least for security and privacy? It would be a shame if we look back in a couple decades to find that we enabled the Internet of Things without the forethought to secure the environment and protect the privacy of all who wish to embrace what it promises. It may be the appropriate time to look at the dilemma in the same way Charles Dudley did in the 1880's.

CONCLUSION

Standards and standardization are one of the mechanisms by which industries, nations, or groups of nations encourage commerce. Historically, standardization has not been managed so much as it has responded to "incidents" and "accidents". Industrial agreements, cross industry relationships and government control have been incorporated into best practices to enhance the process. The growth of the structures and management practices have emerged from particular sets of needs that occurred at particular times. This discussion suggests that the current systems for managing standards may be adequate for many areas of standardization but in areas where there are many players new to standardization or where standards are appearing very rapidly, roadmaps or architectural oversight may be lacking. In these situations, industries or professional societies may not be able to afford the cost of providing structures and government support for the efforts might be warranted. This could include support for research on various approaches, and provision of guidance and architectures for action. In the ICT arena, which is expanding very rapidly, a first step might include a government study of who is currently involved in the standardization process with a focus on the under-documented consortia and working groups currently setting standards. A second step might include an analysis of how well the current efforts are doing in moving toward a secure cyberinfrastructure with appropriate provisions for privacy.

REFERENCES

Baron, J., & Pohlmann, T. (2013). Who Cooperates in Standards Consortia—Rivals Or Complementors? *Journal of Competition Law & Economics*, *9*(4), 905–929. doi:10.1093/joclec/nht034

Baron, J., Ménière, Y., & Pohlmann, T. (2014). Standards, Consortia, and Innovation. *International Journal of Industrial Organization*, *36*, 22–35. doi:10.1016/j.ijindorg.2014.05.004

Bonino, M., & Spring, M. (1991). Standards as Change Agents in the Information Technology Market. *Computer Standards & Interfaces*, *12*(2), 97–107. doi:10.1016/0920-5489(91)90059-9

Cochrane, R. (1974). *Measures for Progress: A History of the National Bureau of Standards*. Washington, DC: U.S. Government Printing Office.

Brunsson, N., Rasche, A., & Seidl, D. (2012). The Dynamics of Standardization: Three Perspectives on Standards. *Organization Studies*, *33*(5-6), 613–632. doi:10.1177/0170840612450120

Buttle, F. (1997). ISO 9000: Marketing motivations and benefits. *International Journal of Quality & Reliability Management*, *14*(9), 936–947. doi:10.1108/02656719710186867

De Vries, H., Verheul, H., & Willemse, H. (2003). Stakeholder identification in IT standardization processes. *Proceedings of the Workshop on Standard Making: A Critical Research Frontier for Information Systems,* 12-14.

Egyedi, T. M. (2010). *On the implications of competing standards. Pros and Cons of Standard Setting*. Stockholm, Sweden: SC Authority.

Egyedi, T. M. (2012) The Impact of Competing Committee Standards: Integrating the Interoperability Perspective. *Proceedings of the 17th EURAS Annual Standardisation Conference*.

Egyedi, T. M., & Enserink, B. (2013). Measuring the benefits of open standards: A contribution to Dutch politics. *Standards Today*, *12*(1), 1–13.

Ernst, D. (2011). *Indigenous Innovation and Globalization: The Challenge for China's Standardization Strategy*. Available at SSRN: http://ssrn.com/abstract=2400855

Jakobs, K. (2002). A proposal for an alternative standards setting process. *IEEE Communications Magazine*, *40*(7), 118–123. doi:10.1109/MCOM.2002.1018016

IEEE. (2014). *Industry Connections Security Group (ICSG), Industry Connections Activity Initiation Document (ICAID)*. Retrieved August 30, 2017, from http://standards.ieee.org/about/sasb/iccom/IC09-001-04_Computer_Security_Group_ICSG.pdf

NIST. (2011). *Cloud Computing Standards Roadmap Working Group*. Special Publication 500-29, July 2011. Retrieved August 30, 2017, from http://www.nist.gov/itl/cloud/upload/NIST_SP-500-291_Jul5A.pdf

Oksala, S., Rutkowski, A., Spring, M. B., & O'Donnell, J. (1996). The Structure of IT Standardization. *StandardView*, *4*(1), 9–22.

Okun-Kozlowicki, J. (2016). *Standards and Regulations: Measuring the Link to Goods Trade*. Office of Standards and Investment Policy International Trade Administration, U.S. Department of Commerce.

OTA. (1990). *1990, Critical Connections: Communications for the Future. U.S. Congress, Office of Technology Assessment*. Washington, DC: U.S. Government Printing Office.

OTA. (1992). *Global Standards: Building Blocks for the Future. U.S. Congress, Office of Technology Assessment*. Washington, DC: U.S. Government Printing Office.

Polyakova, M., & Rubin, G. (2017). Integrated methodology for standard-setting norms of innovative product in the new competitive environment. AIP Conference Proceedings (Vol. 1858, No. 1, p. 040005). AIP Publishing. doi:10.1063/1.4989954

Seck, M., & Evans, D. (2004). Major U.S. Cities Using National Standard Fire Hydrants, One Century After the Great Baltimore Fire. *NISTIR,* 7158. Retrieved August 30, 2017, from http://www.fire.nist.gov/bfrlpubs/fire04/PDF/f04095.pdf

Spring, M. B., & Weiss, M. B. (1999). Intellectual Property Issues in Standardization. In K. Jakobs (Ed.), Information Technology Standards and Standardization: A Global Perspective (pp. 63-79). Hershey, PA: IGI Global.

Spring, M. B., & Weiss, M. B. (1995). Financing the Standards Development Process. In B. Kahin & J. Abbatte (Eds.), *Standards Policy for Information Infrastructure* (pp. 289–320). Cambridge, MA: The MIT Press.

W3C. (2005). *World Wide Web Consortium Process Document*. Retrieved August 30, 2017, from http://www.w3.org/2005/10/Process-20051014/cover.html#toc

Yue, C., Beghin, J., & Jensen, H. (2006). Tariff Equivalent of Technical Barriers to Trade with Imperfect Substitution and Trade Costs. *American Journal of Agricultural Economics*, *88*(4), 947–960. doi:10.1111/j.1467-8276.2006.00908.x

Chapter 14
Measuring the Deployment of Internet Protocols

Tapio Levä
Telia Company, Finland

Antti Riikonen
Verto Analytics Inc., Finland

Juuso Töyli
Aalto University, Finland & University of Turku, Finland

Heikki Hämmäinen
Aalto University, Finland

ABSTRACT

Internet protocols and traffic are the foundation of the internet and reflect the on-going digital transformation in the society. Internet protocols spread to potential adopters through several successive phases of implementation, commercialization, acquisition, and adoption. This protocol deployment process involves several stakeholders and varies depending on the deployment environment and the protocol in question. This complexity and the lack of comprehensive measurement studies call for a further conceptualization of measuring protocol diffusion along the whole deployment process. Therefore, this chapter develops a framework for measuring the deployment of internet protocols, consisting of deployment steps, deployment models, deployment measures, and data sources. The illustrative results indicate that protocol deployment is driven by applications, and show the existence of large deployment gaps between the protocol possession and usage.

INTRODUCTION

The Internet constitutes a unique environment for innovation diffusion due to its global, distributed and loosely regulated nature where control over resources is spread among a multitude of stakeholders with diverse economic goals (Marcus, 2004). The Internet Engineering Task Force (IETF) develops and standardizes Internet protocols as voluntary standards. Diffusion of Internet protocols is a relevant

DOI: 10.4018/978-1-5225-5320-5.ch014

and special example of standards diffusion because the IETF operates a bottom-up marketplace for individual protocol standards. Moreover, the Internet protocols are networked innovations, which exhibit significant network externalities (Katz & Shapiro, 1986). As a result, the diffusion of Internet protocols is a market-based process where the successful alignment of stakeholders' incentives is a key to success (Clark, Wroclawski, Sollins, & Braden, 2005).

A protocol is a software component or feature, which enables applications and services (Jorstad, Dustdar, & Do, 2005). Protocols typically spread to the end users embedded in products, such as applications, operating systems (OS), or devices – thus, diminishing the direct impact of a protocol on the end users' adoption decision (Warma, Levä, Tripp, Ford, & Kostopoulos, 2011) and increasing the impact of supply-side decisions to include the protocol in products (Kivi, Smura, & Töyli, 2012). This is an example of market-pull vs. technology-push (Ende & Dolfsma, 2005). In case of a strong technology-push strategy, a protocol can be acquired by a large population as part of a product bundle, but is possibly only used by few users. Such phenomenon, related to the gap between different adoption events, was introduced by Fichman and Kemerer (1999). On the other hand, protocols and other software features may not even become available to the potential end users due to the decisions of software and hardware vendors (Levä, Komu, Keränen, & Luukkainen, 2013), hindering the diffusion. For example, the decision of Apple not to support Flash in their mobile devices prevented end users from adopting (services based on) it.

Despite the important role of technology providers, the traditional diffusion of innovation theories (Rogers, 2003) and the case studies on protocol diffusion (e.g., Hovav, Patnayakuni, & Schuff, 2004; Ozment & Schechter, 2006; Joseph, Shetty, Chuang, & Stoica, 2007) often limit to modeling and measuring the end user adoption. As Lyytinen and Damsgaard (2001) conclude, this is insufficient for explaining the diffusion of complex, networked technologies, and the focus needs to be widened to cover the critical process features and all key players. In order to overcome too narrow perspective when analyzing the protocols' feasibility, which affects their diffusion, Levä and Suomi (2013) define *protocol deployment as a process, during which a protocol is advanced from the first specification into actual use on the Internet through steps such as implementation, commercialization, acquisition, and adoption of the protocol.* Measuring and analyzing the diffusion during all these steps is essential for understanding the deployment dynamics and identifying the critical success factors of Internet protocols.

Motivated by the special characteristics of Internet protocols and the lack of comprehensive measurement studies on protocol deployment, this chapter develops a framework for measuring the deployment of Internet protocols during the different steps. This is achieved by identifying the deployment models, measures, and data sources of each step. In addition to measuring the deployment levels that are directly linked to the steps, also deployment gaps and delays between these steps are defined and described. The developed framework is then applied to analyze the deployment of 11 protocols in the Finnish mobile market using an extensive longitudinal and cross-sectional data collected from 2003-2012.

The framework builds on the existing literature and the authors' cumulated knowledge on protocol deployment and previous data collection efforts on mobile device diffusion and usage. The data collection for the application of the framework is based on long cooperation with several industry partners and experiences from analyzing the collected datasets. The development of the framework was an iterative process, where the framework was refined based on the data analysis of the case Finland, and vice versa.

The chapter makes a methodological contribution by identifying alternative deployment models of Internet protocols and developing a framework for measuring the deployment. The chapter makes a practical contribution by applying the developed framework to an example market, and analyzing the success of the protocols in each of the deployment steps. The chapter also contributes by combining and comparing different data sources in a novel way.

The rest of the chapter is structured as follows. Next section introduces the background on innovation diffusion and protocol deployment. Then, methods section presents the research process and data that are used later for developing the framework and applying the framework to the Finnish mobile market. Finally, the findings, limitations and future research opportunities are discussed.

BACKGROUND

The innovation diffusion theories tend to define the development and diffusion of innovations as a stepwise process. For example, according to Rogers (2003), the innovation-development process consists of six main stages: 1) needs/problem, 2) basic and applied research, 3) development, 4) commercialization, 5) diffusion and adoption, and 6) consequences. Other closely related process definitions have been suggested, for example, by Tornazky, Fleischer, and Chakrabarti (1990). In a wider industrial context, diffusion of radical innovations, when replacing another innovation, can cause market disruptions (Christensen, 1997). A related disruption framework by Kilkki, Mäntylä, Karhu, Hämmäinen, and Ailisto (2017) covers the spread of a single innovation vertically from scientific community to final societal impacts as well as the spread of innovation horizontally from one industry to another.

As defined by Levä and Suomi (2013), protocol deployment covers the commercialization and adoption stages of Rogers' process. The interplay between the providers commercializing and end users adopting the innovation, known as technology-push vs. market-pull (Zmud, 1984), defines the diffusion of the technology. Prescott and Van Slyke (1997), who analyzed the diffusion of Internet as a collection of various hardware and software technologies, and Ende and Dolfsma (2005), who studied the evolution and diffusion of various computing technologies, conclude that both technology-push and market-pull are important for diffusion but their importance fluctuates across time. In addition to the technology-push from the providers, also governmental mandates may push the technology to the market (Carter, Jambulingam, Gupta, & Melone, 2001). Shin, Kim and Hwang (2015) categorized existing standardization research into demand-side and supply-side literature.

Existing research concerning the commercialization step has focused largely on provider strategies for competing against other providers. Firstly, providers decide on which standards they invest in. Here it is important to decide whether to compete within or between standards (Besen & Farrell, 1994). For example, personal computer (PC) vendors compete within the Windows ecosystem, whereas Apple competes against that with its own operating system. Secondly, the providers of software have various strategies to attract customers, including versioning (Shapiro & Varian, 1998), bundling (Bakos & Brynjolfsson, 1999), and pricing (Lehmann & Buxmann, 2009). These all affect the market offering provided to the end users. Despite the interest towards provider strategies, measurement studies on the commercialization of particular technologies are rare in the existing literature (Adams, Bessant, & Phelps, 2006).

The adoption step can be divided on a high level to pre-adoption, adoption decision, and post-adoption phases. The diffusion of innovation literature (Rogers, 2003) focuses often on the adoption decision traditionally approximated by the number of cumulative purchases. This is a reasonable approach for studying the diffusion of fairly simple consumer goods among individual adopters. However, this approach falls short when studying complex, networked IT innovations (Lyytinen and Damsgaard, 2001) with significant network effects and knowledge barriers (Fichman and Kemerer, 1999). This is particularly true in the organizational context with large number of adoption events that involve different decision-makers. Because of this complex nature of organizational innovation diffusion, the information systems implementation literature (e.g., Cooper & Zmud, 1990; Liker, Fleischer, & Arnsdorf, 1992) has identified large differences in the number of employees that have certain products installed and the number of actual users in organizations. Fichman and Kemerer (1999) call this difference as *assimilation gap* and define it as "*the difference between the pattern of cumulative acquisitions and cumulative deployments of an innovation across a population of potential adopters*". However, they do not restrict the term to only these two particular adoption events.

Shih and Venkatesh (2004) suggest that assimilation gaps can emerge also on adoption of complex consumer innovations. This is supported by some diffusion, adoption, and consumer product choice studies on mobile handsets. Verkasalo (2008) analyzed the behavioral intention to use mobile services and identified that even if there is an intention to use specific services, the intention does not always materialize into actual usage. In mobile handset choice studies the identified main factors affecting product choice decision include price and properties, brand, prior experience and size (Karjaluoto et al., 2005; Liu, 2002; Riquelme, 2001). If the importance of a specific feature is small in the purchase decision, the gap between the possession of such feature and the usage of it could be rather large. However, to the best of our knowledge, these gaps have not been previously studied empirically in the case of Internet protocols.

The protocols are complex, networked innovations typically embedded in other products, such as mobile phones, PCs, or operating systems (Fomin, Su, & Gao, 2011). Due to the hidden nature, the adopters do not necessarily make conscious decision concerning the adoption of the protocol (Warma et al., 2011). As networked innovations, the protocols exhibit network effects since the utility of the protocol depends on the number of potential communication partners (Katz & Shapiro, 1986) and the availability of complementary products (Nair, Chintagunta, & Dubé, 2004). Consequently, the utility of protocols is often too low for early adopters before the critical mass (Rogers, 2003) is achieved, known as the bootstrap problem (Hanseth & Lyytinen, 2010). Multiple authors have proposed solutions for this problem by, for example, introducing converters between the new and the incumbent protocol (Joseph et al., 2007; Jin, Sen, Guérin, Hosanagar, & Zhang, 2008) and suggesting business strategies, such as bundling, coordination and subsidization (Ozment & Schecter, 2006).

Requirement of interoperability is another key characteristic of networked innovations, due to which protocols are typically standardized. The specifications of Internet protocols, known as RFCs (Request For Comments), are voluntary standards developed by the IETF through the process based on "rough consensus and running code" (Alvestrand, 2004). The standards play an important role in mediating different interests and motivations among the various stakeholders (Yoo, Lyytinen, & Yang, 2005). As Jakobs (2017) notes, the end result of standardization depends both on the corporate goals as well as personal interests and social capital of individuals participating in standards development. The role of strong individuals is particularly large in the less hierarchic standardization organization, such as the IETF (Ali-Vehmas, 2016). Standardization also affects the diffusion of protocols, as it encourages deployment by reducing technological uncertainty and lowering costs (Weitzel, Beimborn, & König, 2006). However, standardization does not affect measuring deployment. Therefore, the framework developed in this chapter is applicable for both standardized and proprietary protocols.

The existing measurement studies on protocol deployment focus on measuring the possession and usage of protocols. Eggert (2017) maintains a web site that reports the deployment status of different Internet protocols on the most popular web services and sites. Huston (2017) measures the deployment of IPv6 on the autonomous systems based on the BGP routing tables. Colitti, Gunderson, Kline, and

Refice (2010) use server side measurements to analyze the usage of IPv6 by the clients using Google's services. Internet traffic classification studies (e.g., Thompson, Miller, & Wilder, 1997; Maier, Feldmann, Paxson, & Allman, 2009) measure the traffic volumes generated by different Internet protocols on an aggregate level. Even though all these studies provide valuable insights into the adoption and usage of Internet protocols, they usually focus on a single deployment event, preventing the analysis of the whole deployment process.

METHODS

As was identified in the background section, a need for a holistic conceptualization of the protocol deployment process and its measurement exists. This chapter describes the research process used for developing and applying the framework for measuring protocol deployment, as well as the data sources.

Research Process

The research process consisted of three main parts: framework development, data selection and collection, and application of the framework. First, the framework was developed based on a literature review, the cumulated knowledge from our previous protocol deployment studies, and experiences from several data collection efforts on mobile devices. Then, the appropriate data for the illustrative application of the framework were selected and collected where necessary. Last, the deployment measures identified in the framework were calculated and characterized to show how the framework can be applied.

However, some improvements to the framework were identified in the data collection and application parts, and therefore the framework and its terminology were modified during several iteration rounds. The iteration rounds include an early version of the framework and its application that were presented in a conference paper of Levä and Riikonen (2013). After the feedback from the conference, industry experts, and the new co-authors, the framework was extended and improved. These changes were also taken into account in the refined application part. Figure 1 depicts the iterative nature of the research process.

Figure 1. Research process

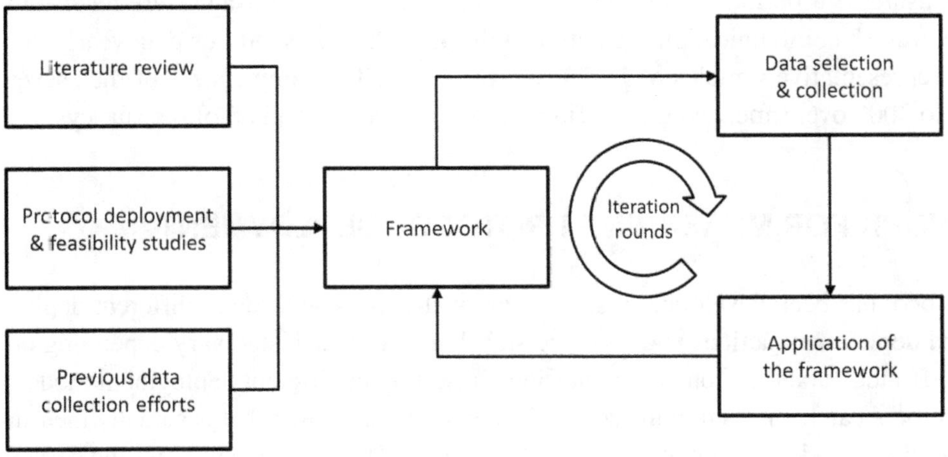

The previous protocol deployment and feasibility studies by the authors of this chapter provided substance knowledge of protocol deployment process (Warma & Levä, 2011; Levä & Suomi, 2013; Levä et al., 2013). The previous data collection efforts providing experience about measuring the device and service usage focused on the Finnish mobile market during 2005-2012 in the context of several research projects. These datasets have been collected in close collaboration with different market stakeholders, including mobile operators, device manufacturers, market research companies, and regulators. Previous results of the analysis of these datasets are reported, for example, by Riikonen et al. (2013) and Kivi et al. (2012). These results also provided substance understanding on the topic area of mobile device and product feature diffusion, of which especially the latter is related to protocol diffusion. Different methods for measuring mobile service usage are discussed in more detail by Smura et al. (2012).

Data

Several primary and secondary datasets were utilized in the research. Information about the pre-installed protocols of different handset models were collected specifically for this article. The data were collected from several sources due to the large number of device models on sale and in use, and the lack of a single, comprehensive data source. These sources included, for example, WURFL device description repository (ScientiaMobile, 2012), manufacturer websites, and other public device information databases. Because Internet protocols were rarely mentioned directly in the used data sources, the existence of protocols in the handset models was identified based on other capabilities. In total, protocol information was found for over 1000 handset models, covering the major models on sale and in use during the measurement period. The protocol data collection process is described in more detail in the Appendix.

Three other data sets were collected about mobile handset sales, installed base of mobile devices, and mobile service usage. Handset sales data were received from the market research company GfK. Data on the model-level unit sales includes a time period from January 2003 to September 2012 and covers roughly 70-90% of the Finnish mobile handset retail market over the measurement period. Data on the installed base of devices using mobile networks were collected from all three Finnish mobile network operators annually in the end of September from 2005 to 2012, consisting of the number of units of each handset model that the operators had identified in their networks. Data collection was based on the charging functionalities of mobile networks and the unique identifiers of handset models called Type Allocation Codes (TAC). The installed base data covers roughly 80-99% of the handset base in Finland over the measurement period. Public survey data on mobile service usage were received from Ficora (2012), the Finnish communication regulation authority, who collected this data yearly during 2006 to 2012 from representative samples of 15-79 year old Finns. The sample sizes of the surveys increased from 1500 to 3000 over time, and the confidence level is over 95% in all of the surveys.

FRAMEWORK FOR MEASURING PROTOCOL DEPLOYMENT

After a protocol has been developed, it spreads gradually into use through different deployment steps. The detailed deployment actions and involved stakeholders of each step vary depending on the chosen deployment strategy and the protocol in question – resulting in different deployment models. The diffusion of a protocol can be measured in each of the deployment steps. What is actually measured depends on the deployment model and the objectives of the inquiry. This section combines the deployment steps, deployment models, deployment measures, and data sources into a framework for measuring protocol diffusion. The framework focuses on the client-side deployment of standardized Internet protocols for individual end users (i.e., consumers). Due to its generality, however, the framework should be applicable also to other software innovations, such as mobile applications, and to the server side deployment.

Deployment Steps

Building on the innovation deployment processes introduced in Section 1, the protocol deployment is divided into four steps. During the first two steps – 1) implementation and 2) commercialization – the providers of software and hardware implement and make the protocol available for end users. During the last two steps – 3) acquisition and 4) adoption – the end users acquire and start using the protocol.

In the implementation step, providers implement the protocol based on the protocol specification developed and standardized by the IETF. As illustrated in Figure 2, at least three protocol implementation locations in the different layers of software can be identified: i) the kernel of an operating system (e.g., TCP/IP protocols), ii) middleware (e.g., OpenSSL for SSL/TLS), and iii) directly in an application utilizing it (e.g., BitTorrent). A new protocol is often first introduced in the higher layers of software hierarchy, from where it moves to the lower layers in the course of time. For example, the Microsoft's operating systems did not have native support for TCP/IP suite before Windows 95 (Young, 2016). If the protocol implementation and the application utilizing it are separated, access to the protocol functionality is provided through an application programming interface (API), such as Sockets API for TCP/IP. This allows the same protocol implementation to be used by multiple applications. The separation also means that the implementation of protocols and applications utilizing them can involve different stakeholders, potentially leading to separate deployment.

In the commercialization step, providers make the protocol available to end users by including it in products that the end users can acquire. These products can be the same software, where the protocol is implemented in, or operating systems and hardware including the higher-level software. For example,

Figure 2. Protocol implementation locations (P) in the software hierarchy

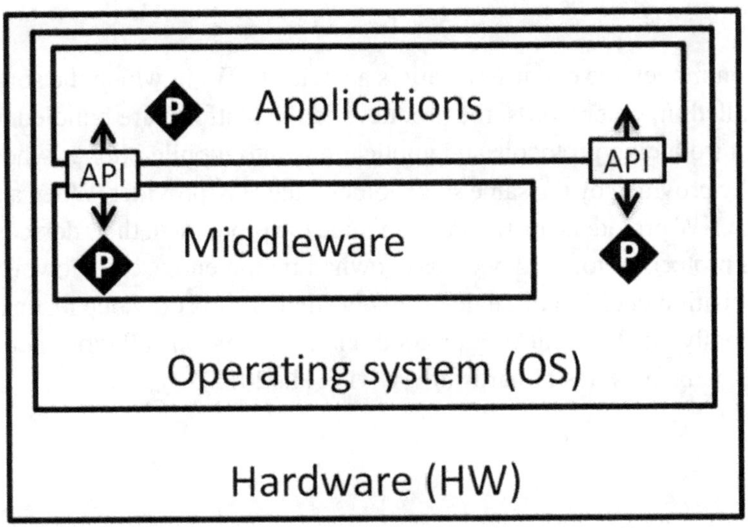

if the protocol is implemented in an application, the commercial product that the end user acquires can either be an application, an operating system containing the application, or hardware containing the operating system, which contains the application. This modularity is visualized in Figure 2 with nested boxes. Consequently, *product is defined being something that end users can acquire (and which contains the protocol)*, whereas *platform is something where the end user installs the product after acquisition.*

In the acquisition step, end users acquire the capability to use the protocol. Acquisition can be divided into 1) the acquisition of the platform where the product can be installed, and 2) the acquisition of the product that contains the protocol. These acquisitions can occur simultaneously if the platform and the product are bundled together. For example, smart phones are always bundled with an operating system, and operating systems contain many applications when sold to the customers. After the acquisition step, the end user possesses the necessary components for using the protocol: hardware, operating system, an application, and the protocol embedded in one of the above-mentioned locations.

In the adoption step, end users start using the protocol. In practice this means that they start to generate protocol traffic by using applications enabled by the protocol. The level of activity required for taking the protocol into use varies from one case to another. On the one hand, the protocol may be used by an operating system service that is started automatically when the device is turned on. On the other hand, the end user may need to enable the protocol manually, as is the case of IPv6 in Windows XP (Colitti et al., 2010).

Deployment Models

The deployment steps describe the process of advancing the protocol from the specifications into actual use on a rather high level. However, the detailed deployment actions and involved stakeholders of each step vary depending on the products in which the protocol is included – resulting in three deployment models: i) pre-installation, ii) update installation, and iii) post-installation. The models are not mutually exclusive and can co-exist depending on the protocol. Identification of the deployment models is relevant, since the dynamics of protocol deployment and the sources for the measurement data differ between the models.

Pre-Installation

In the pre-installation model, the end user acquires a device (HW) in which the software containing the protocol is pre-installed. In other words, the product and the platform are bundled together. This model is used heavily for introducing protocols and applications into mobile phones where the hardware and software are typically provided by the same stakeholder – the HW provider. When a protocol is deployed along this model, the HW providers' decisions to include the protocol in their devices affect significantly the diffusion of the protocol into the device base owned by the end users. However, the impact of the protocol on the acquisition decision of device is likely small as it is only one feature among a multitude of features. Additionally, deployment of a protocol along the pre-installation model is a slow process due to the long replacement cycles of hardware (Kivi et al., 2012).

Post-Installation

In the post-installation model, the end user acquires the software (product) and the device (platform) separately from each other, and then installs the software to the device. Software distribution over the Internet has increased the importance of the post-installation model, which is the predominant model for providing applications to smart phones through the app stores. In the post-installation model, the protocol is potentially a more significant part of the product than in the pre-installation model, thus affecting more the acquisition decision. For example, a decision to download the VLC media player from the web is more likely motivated by an interest towards video streaming over real-time streaming protocol (RTSP) than a decision to purchase a laptop. The model also allows the SW providers to bypass the HW providers, which may speed up protocol deployment.

Update Installation

In the update installation model, the end user acquires the protocol through a SW update. In this model, the product is the SW update and the platform is the previously acquired software. For example, web browsers, smart phone applications, and operating systems are updated regularly. The update installation model resembles the post-installation model, with the distinction that the end user does not acquire a completely new product. Therefore, the acquisition decision is likely smaller than with a new product. The other key difference from the post-installation model is that the updates are often pushed to end users through the update mechanism included in the software, such as Windows Update, so that the end user accepts the installation either separately for each update or once for all the future updates. The update installation model potentially allows fast deployment of new protocols for the owners of the required platform, as the fast diffusion of multipath TCP (Bonaventure, 2013) along with the Apple's iOS 7 OS update (Chitika, 2013) demonstrates.

Deployment Measures and Data Sources

The deployment level of a protocol can be measured in each of the deployment steps. These measures are useful in analyzing the success of protocol deployment. What is actually measured in each step depends on the deployment model and the objectives of the inquiry. On the provider side, the protocol spreads into the different product models, whereas in the end user side the analysis concerns the products owned and used by the end users. In addition to analyzing the deployment levels of each step separately, comparison between them by measuring the deployment gaps and delays can help to identify the bottlenecks of protocol deployment.

Provider Side Measures

Implementation level is a measure for the implementation step. It measures how widely a protocol is implemented for different platforms and supported by applications, i.e., *the number of implementations*. Defining the target population is difficult for the implementation level. One possibility is to calculate the share of certain type of software, such as operating systems, for which an implementation exists. However, more important than measuring the implementation level itself is to identify the implementations and applications utilizing them, so that this information can be used in the later steps. Data sources for analyzing the implementation level include product specifications, device description repositories, such as WURFL (ScientiaMobile, 2012), or source code analysis (Komu, Varjonen, Gurtov, & Tarkoma, 2012).

Availability level is a measure for the commercialization step. It measures how widely a protocol is included in the product models the end user can acquire, i.e., *the share of available product models containing the protocol*. In the pre-installation model, this translates into measuring the share of device models on sale containing the protocol, whereas in the post- and update installation models, the measure is the share of available operating systems or some class of applications, such as web browsers, containing the protocol. This information can be found from the product catalogs, such as the web sites of online shops or the application catalog of app stores. Additionally, data on product sales collected by, for example, market research companies can be used to identify the products on sale.

Availability level alone is not a perfect measure for the success of the commercialization step, since it does not take into account the popularity of the products. Scaling the availability level with the sales volumes of different products can complement the analysis. This measure, called as acquisition rate, provides a linkage between the provider side and the end user side. It is defined as *the share of sold products containing the protocol*. The data on product sales is available from market research companies or directly from the stores selling the products. Availability level and acquisition rate are relevant measures to the protocol developers and providers for finding the optimal commercialization strategies.

End User Side Measures

On the end user side, the diffusion is measured in relation to a specific population. In addition to the typical approach of analyzing the diffusion into a population consisting of end users, one can also analyze the diffusion into a population consisting of products. For example, the user population of the protocol can be measured either as the number of laptops or the number of end users that generate protocol traffic.

Acquisition step has two measures related to the possession of platform and the protocol. Platform possession level measures the share of end users that could acquire (the products containing) the protocol, i.e., *the population possessing the platform*. Due to the lack of separate platform and protocol acquisitions in the pre-installation model, this measure exists only in the post- and update installation models. Platform possession level is a relevant measure to protocol developers that decide in which platforms they implement their protocols, as they are interested in estimating the population they can reach with specific implementations. Protocol possession level measures the share of end users that could use a protocol, i.e., *the population possessing (the product containing) the protocol*. Possession level is a relevant measure to application and content service providers that decide upon which protocols they build their services. Data on cumulative sales of the HW and SW products and the SW updates containing the protocol provide one approximation of the possession levels. However, by using cumulative sales data it is difficult to take the replacement acquisitions and discards into account. Consequently, the number of products in use provides more accurate information on the possession levels. In case of mobile devices, mobile operators' reporting systems are one potential source for these data. Additionally, device models and operating systems can be identified from the network traffic or from the OS vendors' customer databases.

Usage level is a measure for the adoption step. It measures the share of end users that generate protocol traffic, i.e., *the population using the protocol*. When measuring the usage, it is important to clearly define the measure. Usage can namely be measured both on binary (i.e., use or no use) and continuous scale (i.e., rate or variety of use (Shih & Venkatesh, 2004)). Data sources for measuring the usage include end user surveys, client monitoring systems, network traffic measurements, and server-side measurements (Smura, Kivi, & Töyli, 2009). Usage level is a relevant measure to other potential adopters that are interested in communicating with the protocol, Internet service providers who manage their networks, and application service providers whose revenues depend on the usage of their services.

Deployment Gaps and Delays

In addition to measuring the deployment separately for each step, comparison of the measures by analyzing the gaps and delays can be used to identify the bottlenecks and dynamics of protocol deployment. In our terminology the *deployment gap is defined as the relative or absolute difference between two deployment measures at time t.* Similarly, the *deployment delay is defined as the timely difference between two deployment steps reaching a specific deployment level.*

On the provider side, the number of implementations and the number of product models that contain the protocol are not commensurable. Nevertheless, it can still be beneficial for the protocol developers to *compare the implementation level and the availability level* in order to analyze the relative efficiency of protocol commercialization in comparison to the implementation effort. For example, despite a large number of protocol implementations for different OSes, the host identity protocol is not included in any commercial OS due to the OS providers' preference for substitute solutions and reluctance to include unnecessary protocol implementations that increase the maintainable code base, and thus costs (Levä et al., 2013).

Comparison of the availability level to the acquisition rate links the provider side to the end user side through the product models on sale, thus enabling the analysis of technology-push vs. market-pull. This gap reveals the popularity of the products containing the protocol compared to those not containing the protocol over a certain period of time. This is especially relevant comparison in the pre-installation model, where the success of commercialization cannot be analyzed before acquisitions have occurred.

Comparison of the availability and possession levels, for one, allows analysis of the delay between the availability of the protocol in product models and its possession by the end-users. Due to the different populations – product models on sale vs. product base in use – these measures are not directly comparable. This delay differs between the deployment models depending on the replacement and update cycles of hardware and software. Additionally, the gap between availability and possession levels is likely smaller in cases of strong technology-push, e.g., when the protocol comes bundled in hardware or operating systems.

On the end user side, the measures can be calculated as relative shares to the total population and are, therefore, directly comparable. Figure 3 visualizes the end user measures and the gaps and delays between them with example data.

Platform possession gap is calculated as the difference between the total population and the platform possession level. This gap does not exist in the pre-installation model, because the platform and the product are acquired simultaneously. In the post installation and update installation models this gap exists, as there can be a notable difference between the total population and the number of platform owners in the market. The size of the gap depends heavily on the provider side decisions on which platforms the protocol is made available. For example, the lack of Android OS implementation contributes significantly to the platform possession gap in the Finnish mobile market since the share of Android OS in the Finnish mobile handset base was 11% on September 2012 (Riikonen & Smura, 2013).

Figure 3. Deployment levels, gaps, and delays on the end user side

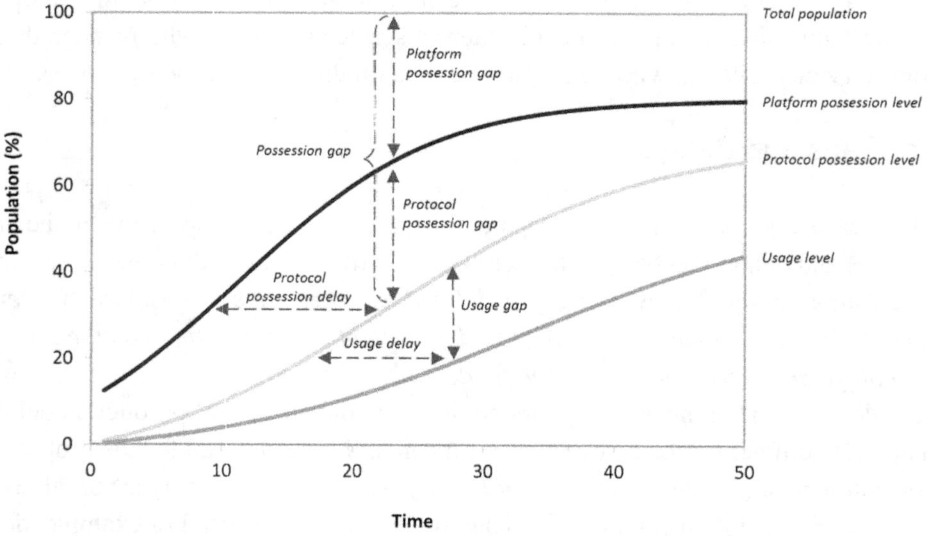

Protocol possession gap (or delay) is calculated as the difference between the platform possession level and the protocol possession level, that is, the difference between the platform owners and the population with the protocol already installed. Again, this gap exists only in the post- and update installation models, where the end users first purchase the platform, and later acquire the protocol itself. The protocol possession gap indicates the difference in the popularity of the platform compared to the popularity of the products. From the protocol developer's perspective, it defines the market reachable with only single acquisition. Platform possession gap and protocol possession gap can be added up to describe the total *possession gap*, which is especially relevant in the pre-installation model.

Usage gap (or delay) is calculated as the difference between the possession level and the usage level, that is, between the end users who have the protocol, but do not use it, and the actual users. The size of this gap depends on the end users' awareness of and interest towards the protocol and the applications utilizing it. The gap is likely smaller, if the protocol has a significant role in the product the end user acquires. For example, Skype application is more likely to be used by an end user who downloads it himself from the web compared with the end user who receives it pre-installed in a device. However, the usage level of Skype may actually be higher in the pre-installation model since the pre-installed application may attract end users who would not download the application themselves. This example demonstrates that the analysis of protocol deployment needs to include both deployment levels and the deployment gaps.

Framework

In order to manage the complexity of protocol deployment, Figure 4 combines the deployment steps, deployment models, deployment measures, and data sources into a framework for measuring protocol deployment. The framework is divided into the provider and end user sides based on the key stakeholder of each step. The measures mapping directly to the steps define the deployment level, i.e., the success of the protocol, in each step. In addition to the deployment levels visible in the figure, the gaps and delays between them are relevant measures for identifying the bottlenecks of protocol deployment. Finally, the selection of the data source for each measure depends on the deployment model under analysis. The deployment models are also relevant since they represent alternative deployment strategies with differing characteristics. The transparent boxes in the post- and update installation models indicate the platform acquisition steps, which do not involve the protocol but need to occur in order to enable the protocol acquisition.

A few practical issues should be taken into account when conducting the measurements according to the framework. For each measure, it is important to identify the population, into which the protocol is diffusing. In order to allow direct comparison, the measures of different steps should be commensurable, i.e., the population should be the same. The population can consist either of i) the products in which the protocol is implemented, installed, and used, or ii) the stakeholders that take these actions. On the provider side, the product models are more relevant than their providers, since they link the two sides together through product sales. On the end user side, however, both the end user population and the device population can be selected. These populations may differ from each other since an end user may use multiple devices, and single device can be used by multiple end users. This needs to be taken into account when combining and comparing the datasets with different populations, such as the possession level based on the number of devices and the usage level based on the number of end users. Variation is also possible inside each measure since deployment levels can be measured both on binary and continuous scales (e.g., use or no use vs. rate of use). Additionally, analyzing both the deployment rates (i.e., the change in the size of the population over time t) and the deployment levels (i.e., the size of the population at time t) can complement the analysis.

Figure 4. Framework for measuring the deployment of Internet protocols

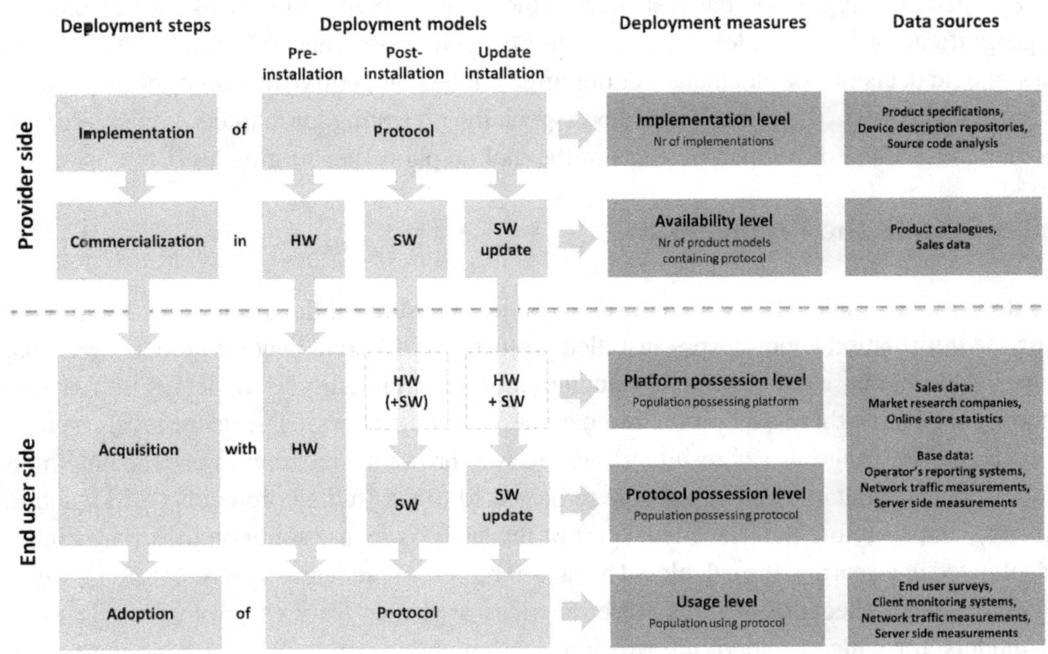

APPLICATION OF THE FRAMEWORK TO THE FINNISH MOBILE MARKET

In this section the developed framework is applied to analyzing the protocol deployment in the Finnish mobile market in 2003-2012 by focusing on the measures, delays, and gaps defined in the framework.

Scope

In total 11 Internet protocols were selected for the analysis. These protocols cover the basic network and transport layer protocols IPv4, TCP, and UDP, widely used security protocol TLS, and the key application layer protocols for email (SMTP, IMAP4, POP3), web browsing (HTTP), video streaming (RTSP, RTP), and voice over IP (SIP). Email protocols POP, IMAP, and SMTP are bundled as a single protocol (EMAIL), because they typically co-exist in the devices they have been included in.

Because of the wide deployment of these protocols in the personal computer market and, therefore, in the server side, this analysis limits to the client side deployment for mobile devices. Additionally, the analysis focuses on the pre-installation model. This is because for the selected protocols the pre-installation model was the dominant deployment model during the measurement period, and because the deployment gaps and delays should be most clearly visible with this model.

The Finnish mobile market is chosen as the focus market. From the viewpoint of the framework, the selected market was appropriate, as data of all the protocol deployment steps in Finland were available from a time period of several years and the authors have substance understanding of the market. From the viewpoint of the analysis results, Finland represents a developed European mobile market (Smura, Kivi & Töyli, 2011) with a population of roughly 5.4 Million in 2013 and some local peculiarities. On the one hand, Finland has long traditions in mobile technologies and was one of the fastest growing mobile markets in the 1990s (Jaakkola, Gabbouj, & Neuvo, 1998). Therefore, 89% of 15-74 year old Finns were using mobile phones already in 2002 (MINTC, 2004). On the other hand, Finland is the home market of the device manufacturer Nokia. This single device vendor dominated the market with over 70% of people using Nokia handsets during the measurement period (Riikonen & Smura, 2013), differing significantly from the global market shares (Cecere, Corrocher, & Battaglia, 2015). This means that the strategic decisions of one manufacturer affected the competitive dynamics of the market during the measurement period. The results should be especially interesting for markets with similar properties or ones that are still less developed in terms of the mobile phone penetration level.

Mapping Data to the Framework

Figure 5 describes the linkages between the data and the measures of the framework. The protocol data, including the information about the pre-installed protocols of different handset models, were mapped to the handset sales and base data to enable calculation of the availability level, acquisition rate, and possession level over time, that is, along the protocol deployment process. The usage levels were received from Ficora's public usage survey results. However, web browsing and email were the only mobile services surveyed in the data that were possible to be linked to the studied protocols (HTTP and EMAIL, respectively), and therefore the usage level part of the analysis focuses only on those two protocols.

The collected datasets and the calculated measures represent the Finnish market well. However, some missing data remains because of unidentified devices or device features (protocols). The over 1000 handset models, for which protocol information was found, cover over 92% of the handset models on

Figure 5. Linkage between the deployment measures and the collected datasets

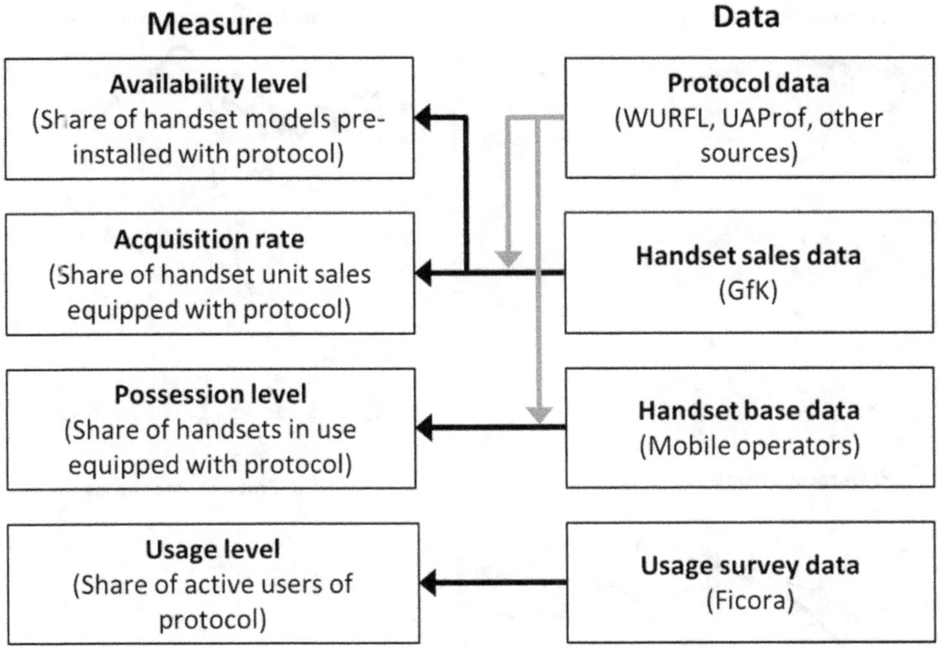

sale, over 99% of the total unit sales, and over 99% of the handset base data depending on the point in time and the protocol. Exceptions are RTP, RTSP, and SIP protocols, for which the protocol information coverage was notably lower, and therefore the findings concerning RTP, RTSP, and SIP are less reliable.

Deployment Measure Analysis

Figure 6 presents a) the availability level, b) acquisition rate, c) possession level, and d) usage level of the selected protocols in Finland from 2003 to 2012. Figures 6a and 6b depict the provider side measures, whereas Figures 6c and 6d show the end user side measures.

On a general level, four groups[1] of protocols can be identified by visual inspection of internally similar diffusion patterns: 1) IPv4 and UDP, 2) TCP and HTTP, 3) TLS and EMAIL, and 4) RTP, RTSP, and SIP. The diffusion patterns of the groups are related to the diffusion of the following mobile services: WAP browsing (group 1), HTML browsing (group 2), email (group 3), and real time communications (group 4). The last group is internally less similar than the other groups, because real time communications protocols can be divided into streaming and VoIP related parts. Even though TLS is important also in securing HTTP browsing, the results indicate that its deployment has been primarily driven by EMAIL.

On the provider side, Figure 6a illustrates the historical protocol availability levels, that is, the share of handset models on sale equipped with the protocols. There are large differences on the availability levels between the protocols and the groups. For example, the availability levels of IPv4 (group 1) were 78%-96% during 2003-2012, whereas for SIP (group 4) the availability level remained below 51% during the same period. Naturally, the availability of protocols in handset models available is dictated mainly by the provider and retailer decisions, which can be based on competitive strategies and differentiation of handset models, customer demand, but also on the use cases and device requirements of the protocols.

Figure 6. Protocol a) availability levels, b) acquisition rates, c) possession levels, and d) usage levels

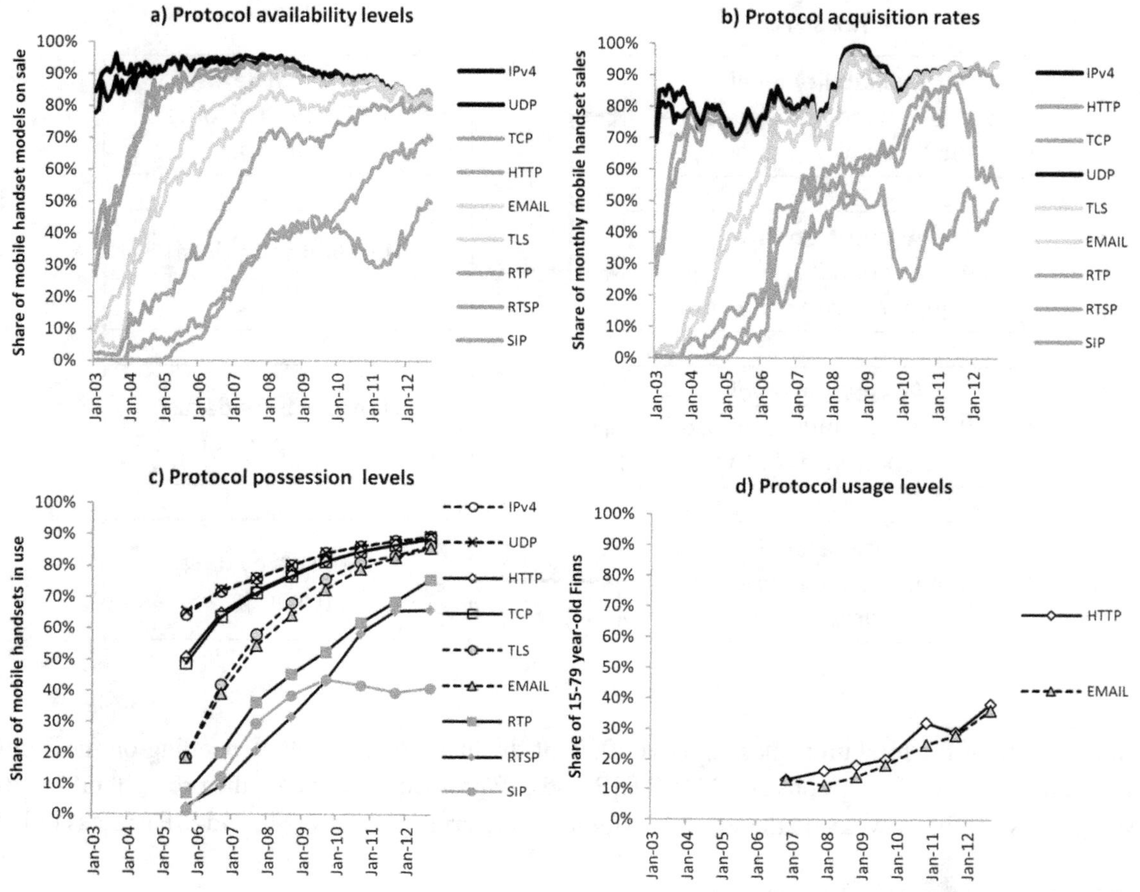

Within the groups the availability levels were close to each other, group 4 (real time communications) with the largest differences. There RTP was identified to have higher availability levels than RTSP and SIP, which is natural since RTP is used with both video streaming and VoIP applications, whereas RTSP and SIP relate to only one of those. It should be noted that the number of models on sale increased during the measurement period from tens in 2003 to a level of roughly 200-300 as of 2007. Therefore, even though the relative availability levels of some protocols stayed rather stable, the absolute number of handset models on sale including the protocols increased considerably.

Figure 6b shows the acquisition rates that measure the share of handsets including the protocol from the monthly handset unit sales. Handset models including protocols from groups 1 and 2 were popular in sales almost throughout the measurement period with protocol acquisition rates of 68%-99% during 2004-2012. The acquisition rates of group 3 reached groups 1 and 2 in 2006, after which the three groups have followed closely each other. For the protocols of group 4 (RTP, RTSP, SIP), Figure 6b shows an abrupt increase in sales in 2006. This peak falls on the point in time when the Finnish regulator allowed bundling of mobile subscriptions and 3G handsets (Tallberg, Hämmäinen, Töyli, & Kamppari, 2007). This change affected sales behavior of 3G devices positively in short term and was discussed in Smura et al. (2011). Another unexpected pattern was the sudden drop for SIP in 2009. One reason for this drop could be that Nokia's Symbian S60 devices with pre-installed SIP were possibly replaced with Apple iPhone and Google Android devices that did not include SIP at that point.

On the end user side, Figure 6c illustrates the protocol possession levels, that is, the share of handsets equipped with each protocol out of all devices using mobile networks. Many of the selected protocols diffused slowly towards 90-95% saturation level during the measurement period. One exception is again SIP, for which the previously identified drop in acquisition rates is also visible in the possession levels of Figure 5c. This supports the presented reasoning that S60 handsets were replaced by handsets without pre-installed SIP.

Figure 6d shows the usage levels for two protocols, EMAIL and HTTP, for which survey data was available. The usage levels of these two protocols followed closely each other from 2008 to 2012, growing slowly from 10-15% to 30-35% during that time period. In general, Figure 6c and Figure 6d indicate large differences in the possession levels and usage levels of Internet protocols. These usage gaps and delays are analyzed in more detail in the next section.

Deployment Gap Analysis

Figure 7 shows the provider side measures - availability level and acquisition rate - for the protocols according to the group. The framework identifies the main criteria for protocol success in the commercialization stage to consist of comparing the availability levels and acquisition rates, as the difference of these measures describes the popularity of the device models in which the protocols were included. Figure 7 indicates that during the initial stages of protocol diffusion, the availability level of several protocols was higher than the capability acquisition rate. For most of the protocols this changed later and the capability rates increased close to or higher than the availability level, meaning that the popularity of the models equipped with the protocols increased. For instance, for EMAIL the acquisition rate reached its availability level in 2006 at a level of about 70%. After this point these two measures followed each other closely. In comparison, for HTTP the same point in time was in 2004, after which the acquisition rate remained higher or close to the availability rate.

The identified gaps could be explained by several factors. For example, the popularity of smartphones and handsets with advanced features, including improved data connectivity, increased generally during the measurement period and several of the selected protocols became basic features also for many feature phone models. On the other hand, new features and protocols are possibly first introduced in high-end or other experimental models, which might explain the lower acquisition rates in the beginning of the measurement period.

When comparing the provider side measures, all protocols from groups 1, 2, and 3 can be argued to be successful in terms of their provider side deployment in the Finnish mobile handset market. These protocols reached availability levels and acquisition rates close to 80-90% by 2012, meaning that they are included in most of the handset models provided by device vendors, and these models include the most popular models in sales. The group 4, and especially RTSP and SIP are exception. For RTSP the acquisition rate stayed below the availability level for long periods until 2012, and for SIP both measures experienced the aforementioned large drop in 2010. Reason for these patterns could be that both RTSP and SIP are advanced application layer protocols, and were mainly included in more high-end models. These protocols also have more competition than most of the other protocols, which have already reached a dominant position. For example, proprietary protocols such as Skype provide competing services to what SIP enables, and therefore, substitution between these protocols might exist.

Figure 7. Provider side measures – availability level and acquisition rate – for protocol groups 1-4 (figures a-d, respectively)

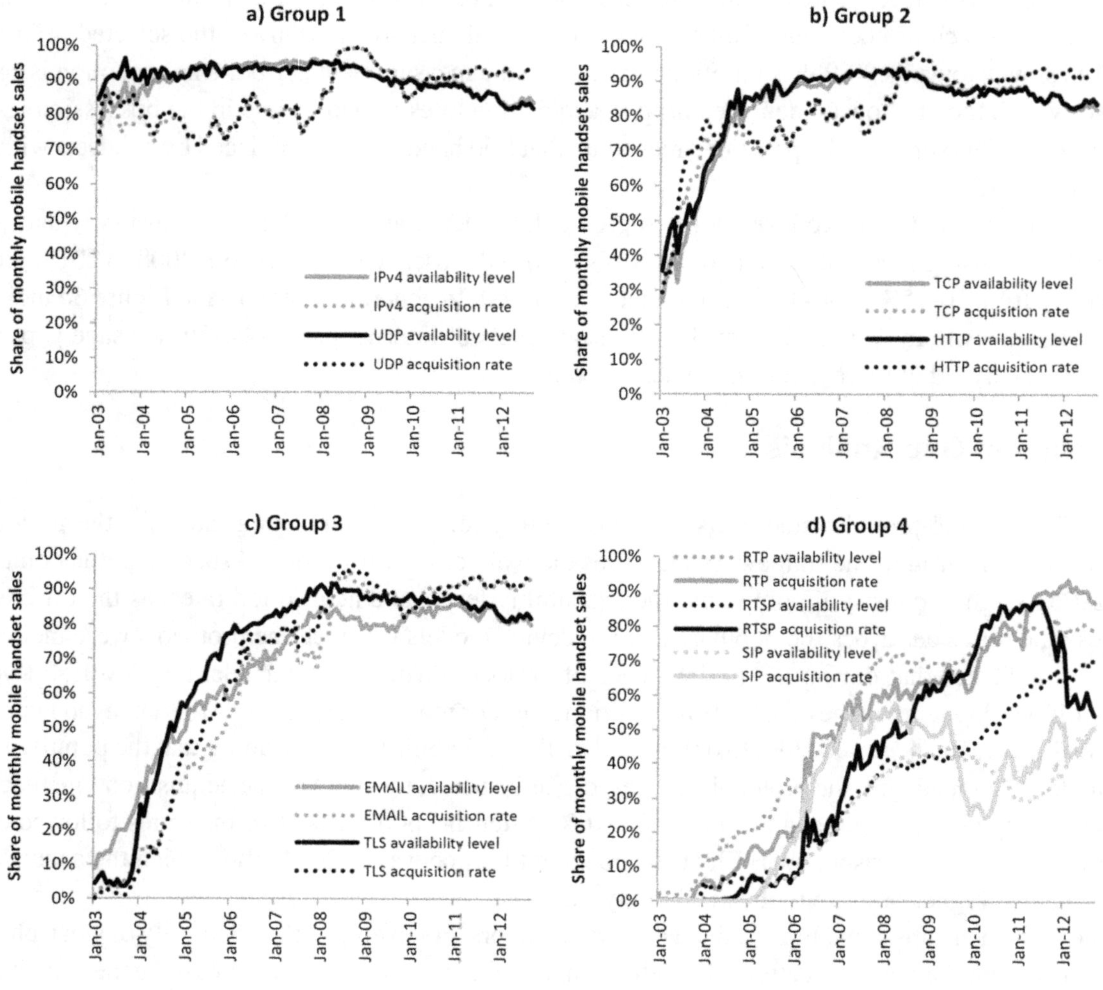

Figure 8 depicts the end user side measures – protocol possession level and usage level – for HTTP and EMAIL protocols, for which usage information was available. In general the provider side success in protocol commercialization is also visible on the end user side, where the strong technology-push by the handset providers resulted in diffusion of the protocols, that is, increase in the possession levels in 2005–2012. However, the typical slowness of the pre-installation model because of the slow replacement rate of mobile handsets is visible in the data. For example, the delay for EMAIL from reaching 80% availability level and acquisition rate to reaching the same possession level was about three years.

It is also visible in the figure that the technology-push did not affect strongly the usage levels of the two protocols, as large usage gaps (and delays) were identified for both. The possession levels of HTTP and EMAIL grew from 39% and 65% in September 2006 to 86% and 89% in September 2012, respectively. At the same time the usage levels increased from 13% for both HTTP and EMAIL to 36% and 38%, respectively. For HTTP, the usage gap remained rather stable at 50–55 percentage units during

Figure 8. End user side measures – possession level and usage level – for HTTP and EMAIL

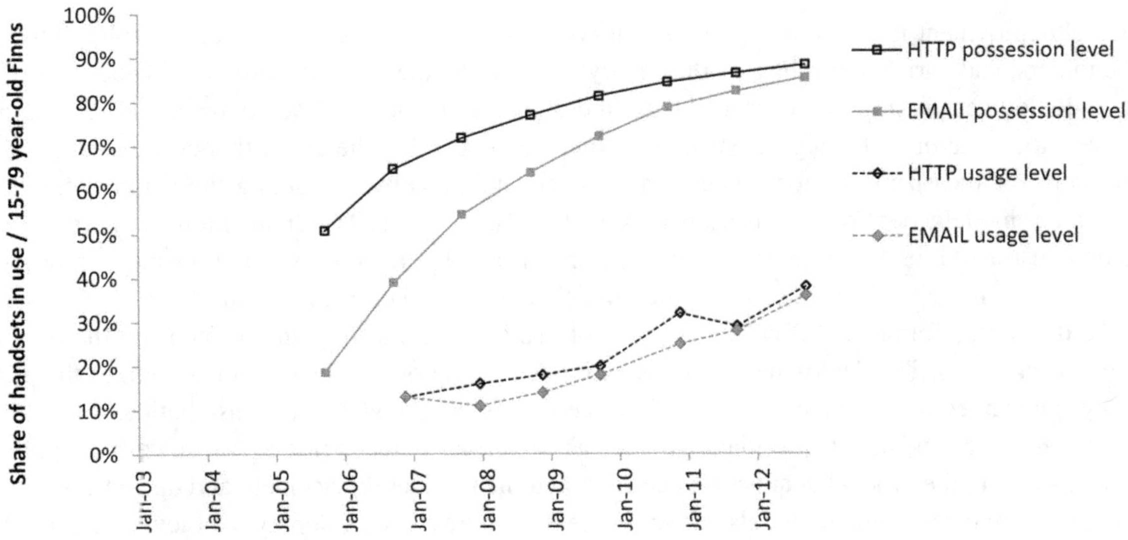

this time period, whereas for EMAIL the usage gap increased notably from 26 to 50 percentage units. A rough estimation of the usage delay for EMAIL at 30% penetration level is about 6 years. This long delay could be caused by two reasons. Either there is a large existing non-user population of the protocol who continue as non-users, or new protocol acquirers simply do not start using the protocol.

These results demonstrate adoption as one deployment step where HTTP and EMAIL protocols have not succeeded. Even though the providers introduced EMAIL in popular models, the popularity of these models was caused by other factors, as the users did not take the protocol in use. The large usage gap indicates that web browsing and email were not important services to the users of mobile handsets especially before 2010. This observation is supported by the Ficora (2012) survey data showing that the interest towards new mobile services, such as email, had only small impact on the decision to acquire a 3G handset in 2006-2009.

Other complementary explanations for the usage gaps could be that the applications utilizing email protocols did not provide good usability or that the pricing of mobile data connectivity prohibited people from using mobile data services. However, strong price competition for mobile data started in the Finnish market in the end of 2007 when USB modems were introduced, after which the mobile data prices in Finland have been low. In addition, the new touch screen –oriented smartphone operating systems were introduced first with the Apple iPhone 3G in 2008. However, the combined share of iOS, Android, and Windows Phone operating systems was still only 21% in September 2012 (Riikonen & Smura, 2013), explaining the slow growth in usage levels to some extent. Age is also a factor affecting the usage of mobile services. For instance, according to Ficora (2012), the usage of mobile phones with older age groups focused on a smaller range of services, and especially on the basic voice calling and text messaging functionalities in 2012. If, however, older users own devices with advanced functionalities, the usage gap will not fully close unless the usage habits of these users change in the future.

DISCUSSION

Protocol deployment is a complex process that consists of several successive steps, involves multiple stakeholders, and varies depending on the deployment environment and the protocol in question. Measuring deployment is important for analyzing the dynamics of protocol deployment and for identifying the critical factors affecting the success of Internet protocols. Therefore, this chapter developed a framework for measuring protocol deployment levels and gaps by combining the deployment steps, deployment models, deployment measures, as well as data sources. The application of the framework was demonstrated by measuring how a set of 11 pre-installed protocols spread into handset models on sale, handsets in use, and actual usage in the Finnish mobile market from 2003 to 2012.

The developed framework divides the protocol deployment into measurable steps on the provider and end users sides. The deployment process varies depending on in which product bundles the protocol is implemented and included, resulting in three deployment models – pre-installation model, post-installation model, and update installation model. The interplay of technology-push and market-pull, the protocol's role in the product acquisition decision, and the product replacement and update cycles may vary between the deployment models, affecting the deployment speed, deployment levels, and the size of deployment gaps. Because the deployment levels define the success of a protocol in each step and the gaps help identifying the bottlenecks of protocol deployment, a complete analysis of the success of a protocol should involve both deployment levels and gaps.

The application of the framework to the Finnish mobile market confirms some of the earlier findings and provides understanding of the dynamics of protocol deployment along the pre-installation model. The protocol deployment is generally driven by applications, such as web browsing, email or video streaming. The results illustrate the slowness of the deployment along the pre-installation model and the differences in the gaps and delays on both the provider and end user sides. For example, the delay from the high provider-driven availability level to the similar end user –related protocol possession level was years in the case of EMAIL protocol. In addition, a high possession level does not necessarily translate into actual use, as the identified large gaps between the protocol possession and usage levels of EMAIL and HTTP show. This result extends the assimilation gap concept proposed by Fichman and Kemerer (1999) from organizational innovation adoption to consumer adoption. It also supports the finding of Warma et al. (2011) that protocols are often acquired unintentionally alongside handset acquisitions motivated by other reasons than the protocols. Consequently, adoption should be studied by measuring the actual usage of a protocol instead of using the acquisition of the protocol-containing hardware and software products as a proxy for adoption.

The measures and gaps identified by the framework are useful for analyzing the dynamics of protocol deployment in the different deployment models. The understanding of the pros and cons of each model helps in choosing the optimal protocol deployment strategies for each protocol case, which is especially important for the technology providers. In addition, researchers seeking to analyze protocol deployment on a holistic level can utilize the framework when planning and conducting their measurements and research. Researchers also benefit from the application of the framework to the Finnish market, demonstrating how the data from the different data sources can be combined. The framework also captures the complexity related to the protocol implementation in the software hierarchy and its relation to the products and platforms end users acquire, which may be useful for researchers working with other modular and systemic innovations.

The identified deployment gaps call for further discussion concerning their reasons and evolution. Firstly, the gap between availability and possession is likely to increase until the critical mass of end users possessing a protocol is reached and the utility of the protocol is high enough. Technology providers often take risk by using the pre-installation model to build the critical mass, making the possession gap a natural phenomenon. Building the critical mass is often asymmetric so that either the client or server side acts first, the other one following. In the analyzed case, for example, the problem of critical mass had already been solved due to the existing deployment on the server side driven by the PC market. However, with completely new, mobile-driven protocols, also the server side deployment should be analyzed in order to take into account the cross-side network effects. Similar asymmetry can also be found between the network and device functionalities, as the introduction of 3G networks before the availability of 3G phones demonstrates.

Secondly, multiple reasons may contribute to the low usage levels, resulting in large usage gaps. In general, the capabilities of mobile devices seem to have evolved faster than people's interest and knowledge of new services, the availability of attractive applications, the usability of devices, the data rates of devices, and the pricing of data subscriptions. A natural way to close the gaps is to solve these issues. For example, the app stores and automatic OS update services have simplified the installation of new applications, whereas the touch screens and modern operating systems have improved the user experience. As rather recent phenomena, these evolutionary steps had not fully affected the usage gaps in the Finnish mobile market in 2012. Still, the usage gaps of some size are likely to exist as long as people use only a part of the functionalities of their mobile phones.

Thirdly, the competition against other incompatible protocols and proprietary applications may also explain the deployment patterns of protocols substituting each other. For example, in the mobile Internet, email faces competition from an increasing number of messaging applications, including the traditional SMS and new instant messaging and social networking applications, such as WhatsApp and Facebook. In case of newer protocols missing the established user base and server side availability built on the PC market, such as SIP, the competition from proprietary applications (e.g., Skype) may affect even the availability and possession levels of the protocols. In the most extreme cases, the competition leads to "standards wars" (Shapiro & Varian, 1999), where the winner takes it all and the loser vanishes from the market.

Finally, the cause and significance of usage gaps are neither obvious nor same in every case. For example, already a small usage gap for a protocol deployed along the post-installation model indicates problems in its attractiveness, whereas in the pre-installation model a many times larger gap may not be a problem. Moreover, the strategies to increase the deployment levels and decrease the related gaps are likely different in these two cases. Consequently, one should not jump to hasty conclusions when analyzing the results of protocol deployment measurements.

The research could be continued to multiple directions. First, the framework should be further validated and improved. Because the framework has been mainly created for analyzing the client-side deployment of a particular type of protocols – voluntary standards developed by the IETF, the application of the framework to the server-side and to other types of protocols (or even other software components, such as mobile games) requires testing and validation. A deeper analysis of the modularity of software – the relationship between software components, products, and platforms over time – could also help to improve the framework. Because the framework does not explain the motives behind the stakeholders' deployment decisions, the framework should be accompanied with suitable analytical tools for analyzing stakeholder incentives.

Second, the application of the framework could be extended to cover additional data sources and protocols. For example, conducting network traffic measurements would provide data on usage levels for additional protocols and would enable the analysis of frequency and usage volumes. The introduction of automatic operating system update models and convenient distribution of applications through app stores has contributed notably to the success of iOS and Android (Campbell-Kelly, Carcia-Swartz, Lam, & Yang, 2015), and increased the role of post- and update-installation models. Studying a case where these models were more relevant, would allow comparing the deployment levels and gaps of different deployment models. Likewise, the comparative analysis of multi-device usage, i.e., the same person using multiple types of devices as pointed out by Finley and Soikkeli (2016), would allow further understanding the deployment linkages between device types.

Last, the mobile market has evolved significantly after 2012. Share of Android and iOS operating systems has increased (Vesselkov & Hämmäinen, 2016) and new protocols such as QUIC, WebRTC and Multipath TCP have become available. Therefore, continuing the Finland case study beyond 2012 could introduce new findings. Moreover, comparison of different markets, either geographic or device markets (mobile handsets vs. PCs) could help to explain the role of deployment environment on protocol deployment.

ACKNOWLEDGMENT

The authors wish to thank Miika Komu, Jörg Ott, Pasi Sarolahti, Varun Singh, Henna Suomi, and the anonymous reviewers. The work has been supported by the MoMIE research project.

REFERENCES

Adams, R., Bessant, J., & Phelps, R. (2006). Innovation management measurement: A review. *International Journal of Management Reviews*, *8*(1), 21–47. doi:10.1111/j.1468-2370.2006.00119.x

Ali-Vehmas, T. (2016). Complex Network Perspective on Collaboration in the ICT Standardization. *International Journal of Standardization Research*, *14*(2), 33–64. doi:10.4018/IJSR.2016070103

Alvestrand, H. (2004). The role of the standards process in shaping the internet. *Proceedings of the IEEE*, *92*(9), 1371–1374. doi:10.1109/JPROC.2004.832973

Bakos, Y., & Brynjolfsson, E. (1999). Bundling information goods: Pricing, profits, and efficiency. *Management Science*, *45*(12), 1613–1630. doi:10.1287/mnsc.45.12.1613

Besen, S. M., & Farrell, J. (1994). Choosing how to compete: Strategies and tactics in standardization. *The Journal of Economic Perspectives*, *8*(2), 117–131. doi:10.1257/jep.8.2.117

Bonaventure, O. (2013, September 18). *Apple seems to also believe in Multipath TCP*. Retrieved from http://perso.uclouvain.be/olivier.bonaventure/blog/html/2013/09/18/mptcp.html

Campbell-Kelly, M., Garcia-Swartz, D., Lam, R., & Yang, Y. (2015). Economic and business perspectives on smartphones as multi-sided platforms. *Telecommunications Policy*, *39*(8), 717–734. doi:10.1016/j.telpol.2014.11.001

Carter, F. J. Jr, Jambulingam, T., Gupta, V. K., & Melone, N. (2001). Technological innovations: A framework for communicating diffusion effects. *Information & Management*, *38*(5), 277–287. doi:10.1016/S0378-7206(00)00065-3

Cecere, G., Corrocher, N., & Battaglia, R. D. (2015). Innovation and competition in the smartphone industry: Is there a dominant design? *Telecommunications Policy*, *39*(3), 162–175. doi:10.1016/j.telpol.2014.07.002

Chitika. (2013, September 19). *UPDATE: iOS 7 on 52% of iOS devices one week after public release*. Retrieved from http://www.chitika.com/insights/2013/ios-7-adoption

Christensen, C. M. (1997). *The innovator's dilemma: When new technologies cause great firms to fail*. Harvard Business Review Press.

Clark, D. D., Wroclawski, J., Sollins, K., & Braden, R. (2005). Tussle in cyberspace: Defining tomorrow's Internet. *IEEE/ACM Transactions on Networking*, *13*(3), 462–475. doi:10.1109/TNET.2005.850224

Colitti, L., Gunderson, S. H., Kline, E., & Refice, T. (2010). Evaluating IPv6 adoption in the Internet. In Proc. Passive and Active Measurement (pp. 141–150). Springer Berlin Heidelberg. doi:10.1007/978-3-642-12334-4_15

Cooper, R. B., & Zmud, R. W. (1990). Information technology implementation research: A technological diffusion approach. *Management Science*, *36*(2), 123–139. doi:10.1287/mnsc.36.2.123

Eggert, L. (2017). *Deployment trends of IPv6, SIP, XMPP, DNSSEC, DKIM, SPF, and DMARC*. Retrieved from http://eggert.org/

Ende, J., & Dolfsma, W. (2005). Technology-push, demand-pull and the shaping of technological paradigms – Patterns in the development of computing technology. *Journal of Evolutionary Economics*, *15*(1), 83–99. doi:10.1007/s00191-004-0220-1

Fichman, R. G., & Kemerer, C. F. (1999). The illusory diffusion of innovation: An examination of assimilation gaps. *Information Systems Research*, *10*(3), 255–275. doi:10.1287/isre.10.3.255

Ficora. (2012). *Consumer survey on communications services*. Retrieved from https://www.viestintavirasto.fi/en/aboutthesector/studiesandsurveys/viestintapalvelujenkuluttajatutkimus.html

Finley, B., & Soikkeli, T. (2016). Multidevice mobile sessions: A first look. *Pervasive and Mobile Computing*, *39*, 267–283. doi:10.1016/j.pmcj.2016.11.001

Fomin, V. V., Su, J., & Gao, P. (2011). Indigenous standard development in the presence of dominant international standards: The case of the AVS standard in China. *Technology Analysis and Strategic Management*, *23*(7), 745–758. doi:10.1080/09537325.2011.592270

Gabin, F., Kampmann, M., Lohmar, T., & Priddle, C. (2010, November). 3GPP mobile multimedia streaming standards. *IEEE Signal Processing Magazine*, 134–138.

Hanseth, O., & Lyytinen, K. (2010). Information infrastructures: The case of building Internet. *Journal of Information Technology*, *25*(1), 1–19. doi:10.1057/jit.2009.19

Hovav, A., Patnayakuni, R., & Schuff, D. (2004). A model of Internet standards adoption: The case of IPv6. *Information Systems Journal, 14*(3), 265–294. doi:10.1111/j.1365-2575.2004.00170.x

Huston, G. (2017). *IPv6 reports*. Retrieved from http://bgp.potaroo.net/index-v6.html

Jaakkola, H., Gabbouj, M., & Neuvo, Y. (1998). Fundamentals of technology diffusion and mobile phone case study. *Circuits, Systems, and Signal Processing*, *17*(3), 421–448. doi:10.1007/BF01202301

Jakobs, K. (2017). Two dimensions of success in ICT standardization – A review. *ICT Express*, *3*(2), 85–89. doi:10.1016/j.icte.2017.05.008

Jin, Y., Sen, S., Guérin, R., Hosanagar, K., & Zhang, Z.-L. (2008). Dynamics of competition between incumbent and emerging network technologies. *Proceedings of the 3rd international workshop on Economics of networked systems*, 49–54. doi:10.1145/1403027.1403039

Jorstad, I., Dustdar, S., & Do, T. (2005). An analysis of current mobile services and enabling technologies. *International Journal of Ad Hoc and Ubiquitous Computing*, *1*(1), 92–102. doi:10.1504/IJAHUC.2005.008023

Joseph, D., Shetty, N., Chuang, J., & Stoica, I. (2007). Modeling the adoption of new network architectures. *Proceedings of International Conference on Emerging Networking Experiments and Technologies (CoNEXT'07)*. doi:10.1145/1364654.1364661

Karjaluoto, H., Karvonen, J., Kesti, M., Koivumaki, T., Manninen, M., Pakola, J., & Salo, J. (2005). Factors affecting consumer choice of mobile phones: Two studies from Finland. *Journal of Euromarketing*, *14*(3), 59–82. doi:10.1300/J037v14n03_04

Katz, M. L., & Shapiro, C. (1986). Technology adoption in the presence of network externalities. *Journal of Political Economy*, *94*(4), 822–841. doi:10.1086/261409

Kilkki, K., Mäntylä, M., Karhu, K., Hämmäinen, H., & Ailisto, H. (2017). *A disruption framework*. Manuscript submitted for publication.

Kivi, A., Smura, T., & Töyli, J. (2012). Technology product evolution and the diffusion of new product features. *Technological Forecasting and Social Change*, *79*(1), 107–126. doi:10.1016/j.techfore.2011.06.001

Komu, M., Varjonen, S., Gurtov, A., & Tarkoma, S. (2012). Sockets and beyond: Assessing the source code of network applications. *Linux Symposium*, 7–24.

Lehmann, S., & Buxmann, P. (2009). Pricing strategies of software vendors. *Business & Information Systems Engineering*, *1*(6), 452–462. doi:10.1007/s12599-009-0075-y

Levä, T., Komu, M., Keränen, A., & Luukkainen, S. (2013). Adoption barriers of network layer protocols: The case of host identity protocol. *Computer Networks*, *57*(10), 2218–2232. doi:10.1016/j.comnet.2012.11.024

Levä, T., & Riikonen, A. (2013). Diffusion of Internet protocols – Measurement framework and empirical analysis of the Finnish mobile market. *Proc. of the 8th IEEE Conference on Standardisation and Innovation in Information Technology (SIIT 2013)*. doi:10.1109/SIIT.2013.6774578

Levä, T., & Suomi, H. (2013). Techno-economic feasibility analysis of Internet protocols: Framework and tools. *Computer Standards & Interfaces*, *36*(1), 76–88. doi:10.1016/j.csi.2013.07.011

Liker, J. K., Fleischer, M., & Arnsdorf, D. (1992). Fulfilling the promises of CAD. *Sloan Management Review*, *33*(3), 74–86.

Liu, C.-M. (2002). The effects of promotional activities on brand decision in the cellular telephone industry. *Journal of Product and Brand Management*, *11*(1), 42–51. doi:10.1108/10610420210419540

Lyytinen, K., & Damsgaard, J. (2001). What's wrong with the diffusion of innovation theory? The case of a complex and networked technology. *IFIP TC8 WG8.1 Fourth Working Conference on Diffusing Software Products and Process Innovations*, 173–190.

Maier, G., Feldmann, A., Paxson, V., & Allman, M. (2009). On dominant characteristics of residential broadband internet traffic. In *Proceedings of the 9th ACM SIGCOMM conference on Internet measurement conference (IMC '09)*. ACM. doi:10.1145/1644893.1644904

Marcus, J. (2004). Evolving core capabilities of the Internet. *Journal on Telecommunications & High Technology Law*, *3*, 121–161.

MINTC. (2004). *Mobile services market in Finland 2003, Ministry of Transport and Communications*. Retrieved from http://urn.fi/URN:ISBN:951-723-710-3

Nair, H., Chintagunta, P., & Dubé, J. (2004). Empirical analysis of indirect network effects in the market for personal digital assistants. *Quantitative Marketing and Economics*, *2*(1), 23–58. doi:10.1023/B:QMEC.0000017034.98302.44

Open Mobile Alliance. (2002). *WAP 2.0 - Technical White Paper.* Author.

Open Mobile Alliance. (2006). *User Agent Profile V2.0.* Author.

Ozment, A., & Schechter, S. E. (2006). Bootstrapping the adoption of Internet security protocols. *Proceedings of the Fifth Workshop on the Economics of Information Security.*

Prescott, M. M., & Van Slyke, C. (1997). Understanding the Internet as an innovation. *Industrial Management & Data Systems*, *97*(3), 119–124. doi:10.1108/02635579710173185

Riikonen, A., & Smura, T. (2013). *Mobile handset population in Finland 2005-2012*. Technical report. Retrieved from http://momie.comnet.aalto.fi/wp-uploads/Mobile_Handset_Population_2005-2012.pdf

Riikonen, A., Smura, T., Kivi, A., & Töyli, J. (2013). Diffusion of mobile handset features: Analysis of turning points and stages. *Telecommunications Policy*, *37*(6–7), 563–572. doi:10.1016/j.telpol.2012.07.011

Riquelme, H. (2001). Do consumers know what they want? *Journal of Consumer Marketing*, *18*(5), 437–448. doi:10.1108/07363760110398772

Rogers, E. M. (2003). *Diffusion of Innovations* (5th ed.). New York: Free Press.

ScientiaMobile. (2012). *WURFL repository v. 2.3.4*. Retrieved from http://wurfl.sourceforge.net

Shapiro, C., & Varian, H. R. (1998, November). Versioning: The smart way to sell information. *Harvard Business Review*, 106–114.

Shih, C., & Venkatesh, A. (2004). Beyond adoption: Development and application of a use-diffusion model. *Journal of Marketing*, *68*(1), 59–72. doi:10.1509/jmkg.68.1.59.24029

Shin, D. H., Kim, H., & Hwang, J. (2015). Standardization revisited: A critical literature review on standards and innovation. *Computer Standards & Interfaces*, *38*, 152–157. doi:10.1016/j.csi.2014.09.002

Smura, T., Kivi, A., & Töyli, J. (2009). A framework for analysing the usage of mobile services. *Info*, *11*(4), 53–67. doi:10.1108/14636690910970973

Smura, T., Kivi, A., & Töyli, J. (2011). Mobile data services in Finland: Usage of networks, devices, applications and content. *International Journal of Electronic Business*, *9*(1), 138–157. doi:10.1504/IJEB.2011.040359

Tallberg, M., Hämmäinen, H., Töyli, J., Kamppari, S., & Kivi, A. (2007). Impacts of handset bundling on mobile data usage: Case Finland. *Telecommunications Policy*, *31*(10), 648–659. doi:10.1016/j.telpol.2007.08.007

Thompson, K., Miller, G. J., & Wilder, R. (1997). Wide-area Internet traffic patterns and characteristics. *IEEE Network*, *11*(6), 10–23. doi:10.1109/65.642356

Tornatzky, L. G., Fleischer, M., & Chakrabarti, A. K. (1990). *The processes of technological innovation* (Vol. 273). Lexington, MA: Lexington Books.

Verkasalo, H. (2008). Dynamics of mobile service adoption. *International Journal of E-Business Research*, *4*(3), 40–63. doi:10.4018/jebr.2008070103

Vesselkov, A., & Hämmäinen, H. (2016). *Mobile handset population in Finland 2005-2015*. Technical report. Retrieved from http://metsahovi.aalto.fi/en/midcom-serveattachmentguid-1e5c68b3bf94696c68b11e5acd70192aed101490149/mobile_handset_population_2005-2015.pdf

Warma, H., Levä, T., Tripp, H., Ford, A., & Kostopoulos, A. (2011). Dynamics of communication protocol diffusion: The case of Multipath TCP. *NETNOMICS: Economic Research and Electronic Networking*, *12*(2), 133–159. doi:10.1007/s11066-011-9063-8

Weitzel, T., Beimborn, D., & Konig, W. (2006). A unified economic model of standard diffusion: The impact of standardization cost, network effects, and network topology. *Management Information Systems Quarterly*, *30*(1), 489–514.

Yoo, Y., Lyytinen, K., & Yang, H. (2005). The role of standards in innovation and diffusion of broadband mobile services: The case of South Korea. *The Journal of Strategic Information Systems*, *14*(3), 323–353. doi:10.1016/j.jsis.2005.07.007

Young, W. (2016, March 24). *The History of Winsock*. Retrieved from http://tangentsoft.net/wskfaq/articles/history.html

Zmud, R. (1984). An examination of push-pull theory applied to process innovation in knowledge work. *Management Science*, *30*(6), 727–738. doi:10.1287/mnsc.30.6.727

APPENDIX: COLLECTION PROCESS OF PROTOCOL DATA

The protocol data were collected from several sources. First, a script was created to fetch the data from the WURFL repository (ScientiaMobile, 2012) containing information about device capabilities and features for a variety of mobile devices. The capabilities included the URLs of UAProf files that describe the capabilities of a mobile handset model in an XML format standardized by the Open Mobile Alliance (OMA) (2006). Second, these UAProf files were downloaded and parsed programmatically to complement the WURFL data. Finally, the missing information was collected manually from the handset specifications available in the handset vendors' web pages and in online databases, such as GSMArena (www.gsmarena.com).

Internet protocols were rarely mentioned directly in the used data sources. As a consequence, the existence of protocols in the handset models was identified based on other capabilities (Table 1). For example, the inclusion of wireless application protocol (WAP) version 1.x into a device model indicates that the model is IPv4 and UDP capable, whereas version 2.x adds support for TCP, TLS and HTTP (Open Mobile Alliance, 2002). Similarly, support for OMA packet-switched streaming (PSS) standard translates into RTP and RTSP support (Gabin, Kampmann, Lohmar, & Priddle, 2010), certain Java APIs (MIDP, SIP API) and applications (email, VoIP, push to talk (PoC), video streaming) indicate support for other protocols, and some protocols are included as features in operating systems. Finally, application layer protocols typically run over certain lower layer protocols. Consequently, the availability of application layer protocols indicates the availability of lower layer protocols, whereas the lack of lower layer protocols translates into the lack of application layer protocols.

Table 1. Rules for identifying the existence of protocols in the handset models

Protocol	Features Indicating Protocol Support (and Non-Support)
EMAIL - *SMTP (* Simple Mail Transfer Protocol, RFC 5321*)* - *IMAP4 (* Internet Message Access Protocol v4, RFC 3501*)* - *POP3 (* Post Office Protocol v3, RFC 1939*)*	1) Email application **OR** 2) UAProf: <prf:Email-URI-Schemes> = POP or IMAP or SMTP NO SUPPORT: 1 ≠ true **AND** 2 ≠ true
HTTP (HyperText Transfer Protocol, RFC 2616)	1) HTML browser **OR** 2) WURFL: <j2me_midp_1_0> or <j2me_midp_2_0> = true **OR** 3) WURFL: <j2me_http> = true **OR** 4) WAP version = 2.x NO SUPPORT: 1 ≠ true **AND** 2 ≠ true **AND** 3 ≠ true **AND** 4 ≠ 2.x
TLS (Transport Layer Security, RFC 5246)	1) WURFL: <j2me_https> = true **OR** 2) UAProf: <prf:SecuritySupport> = SSL/TLS **OR** 3) WAP version = 2.x NO SUPPORT: (1 ≠ true **AND** 2 ≠ SSL/TLS **AND** 3 ≠ 2.x) **OR** *IPv4* = false
SIP (Session Initiation Protocol, RFC 3261)	1) JSR-180 SIP API for J2ME **OR** 2) PoC (Push-to-talk over Cellular) application **OR** 3) VoIP application, including also FaceTime **OR** 4) OS = Android 2.3 --> **OR** Symbian v9.1 / S60 3E --> **OR** Windows Mobile 6.0 --> NO SUPPORT: (2 ≠ true **AND** 3 ≠ true) **OR** IPv4 = false **OR** WAP version = 1.x **OR** OS = (Windows Phone **OR** Samsung Bada **OR** Blackberry)
RTSP (Real Time Streaming Protocol, RFC 2326)	1) WURFL: <streaming_real_media> = true **OR** 2) UAProf: <rdf:Description rdf:ID="Streaming"> = exists **OR** 3) OS = Nokia S40 3E FP2 --> **OR** Symbian v8.1 / S60 2E FP3 --> **OR** Blackberry 4.3 --> **OR** Android **OR** Samsung Bada 1.0 --> NO SUPPORT: RTP = false **OR** IPv4 = false **OR** WAP version = 1.x **OR** OS = (iOS **OR** Windows Phone)
RTP (Real-time Transport Protocol, RFC 3550)	1) *SIP* **OR** 2) *RTSP* **OR** 3) Video streaming support **OR** 4) WURFL: <streaming_video> = true **OR** 5) UAProf: <pss6:RtpProfiles> = RTP **OR** 6) UAProf: <rdf:Description rdf:ID="Streaming"> = exists NO SUPPORT: (1 ≠ true **AND** 3 ≠ true) **OR** IPv4 = false **OR** WAP version = 1.x
TCP (Transport Control Protocol, RFC 793)	1) *EMAIL* **OR** 2) HTML browser **OR** 3) WURFL: <j2me_socket> = true **OR** 4) WAP version = 2.x NO SUPPORT: 1 ≠ true **AND** 2 ≠ true **AND** 3 ≠ true **AND** 4 ≠ 2.x
UDP (User Datagram Protocol, RFC 768)	1) *RTP* **OR** 2) *SIP* **OR** 3) WURFL: <j2me_udp> = true **OR** 4) WAP version = 1.x or 2.x NO SUPPORT: IPv4 = false
IPv4 (Internet Protocol, version 4, RFC 791)	1) EMAIL **OR** 2) HTTP **OR** 3) GPRS **OR** 4) WAP version = 2.x NO SUPPORT: 3 = false **AND** 4 = 0 (i.e., no WAP)

[1] One reason for the emergence of the groups is that the identification of the protocols' existence on the mobile handsets is based on the dependencies between different protocols (e.g., EMAIL runs over TCP) and common features (e.g., WAP 2.x indicates existence of both HTTP and TCP).

Compilation of References

3GPP. (2002). *Technical Specification Group Services*. Retrieved from http://www.3gpp.org/ftp/tsg_sa/TSG_SA/TSGS_17/Docs/PDF/SP-020418.pdf

Abdullah, L. M., & Verner, J. M. (2012). Analysis and application of an outsourcing risk framework. *Journal of Systems and Software*, *85*(8), 1930–1952. doi:10.1016/j.jss.2012.02.040

Abernathy, W. J., & Utterback, J. M. (1978). Patterns of innovation in industry. *Technology Review*, *80*(7), 40–47.

ABIresearch. (2013). *Standard leadership within the 3GPP*. Retrieved July 29, 2017, from https://www.abiresearch.com/market-research/product/1016671-standards-leadership-within-the-3gpp/

acatech (Eds.). (2011). *Cyber-Physical Systems – Driving force for innovation in mobility, health, energy and production*. Retrieved from http://www.acatech.de/fileadmin/user_upload/Baumstruktur_nach_Website/Acatech/root/de/Publikationen/Stellungnahmen/acatech_POSITION_CPS_Englisch_WEB.pdf

Acemoglu, D., Gancia, G., & Zilibotti, F. (2012). Competing Engines of Growth: Innovation and Standardization. *Journal of Economic Theory*, *147*(2), 570–601.e3. doi:10.1016/j.jet.2010.09.001

Adams, R., Bessant, J., & Phelps, R. (2006). Innovation management measurement: A review. *International Journal of Management Reviews*, *8*(1), 21–47. doi:10.1111/j.1468-2370.2006.00119.x

Addy, R. (2007). *Effective IT Service Management: To ITIL and Beyond!* Berlin: Springer.

Adler, P. (2005). The evolving object of software development. *Organization*, *12*(3), 401–435. doi:10.1177/1350508405051277

Adler, P. S., McGarry, F. E., Irion-Talbot, W. B., & Binney, D. J. (2005a). Enabling process discipline: Lessons from the journey to CMM Level 5. *MIS Quarterly Executive*, *4*(1), 215–227.

Adner, R., & Kapoor, R. (2010). Value creation in innovation ecosystems: How the structure of technological interdependence affects firm performance in new technology generations. *Strategic Management Journal*, *31*(3), 306–333. doi:10.1002/smj.821

Agarwal, R., & Bayus, B. L. (2002). The market evolution and sales takeoff of product innovations. *Management Science*, *48*(8), 1024–1041. doi:10.1287/mnsc.48.8.1024.167

AIOTI (Ed.). (2015a). *IoT LSP Standard Framework Concepts – Release 2.0*. Retrieved from https://ec.europa.eu/digital-agenda/sites/digital-agenda/files/discussions/aioti_wg3_sdos_alliances_landscape_-_iot_lsp_standard_framework_concepts_-_release_2_0.pdf

Alfred, G., & Warner, J. F. F. (2006). Managing Uncertainty in a Formal Standards-Based Industry: A Real Options Perspective on Acquisition Timing. *Journal of Management*, *32*(2), 279–298. doi:10.1177/0149206305280108

Aliprandi, S. (2011). Interoperability And Open Standards: The Key To True Openness And Innovation. *International Free and Open Source Software Law Review*, *3*(1), 5–24. doi:10.5033/ifosslr.v3i1.53

Ali-Vehmas, T. (2016). Complex Network Perspective on Collaboration in the ICT Standardization. *International Journal of Standardization Research*, *14*(2), 33–64. doi:10.4018/IJSR.2016070103

Ali-Vehmas, T., & Casey, T. R. (2012). Evolution of wireless access provisioning: A systems thinking approach. *Competition and Regulation in Network Industries*, *13*(4), 333–361. doi:10.1177/178359171201300402

Al-Karaghouli, W., Al Azri, A., & Al Salti, Z. (n.d.). *E-Government and EIS Change Management and Critical Success Factors: An Omani Success Story*. Academic Press.

Allen, D. W. E. (2017). *Blockchain Innovation Commons*. Retrieved September 24th, 2017, from https://ssrn.com/abstract=2919170

Allison, I., & Merali, Y. (2007). Software process improvement as emergent change: A structurational analysis. *Information and Software Technology*, *49*(6), 668–681. doi:10.1016/j.infsof.2007.02.003

Alreemy, Z., Chang, V., Walters, R., & Wills, G. (2016). Critical success factors (CSFs) for information technology governance (ITG). *International Journal of Information Management*, *36*(6), 907–916. doi:10.1016/j.ijinfomgt.2016.05.017

Alvestrand, H. (2004). The role of the standards process in shaping the internet. *Proceedings of the IEEE*, *92*(9), 1371–1374. doi:10.1109/JPROC.2004.832973

Ambrosus. (2017). *Ambrosus Whitepaper*. Retrieved September 24th, 2017, from https://ambrosus.com/assets/Ambrosus-White-Paper-V8-1.pdf

Amid, A., Moalagh, M., & Ravasan, A. Z. (2012). Identification and classification of ERP critical failure factors in Iranian industries. *Information Systems*, *37*(3), 227–237. doi:10.1016/j.is.2011.10.010

Andersen, N. (2016). *Blockchain Technology - A game-changer in accounting?* Retrieved September 24th, 2017, from https://www2.deloitte.com/content/dam/Deloitte/de/Documents/Innovation/Blockchain_A%20game-changer%20in%20accounting.pdf

Anderson, C. (2006). *The long tail: Why the future of business is selling less of more*. New York, NY: Hyperion Books.

Anderson, P. C., & Tushman, M. L. (1990). Technological Discontinuities and Dominant Designs: A Cyclical Model of Technological Change. *Administrative Science Quarterly*, *35*(4), 604–633. doi:10.2307/2393511

ANSI. (n.d.a). *About Ansi*. Retrieved from https://www.ansi.org/about_ansi/overview/overview.aspx?menuid=1

ANSI. (n.d.b). *ANSI Standards*. Retrieved from https://share.ansi.org/Shared%20Documents/Standards%20Activities/American%20National%20Standards/ANSI%20Accredited%20Standards%20Developers/AUG2016ASD.pdf

AOL. (1998). *America Online, Inc. to Acquire Netscape Communications Corporation in Stock Transaction Valued at $4.2 Billion*. Retrieved September 24th, 2017, from http://www.timewarner.com/newsroom/pressreleases/1998/11/America_Online_Inc_to_Acquire_Netscape_Communications_11-24-1998.php

Apple Inc. and NeXT Software Inc. v. Motorola Inc. and Motorola Mobility, Inc., No. 11-cv-08540 (N.D. Ill. 2012) (2012).

Apple Inc. v. Motorola Inc., No. 11-cv-0017 (W.D.Wisc. 2012) (2012).

Archer, K. J., & Lemeshow, S. (2006). Goodness-of-Fit Test for a Logistic Regression Model Fitted Using Survey Sample Data. *The Stata Journal*, *6*(1), 97–105.

Arthur, W. B. (1989). Competing Technologies, Increasing Returns, and Lock-in by Historical Events. *Economic Journal (London), 99*(394), 116–131. doi:10.2307/2234208

ATIS. (n.d.). *About our Work.* Retrieved from http://www.atis.org/01_about/our_work.asp

Autio, E. (2005). Creative tension: The significance of Ben Oviatt's and Patricia McDougall's article 'toward a theory of international new ventures'. *Journal of International Business Studies, 36*(1), 9–19. doi:10.1057/palgrave.jibs.8400117

Autio, E., Sapienza, H. J., & Almeida, J. G. (2000). Effects of age at entry, knowledge intensity, and imitability on international growth. *Academy of Management Journal, 43*(5), 909–924. doi:10.2307/1556419

Axios-Systems. (2008). *ITIL Adoption Surges Despite Confusion.* Author.

Baird, S. (2007). The Government at the Standards Bazaar. *Stanford Law & Policy Review, 18*(1), 35–100.

Bakos, Y., & Brynjolfsson, E. (1999). Bundling information goods: Pricing, profits, and efficiency. *Management Science, 45*(12), 1613–1630. doi:10.1287/mnsc.45.12.1613

Baldini, G., Holland, O., Stavroulaki, V., Tsagkaris, K., Demestichas, P., Polydoros, A., & Allen, D. et al. (2013). The evolution of cognitive radio technology in Europe: Regulatory and standardization aspects. *Telecommunications Policy, 37*(2-3), 96–107. doi:10.1016/j.telpol.2012.07.003

Ballard, M. (2012). Portugal's prescribed open standards - full list. *Computer Weekly.* Retrieved from http://www.computerweekly.com/blogs/public-sector/2012/11/-tablemytable-border-collapsec.html

Bansler, J., & Havn, E. (2004). Technology-use mediation. Making sense of electronic communication in an organizational context. *Scandinavian Journal of Information Systems, 16*, 57–84.

Barabási, A.-L., & Bonabeau, E. (2003). Scale-Free Networks. *Scientific American*, (May): 50–59. PMID:12701331

Barley, S. (1986). Technology as an occasion for structuring: Evidence from observations of ct scanners and the social order of radiology departments. *Administrative Science Quarterly, 31*(1), 78–108. doi:10.2307/2392767 PMID:10281188

Barley, S., & Tolbert, P. (1997). Institutionalization and structuration: Studying the links between action and institution. *Organization Studies, 18*(1), 93–117. doi:10.1177/017084069701800106

Baron, J., Meniere, Y., & Pohlmann, T. (2014). Standards, consortia, and innovation. *International Journal of Industrial Organization, 36*, 22–35. doi:10.1016/j.ijindorg.2014.05.004

Baron, J., & Pohlmann, T. (2013). Who Cooperates in Standards Consortia—Rivals Or Complementors? *Journal of Competition Law & Economics, 9*(4), 905–929. doi:10.1093/joclec/nht034

Barry, A. (2000). Coordinating Technology: Studies in the International Standardization of Telecommunications [review]. *Technology and Culture, 41*(1), 153–154. doi:10.1353/tech.2000.0003

Bar, T., & Leiponen, A. E. (2014). Committee composition and networking in standard setting: The case of wireless telecommunications. *Journal of Economics & Management Strategy, 23*(1), 1–23. doi:10.1111/jems.12044

Bauer, J. M., Ha, I. S., & Saugstrup, D. (2007). Mobile Television: Challenges of Advanced Service Design. *Communications of the Association for Information Systems, 20*, 621–631.

BCG. (2015). *The Mobile Revolution – How Mobile Technologies Drive a Trillion Dollar Impact.* The Boston Consultancy Group.

Bekkers, R., Bongard, R., & Nuvolari, A. (2011). An empirical study on the determinants of essential patent claims in compatibility standards. *Research Policy, 40*(7), 1001–1015. doi:10.1016/j.respol.2011.05.004

Bekkers, R., & Updegrove, A. (2012). *A study of IPR policies and practices of a representative group of Standards Setting Organizations worldwide, US National Academies of Science, Board of Science, Technology, and Economic Policy*. STEP.

Bekkers, R., & Updegrove, A. (2013). *A study of IPR policies and practices of a representative group of Standards Setting Organizations worldwide (updated version)*. Washington, DC: National Academies of Science. Retrieved from http://www.nap.edu/catalog.php?record_id=18510

Bekkers, R., Verspagen, B., & Smits, J. (2002). Intellectual property rights and standardization: The case of GSM. *Telecommunications Policy*, *26*(3-4), 171–188. doi:10.1016/S0308-5961(02)00007-1

Bekkers, R., & West, J. (2009). The limits to IPR standardization policies as evidenced by strategic patenting in UMTS. *Telecommunications Policy*, *33*(1-2), 80–97. doi:10.1016/j.telpol.2008.11.003

Berman, B. (2013). *Public IP companies (PIPCOs) – a business model whose time has come*. Intellectual Asset Management Magazine, March/April.

Besen, S. M., & Farrell, J. (1994). Choosing How to Compete: Strategies and Tactics in Standardization. *Journal of Economic Perspectives*, *8*(2), 117–131.

Besen, S. M., & Farrell, J. (1994). Choosing how to compete: Strategies and tactics in standardization. *The Journal of Economic Perspectives*, *8*(2), 117–131. doi:10.1257/jep.8.2.117

Bessen, J., & Meurer, M. (2009). *Patent Failure: How Judges, Bureaucrats, and Lawyers Put Innovators at Risk*. Princeton University Press. doi:10.1515/9781400828692

Bird, G. B. (1998). The Business Benefit of Standards. *StandardView*, *5*(2), 76–80. doi:10.1145/301688.301691

Blind, K. (2004). The economics of standards. Theory, evidence, policy. Cheltenham, UK: Academic Press.

Blind, K. (2013). *The impact of standardization and standards on innovation*. Retrieved from http://www.innovation-policy.org.uk/share/14_The%20Impact%20of%20Standardization%20and%20Standards%20on%20Innovation.pdf

Blind, K., & Grupp, H. (2000). *Standards Statistics as New Indicators for the Diffusion of Technology*. Paper presented to the International JA Schumpeter Conference, Manchester, UK.

Blind, K. (2004). *The Economics Of Standards; Theory, Evidence, Policy*. Cheltenham, UK: Edward Elgar Publishing.

Blind, K. (2008). The Influence of Companies' Patenting Motives on their Standardisation Strategies. EURAS Proceedings 2008 Aachener Beiträge zur Informatik, 40.

Blind, K. (2011). An economic analysis of standards competition: The example of the ISO ODF and OOXML standards. *Telecommunications Policy*, *35*(4), 373–381. doi:10.1016/j.telpol.2011.02.007

Blind, K., Bekkers, R., Dietrich, Y., Iversen, E., Köhler, F., Müller, B., & Verweijen, J. et al. (2011). *Study on the Interplay between standards and Intellectual Property Rights (IPRs) Final Report. Publications Office of the European Union*. European Union.

Blind, K., de Vries, H. J., & Mangelsdorf, A. (2012). External Knowledge Sourcing and Involvement in Standardization - Evidence from the Community Innovation Survey. *IEEE International Technology Management Conference*, 1–9. doi:10.1109/ITMC.2012.6306399

Blind, K., & Gauch, S. (2008). Trends in ICT standards: The relationship between European standardisation bodies and standards consortia. *Telecommunications Policy*, *32*(7), 503–513. doi:10.1016/j.telpol.2008.05.004

Blind, K., & Jungmittag, A. (2008, February). The impact of patents and standards on macroeconomic growth: A panel approach covering four countries and 12 sectors. *Journal of Productivity Analysis*, *29*(1), 51–60. doi:10.1007/s11123-007-0060-8

Blind, K., & Mangelsdorf, A. (2013). Alliance Formation of SMEs: Empirical Evidence From Standardization Committees. *IEEE Transactions on Engineering Management*, *60*(1), 148–156. doi:10.1109/TEM.2012.2192935

Blind, K., & Mangelsdorf, A. (2016). Motives to standardize: Empirical evidence from Germany. *Technovation*, *48-49*(1), 13–24. doi:10.1016/j.technovation.2016.01.001

Blind, K., & Thumm, N. (2004). Interrelation between patenting and standardisation strategies: Empirical evidence and policy implications. *Research Policy*, *33*(10), 1583–1598. doi:10.1016/j.respol.2004.08.007

Bluetooth. (n.d.). *Our History*. Retrieved from https://www.bluetooth.com/media/our-history

Boiral, O. (2003). ISO 9000: Outside the iron cage. *Organization Science*, *14*(6), 720–737. doi:10.1287/orsc.14.6.720.24873

Boldrin, M., & Levine, D. (2013, Winter). The Case Against Patents. *The Journal of Economic Perspectives*, *27*(1), 3–22. doi:10.1257/jep.27.1.3

Bonaventure, O. (2013, September 18). *Apple seems to also believe in Multipath TCP*. Retrieved from http://perso.uclouvain.be/olivier.bonaventure/blog/html/2013/09/18/mptcp.html

Bonino, M., & Spring, M. (1991). Standards as Change Agents in the Information Technology Market. *Computer Standards & Interfaces*, *12*(2), 97–107. doi:10.1016/0920-5489(91)90059-9

Borenstein, N., & Blake, J. (2011). Cloud computing standards: Where's the beef? *IEEE Internet Computing*, *15*(3), 74–78. doi:10.1109/MIC.2011.58

Bouchoux, D. (2009). *Intelectual Property for Paralegals – The Law of Trademarks, Copyrights, Patents and Industry secrets*. Cengage Learning.

Boudreau, K. (2007). *Too Many Complementors?* Evidence on Software Developers.

Bousquet, F., Fomin, V. V., & Drillon, D. (2009). Standardization: a Major Tool for Competitive Intelligence. In *14th EURAS Annual Standardisation Conference "Standardisation and Corporate Intelligence"* (pp. 27–36). Cergy, France: Wissenschaftsverlag Mainz in Aachen.

Braa, J., & Hedberg, C. (2002). The Struggle for District-Based Health Information Systems in South Africa. *The Information Society*, *18*(2), 113–127. doi:10.1080/01972240290075048

Brandenburger, A., & Nalebuff, B. J. (1997). *Co-opetition*. New York: Currency Doubleday.

Brandes, F. (2009). The UK technology foresight programme: An assessment of expert estimates. *Technological Forecasting and Social Change*, *76*(7), 869–879. doi:10.1016/j.techfore.2009.01.005

Brismark, G., & Alfalahi, K. (2008). Patent strategies – a fork in the road toward 4G. *Ericsson Business Review*, (3), 62-65.

Brouwer & Kleinknecht. (1999). Innovative output, and a firm's propensity to patent: An exploration of CIS micro data. *Research Policy*, *28*(6), 615-624.

Brown, G. I. (1998). *The Big Bang; History of Explosives*. Sutton Publishing Limited.

Brunsson, N. (2000a). Organizations, Markets, and Standardization. In N. Brunsson & B. Jacobsson (Eds.), *A World of Standards* (pp. 21–40). Oxford, UK: Oxford University Press.

Brunsson, N. (2000b). Standardization and uniformity. In N. Brunsson & B. Jacobsson (Eds.), *A World of Standards* (pp. 138–151). Oxford, UK: Oxford University Press.

Brunsson, N., & Jacobsson, B. (2000c). *A World of Standards*. Oxford, UK: Oxford University Press.

Brunsson, N., & Jacobsson, B. (2000c). The contemporary expansion of standardization. In N. Brunsson & B. Jacobsson (Eds.), *A World of Standards* (pp. 1–19). Oxford, UK: Oxford University Press.

Brunsson, N., & Jacobsson, B. (2000d). Following standards. In N. Brunsson & B. Jacobsson (Eds.), *A World of Standards* (pp. 127–138). Oxford, UK: Oxford University Press.

Brunsson, N., Rasche, A., & Seidl, D. (2012). The Dynamics of Standardization: Three Perspectives on Standards in Organization Studies. *Organization Studies*, *33*(5-6), 613–632. doi:10.1177/0170840612450120

Brunswicker, S.; Rodriguez, J. A. & Wareham, J. D. (2015). Governing Standardization through Formal and Informal Standard Development Organizations. *Academy of Management Proceedings 2015.*

Brusoni, S., & Corrocher, N. (2006). *The development and commercial success of SMS: standard committees as engine of knowledge-based entrepreneurship*. Deliverable 13, Work Package 2A of project Knowledge-based Entrepreneurship: Innovation, Networks and Systems. European Commission Sixth Framework Programme (2002-2006).

Brykman, L. (2011). Better Standards to Enhance European Competitiveness. In *Proceedings of the 7th biennial Standardisation and Innovation in Information Technology conference (SIIT 2011)* (pp. 193–194). Berlin: Mainz Publishers.

Brzezinski, K. M. (2007). Network Operators' Requirements and the Structure of Telecommunications Standards. *International Journal of IT Standards and Standardization Research*, *5*(1), 103–117. doi:10.4018/jitsr.2007010102

BS15000-Associates-Group. (2012). *The ISO 20000 (BS15000 / BS 15000)*. ITSM Standard.

Bundesgerichtshof. (2009, May 6). *Orange-Book-Standard*. KZR 39/06.

Buterin, V. (2014). *A next-generation smart contract and decentralized application platform.* Retrieved September 24, 2017, from http://www.the-blockchain.com/docs/Ethereum_white_paper-a_next_generation_smart_contract_and_decentralized_application_platform-vitalik-buterin.pdf

Buttle, F. (1997). ISO 9000: Marketing motivations and benefits. *International Journal of Quality & Reliability Management*, *14*(9), 936–947. doi:10.1108/02656719710186867

BylawsIEEE (2015).

Cabral, L., & Salant, D. (2014). Evolving technologies and standards regulation. *International Journal of Industrial Organization*, *36*, 48–56. doi:10.1016/j.ijindorg.2013.07.006

Campbell-Kelly, M., & Garcia-Swartz, D. D. (2013). The history of the internet: The missing narratives. *Journal of Information Technology*, *28*(1), 18–33. doi:10.1057/jit.2013.4

Campbell-Kelly, M., Garcia-Swartz, D., Lam, R., & Yang, Y. (2015). Economic and business perspectives on smartphones as multi-sided platforms. *Telecommunications Policy*, *39*(8), 717–734. doi:10.1016/j.telpol.2014.11.001

Caragliu, A., Del Bo, C., & Nijkamp, P. (2011). Smart cities in Europe. *Journal of Urban Technology*, *18*(2), 65–82. do i:10.1080/10630732.2011.601117

Carey, J., & Moss, M. L. (1985). The diffusion of telecommunication technologies. *Telecommunications Policy*, *6*(2), 145–158. doi:10.1016/0308-5961(85)90038-2

Carlile, P. R. (2002). A Pragmatic View of Knowledge and Boundaries: Boundary Objects in New Product Development. *Organization Science*, *13*(4), 442–455. doi:10.1287/orsc.13.4.442.2953

Carlsson, B., Acs, Z. J., Audretsch, D. B., & Braunerhjelm, P. (2009). Knowledge creation, entrepreneurship, and economic growth: A historical review. *Industrial and Corporate Change*, *18*(6), 1193–1229. doi:10.1093/icc/dtp043

Carter, F. J. Jr, Jambulingam, T., Gupta, V. K., & Melone, N. (2001). Technological innovations: A framework for communicating diffusion effects. *Information & Management*, *38*(5), 277–287. doi:10.1016/S0378-7206(00)00065-3

Catalini, C., & Gans, J. S. (2016). *Some Simple Economics of the Blockchain*. MIT Sloan School of Management. doi:10.3386/w22952

Cater-Steel, A., & Tan, W.-G. (2005). *Implementation of IT Infrastructure Library (ITIL) in Australia: Progress and success factors*. Paper presented at the IT Governance International Conference, Auckland, New Zealand.

Cater-Steel, A., M, T., & Tan, W.-G. (2006). *Transforming IT Service Management – the ITIL Impact*. Paper presented at the 17th Australasian conference on information systems, Adelaide, Australia.

Cater-Steel, A., Tan, W.-G., & Toleman, M. (2006). *Challenge of adopting multiple process improvement frameworks*. Paper presented at the 14th European Conference on Information Systems, Goteborg, Sweden.

Cater-Steel, A., & Toleman, M. (2009). Education for IT service management standards. *International Journal of IT Standards and Standardization Research*, *5*(2), 27–41. doi:10.4018/jitsr.2007070103

Cater-Steel, A., Toleman, M., & Tan, W.-G. (2006). Transforming IT service management-the ITIL impact. *Proceedings of the 17th Australasian Conference on Information Systems (ACIS 2006).*

Cavusgil, S. T., & Knight, G. (2009). *Born Global Firms: A New International Enterprise*. doi:10.4128/9781606490136

CCSA. (n.d.). *About*. Retrieved from http://www.ccsa.org.cn/english/about.php

CEC. (1996). *Communication from the Commission to the Council and the Parliament on "Standardization and the Global Information Society: The European Approach"*. COM(96) 359 final. Retrieved from http://eur-lex.europa.eu/LexUriServ/LexUriServ.do?uri=COM:1996:0359:FIN:EN:PDF

CEC. (2012). *Unleashing the Potential of Cloud Computing in Europe*. Communication from European Commission COM(2012) 529 final. Retrieved from http://eur-lex.europa.eu/LexUriServ/LexUriServ.do?uri=COM:2012:0529:FIN:EN:PDF

Cecere, G., Corrocher, N., & Battaglia, R. D. (2015). Innovation and competition in the smartphone industry: Is there a dominant design? *Telecommunications Policy*, *39*(3), 162–175. doi:10.1016/j.telpol.2014.07.002

CEN-CENELEC. (2012). *CEN-CENELEC Position Paper on Horizon 2020*. Retrieved September 27, 2017, from http://www.cencenelec.eu/News/Policy_Opinions/PolicyOpinions/PPhorizon2020.pdf

CEPT. (n.d.). Retrieved from http://www.cept.org/

CERN. (n.d.). *About CERN*. Retrieved from http://home.cern/about

Charles, I. (2016). *Ericsson's offer of SEP terms to TCL held to adhere to FRAND standards by US courts*. Retrieved July 29, 2017, from https://spicyip.com/2016/08/ericssons-offer-of-sep-terms-to-tcl-held-to-adhere-to-frand-standards-by-u-s-court.html

Chiao, B., Lerner, J., & Tirole, J. (2007). The Rules of Standard-Setting Organizations: An Empirical Analysis. *The Rand Journal of Economics*, *38*(4), 905–930. doi:10.1111/j.0741-6261.2007.00118.x

Chin, W. W., Gopal, A., & Salisbury, W. D. (1997). Advancing the theory of adaptive structuration: The development of a scale to measure faithfulness of appropriation. *Information Systems Research*, *8*(4), 342–367. doi:10.1287/isre.8.4.342

Chitika. (2013, September 19). *UPDATE: iOS 7 on 52% of iOS devices one week after public release*. Retrieved from http://www.chitika.com/insights/2013/ios-7-adoption

Choi, H., Kim, S.-H., & Lee, J. (2010). Role of network structure and network effects in diffusion of innovations. *Industrial Marketing Management*, *39*(1), 170–177. doi:10.1016/j.indmarman.2008.08.006

Chou, D. C., & Chou, A. Y. (2009). Information systems outsourcing life cycle and risks analysis. *Computer Standards & Interfaces*, *31*(5), 1036–1043. doi:10.1016/j.csi.2008.09.032

Chowdry, A. (2014). Apple And Samsung Drop Patent Disputes Against Each Other Outside Of the U.S. *Forbes*. Retrieved July 29, 2017, from https://www.forbes.com/sites/amitchowdhry/2014/08/06/apple-and-samsung-drop-patent-disputes-against-each-other-outside-of-the-u-s/#6dd91db5418c

Christensen, C. M. (1997). *The innovator's dilemma: When new technologies cause great firms to fail*. Harvard Business Review Press.

Christensen, C. M., Suarez, F. F., & Utterback, J. M. (1998). Strategies for Survival in Fast-Changing Industries. *Management Science*, *44*(12), 207–220. doi:10.1287/mnsc.44.12.S207

Chronopoulos, A. (2009). Patenting Standards - A case for US Antitrust Law or a Call for Recognizing Immanent Public Policy Limitations to the Exploitation Rights Conferred by the Patent Act? *International Review of Intellectual Property and Competition Law*, *40*(7), 782–816.

Clark, J. (2016). *Convergence, Collaboration and Smart Shopping in Open Standards and Open Source*. In Open Source and Standards in 5G, ITU / NGMN Joint Workshop on OS, San Diego, CA. Retrieved from http://www.itu.int/en/ITU-T/Workshops-and-Seminars/itu-ngmn/Documents/Abstracts_and_Presentations/Jamie_Clark_v2.pdf

Clark, D. D., Wroclawski, J., Sollins, K., & Braden, R. (2005). Tussle in cyberspace: Defining tomorrow's Internet. *IEEE/ACM Transactions on Networking*, *13*(3), 462–475. doi:10.1109/TNET.2005.850224

Clark, K. B. (1985). The interaction of design hierarchies and market concepts in technological evolution. *Research Policy*, *14*(5), 235–251. doi:10.1016/0048-7333(85)90007-1

Clear, T., & MacDonell, S. G. (2011). Understanding technology use in global virtual teams: Research methodologies and methods. *Information and Software Technology*, *53*(9), 994–1011. doi:10.1016/j.infsof.2011.01.011

Cochrane, R. (1974). *Measures for Progress: A History of the National Bureau of Standards*. Washington, DC: U.S. Government Printing Office.

Coindesk. (2017). *State of blockchain Q2 2017*. Retrieved September 24th, 2017, from https://media.coindesk.com/uploads/2017/09/state_of_blockchain_q2_2017.pdf

Colitti, L., Gunderson, S. H., Kline, E., & Refice, T. (2010). Evaluating IPv6 adoption in the Internet. In Proc. Passive and Active Measurement (pp. 141–150). Springer Berlin Heidelberg. doi:10.1007/978-3-642-12334-4_15

Constant, E. W. II. (1980). *The Origins of the Turbojet Revolution*. Baltimore, MD: The Johns Hopkins University Press.

Contreras, J. L. (2015). *CSIRO v. Cisco: The Convergence of RAND and Non-RAND Royalties for Standards-Essential Patents*. Available at: https://papers.ssrn.com/sol3/papers.cfm?abstract_id=2749653

Contreras, J. L. (2017). *Technical Standards, Standards-Setting Organizations and Intellectual Property: A Survey of the Literature (with an Emphasis on Empirical Approaches).* Utah Law Faculty Scholarship 11. Retrieved from http://dc.law.utah.edu/scholarship/11

Contreras, J. L. (2013). Fixing FRAND: A Pseudo-Pool Approach to Standards-Based Patent Licensing, *79. Antitrust Law Journal*, 47.

Contreras, J. L. (2015). A Brief History of FRAND: Analyzing Current Debates in Standard Setting and Antitrust Through a Historical Lens. *Antitrust Law Journal*, *80*(1), 39–120.

Contreras, J. L., & Gilbert, R. J. (2015). (Forthcoming). A Unified Framework for RAND and Other Reasonable Royalties. *Berkeley Technology Law Journal.*

Cooper, R. B., & Zmud, R. W. (1990). Information technology implementation research: A technological diffusion approach. *Management Science*, *36*(2), 123–139. doi:10.1287/mnsc.36.2.123

Court of Justice of the European Union (2015, July 16). *Huawei v. ZTE Corp. and ZTE Deutschland GmbH.* C-170/13.

Crespo, J., Suire, R., & Vicente, J. (2014). Lock-in or lock-out? How structural properties of knowledge networks affect regional resilience. *Journal of Economic Geography*, *14*(1), 199–219. doi:10.1093/jeg/lbt006

Crespo, J., Suire, R., & Vicente, J. (2016). Network structural properties for cluster long-run dynamics: Evidence from collaborative R&D networks in the European mobile phone industry. *Industrial and Corporate Change*, *25*(2), 261–282. doi:10.1093/icc/dtv032

Critical Values for the Two-sample Kolmogorov-Smirnov Test. (n.d.). Retrieved from https://www.webdepot.umontreal.ca/Usagers/angers/MonDepotPublic/STT3500H10/Critical_KS.pdf

Crocker, D. (1993). Making standards the IETF way. *StandardView*, *1*(1), 48–56. doi:10.1145/174683.174689

Crosby, M., Nachiappan, Pattanayak, P., Verma, S., & Kalyanaraman, V. (2015). *Blockchain Technology: Beyond Bitcoin.* Sutardja Center for Entrepreneurship & Technology Technical Report.

CSC. (2013). *Cloud Standards Coordination.* Final Report. Retrieved from http://ec.europa.eu/information_society/newsroom/cf/dae/document.cfm?doc_id=3988

CSIRO v. Cisco (Fed. Circ. 2015) (2015).

Curran, P. D. (2003, Summer). Standard-Setting Organizations: Patents, Price Fixing and Per Se Legality. *The University of Chicago Law Review. University of Chicago. Law School*, *70*(3), 983–1009. doi:10.2307/1600663

Cusumano, M. A , & Gawer, A. (2002). The Elements of Platform Leadership. *MIT Sloan Management Review*, *43*(3), 51–58.

Dardailler, D., Archer, P., & Wenning, R. (2012). *W3C Response to UK Cabinet Office Open Standards Consultation.* Retrieved from http://www.w3.org/2012/04/openstandards.html

DARPA. (n.d.). *History and Timeline.* Retrieved from http://www.darpa.mil/about-us/darpa-history-and-timeline

David, P. A. (1985). Clio and the Economy of QWERTY. *The American Economic Review*, *75*(2), 332–337.

David, P. A., & Greenstein, S. (1990). The Economics Of Compatibility Standards: An Introduction To Recent Research 1. *Economics of Innovation and New Technology*, *1*(1-2), 3–41. doi:10.1080/10438599000000002

Davidson, S., De Filippi, P., & Potts, J. (2016). *Economics of blockchain.* Retrieved September 24th, 2017, from http://ssrn.com/abstract=2744751

Davidson, E., & Chiasson, M. (2005). Contextual influences on technology use mediation: A comparative analysis of electronic medical record systems. *European Journal of Information Systems, 14*(1), 6–18. doi:10.1057/palgrave.ejis.3000518

Davidson, E., & Chismar, W. (2007). The interaction of institutionally triggered and technology-triggered social structure change: An investigation of computerized physician order entry. *Management Information Systems Quarterly, 31*(4), 739–758.

Davies, A., & Hobday, M. (2005). *The Business of Projects: Managing Innovation in Complex Products and Systems.* Cambridge University Press. doi:10.1017/CBO9780511493294

Davis, J., Edgar, T., Porter, J., Bernaden, J., & Sarli, M. (2012). Smart manufacturing, manufacturing intelligence and demand-dynamic performance. *Computers & Chemical Engineering, 47*, 145–156. doi:10.1016/j.compchemeng.2012.06.037

de Sousa Pereira, R. F., & da Silva, M. M. (2010). *A maturity model for implementing itil v3.* Paper presented at the Services (SERVICES-1), 2010 6th World Congress on.

De Vries, H. J. (2006). The paradox of standardization and innovation. *ISO Focus, 3*(2).

de Vries, H. J. (1997). Standardization - What's in a name? *Terminology, 4*(1), 55–83. doi:10.1075/term.4.1.05vri

de Vries, H. J. (2006). IT Standards Typology. In K. Jakobs (Ed.), *Advanced Topics in Information Technology Standards and Standardization Research* (Vol. 1, pp. 1–26). Hershey, PA: Idea Group Publishing. doi:10.4018/978-1-59140-938-0.ch001

De Vries, H. J., de Ruijter, J. P. M., & Argam, N. (2011). Dominant design or multiple designs: The flash memory card case. *Technology Analysis and Strategic Management, 23*(3), 249–262. doi:10.1080/09537325.2011.550393

De Vries, H., Verheul, H., & Willemse, H. (2003). Stakeholder identification in IT standardization processes. *Proceedings of the Workshop on Standard Making: A Critical Research Frontier for Information Systems,* 12-14.

Decker, S., & King, I. (2015). Qualcomm Says It Won't Follow New Wi-Fi Rules on Patents. *Bloomberg.* Retrieved July 29, 2017, from https://www.bloomberg.com/news/articles/2015-02-11/qualcomm-says-new-wi-fi-standard-rules-unfair-may-not-take-part

Delaere, S., & Ballon, P. (2008). Multi-Level Standardization and Business Models for Cognitive Radio: The Case of the Cognitive Pilot Channel. In *3rd IEEE Symposium on New Frontiers in Dynamic Spectrum Access Networks (DySPAN'08).* Chicago, IL: IEEE. doi:10.1109/DYSPAN.2008.67

Delcamp, H. R., & Leiponen, A. E. (2014). Innovating Standards Through Informal Consortia: The Case of Wireless Telecommunications. *International Journal of Industrial Organization, 36*, 36–47. doi:10.1016/j.ijindorg.2013.07.004

Deloitte. (2016). *Connecting Global FinTech: Hub Review 2016.* Retrieved September 24, 2017, from https://www2.deloitte.com/content/dam/Deloitte/uk/Documents/Innovation/deloitte-uk-connecting-global-fintech-hub-federation-innotribe-innovate-finance.pdf

Denicolo, V., Geradin, D., Layne-Farrar, A., & Padilla, J. (2008). Revisiting Injunctive Relief – Interpreting eBay in High-Tech Industries with Non-Practicing Patent Holders. *Journal of Competition Law & Economics, 4*(3), 571–608. doi:10.1093/joclec/nhn028

DeSanctis, G., & Poole, M. S. (1994). Capturing the complexity in advanced technology use: Adaptive structuration theory. *Organization Science, 5*(2), 121–147. doi:10.1287/orsc.5.2.121

Di Minin, A., & Bianchi, M. (2011). Safe nests in global nets: Internationalization and appropriability of R&D in wireless telecom. *Journal of International Business Studies, 42*(7), 910–934. doi:10.1057/jibs.2011.16

Dinh, T. T. A., Liu. R., Zhang, M., Chen, G., Ooi, B. C., & Wang, J. (2017). *Untangling Blockchain: A Data Processing View of Blockchain Systems.* arXiv preprint arXiv:1708.05665

Directive 2004/48 of the European Parliament and the Council of 29 April 2004 on the Enforcement of Intellectual Property Rights.

Djelic, M. L., & den Hond, F. (2014). Introduction: Multiplicity and plurality in the world of standards. *Business and Politics, 16*(1), 67–77. doi:10.1515/bap-2013-0034

Dorsey, E., & McGuire, M. R. (2013). How the Google Consent Order Alters the Process and Outcomes of FRAND Bargaining. *George Mason Law Review, 20*, 979.

DTI. (2005). *The Empirical Economics Of Standards.* DTI Economics Paper 12. London: Department of Trade and Industry. Retrieved from www.bis.gov.uk/files/file9655.pdf

Dumez, H. (2013). *Méthodologie de la recherche qualitative - Les 10 questions clés de la démarche compréhensive.* Paris: VUIBERT.

eBay v. MerckExchange, L.L.C., 547 U.S.388 (2006).

EC. (2007). *Antitrust: Commission confirms sending a Statement of Objections to Rambus.* MEMO/07/330, European Commission.

EC. (2009). *Antitrust: Commission accepts commitments from Rambus lowering memory chip royalty rates.* IP/09/1897, European Commission.

EC. (2012). *Implementing FRAND standards in Open Source: Business as usual or mission impossible?* Report summary from a workshop organised by the European Commission (EC) and the European Patent Office.

EC. (2013a). *European Commission: Press Release, Digital Agenda: Open standards would save public sector €1 billion a year.* Retrieved from http://europa.eu/rapid/press-release_IP-13-602_en.htm

EC. (2013b). *Commission staff working document: Guide for the procurement of standards-based ICT — Elements of Good Practice.* SWD(2013) 224 final, European Commission. Retrieved from http://ec.europa.eu/information_society/newsroom/cf/dae/document.cfm?doc_id=2326

EC. (2016). *ICT Standardisation Priorities for the Digital Single Market.* Communication from the Commission to the European Parliament, the Council, the European Economic and Social Committee and the Committee of the regions, COM(2016) 176 final, European Commission.

EC. (Ed.). (2006). *European Technology Platform SmartGrids. Vision and Strategy for Europe's Electricity Networks of the Future.* Retrieved from https://ec.europa.eu/research/energy/pdf/smartgrids_en.pdf

Economides, N. (1996). Network Externalities, Complementarities and Invitations to Enter. *European Journal of Political Economy, 12*(2), 211–233. doi:10.1016/0176-2680(95)00014-3

ECSIP. (2014). Patents and Standards: A modern framework for IPR-based standardization. Final Report: A study prepared for the European Commission Directorate-General for Enterprise and Industry, European Competitiveness and Sustainable Industrial Policy Consortium, Ref. Ares(2014)917720 – 25/03/2014.

Eggert, L. (2017). *Deployment trends of IPv6, SIP, XMPP, DNSSEC, DKIM, SPF, and DMARC.* Retrieved from http://eggert.org/

Egyedi, T. & Enserink, B. (2013). Measuring the Benefits of Open Standards: A Contribution to Dutch Politics. *Standards Today, 12*(1).

Egyedi, T. (2007). Standard-compliant, but incompatible?! *Computer Standards & Interfaces*, *29*(6), 605–613. doi:10.1016/j.csi.2007.04.001

Egyedi, T. M. (2008). An implementation perspective on sources of incompatibility and standards' dynamics. In T. M. Egyedi & K. Blind (Eds.), *The Dynamics of Standards* (pp. 28–43). Cheltenham, UK: Edward Elgar.

Egyedi, T. M. (2010). *On the implications of competing standards. Pros and Cons of Standard Setting*. Stockholm, Sweden: SC Authority.

Egyedi, T. M. (2012) The Impact of Competing Committee Standards: Integrating the Interoperability Perspective. *Proceedings of the 17th EURAS Annual Standardisation Conference.*

Egyedi, T. M., & Enserink, B. (2013). Measuring the benefits of open standards: A contribution to Dutch politics. *Standards Today*, *12*(1), 1–13.

Egyedi, T. M., & Hawkins, R. (2011). Standards Consortia. In W. Hesser, A. Feilzer, & H. de Vries (Eds.), *Standardization in Companies and Markets. Helmut Schmidt University Germany.*

Egyedi, T. M., & Sherif, M. H. (2010). Standards Dynamics through an Innovation Lens: Next Generation Ethernet Networks. *IEEE Communications Magazine*, *48*(October), 166–171. doi:10.1109/MCOM.2010.5594692

Einhorn, B. (2015). The Cheap Phones Quiety Winning the U.S. *Bloomberg*. Retrieved July 29, 2017, from https://www.bloomberg.com/news/articles/2015-09-03/zte-s-cheap-phones-quietly-winning-the-u-s-

Elhauge, E. (2008). Do Patent Holdup and Royalty Stacking Lead to Systematically Excessive Royalties? *Journal of Competition Law & Economics*, *4*(3), 3. doi:10.1093/joclec/nhn027

Elliott, A. (2017). *B3i moves blockchain to market with working reinsurance prototype*. Retrieved September 24th, 2017, from https://www.zurich.com/en/knowledge/articles/2017/09/b3i-moves-blockchain-to-market-with-working-reinsurance-prototype

Elmer-DeWitt, P. (2016). How Apple Sucks the Profit Out of Mobile Phones. *Fortune*. Retrieved July 29, 2017, from http://fortune.com/2016/02/14/apple-mobile-profit-2015/

Ende van den, J., Kaa van de, G., Uijl den, S., & Vries de, H. J. (2012). The paradox of standard flexibility: The effects of co-evolution between standard and interorganizational network. *Organization Studies, 33*(5–6), 705–736. doi:10.1177/0170840612443625

Ende, J., & Dolfsma, W. (2005). Technology-push, demand-pull and the shaping of technological paradigms – Patterns in the development of computing technology. *Journal of Evolutionary Economics*, *15*(1), 83–99. doi:10.1007/s00191-004-0220-1

EP. (Ed.). (2016). *Ethical Aspects of Cyber-Physical Systems – Scientific Foresight study*. Retrieved from http://www.europarl.europa.eu/RegData/etudes/STUD/2016/563501/EPRS_STU(2016)563501_EN.pdf

Epstein, R. A., Kieff, F. S., & Spulber, D. F. (2012). The FTC, IP, and SSOs: Hold-Up Replacing Private Coordination. *Journal of Competition Law & Economics*, *8*(1), 1–46. doi:10.1093/joclec/nhs002

Ericsson v. D-Link, 773 F.3d (Fed. Circ. 2014) (2014).

Ernst, D. (2011). *Indigenous Innovation and Globalization: The Challenge for China's Standardization Strategy*. Available at SSRN: http://ssrn.com/abstract=2400855

Ernst, D. (2013). *Standards, innovation, and latecomer economic development: A conceptual framework*. East-West Center Working Paper, Economics Series no. 134. Retrieved from http://scholarspace.manoa.hawaii.edu/bitstream/handle/10125/30723/econwp134.pdf?sequence=1

Estevadeordal, A., & Goodman, L. W. (2017). 21st-century cooperation, regional public goods, and sustainable development. In 21st Century Cooperation: Regional Public Goods, Global Governance, and Sustainable Development. Routledge.

ETsI. (n.d.). Retrieved from http://www.etsi.org/about/what-we-are

EU. (2004). *European Interoperability Framework for pan-European eGovernment Services, Version 1.0*. European Commission. Retrieved from http://ec.europa.eu/idabc/servlets/

EU. (2008). *Communication from the commission to the council, the European parliament and the European economic and social committee n 133. Towards an Increased Contribution from Standardisation to Innovation in Europe*. Brussels: Commission of the European Communities. Retrieved November 27, 2015, from http://eur-lex.europa.eu/LexUriServ/LexUriServ.do?uri=COM:2008:0133:FIN:EN:PDF

EU. (2012). *Regulation (EU) No 1025/2012 of the European Parliament and of the Council on European standardization*. Retrieved from http://eur-lex.europa.eu/LexUriServ/LexUriServ.do?uri=OJ:L:2012:316:0012:0033:EN:PDF

European Commission. (2007). *Final report of the study on the specific policy needs for ICT standardization*. European Commission. Retrieved from http://www.epractice.eu/en/library/281182

European Commission. (2009a). *Radio Spectrum Policy Group Report On "Cognitive Technologies". Final Draft* (No. Electronic Communications Policy). Brussels: Information Society and Media Directorate-General. Retrieved from http://rspg.groups.eu.int/_documents/documents/meeting/rspg20/rspg09_299.pdf

European Commission. (2009b). *White Paper: Modernising ICT Standardisation in the EU – The Way Forward.* (White Paper No. COM(2009) 324 final). European Commission. Retrieved from http://eur-lex.europa.eu/LexUriServ/LexUriServ.do?uri=CELEX:52009DC0324:EN:NOT

European Commission. (2014). *Patents and Standards: A modern framework for IPR-based standardization*. European Commission – Directorate-General for Enterprise and Industry.

European Commission. (2014a, April 29). *Antitrust Procedure, Case AT-39939-Samsung-Enforcement of UMTS Standard Essential Patents*. Author.

European Commission. (2014b, April 29). *Antitrust Procedure, Case AT-39985-Motorola-Enforcement of GPRS Standard Essential Patents*. Author.

European Commission. (2016). *Communication from the Commission to the European Parliament, the Council, the European Economic and Social Committee and the Committee of the Regions/ ICT Standardisation Priorities for the Digital Single Market*. COM (2016) 176 final.

European Commission. (2017). *Roadmap for a Communication for Standard Essential Patents for a European Digitalized Economy*. Author.

Eurostat. (2009). *High-Technology' and 'Knowledge Based Services' Aggregations Based on NACE Rev. 2*. Retrieved November 27, 2015, from http://epp.eurostat.ec.europa.eu/cache/ITY_SDDS/Annexes/htec_esms_an3.pdf

Ewing, T. (2012). Indirect Exploitation of Intellectual Property Rights by Corporations and Investors: IP Privateering and Modern Letters of Marque and Reprisal. *Hastings Science and Technology Law Journal, 4*.

Farhangi, H. (2010). The path of the smart grid. *IEEE Power & Energy Magazine*, *8*(1), 18–28. doi:10.1109/MPE.2009.934876

Farrell, J., & Klemperer, P. (2007). Chapter 31 Coordination and Lock-In: Competition with Switching Costs and Network Effects. Handbook of Industrial Organization. doi:10.1016/S1573-448X(06)03031-7

Farrell, J., Hayes, J., Shapiro, C., & Sullivan, T. (2007). Standard Setting, Patents, and Hold-up. *Antitrust Law Journal*, *74*, 3.

Farrell, J., Hayes, J., Shapiro, C., & Sullivan, Th. (2007). Standard Setting, Patents, and Hold-Up. *Antitrust Law Journal*, *74*(3), 603–670.

Farrell, J., & Saloner, G. (1988). Coordination Through Committees and Markets. *The Rand Journal of Economics*, *19*(2), 235. doi:10.2307/2555702

Fettweis, G. (2014). *The Tactile Internet.* TU-T Technology Watch Report. Retrieved from http://www.itu.int/dms_pub/itu-t/oth/23/01/T23010000230001PDFE.pdf

Fichman, R. G., & Kemerer, C. F. (1999). The illusory diffusion of innovation: An examination of assimilation gaps. *Information Systems Research*, *10*(3), 255–275. doi:10.1287/isre.10.3.255

Ficora. (2012). *Consumer survey on communications services*. Retrieved from https://www.viestintavirasto.fi/en/about-thesector/studiesandsurveys/viestintapalvelujenkuluttajatutkimus.html

Finley, B., & Soikkeli, T. (2016). Multidevice mobile sessions: A first look. *Pervasive and Mobile Computing*, *39*, 267–283. doi:10.1016/j.pmcj.2016.11.001

Fischer, R., Janiesch, C., Strach, J., Bieber, N., Zink, W., & Tai, S. (2013). Eine Bestandsaufnahme von Standardisierungs-potentialen und-lücken im Cloud Computing. *Proc. 11th International Conference on Wirtschaftsinformatik*. Retrieved from http://www.wi2013.de/proceedings/WI2013%20-%20Track%209%20-%20Fischer.pdf

Fitzgerald, A., & Pappalardo, K. (2009). Moving Towards Open Standards. *SCRIPted*, *6*(2), 467–483.

Fleck, J. (1994). Learning by trying: The implementation of configurational technology. *Research Policy*, *23*(6), 637–652. doi:10.1016/0048-7333(94)90014-0

Fomin, V. V., & Keil, T. (2000). Standardization: Bridging the Gap Between Economic and Social Theory. In *The Twenty First International Conference on Information Systems (ICIS 2000)* (pp. 206–217). Brisbane, Australia: AIS. Retrieved from http://aisel.aisnet.org/icis2000/20

Fomin, V. V., Keil, T., & Lyytinen, K. J. (2003). Theorizing about Standardization: Integrating Fragments of Process Theory in Light of Telecommunication Standardization Wars. *Sprouts: Working Papers on Information Environments, Systems and Organization, 3.*

Fomin, V. V., Medeišis, A., & Vitkutė-Adžgauskienė, D. (2011). The Role of Policy in the Development of Cognitive Radio Systems: Co-Evolutionary Perspective. In *The 7th International Conference on Standardisation and Innovation in Information Technology (SIIT 2011)* (pp. 79–90). Berlin: Mainz Publishing.

Fomin, V. V., Medeišis, A., & Vitkutė-Adžgauskienė, D. (2012). Pre-Standardization of Cognitive Radio Systems. *International Journal of IT Standards and Standardization Research*, *10*(1), 1–16. doi:10.4018/jitsr.2012010101

Fomin, V. V., Pedersen, M. K., & de Vries, H. J. (2008). Open Standards and Government Policy: Results of a Delphi Survey. *Communications of the Association for Information Systems*, *22*(1), 25.

Fomin, V. V., Su, J., & Gao, P. (2011). Indigenous standard development in the presence of dominant international standards: The case of the AVS standard in China. *Technology Analysis and Strategic Management, 23*(7), 745–758. doi:10.1080/09537325.2011.592270

Fransman, M. (2010). *The New ICT Ecosystem: Implications for Policy and Regulation.* Cambridge, UK: Cambridge University Press. doi:10.1017/CBO9780511676130

Fraunhofer. (2011). *Study on the Interplay between Patents and Standards.* Retrieved July 29, 2017 from http://www.iplytics.com/download/docs/studies/ipr_study_final_report_en.pdf

Fried, A. (2010). Performance measurement systems and their relation to strategic learning: A case study in a software-developing organization. *Critical Perspectives on Accounting, 21*(2), 118–133. doi:10.1016/j.cpa.2009.08.007

Fried, A., Gey, R., Pretorius, A., & Günther, L. (2013). Decoupling from standards - process management and technical innovation in software development organisations. *International Journal of Innovation Management, 17*(4), 1–34. doi:10.1142/S1363919613500126

Friedel, R. (1983). *Pioneer Plastic.* Madison, WI: University of Wisconsin Press.

Friedrich, J. (2011). Making Innovation Happen: The Role of Standards and Openness in an Innovation-Friendy Ecosystem, In *Proceedings of 7th International Conference on Standardization and Innovation in Information Technology (SIIT).* IEEE Computer Society. doi:10.1109/SIIT.2011.6083609

Froeb, L., & Shor, M. (2015). Innovators, Implementers, and Two-Sided Hold-Up. *The Antitrust Source, 205*, 1-10.

FSF. (2017). *Free Software Foundation.* Retrieved from www.fsf.org

Funk, J. L. (2003). Standards, dominant designs and preferential acquisition of complementary assets through slight information advantages. *Research Policy, 32*(8), 1325–1341. doi:10.1016/S0048-7333(02)00121-X

Funk, J. L., & Methe, D. T. (2001). Market- and committee-based mechanisms in the creation and diffusion of global industry standards: The case of mobile communication. *Research Policy, 30*(4), 589–610. doi:10.1016/S0048-7333(00)00095-0

Furusten, S. (2000). The knowledge base of standards. In N. Brunsson & B. Jacobsson (Eds.), *A World of Standards* (pp. 71–85). Oxford, UK: Oxford University Press.

Gabin, F., Kampmann, M., Lohmar, T., & Priddle, C. (2010, November). 3GPP mobile multimedia streaming standards. *IEEE Signal Processing Magazine*, 134–138.

Galbraith, C. S., DeNoble, A. F., Ehrlich, S. B., & Kline, D. M. (2007). Can experts really assess future technology success? A neural network and Bayesian analysis of early stage technology proposals. *The Journal of High Technology Management Research, 17*(2), 125–137. doi:10.1016/j.hitech.2006.11.002

Galbraith, C. S., Ehrlich, S. B., & DeNoble, A. F. (2006). Predicting Technology Success: Identifying Key Predictors and Assessing Expert Evaluation for Advanced Technologies. *The Journal of Technology Transfer, 31*(6), 673–684. doi:10.1007/s10961-006-0022-8

Galetovic, A., & Gupta, K. (2016). *Royalty Stacking and Standard Essential Patents: Theory and Evidence from the World Mobile Wireless Industry.* Academic Press.

Galetovic, A., Haber, S., & Levine, R. (2015). *An Empirical Examination of Patent Hold-Up.* Nat'l Bureau of Econ. Research, Working Paper No. 21090.

Galetovic, A., Haber, S., & Zaretzki, L. (2016). *A New Dataset on Mobile Phone Patent License Royalties: September 2016 Update.* Hoover IP2 Working Paper 16011.

Galetovic, A., & Haber, S. (2017). The Fallacies of Patent-Holdup Theory. *Journal of Competition Law & Economics, 13*(1), 1–44. doi:10.1093/joclec/nhx006

Galetovic, A., Harber, S., & Levine, R. (2015). An Empirical Examination of Patent Holdup. *Journal of Competition Law & Economics, 11*(3), 549–578. doi:10.1093/joclec/nhv024

Gallagher, S. (2007). The complementary role of dominant designs and industry standards. *IEEE Transactions on Engineering Management, 54*(2), 371–388. doi:10.1109/TEM.2007.893991

Galup, S. D., Dattero, R., Quan, J. J., & Conger, S. (2009). An overview of IT service management. *Communications of the ACM, 52*(5), 124–127. doi:10.1145/1506409.1506439

Gamalielsson, J., Lundell, B., Feist, J., Gustavsson, T., & Landqvist, F. (2015). On organisational influences in software standards and their open source implementations. *Information and Software Technology, 67*(Nov), 30–43. doi:10.1016/j.infsof.2015.06.006

Gandal, N., Salant, D., & Waverman, L. (2003). Standards in wireless telephone networks. *Telecommunications Policy, 27*(5–6), 325–332. doi:10.1016/S0308-5961(03)00026-0

Garas, A., Tomasello, M. V., & Schweitzer, F. (2014). *Selection rules in alliance formation: strategic decisions or abundance of choice?* Retrieved from http://arxiv.org/abs/1403.3298

Garcia, R., & Calantone, R. (2002). A critical look at technological innovation typology and innovativeness terminology: A literature review. *Journal of Product Innovation Management, 19*(2), 110–132. doi:10.1016/S0737-6782(01)00132-1

Gawer, A. (2009). Platforms, Markets and Innovation. Cheltenham, UK: Edward Elgar.

Gawer, A. (2011). *Platforms: Strategy, Leadership and Organization – An overview*. Keynote presented at EuroMOT 2011, Tampere, Finland. Retrieved from https://s3-eu-west-1.amazonaws.com/esrc-files/outputs/T9DpGeKdokaDkureK8C-Fw/oJdW7e8cTUK7qU0Uczx-jw.pdf

Gawer, A. (2002). *Platform Leadership: How Intel, Microsoft, and Cisco Drive Industry Innovation*. Boston: Harvard Business School Press.

Gawer, A. (2010). The Organization of Technological Platforms. *Research in the Sociology of Organizations, 29*, 287–296. doi:10.1108/S0733-558X(2010)0000029022

Gawer, A., & Cusumano, M. (2002). *Platform leardership: How Intel, Microsoft and Cisco drive industry innovation*. Harvard Business School Press.

Gawer, A., & Cusumano, M. (2014). Platforms and ecosystem innovation. *Journal of Product Innovation Management, 31*(3), 417–433. doi:10.1111/jpim.12105

Gawer, A., & Cusumano, M. A. (2008). How Firms Become Platform Leaders. *MIT Sloan Management Review, 49*(2), 28–35.

Gawer, A., & Henderson, R. (2007). Platform Owner Entry and Innovation in Complementary Markets, Evidence from Intel. *Journal of Economics & Management Strategy, 16*(1), 1–34. doi:10.1111/j.1530-9134.2007.00130.x

Gaynor, M., & Bradner, S. (2001). The real options approach to standardization. *Proceedings of the 34th annual Hawaii International Conference On System Sciences*. doi:10.1109/HICSS.2001.926526

Gençer, M. (2007). Structure and collaborative aspects of internet standards. *Proceedings - IEEE Symposium on Computers and Communications*, 733–739. doi:10.1109/ISCC.2007.4381499

Gençer, M. (2012). The evolution of IETF standards and their production. *International Journal of IT Standards and Standardization Research, 10*(1), 17–33. doi:10.4018/jitsr.2012010102

Geradin, D., & Rato, M. (2007). Can Standard-Setting Lead to Exploitative Abuse: A Dissonant View on Patent Hold-Up, Royalty Stacking and the Meaning of Frand. *European Competition Journal, 3.*

Geradin, D., Layne-Farrar, A., & Padilla, A. J. (2007). *Royalty Stacking in High Tech Industries: Separating Myth From Reality*. CEMFI Working Paper No.071. Retried July 29, 2017, from ftp://ftp.cemfi.es/wp/07/0701.pdf

Gershon, R. A. (2003). Cellular Telephony. Encyclopedia of International Media and Communications, 1, 175-187.

Gerst, M. H. (2003). The role of standardisation in the context of e-collaboration: a snap shot. *The 3rd Conference on Standardization and Innovation in Information Technology*, 113–119. doi:10.1109/SIIT.2003.1251200

Gey, R. (2011). Appropriation of Software Process Improvement Standards: An Empirical Study in Software Development Organizations. *Proceedings of the 8th International Conference on Intellectual Capital, Knowledge Management & Organisational Learning*, 182–189.

Ghosh, R. A. (2005). *An Economic Basis for Open Standards*. FLOSSPOLS. Retrieved from http://www.flosspols.org/deliverables/FLOSSPOLS-D04-openstandards-v6.pdf

Giddens, A. (1984). *The constitution of society: Outline of the theory of structuration*. Cambridge, UK: Polity Press.

Ginsburg, D.H., Wrong-Ervin, K., & Wright, J.D. (2015). The Troubling Use of Antitrust to Regulate FRAND Licensing. *CPI Antitrust Chronicle, 2015*(1), 2-8.

Gioia, D., & Poole, P. (1984). Scripts in organizational behavior. *Academy of Management Review, 9*(3), 449–459.

Global Certification Forum. (2014). Retrieved from http://www.globalcertificationforum.org/images/downloads/GCF-WP-GCF_Certification_June2014.pdf

Gnyawali, D. R., & Madhavan, R. (2001). Cooperative Networks and Competitive Dynamics: A Structural Embeddedness Perspective. *Academy of Management Review, 26*(3), 431–445.

Golden, J. (2013). Patent Privateers: Private Enforcement's Historical Survivors. *Harvard Journal of Law & Technology*, 26.

Granstrand, O. (1999). *The Economics and Management of Intellectual Property*. Edward Elgar.

Greene, W. H. (2008). Econometric Analysis (6th ed.). Prentice Hall.

Greene, W. H. (2008). *Econometric Analysis* (6th ed.). Upper Saddle River, NJ: Prentice Hall.

Grindley, P. (1995). *Standards*. New York: Strategy and Policy.

Grindley, P., & Teece, D. (1997, Winter). Managing Intellectual Capital: Licensing and Cross-Licensing in Semiconductors and Electronics. *California Management Review, 39*(2), 2. doi:10.2307/41165885

Groenfeldt, T. (2016). *Stellar Makes Sending Money Internationally As Easy As Email?* Retrieved September 24th, 2017, from https://www.forbes.com/sites/tomgroenfeldt/2016/12/07/stellar-makes-sending-money-internationally-as-easy-as-email/#288ad1f65078

Grøtnes, E. (2008). Strategies for Influencing the Standardization Process: Examples from Within. In *16th European Conference on Information Systems*. Association for Information Systems.

Gunasekaran, V., & Harmantzis, F. C. (2008). Towards a Wi-Fi ecosystem: Technology integration and emerging service models. *Telecommunications Policy, 32*(3–4), 163–181. doi:10.1016/j.telpol.2008.01.002

Gunes, V., Peter, S., Givargis, T., & Vahid, F. (2014). A Survey on Concepts, Applications, and Challenges in Cyber-Physical Systems. *Transactions on Internet and Information Systems (Seoul)*, *8*(12), 4242–4268.

Gungor, V. C., Lu, B., & Hancke, G. P. (2010). Opportunities and challenges of wireless sensor networks in smart grid. *Industrial Electronics. IEEE Transactions on*, *57*(10), 3557–3564.

Gupta, K., & Snyder, M. (2014). *Smart Phone Litigation and Standard Essential Patents*. Hoover IP² Working Paper Series No.14006.

Habets, M. J. M. (2012). *Complementors: their effects on new value creation by SMEs in the Dutch printing industry*. Universiteit Twente. doi:10.3990/1.9789036534529

Hacklin, F., Battistini, B., & von Krogh, G. (2013). Strategic Choices in Converging Industries. *MIT Sloan Management Review*, *55*(1), 65–73.

Hagedoorn, J. (2002). Inter-firm R&D partnerships: An overview of major trends and patterns since 1960. *Research Policy*, *31*(4), 477–492. doi:10.1016/S0048-7333(01)00120-2

Hanneman, R. A., & Riddle, M. (2005). *Introduction to social network methods. Riverside, CA: University of California*. Retrieved from http://faculty.ucr.edu/~hanneman/

Hanseth, O., & Lyytinen, K. (2010). Information infrastructures: The case of building Internet. *Journal of Information Technology*, *25*(1), 1–19. doi:10.1057/jit.2009.19

Hanseth, O., Monteiro, E., & Hatling, M. (1996). Developing information infrastructure: The tension between standardization and flexibility. *Science, Technology & Human Values*, *21*(4), 407–426. doi:10.1177/016224399602100402

Haug, T. (2002). A commentary on standardization practices: Lessons from the NMT and GSM mobile telephone standards histories. *Telecommunications Policy*, *26*(3–4), 101–107. doi:10.1016/S0308-5961(02)00003-4

Hawkins, R., & Ballon, P. (2007). When standards become business models: Reinterpreting "failure" in the standardization paradigm. *Info*, *9*(5), 20–30. doi:10.1108/14636690710816426

Heiden, B., & Petit, N. (2017). *Patent Trespass and the Royalty Gap: Exploring the Nature and Impact of Patent Holdout*. Available at https://papers.ssrn.com/sol3/papers.cfm?abstract_id=2981577

Heiden, B. (2016). The viability of FRAND: How the seminal landmark Microsoft ruling could impact the value of standard essential patents and the future of telecom standards. *Telecommunications Policy*, *40*(9), 870–887. doi:10.1016/j.telpol.2016.03.004

Heiden, B., & Andreasson, J. (2016). Reevaluating Patent Damages in the Knowledge Economy: The Intellectual Value Chain and the Royalty Base for Standard-Essential Patents. *Criterion J. on Innovation*, *1*, 229.

Henriksen, D., Nicolajsen, H., & Pors, J. (2002). Towards variation or uniformity? Comparing technology-use mediations of web-based groupware. *Proceedings of the European Conference on Information Systems 2002*, 1174–1184.

High Court of Delhi at New Delhi. (2015, March 31). *Telefonaktiebolaget LM Ericsson v. Intex Technologies Limited.*

High Court of Delhi at New Delhi. (2015, September 2). *Telefonaktiebolaget LM Ericsson v. M/S Best IT World Private Limited (iBall).*

High Court of England and Wales (2017, April 5). *Unwired Planet International Ltd. v. Huawei Technologies Co. Ltd, Huawei Technologies (UK) Co. Ltd v. Unwired Planet LLC.* EWCH 711 (Pat).

High Court of England and Wales. (2012, May 10). *Nokia Corp. v. IPCom GmbH & Co. KG*, [2012] EWHC 1446 (Pat.).

High Court of England and Wales. (2015, January 30). *Vringo Infrastructure Inc. and ZTE Limited.* EWCH 214 (Pat).

Hillebrand, F. (2013). The creation of standards for global mobile communication: GSM and UMTS standardization from 1982 to 2000. *IEEE Wireless Communications, 20*(5), 24–33. doi:10.1109/MWC.2013.6664470

Ho, J. (2017). *A Systems Perspective on Standardisation in Technological Innovation: A Conceptual Framework and a Process Model Supporting Strategic Policy Foresight* (Doctoral thesis). University of Cambridge, UK. doi:10.17863/CAM.8705

Ho, J.-Y., & Sullivan, E. (2016). The Multi-dimensional Nature of Standardisation in Support of Innovation: A Systematic Analysis of the History of Photovoltaic Technology. *EURAS Proceedings 2016.*

Hobday, M. (1998). Product complexity, innovation and industrial organisation. *Research Policy, 26*(6), 689–710. doi:10.1016/S0048-7333(97)00044-9

Hochstein, A., Tamm, G., & Brenner, W. (2005). *Service-oriented IT-Management: Benefit, Cost and Success Factors.* Paper presented at the 13th European Conference on Information Systems, Regensburg.

Hoel, T., Hollins, P. A., & Pawlowski, J. M. (2010). On the Status of Learning Technology Specifications and Standards. *International Journal of IT Standards and Standardization Research, 8*(2), i–vi.

Holtbruegge, D., & Wessely, B. (2009). Initiating forces and success factors of born global firms. *European Journal of International Management, 3*(2), 232–260. doi:10.1504/EJIM.2009.024324

Hovav, A., Patnayakuni, R., & Schuff, D. (2004). A model of Internet standards adoption: The case of IPv6. *Information Systems Journal, 14*(3), 265–294. doi:10.1111/j.1365-2575.2004.00170.x

HP. (2003). *HP IT Service Management (ITSM); Transforming IT organizations into service providers.* Retrieved August 02 2012 from http://www.hp.com/hpinfo/newsroom/press_kits/2007/businesstechnology/wp_it_transformation.pdf

HP. (2015). *HP Response to Consultation on Patents and Standards.* Hewlett-Packard.

Huang, S.-J., Wu, M.-S., & Chen, L.-W. (2012). Critical success factors in aligning IT and business objectives: A Delphi study. *Total Quality Management & Business Excellence*, 1-22.

Huang, Z., & Palvia, P. (2001). ERP implementation issues in advanced and developing countries. *Business Process Management Journal, 7*(3), 276–284. doi:10.1108/14637150110392773

Hubbert, E. (2010). *Market Overview: IT Service Management Support Tools. An Overview Of The Modern Service Desk And Other Technologies.* Forrester Research, Inc.

Hughes, T. P. (1993). The Evolution of Large Technological Systems. In *The social construction of technological systems: New directions in the sociology and history of technology* (pp. 51–82). Cambridge, MA: MIT Press.

Humphrey, W. (1992). Introduction to software process improvement. Technical report, Software Engineering Institute, Carnegie-Mellon University.

Huston, G. (2017). *IPv6 reports.* Retrieved from http://bgp.potaroo.net/index-v6.html

Huurdeman, A. A. (2003). *The Worldwide History of Telecommunications.* Hoboken, NJ: Wiley-Interscience. doi:10.1002/0471722243

IBM-Global-Services. (2004). *IBM and the IT Infrastructure Library. How IBM supports ITIL and provides ITIL-based capabilities and solutions.* Author.

Iden, J., & Langeland, L. (2010). Setting the Stage for a Successful ITIL Adoption: A Delphi Study of IT Experts in the Norwegian Armed Forces. *Information Systems Management, 27*(2), 103-112. doi: 10.1080/10580531003708378

Iden, J., Steindal, F., & Stokke, B. (2007). The Implementation of IT Infrastructure Library (ITIL) in Norway: Progress, success factors and benefits. In L. Habib (Ed.), *Norsk konferanse for organisasjoners bruk av informasjonsteknologi. NOKOBIT 2007* (pp. 43–56). Trondheim, Norway: Tapir Akademisk Forlag.

IEEE. (2014). *Industry Connections Security Group (ICSG), Industry Connections Activity Initiation Document (ICAID).* Retrieved August 30, 2017, from http://standards.ieee.org/about/sasb/iccom/IC09-001-04_Computer_Security_Group_ICSG.pdf

IEEE. (n.d.). *History*. Retrieved from https://www.ieee.org/about/ieee_history.html

ILNAS/ANEC. (2012). *Standards analysis ICT sector Luxembourg*. Retrieved November 27, 2015, from http://www.ilnas.public.lu/fr/publications/confiance-numerique/etudes-nationales/standards-analysis-ict-v1-0.pdf

In re Innovatio IP Ventures L.L.C, No. 2303, 2013 WL 5593609 (N.D. Ill. 2013) (2013).

Insight, C. B. (2017). *Deals To Bitcoin & Blockchain Startups Fall Below 2014 Levels.* Retrieved September 24th, 2017, from https://www.cbinsights.com/research/bitcoin-blockchain-startup-funding

Intellectual Property Rights PolicyETSI (2017).

International Trade Commission. (2013c, June 5). *In the Matter of Certain Electronic Devices, Including Wireless Communication Devices, Portable Music and Data Processing Devices, and Tablet Computer.* Inv. 337-TA-794.

International Trade Commission. (2014, June 13). *In the Matter of Certain Wireless Devices With 3G and/or 4G Capabilities, and Components Thereof.* Inv. 337-TA-868, Initial Determination.

International Trade Commission. (2015, April 27). *In the Matter of Certain 3G Mobile Handsets and Components Thereof.* Inv. 337-TA-613.

Internet Society. (n.d.). *History*. Retrieved from http://www.internetsociety.org/history

Internet World Stats. (n.d.). *Marketing*. Retrieved from http://www.internetworldstats.com/emarketing.htm

IoT-A. (Ed.). (2013). *Internet of Things – Architecture – Deliverable D1.5 – Final architectural reference model for the IoT v3.0*. Retrieved from http://www.meet-iot.eu/deliverables-IOTA/D1_3.pdf

Isaak, J. (2006). The Role of Individuals and Social Capital in POSIX Standardization. *International Journal of IT Standards and Standardization Research, 4*(1), 1–23. doi:10.4018/jitsr.2006010101

ISACA. (2012). *Control Objectives for Information and related Technologies*. COBIT.

ISO. (2007). *ISO/IEC/ITU common patent policy*. Retrieved from http://isotc.iso.org/livelink/livelink/fetch/2000/2122/3770791/Common_Policy.htm

ISO. (2012). *Guidelines for Implementation of the Common Patent Policy for ITU-T/ITU-R/ISO/IEC, Revision 1*. Retrieved from http://isotc.iso.org/livelink/livelink/fetch/2000/2122/3770791/Common_Guidelines_01_March_07.pdf

ISO. (2015a). *Standards: What is a standard?* Retrieved from http://www.iso.org/iso/home/standards.htm

ISO. (2015b). *Patent declarations submitted to ISO*. Retrieved from http://isotc.iso.org/livelink/livelink/ 13622347/Patents_database.xls?func=doc.Fetch&nodeId=13622347

ISO. (2015c). *ISO Standards and Patents*. Retrieved from http://www.iso.org/iso/standards_development/patents

ISO. (2015d). *Guidelines for Implementation of the Common Patent Policy for ITU-T/ITU-R/ISO/IEC, Revision 2*. Retrieved from http://www.itu.int/dms_pub/itu-t/oth/04/04/T04040000010004PDFE.pdf

ISO. (Ed.). (2011). *Economic benefits of standards – International case studies – Volume 1*. Retrieved 27 April 2015, from http://www.iso.org/iso/home/store/publication_item.htm?pid=PUB100288

ISO. (Ed.). (2012). *Economic benefits of standards – International case studies – Volume 2*. Retrieved 27 April 2015, from http://www.iso.org/iso/home/store/publication_item.htm?pid=PUB100311)

ISO. (n.d.). *Standards in our world*. Retrieved from http://www.iso.org/sites/ConsumersStandards/1_standards.html

ISO/IEC. (2003). *ISO/IEC 15504- 2:2004 Software Engineering – Process Assessment – Part 2: Performing an Assessment*. Author.

ITU. (2009). *Definitions of Software Defined Radio (SDR) and Cognitive Radio System (CRS)* (No. ITU-R SM.2152). ITU-R. Retrieved from http://www.itu.int/dms_pub/itu-r/opb/rep/R-REP-SM.2152-2009-PDF-E.pdf

ITU. (2012). *Recommendation Y.2016 – Overview of the Internet of things: Next Generation Networks – Frameworks and functional architecture model*. Retrieved from https://www.itu.int/rec/dologin_pub.asp?lang=e&id=T-REC-Y.2060-201206-I!!PDF-E&type=items

Iversen, E. (1999). *Standardization and Intellectual Property Rights: ETSI's controversial search for new IPR-procedure*. Paper presented at the IEEE Conference on Standardization and Innovation, Aachen, Germany.

Iversen, E. (1999) Standardization and Intellectual Property Rights: ETSI's controversial search for new IPR-procedures. *IEEE Conference on Standardisation and Innovation in Information Technology*, Aachen, Germany.

Iversen, E. J. (2001). Patenting and Voluntary Standards: Tensions Between the Domains of Proprietary Assets and "Public Goods" in the Innovation of Network Technologies. *Science Studies*, *14*(2), 66–82.

Iversen, E. J., Vedel, T., & Werle, R. (2004, Summer). Standardization and the Democratic Design of Information and Communication Technology. *Knowledge, Technology & Policy*, *17*(2), 104–126. doi:10.1007/s12130-004-1027-y

Jaakkola, H., Gabbouj, M., & Neuvo, Y. (1998). Fundamentals of technology diffusion and mobile phone case study. *Circuits, Systems, and Signal Processing*, *17*(3), 421–448. doi:10.1007/BF01202301

Jacob, R. (2014). *FRAND: A Legal Analysis*. Retrieved July 29, 2017, from https://ec.europa.eu/jrc/sites/jrcsh/files/jrc_is_20141027_-_patentsws-04.frand_.robinjacob.pdf

Jaffe, A., & Lerner, J. (2007). *Innovation and Its Discontents: How Our Broken Patent System is Endangering Innovation and Progress, and What to Do About It*. Princeton University Press.

Jakobs, K. & Mora, M. (2008). Co-ordinating Rule Setters – Co-operation in ICT Standards Setting. *CONF-IRM 2008 Proceedings*, Paper 17.

Jakobs, K. (2003). Information technology standards, standards setting and standards research. *Cotswolds Conference, Stanhope Center for Communication Policy Research*, 8–9.

Jakobs, K. (2003). *Information Technology Standards, Standards Setting and Standards Research: Mapping the Universe*. Retrieved from http://www-i4.informatik.rwth-aachen.de/~jakobs/Papers/stanhope.pdf

Jakobs, K. (2008). ICT standardisation - co-ordinating the diversity. *International Telecommunication Union - Proceedings of the 1st ITU-T Kaleidoscope Academic Conference, Innovations in NGN, K-INGN*. doi:10.1109/KINGN.2008.4542257

Jakobs, K. (2012). Pitfalls in Standards Making. In *EURAS Proceedings 2012: "Standards and Innovation"* (pp. 211–226). Banska Bystrica, Slovakia: The EURAS Board Series.

Jakobs, K., & Wehrle, K. (2017). *Cyber-Physical Systems – Future Standardization for Europe (CPS-FUSE)*. Unpublished Final Project Report.

Jakobs, K. (2002). A proposal for an alternative standards setting process. *IEEE Communications Magazine*, *40*(7), 118–123. doi:10.1109/MCOM.2002.1018016

Jakobs, K. (2006). Shaping user-side innovation through standardisation: The example of ICT. *Technological Forecasting and Social Change*, *73*(1), 27–40. doi:10.1016/j.techfore.2005.06.007

Jakobs, K. (2017). Two dimensions of success in ICT standardization – A review. *ICT Express*, *3*(2), 85–89. doi:10.1016/j.icte.2017.05.008

Jakobs, K. (Ed.). (2009). *ICT Standardization for E-Business Sectors: Integrating Supply and Demand Factors*. Hershey, PA: IGI Global. doi:10.4018/978-1-60566-320-3

Jakobs, K., & Blind, K. (2011). ICT Standardisation Policy in Europe – Recent Past, Presence, and Future(?). In *Proceedings of the 7th biennial Standardisation and Innovation in Information Technology conference (SIIT 2011)* (pp. 101–114). Berlin: Mainz Publishers. doi:10.1109/SIIT.2011.6083611

Jakobs, K., Procter, R., & Williams, R. (2001). The Making of Standards : Looking Inside the Work Groups. *IEEE Communications Magazine*, *39*(4), 102–107. doi:10.1109/35.917511

Janneck, M., & Finck, M. (2006). Appropriation and mediation of technology use in stable self-organised online communities. *Web Based Communities, Proceedings of IADIS International Conference*, 149–156.

Jiang, H., Zhao, S., Li, Z., & Chen, Y. (2015). Interaction between technology standardization and technology development: a coupling effect study. *Information Technology and Management*. DOI 10.1007/s10799-015-0215-7

Jin, Y., Sen, S., Guérin, R., Hosanagar, K., & Zhang, Z.-L. (2008). Dynamics of competition between incumbent and emerging network technologies. *Proceedings of the 3rd international workshop on Economics of networked systems*, 49–54. doi:10.1145/1403027.1403039

Jones, M., Coviello, N., & Tang, Y. K. (2011). International entrepreneurship research (1989–2009): A domain ontology and thematic analysis. *Journal of Business Venturing*, *26*(6), 632–659. doi:10.1016/j.jbusvent.2011.04.001

Jorstad, I., Dustdar, S., & Do, T. (2005). An analysis of current mobile services and enabling technologies. *International Journal of Ad Hoc and Ubiquitous Computing*, *1*(1), 92–102. doi:10.1504/IJAHUC.2005.008023

Joseph, D., Shetty, N., Chuang, J., & Stoica, I. (2007). Modeling the adoption of new network architectures. *Proceedings of International Conference on Emerging Networking Experiments and Technologies (CoNEXT'07)*. doi:10.1145/1364654.1364661

Kaiser, H. F. (1974). An index of factorial simplicity. *Psychometrika*, *39*(1), 31–36. doi:10.1007/BF02291575

Kammarkollegiet (2013). Programvaror och tjänster 2013, Statens Inköpscentral, Kammarkollegiet, 5 February, Dnr. 96-40-2013.

Kammarkollegiet (2014a). Programvaror och tjänster 2014 - Kontorsstöd, Dnr 96-33-2014, 2014-09-11.

Kammarkollegiet (2014b). Programvaror och tjänster 2014 - Grundläggande it, Dnr 96-34-2014, 2014-09-11.

Kammarkollegiet. (2016). *Open IT-standards, National Procurement Services (Kammarkollegiet), 7 March, Dnr 96-38-2014*. Retrieved from http://www.avropa.se/globalassets/open-it-standards.pdf

Kane, E. (2017). *Is Blockchain a General Purpose Technology?* Retrieved September 24th, 2017, https://papers.ssrn.com/sol3/papers.cfm?abstract_id=2932585

Karjaluoto, H., Karvonen, J., Kesti, M., Koivumaki, T., Manninen, M., Pakola, J., & Salo, J. (2005). Factors affecting consumer choice of mobile phones: Two studies from Finland. *Journal of Euromarketing*, *14*(3), 59–82. doi:10.1300/J037v14n03_04

Katusic, D., Weber, M., Bojic, I., Jezic, G., & Kusek, M. (2012). Market, standardization, and regulation development in Machine-to-Machine communications. *20th International Conference on Software, Telecommunications and Computer Networks (SoftCOM)*, 1–7.

Katz, M. L., & Shapiro, C. (1985). Network Externalities, Competition, and Compatibility. *The American Economic Review*, *75*(3), 424–440.

Katz, M. L., & Shapiro, C. (1986). Technology adoption in the presence of network externalities. *Journal of Political Economy*, *94*(4), 822–841. doi:10.1086/261409

Katznelson, R. (2015). *Perilous Deviations from FRAND-Harmony-Operational Pitfalls of the 2015 IEEE Patent Policy*. Paper presented at the 9th International Conference on Standardization and Innovation in Information Technology, Sunnyvale, CA. doi:10.1109/SIIT.2015.7535599

Katznelson, R. (2016). *Decline in non-duplicative Letters of Assurance (LOA) from Product/System Companies for IEEE standards*. Retrieved July 29, 2017, from http://www.iam-media.com/files/LOAs-March-2016_stamped.pdf

Kay, A. L. (2002). *German Jet Engine and Gas Turbine Development 1930-1945*. Shrewsbury: Airlife Publishing Ltd.

Kilkki, K., Mäntylä, M., Karhu, K., Hämmäinen, H., & Ailisto, H. (2017). *A disruption framework*. Manuscript submitted for publication.

King, I. (2016). Qualcomm Fined $975 Million in China, Sets Licensing Rates. *Bloomberg*. Retrieved July 29, 2017, from https://www.theglobeandmail.com/report-on-business/international-business/qualcomm-fined-975-million-in-china/article22883526/

King, M. (2006). *Standards and Innovation* (Master of Science Dissertation). School of Social Sciences, University of Manchester.

Kivi, A., Smura, T., & Töyli, J. (2012). Technology product evolution and the diffusion of new product features. *Technological Forecasting and Social Change*, *79*(1), 107–126. doi:10.1016/j.techfore.2011.06.001

Kiviat, T. I. (2015). Beyond Bitcoin: Issues in Regulating Blockchain Tranactions. *Duke Law Journal*, *65*, 569.

Klein, B., Crawford, R. G., & Alchian, A. A. (1978). Vertical Integration, Appropriable Rents, and Competitive Contracting Processes. *The Journal of Law & Economics*, *21*(2), 297–326. doi:10.1086/466922

Knight, W. (2017). The Technology Behind Bitcoin Is Shaking Up Much More Than Money. *MIT Technology Review*. Retrieved September 24th, 2017, from https://www.technologyreview.com/s/604148/the-technology-behind-bitcoin-is-shaking-up-much-more-than-money/

Knight, G., Madsen, T. K., & Servais, P. (2004). An inquiry into born-global firms in Europe and the USA. *International Marketing Review*, *21*(6), 645–665. doi:10.1108/02651330410568060

Koch, C. (2017). Standardization in Energing Technologies: The Case of Additive Manufacturing. *EURAS Proceedings 2017*, T117-133.

Kohlbacher, F. (2006). The use of qualitative content analysis in case study research. *Forum Qualitative Social Research, 7*(1).

Komu, M., Varjonen, S., Gurtov, A., & Tarkoma, S. (2012). Sockets and beyond: Assessing the source code of network applications. *Linux Symposium*, 7–24.

Konstantinos, C., & Devetsikiotis, M. (2016). Blockchains and Smart Contracts for the Internet of Things. *IEEE Access: Practical Innovations, Open Solutions, 4*, 2292–2303. doi:10.1109/ACCESS.2016.2566339

Krechmer, K. (1998). The Principles of Open Standards. *Standards Engineering Society: WorldStandards Day Paper Competition.*

Krechmer, K. (2000). *The Fundamental Nature of Standards: Economics Perspective*. Paper presented at the International JA Schumpeter Conference, Manchester, UK.

Krechmer, K. (1996). Technical standards: Foundations of the future. *StandardView, 4*(1), 4–8. doi:10.1145/230871.230872

Krechmer, K. (2001). The Need for Openness in Standards. *IEEE Computer, 34*(6), 100–101.

Krechmer, K. (2005). The Meaning of Open Standards, In *Proceedings of the 38th Hawaii International Conference on System Sciences*. IEEE Computer Society. doi:10.1109/HICSS.2005.605

Krechmer, K. (2005). The Meaning of Open Standards. *Hawaii International Conference on System Sciences.*

Krechmer, K. (2006). Open Standards Requirements. *International Journal of IT Standards and Standardization Research, 4*(1), 1. doi:10.4018/jitsr.2006010103

Krippendorff, K. (2004). *Content analysis: An introduction to its methodology*. Thousand Oaks, CA: Sage Publications.

Kumbakara, N. (2008). Managed IT services: The role of IT standards. *Information Management & Computer Security, 16*(4), 336–359. doi:10.1108/09685220810908778

Kwak, J. Y., Lee, H., & Chung, D. B. (2012). The evolution of alliance structure in China's mobile telecommunication industry and implications for international standardization. *Telecommunications Policy, 36*(10–11), 966–976. doi:10.1016/j.telpol.2012.07.017

Kwak, J., Lee, H., & Fomin, V. V. (2011). The governmental coordination of conflicting interests in standardisation: Case studies of indigenous ICT standards in China and South Korea. *Technology Analysis and Strategic Management, 23*(7), 789–806. doi:10.1080/09537325.2011.592285

Lacohée, H., Wakeford, N., & Pearson, I. (2003). A social history of the mobile telephone with a view of its future. *BT Technology Journal, 21*(3), 203–211. doi:10.1023/A:1025187821567

Landgericht Düsseldorf. (2015, March 26). *France Brevet vs.* HTC, 4b O 140/13.

Landgericht Düsseldorf. (2016, March 31). *Saint Lawrence v.* Vodafone, 4a O 73/14.

Landgericht Mannheim. (2015, November 27). *Saint Lawrence v. Deutsche Telekom.* 2 O 106/14.

LaserDynamics, Inc., v. Quanta Computer, Inc., 694 F.3d 51, 67 (Fed. Circ. 2012) (2012).

Lavallée, M., & Robillard, P. N. (2012). The impacts of software process improvement on developers: a systematic review. *Proceedings of the 2012 International Conference on Software Engineering*, 113-122. doi:10.1109/ICSE.2012.6227201

Lawrence, T., & Phillips, N. (1998). Commentary: Separating play and critique. *Journal of Management Inquiry*, *7*(2), 154–160. doi:10.1177/105649269872010

Layne-Farrar, A. (2014). *Patent Holdup and Royalty Stacking Theory and Evidence: Where do we Stand After 15 Years of History*. Available at http://www.oecd.org/officialdocuments/publicdisplaydocumentpdf/?cote=DAF/COMP/WD%282014%29%2084&doclanguage=en

Layne-Farrar, A. (2016, February). Why Patent Holdout is Not Just a Fancy Name for Plain Old Patent Infringement. *CPI,* 1-4.

Layne-Farrar, A., & Wong-Ervin, K. W. (2015). An Analysis of the Federal Circuit's Decision in Ericsson v. D-Link. *CPI Antitrust Chronicle*, (1).

Layne-Farrar, A., Padilla, A. J., & Schmalensee, R. (2007). Pricing Patents for Licensing in Standard-Setting Organizations: Making Sense of FRAND Commitments. *Antitrust Law Journal*, *74*, 3.

Le Borne, F., Treat, D., Dimidschstein, F., & Brodersen, C. (2016). *SWIFT on distributed ledger technologies - Delivering an industry- standard platform through community collaboration.* Retrieved September 24th, 2017, from http://www.ameda.org.eg/files/SWIFT_DLTs_position_paper_FINAL1804.pdf

Lea, G., & Hall, P. (2004). Standards and intellectual property rights: An economic and legal perspective. *Information Economics and Policy*, *16*(1), 67–89. doi:10.1016/j.infoecopol.2003.09.005

Lee, J., O'Neal, D. E., Pruett, M. W., & Thomas, H. (1995). Planning for Dominance: A Strategic Perspective on the Emergence of a Dominant Design. *R & D Management*, *25*(1), 3–15. doi:10.1111/j.1467-9310.1995.tb00896.x

Lehmann, S., & Buxmann, P. (2009). Pricing strategies of software vendors. *Business & Information Systems Engineering*, *1*(6), 452–462. doi:10.1007/s12599-009-0075-y

Lehr, W. H., & McKnight, L. W. (2003). Wireless Internet access: 3G vs. WiFi? *Telecommunications Policy*, *27*(5-6), 351–370. doi:10.1016/S0308-5961(03)00004-1

Leiner, B. M., Cerf, V. G., Clark, D. D., Kahn, R. E., Kleinrock, L., Lynch, D. C., & Wolf, S. et al. (2009). A Brief History of the Internet. *Computer Communication Review*, *39*(5), 22–31. doi:10.1145/1629607.1629613

Leiponen, A. E. (2008). Competing Through Cooperation: The Organization of Standard Setting in Wireless Telecommunications. *Management Science*, *54*(11), 1904–1919. doi:10.1287/mnsc.1080.0912

Lemley, M. A. (2002). Intellectual Property Rights and Standard-Setting Organizations. *California Law Review, 90*(6), 1889-1980. Retrieved from http://www.jstor.org/stable/3481437

Lemley, M. A., & Shapiro, C. (2013). A Simple Approach to Setting Reasonable Royalties for Standard-Essential Patents. *Berkeley Tech. L.J., 28.*

Lemley, M. A. (2002). Intellectual Property Rights and Standard-setting organizations. *California Law Review*, *90*(6), 1889–1990. doi:10.2307/3481437

Lemley, M. A. (2007). Ten Things to Do about Patent Holdup of Standards (and One Not To). *Boston College Law Review. Boston College. Law School*, 48.

Lemley, M. A., & Shapiro, C. (2007). Patent Holdup and Royalty Stacking. *Texas Law Review*, *85*, 1991–2050.

Lemley, M., & Melamed, D. (2013). Missing the Forest for the Trolls. *Columbia Law Review*, *113*, 8.

Lemley, M., & Shapiro, C. (2007). Patent Holdup and Royalty Stacking. *Texas Law Review*, *85*, 1991–2049.

Lemstra, W., & Hayes, V. (2009). License-exempt: Wi-Fi complement to 3G. *Telematics and Informatics*, *26*(3), 227–239. doi:10.1016/j.tele.2008.11.003

Lemstra, W., Hayes, V., & Groenewegen, J. (2011). *The Innovation Journey of Wi-Fi. The Innovation Journey of Wi-Fi: The Road to Global Success*. Cambridge University Press. doi:10.1017/CBO9780511666995

Leonardi, P. (2007). Activating the informational capabilities of information technology for organizational change. *Organization Science*, *18*(5), 813–831. doi:10.1287/orsc.1070.0284

Leonardi, P. M., Bailey, D. E., Diniz, E. H., Sholler, D., & Nardi, B. (2016). Multiplex Appropriation in Complex Systems Implementation: The Case of Brazil's Correspondent Banking System. *Management Information Systems Quarterly*, *40*(2), 461–473. doi:10.25300/MISQ/2016/40.2.10

Leonardi, P., & Barley, S. (2010). What's under construction here? Social action, materiality, and power in constructivist studies of technology and organizing. *The Academy of Management Annals*, *4*(1), 1–51. doi:10.1080/19416521003654160

Lerner, J. (2002). 150 Years of Patent Protection. *The American Economic Review, 92*(2), 221-225. Retrieved from http://www.jstor.org/stable/3083406

Lerner, J. (2009). The Empirical Impact of Intellectual Property Rights on Innovation: Puzzles and Clues. *The American Economic Review, 99*(2), 343-348. Retrieved from http://www.jstor.org/stable/25592422

Lerner, J., & Tirole, J. (2006). Model of Forum Shopping. *The American Economic Review, 96*(4), 1091-111. Retrieved from http://www.jstor.org/stable/30034332

Lerner, J., & Layne-Farrar, A. (2011). To Join or Not to Join: Examining Patent Pool Participation and Rent Sharing Rules. *International Journal of Industrial Organization*, *29*(2), 294–303. doi:10.1016/j.ijindorg.2010.08.006

Lerner, J., & Tirole, J. (2006). A Model of Forum Shopping, with Special Reference to Standard Setting Organizations. *The American Economic Review*, *96*(4), 1091–1113. doi:10.1257/aer.96.4.1091

Lerner, J., & Tirole, J. (2015). Standard-essential patents. *Journal of Political Economy*, *123*(3), 547–586. doi:10.1086/680995

Levä, T., Komu, M., Keränen, A., & Luukkainen, S. (2013). Adoption barriers of network layer protocols: The case of host identity protocol. *Computer Networks*, *57*(10), 2218–2232. doi:10.1016/j.comnet.2012.11.024

Levä, T., & Riikonen, A. (2013). Diffusion of Internet protocols – Measurement framework and empirical analysis of the Finnish mobile market. *Proc. of the 8th IEEE Conference on Standardisation and Innovation in Information Technology (SIIT 2013)*. doi:10.1109/SIIT.2013.6774578

Levä, T., & Suomi, H. (2013). Techno-economic feasibility analysis of Internet protocols: Framework and tools. *Computer Standards & Interfaces*, *36*(1), 76–88. doi:10.1016/j.csi.2013.07.011

Levina, N., & Vaast, E. (2005). The Emergence of Boundary Spanning Competence in Practice: Implications for Implementation and Use of Information Systems. *Management Information Systems Quarterly*, *29*(2), 335–363.

Lewis, G. A. (2013). The role of standards in cloud-computing interoperability. In *Proc. 46th Hawaii International Conference on Systems Sciences*. IEEE. Retrieved from http://repository.cmu.edu/cgi/viewcontent.cgi?article=1679&context=sei

Liker, J. K., Fleischer, M., & Arnsdorf, D. (1992). Fulfilling the promises of CAD. *Sloan Management Review*, *33*(3), 74–86.

Lim, A. S. (2003). *Inter-firm Alliances during Pre-standardization in ICT* (Working paper No. 03.03). Eindhoven Centre for Innovation Studies, Eindhoven University of Technology.

Lindqvist, M. (1991). *Infant multinationals. The internationalisation of young, technology-based Swedish firms* (Dissertation). Stockholm School of Economics.

Lin, I. C., & Liao, T. C. (2017). A Survey of Blockchain Security Issues and Challenges. *International Journal of Network Security, 19*(5), 653–659.

Liu, C.-M. (2002). The effects of promotional activities on brand decision in the cellular telephone industry. *Journal of Product and Brand Management, 11*(1), 42–51. doi:10.1108/10610420210419540

Lloyd, R. (2015). Ericsson and Nokia the latest to confirm that they will not license under the new IEEE patent policy. *IAM*. Retrieved July 29, 2017, from http://www.iam-media.com/blog/detail.aspx?g=d07d0bde-ebd6-495a-aa72-4eecb9dac67d

Lopez Research LLC. (2015). *An Introduction to the Internet of Things (IoT). Part 1 of the IoT Series*. Retrieved July 29, 2017, from http://www.cisco.com/c/dam/en_us/solutions/trends/iot/introduction_to_IoT_november.pdf

Lucio-Nieto, T., Colomo-Palacios, R., Soto-Acosta, P., Popa, S., & Amescua-Seco, A. (2012). Implementing an IT service information management framework: The case of COTEMAR. *International Journal of Information Management, 32*(6), 589–594. doi:10.1016/j.ijinfomgt.2012.08.004

Lu, L., & Yang, C. (2004). The R&D and Marketing Cooperation across new Product Development Stages. *Industrial Marketing Management, 33*(7), 593–605. doi:10.1016/j.indmarman.2003.11.002

Lundell, B. (2011). e-Governance in public sector ICT-procurement: what is shaping practice in Sweden? *European Journal of ePractice, 12*(6), 66-78.

Lundell, B. (2012). Why do we need Open Standards? In M. Orviska & K. Jakobs (Eds.), *Proceedings 17th EURAS Annual Standardisation Conference 'Standards and Innovation'* (pp. 227-240). The EURAS Board Series.

Lundell, B., & Gamalielsson, J. (2013). Usage of Open Standards and Open Source in Swedish schools: On Promotion of Openness and Transparency, In E. Petrinja et al. (Eds.), Open Source Software: Quality Verification (pp. 207-221). IFIP Advances in Information and Communication Technology.

Lundell, B., & Gamalielsson, J. (2017). On the potential for improved standardisation through use of open source work practices in different standardisation organisations: How can open source-projects contribute to development of IT-standards? In *Digitalisation: Challenge and Opportunity for Standardisation: Proceedings of the 22nd EURAS Annual Standardisation Conference*. EURAS Contributions to Standardisation Research.

Lundell, B., Gamalielsson, J., & Tengblad, S. (2016). IT-standarder, inlåsning och konkurrens: En analys av policy och praktik inom svensk förvaltning. *Uppdragsforskningsrapport, 2016*(2). (in Swedish)

Lundell, B., Gamalielsson, J., Tengblad, S., Hooshyar Yousefi, B., Fischer, T., Johansson, G., . . . Lönroth, E. (2017). Addressing lock-in, interoperability, and long-term maintenance challenges through Open Source: How can companies strategically use Open Source? In *The 13th International Conference on Open Source Systems (OSS 2017), IFIP AICT 496*. Springer.

Lundell, B., Marr, D., Opie, E., Piana, C., & van Rooijen, A. (2016b). *Panel discussion @ Workshop on Open source and standardization: Legal interactions*. ETSI, Sophia Antipolis, France.

Lundell, B., Lings, B., & Syberfeldt, A. (2011). Practitioner perceptions of Open Source software in the embedded systems area. *Journal of Systems and Software, 84*(9), 1540–1549. doi:10.1016/j.jss.2011.03.020

Lundell, B., & van der Linden, F. (2013). Open Source Software as Open Innovation: Experiences from the Medical Domain. In J. S. Z. Eriksson et al. (Eds.), *Managing open innovation technologies* (pp. 3–16). Berlin: Springer. doi:10.1007/978-3-642-31650-0_1

Lu, Y., Morris, K. C., & Frechette, S. (2015). Standards landscape and directions for smart manufacturing systems. In *2015 IEEE International Conference on Automation Science and Engineering (CASE)* (pp. 998-1005). IEEE. doi:10.1109/CoASE.2015.7294229

Lyytinen, K., & Damsgaard, J. (2001). What's wrong with the diffusion of innovation theory? The case of a complex and networked technology. *IFIP TC8 WG8.1 Fourth Working Conference on Diffusing Software Products and Process Innovations*, 173–190.

Lyytinen, K. J., Keil, T., & Fomin, V. V. (2008). A framework to build process theories of anticipatory Information and Communication Technology (ICT) Standardizing. *International Journal of IT Standards and Standardization Research*, *6*(1), 1–38. doi:10.4018/jitsr.2008010101

Lyytinen, K., & King, J. L. (2002). Around the cradle of the wireless revolution: The emergence and evolution of cellular telephony. *Telecommunications Policy*, *26*(3–4), 97–100. doi:10.1016/S0308-5961(02)00002-2

MacCallum, R. C., Widaman, K. F., Preacher, K. J., & Hong, S. (2001). Sample size in factor analysis: The role of model error. *Multivariate Behavioral Research*, *36*(4), 611–637. doi:10.1207/S15327906MBR3604_06 PMID:26822184

MacDonald, T. J., Allen, D. W. E., & Potts, J. (2016). Blockchains and the Boundaries of Self-Organized Economies: Predictions for the Future of Banking. In P. Tasca, T. Aste, L. Pelizzon, & N. Perony (Eds.), *Banking Beyond Banks and Money. New Economic Windows*. Cham: Springer. doi:10.1007/978-3-319-42448-4_14

Mack, E. (2013). *The Bitcoin Pizza Purchase That's Worth $7 Million Today*. Retrieved September 24th, 2017, from https://www.forbes.com/sites/ericmack/2013/12/23/the-bitcoin-pizza-purchase-thats-worth-7-million-today/#24ce03862509

Maier, G., Feldmann, A., Paxson, V., & Allman, M. (2009). On dominant characteristics of residential broadband internet traffic. In *Proceedings of the 9th ACM SIGCOMM conference on Internet measurement conference (IMC '09)*. ACM. doi:10.1145/1644893.1644904

Majchrzak, A., Jarvenpaa, S. L., & Bagherzadeh, M. (2015). A Review of Interorganizational Collaboration Dynamics. *Journal of Management*, *41*(5), 1338–1360. doi:10.1177/0149206314563399

Malinova, K., & Park, A. (2016). *Market Design for Trading with Blockchain Technology*. RMIT University.

Mallinson, K. (2015). Cumulative mobile-SEP royalty payments no more than around 5% of mobile handset revenues. *IP Finance*.

Mallison, K. (2013). *Absurd (F)RAND licensing-rate determinations for SEPs*. Retrieved July 29, 2017, from http://www.ip.finance/2013/11/absurd-frand-licensing-rate.html

Mallison, K. (2014). *Stacking the Deck in Analysis of Smartphone Patent Licensing Costs*. Retrieved July 29, 2017, from http://www.wiseharbor.com/pdfs/Mallinson%20on%20Intel's%20Smartphone%20Royalty%20Stack%2019Sept2014.pdf

Mallison, K. (2015). *Cumulative mobile-SEP royalty payments no more than around 5% of mobile handset revenues*. Retrieved July 29, 2017, from http://www.wiseharbor.com/pdfs/Mallinson%20on%20cumulative%20mobile%20SEP%20royalties%20for%20IP%20Finance%202015Aug19.pdf

Mallison, K. (2016). *"Patent holdup" allegations encourage SEP free-riders*. Retrieved July 29, 2017, from http://www.wiseharbor.com/pdfs/Mallinson%20on%20Holdup%20and%20Holdout%20for%20IP%20Finance%2016%20Aug%202016.pdf

Marcus, J. (2004). Evolving core capabilities of the Internet. *Journal on Telecommunications & High Technology Law*, *3*, 121–161.

Marimon, F., Heras, I., & Casadesús, M. (2009). ISO 9000 and ISO 14000 standards: A projection model for the decline phase. *Total Quality Management*, *20*(1), 1–21. doi:10.1080/14783360802614257

Markham, S. K., Ward, S. J., Aiman-Smith, L., & Kingon, A. I. (2010). The Valley of Death as Context for Role Theory in Product Innovation. *Journal of Product Innovation Management*, *27*(3), 402–417. doi:10.1111/j.1540-5885.2010.00724.x

Marrone, M., Kiessling, M., & Kolbe, L. M. (2010). *Are we really innovating? An exploratory study on Innovation Management and Service Management.* Paper presented at the Management of Innovation and Technology (ICMIT), 2010 IEEE International Conference on.

Martinez-Romo, J., Robles, G., Gonzalez-Barahona, J. M., & Ortuño-Perez, M. (2008). Using Social Network Analysis Techniques to Study Collaboration between a FLOSS Community and a Company. *IFIP International Federation for Information Processing*, *275*, 171–186. doi:10.1007/978-0-387-09684-1_14

Maskus, K., & Merrill, S. A. (Eds.). (2013). *Patent Challenges for Standard-Setting in the Global Economy: Lessons from Information and Communication Technology*. National Academy of Sciences.

Mattila, J. (2016). *The Blockchain Phenomenon – The Disruptive Potential of Distributed Consensus Architectures.* BRIE Working Paper 2016-1.

Mattli, W. (2001). The politics and economics of international institutional standards setting: An introduction. *Journal of European Public Policy*, *8*(3), 328–344. doi:10.1080/13501760110056004

Maurer, N., & Haber, S. (2017). *Patent Trolls or Patent Elves? Evidence from Publicly-Traded Patent Assertion Entities.* Hoover IP2 Working Paper 17003.

McDougall, P. P., & Oviatt, B. M. (2000). International entrepreneurship: The intersection of two research paths. *Academy of Management Journal*, *43*(5), 902–906. doi:10.2307/1556418

McDougall, P. P., & Oviatt, B. M. (2003). Some Fundamental Issues in International Entrepreneurship. *Entrepreneurship Theory and Practice*, (July): 2003.

McNaughton, B., Ray, P., & Lewis, L. (2010). Designing an evaluation framework for IT service management. *Information & Management*, *47*(4), 219–225. doi:10.1016/j.im.2010.02.003

McNaughton, R. B. (2000). Determinants of time-span to foreign market entry. *Journal of Euromarketing*, *9*(2), 99–112. doi:10.1300/J037v09n02_06

Medeišis, A. (2014). European Regulatory Developments related to CR. In Cognitive Radio Policy and Regulation Techno-Economic Studies to Facilitate Development of Cognitive Radio. Academic Press.

Medeišis, A., & Holland, O. (Eds.). (2014). *Cognitive Radio Policy and Regulation Techno-Economic Studies to Facilitate Development of Cognitive Radio*. Springer. doi:10.1007/978-3-319-04022-6

Mesquida, A. L., Mas, A., Amengual, E., & Calvo-Manzano, J. A. (2011). IT Service Management Process Improvement based on ISO/IEC 15504: A systematic review. *Information and Software Technology*, *54*(3), 239–247. doi:10.1016/j.infsof.2011.11.002

Microsoft, Corp. v. Motorola, Inc., No. 10-1823 (W.D. Wash. 2013) (2013)

Microsoft, Corp. v. Motorola, Inc., No. 14-35393 (9th Circ., 2015) (2015).

Microsoft-Inc. (2012). *Microsoft Operations Framework*. MOF.

Miles, R. E., Snow, C. C., Meyer, A. D., & Coleman, H. J. (1978). Organizational strategy, structure, and process. *Academy of Management Review, 3*(3), 546–562. PMID:10238389

Milinkovich, M. (2015). *How the Eclipse community works.* Keynote presentation @ the 11th International Conference on Open Source Systems, Florence, Italy.

MINTC. (2004). *Mobile services market in Finland 2003, Ministry of Transport and Communications.* Retrieved from http://urn.fi/URN:ISBN:951-723-710-3

Mione, A. (1994). *Les normes, instrument de la stratégie marketing de l'entreprise* (Thèse de doctorat). Université de Montpellier.

Mitchell, I. G., & Mason, S. (2010). Compatibility Of The Licensing Of Embedded Patents With Open Source Licensing Terms. *International Free and Open Source Software Law Review, 3*(1), 25–58. doi:10.5033/ifosslr.v3i1.57

Moser, P. (2005). How Do Patent Laws Influence Innovation? Evidence from Nineteenth-Century World's Fairs. *The American Economic Review, 95*(4), 1214-1236. Retrieved from http://www.jstor.org/stable/4132712

Mougayar, W. (2017). *Why the Blockchain Still Lacks Mass Understanding.* Retrieved September 24th, 2017, from http://www.huffingtonpost.com/entry/why-the-blockchain-still-lacks-mass-understanding_us_587132afe4b08052400ee2f8

Muhr, T. (1991). Atlas/ti - a prototype for the support of text interpretation. *Qualitative Sociology, 14*(4), 349–371. doi:10.1007/BF00989645

Müller, S. D., Mathiassen, L., & Balshøj, H. H. (2010). Software Process Improvement as organizational change: A metaphorical analysis of the literature. *Journal of Systems and Software, 83*(11), 2128–2146. doi:10.1016/j.jss.2010.06.017

Nair, H., Chintagunta, P., & Dubé, J. (2004). Empirical analysis of indirect network effects in the market for personal digital assistants. *Quantitative Marketing and Economics, 2*(1), 23–58. doi:10.1023/B:QMEC.0000017034.98302.44

Nakamoto, S. (2008). *Bitcoin: A peer-to-peer electronic cash system.* Retrieved September 24, 2017, from https://people.eecs.berkeley.edu/~raluca/cs261-f15/readings/bitcoin.pdf

Nalebuff, B. J. (1997). *Co-Opetition.* New York: Crown Business.

Narayanan, A., Bonneau, J., Felten, E., Miller, A., & Goldfeder, S. (2016). *Bitcoin and Cryptocurrency Technologies: A Comprehensive Introduction.* Princeton University Press.

NASDAQ. (2008). *Company List.* Retrieved September 16, 2013, from http://www.nasdaq.com/asp/symbols.asp?exchange=Q

Navigant. (Eds.). (2011). *'Smart City' report.* Navigant Research.

Nazier, D. (2014). *Infamous Wi-Fi Patent Troll Settles For Peanuts.* Electronic Frontier Foundation. Retrieved from www.eff.org

Ndlovu, S. L., & Kyobe, M. E. (2016). *Challenges of CoBIT 5 IT Governance Framework Migration.* Paper presented at the CONF-IRM.

Neate, R. (2017). *Bitcoin value plummets after China orders trading in currency to cease.* Retrieved September 24th, 2017, from https://www.theguardian.com/technology/2017/sep/15/bitcoin-value-plummets-after-china-orders-trading-in-currency-to-cease

Nemec, J., Orviska, M., & Hudson, J. (2012). Standardization and the European Standards Organizations. In *EURAS Proceedings 2012: "Standards and Innovation"* (pp. 281–294). Banska Bystrica, Slovakia: The EURAS Board Series.

Newman, M. E. J. (2001). The structure of scientific collaboration networks. *Proceedings of the National Academy of Sciences of the United States of America*, *98*(2), 404–409. doi:10.1073/pnas.98.2.404 PMID:11149952

Nfuka, E. N., & Rusu, L. (2011). Critical Success Factors Framework For Implementing Effective IT Governance In Public Sector Organizations In A Developing Country. *AMCIS 2011 Proceedings.*

Ngwenyama, O., & Nielsen, P. A. (2003). Competing values in software process improvement: An assumption analysis of CMM from an organizational culture perspective. *IEEE Transactions on Engineering Management*, *50*(1), 100–112. doi:10.1109/TEM.2002.808267

Niazi, M., Wilson, D., & Zowghi, D. (2005). A maturity model for the implementation of software process improvement: An empirical study. *Journal of Systems and Software*, *74*(2), 155–172. doi:10.1016/j.jss.2003.10.017

Nickerson, J. V., & zur Muehlen, M. (2006). The Ecology Of Standards Processes: Insights From Internet Standard Making. *MIS Quarterly*, *30*, 467–488.

NIST. (2011). *Cloud Computing Standards Roadmap Working Group*. Special Publication 500-29, July 2011. Retrieved August 30, 2017, from http://www.nist.gov/itl/cloud/upload/NIST_SP-500-291_Jul5A.pdf

NIST. (2011). *NIST Cloud Computing Standards Roadmap*. Special Publication 500-291. Retrieved from http://www.nist.gov/customcf/get_pdf.cfm?pub_id=909024

NIST. (Ed.). (2013). *Strategic R&D Opportunities for 21st Century Cyber-Physical Systems*. Retrieved from http://www.nist.gov/el/upload/12-Cyber-Physical-Systems020113_final.pdf

NIST. (Ed.). (2016). *Framework for Cyber-Physical Systems, Release 1.0*. Retrieved from https://s3.amazonaws.com/nist-sgcps/cpspwg/files/pwgglobal/CPS_PWG_Framework_for_Cyber_Physical_Systems_Release_1_0Final.pdf

NOC. (2007). *The Netherlands in Open Connection: An action plan for the use of Open Standards and Open Source Software in the public and semi-public sector*. Ministry of Economic Affairs. Retrieved from https://www.ictu.nl/archief/noiv.nl/files/2009/12/Action_plan_english.pdf

Novak, L. L., Anders, S., Gadd, C. S., & Lorenzi, N. M. (2012). Mediation of adoption and use: A key strategy for mitigating unintended consequences of health IT implementation. *Journal of the American Medical Informatics Association*, *19*(6), 1043–1049. doi:10.1136/amiajnl-2011-000575 PMID:22634157

Novak, L., Brooks, J., Gadd, C., Anders, S., & Lorenzi, N. (2012a). Mediating the intersections of organizational routines during the introduction of a health IT system. *European Journal of Information Systems*, *21*(5), 552–569. doi:10.1057/ejis.2012.2 PMID:24357898

NOVFS. (2011). *Open Document Standards for the Government Guide*. Programma Nederland Open in Verbinding & Forum Standaardisatie. Retrieved from https://www.forumstandaardisatie.nl/fileadmin/ os/documenten/Handreiking_ODF_Engelse_versie.pdf

Nunnally, J. (1978). *Psychometric theory*. McGraw-Hill.

Odell, M. (2009). European Public Sector Award. Public speech by Swedish minister, Maastricht.

Ohlhausen, M. K. (2015). *Antitrust Oversight of Standard-Essential Patents: The Role of Injunctions*. Paper presented at the Antitrust Forum, Beijing, China.

Ohlhausen, M. K. (2017). *The Elusive Role of Competition in the Standard-Setting Antitrust Debate*. Retrieved July 29, 2017, from https://www.ftc.gov/public-statements/2017/07/elusive-role-competition-standard-setting-antitrust-debate

Okamura, K., Orlikowski, W., Fujimoto, M., & Yates, J. (1994). Helping CSCW applications succeed: the role of mediators in the context of use. *Proceedings of the 1994 ACM conference on computer supported cooperative work*, 55-65. doi:10.1145/192844.192871

Oksala, S., Rutkowski, A., Spring, M. B., & O'Donnell, J. (1996). The Structure of IT Standardization. *StandardView*, *4*(1), 9–22.

Okun-Kozlowicki, J. (2016). *Standards and Regulations: Measuring the Link to Goods Trade.* Office of Standards and Investment Policy International Trade Administration, U.S. Department of Commerce.

Olleros, F. (1986). Emerging Industries and the Burnout of Pioneers. *Journal of Product Innovation Management*, *1*(1), 5–18. doi:10.1016/0737-6782(86)90039-1

Open Mobile Alliance. (2002). *WAP 2.0 - Technical White Paper.* Author.

Open Mobile Alliance. (2006). *User Agent Profile V2.0.* Author.

Orlikowski, W. (1992). The duality of technology: Rethinking the concept of technology in organizations. *Organization Science*, *3*(3), 398–427. doi:10.1287/orsc.3.3.398

Orlikowski, W. (1996). Improvising organizational transformation over time: A situated change perspective. *Information Systems Research*, *7*(1), 63–92. doi:10.1287/isre.7.1.63

Orlikowski, W., Yates, J., Okamura, K., & Fujimoto, M. (1995). Shaping electronic communication: The metastructuring of technology in the context of use. *Organization Science*, *6*(4), 423–444. doi:10.1287/orsc.6.4.423

Ortmann, G. (2010). On drifting rules and standards. *Scandinavian Journal of Management*, *26*(2), 204–214. doi:10.1016/j.scaman.2010.02.004

Ortt, J. R., & Suprapto, M. (2011). The Role of Strategic Niches in Creating Large-scale Applications for High-tech Products. *International Association for Management of Technology Conference Proceedings.*

Ortt, J. R., Shah, C. M., & Zegveld, M. A. (2007). Strategies to Commercialize Breakthrough Technologies. *International Association for Management of Technology Conference Proceedings.*

Ortt, J. R. (2010). Understanding the Pre-diffusion Phases. In J. Tidd (Ed.), *Gaining Momentum Managing the Diffusion of Innovations* (pp. 47–80). London: Imperial College Press. doi:10.1142/9781848163553_0002

Ortt, J. R., & Delgoshaie, N. (2008). Why does it take so long before the diffusion of new high-tech products takes off? In B. Abu-Hijleh, M. Arif, T. Khalil, & Y. Hosni (Eds.), *Creating & Managing a Knowledge Economy* (pp. 2–15). Dubai: The Britisch University in Dubai.

Ortt, J. R., & Schoormans, J. P. L. (2004). The Pattern of Development and Diffusion of Breakthrough Communication Technologies. *European Journal of Innovation Management*, *7*(4), 292–302. doi:10.1108/14601060410565047

Ortt, J. R., & Suprapto, M. (2010). Niches Prior to Large-scale Application of Radically New High-Tech Products. *ERSCP-EMSU Conference proceedings.*

Orzeck, K. (2015). TCL Wins Anti-Suit Injunction in Ericsson Patent Row. *Law360.* Retrieved July 29, 2017, from https://www.law360.com/articles/678680/tcl-wins-anti-suit-injunction-in-ericsson-patent-row

O'Shields, R. (2017). Smart Contracts: Legal Agreements for the Blockchain. *Banking Institute Journal*, *21*(1).

OSI. (2017). *Open Source Initiative*. Retrieved from opensource.org

Ostrom, E. (1990). *Governing the Commons – The Evolution of Institutions for Collective Action*. Cambridge University Press. doi:10.1017/CBO9780511807763

OTA. (1990). *1990 Critical Connections: Communications for the Future. U.S. Congress, Office of Technology Assessment*. Washington, DC: U.S. Government Printing Office.

OTA. (1992). *Global Standards: Building Blocks for the Future, TCT-512*. Washington, DC: Congress of the United States, Office of Technology Assessment. Retrieved from http://www.princeton.edu/~ota/disk1/1992/9220/9220.PDF

OTA. (1992). *Global Standards: Building Blocks for the Future. U.S. Congress, Office of Technology Assessment*. Washington, DC: U.S. Government Printing Office.

Othman, M. F. I., Chan, T., & Foo, E. (2011). *IT governance adoption in Malaysia: a preliminary investigation*. Academic Press.

Oviatt, B. M., & McDougall, P. P. (1994). Toward a Theory of International New Ventures. *Journal of International Business Studies*, *25*(1), 45–64. doi:10.1057/palgrave.jibs.8490193

Oviatt, B. M., & McDougall, P. P. (1997). Challenges for internationalisation process theory: The case of international new ventures. *Management International Review*, *2*(Special Issue), 85–99.

Ozment, A., & Schechter, S. E. (2006). Bootstrapping the adoption of Internet security protocols. *Proceedings of the Fifth Workshop on the Economics of Information Security*.

Padilla, J., Geradin, D., & Layne-Farrar, A. (2007). *Royalty Stacking In High Tech Industries: Separating Myth From Reality*. Working Paper, CEMFI.

Paoli, M., & Prencipe, A. (1999). *The role of knowledge in complex product systems; some empirical evidence from the aero engine industry. Journal of Management and Governance*.

Peak, D., Guynes, C. S., & Kroon, V. (2005). Information technology alignment planning—A case study. *Information & Management*, *42*(5), 635–649. doi:10.1016/j.im.2004.02.009

Pech, R. J. (2003). Memetics and innovation: Profit through balanced meme management. *European Journal of Innovation Management*, *6*(2), 111–117. doi:10.1108/14601060310475264

Pentheroudakis, C., & Baron, J. A. (2017). *Licensing Terms for Standard Essential Patents: A Comprehensive Analysis of Cases*. JRC Science for policy report, European Commission.

Pereira, A. A., Jr. (2015, Fall). TCL v. Ericsson: Brazil's First Standard Essential Patent Antitrust Investigation. *Newsletter of the Antitrust Section's Intellectual Property Committee,* 18-27.

Permanent Representatives Committee. (2016). Draft Council conclusions on the "Digital Single Market Technologies and Public Services Modernisation" package. Statement by the United Kingdom, Estonia, Belgium, Slovenia, Poland, Latvia and Malta, 8735/16 ADD 1, 26 May, Brussels.

Peters, G., Panayi, E., & Chapelle, A. (2015). *Trends in crypto-currencies and blockchain technologies: A monetary theory and regulation perspective*. arXiv:1508.04364v1

Petersen, S. S., Blind, K., & Stephan, M. (2012). The Interrelation of Strategic Patenting and Standardization in the Context of the Resource-Based View of the Firm. *EURAS Proceedings 2012 AachenerBeiträgezurInformatik*.

Petit, N. (2015). Huawei v. ZTE: Judicial Conservatism at the Patent-Antitrust Section. *CPI Antitrust Chronicle, 12*(2).

Petit, N. (2016). *The Smallest Salable Patent Practicing Unit ("SSPPU") Experiment, General Purpose, Technologies and the Coase Theorem*. Retrieved from https://papers.ssrn.com/sol3/papers.cfm?abstract_id=2734245

Petrusson, U. (2005). *Intellectual Property & Entrepreneurship: Creating Wealth in an Intellectual Value Chain*. Gothenburg: Center for Intellectual Property.

Petrusson, U., & Heiden, B. (2009). *Assets, Property and Capital in a Globalized Intellectual Value Chain. In From Assets to Profits*. John Wiley & Sons.

Pierce, J., & Medzmaraishvili, M. (2015). Standard Essential Patent and Injunctions – A Question of Imbalance? *EURAS Proceedings 2015*, 305-321.

Pohlmann, T. (2014). The Evolution of ICT Standards Consortia. *The DigiWorld Economic Journal*, (95), 17–40.

Pollard, C., & Cater-Steel, A. (2009). Justifications, Strategies, and Critical Success Factors in Successful ITIL Implementations in U.S. and Australian Companies: An Exploratory Study. *Information Systems Management*, *26*(2), 164–175. doi:10.1080/10580530902797540

Polyakova, M., & Rubin, G. (2017). Integrated methodology for standard-setting norms of innovative product in the new competitive environment. AIP Conference Proceedings (Vol. 1858, No. 1, p. 040005). AIP Publishing. doi:10.1063/1.4989954

Porter, M. E., & Kramer, M. R. (2011). Creating shared value. *Harvard Business Review*, *89*(1/2), 62–77.

Potgieter, B., Botha, J., & Lew, C. (2005). *Evidence that use of the ITIL framework is effective*. Paper presented at the 18th annual conference of the national advisory committee on computing qualifications, Tauranga.

Powerledger. (2017). *Powerledger White paper*. Retrieved September 24th, 2017, from https://tge.powerledger.io/media/Power-Ledger-Whitepaper-v3.pdf

Power, M. (1997). *The audit society: rituals of verification*. Oxford, UK: Oxford University Press.

Prescott, M. M., & Van Slyke, C. (1997). Understanding the Internet as an innovation. *Industrial Management & Data Systems*, *97*(3), 119–124. doi:10.1108/02635579710173185

Quinones, P.-A., Teasley, S. D., & Lonn, S. (2013). Appropriation by unanticipated users: looking beyond design intent and expected use. *Proceedings of the 2013 conference on computer supported cooperative work*, 1515-1526. doi:10.1145/2441776.2441949

Rader, R.R. (2014). Chief Judge, dissenting-in-part in U.S. Court of Appeals for the Federal Circuit, *Apple, Inc. and NeXT Software, Inc., v. Motorola Inc. and Motorola Mobility, Inc.*

Rauber, J. (2014). Why do Firms join SSOs - Exploring the Influence of Firm Characteristics using Survival Analyses An empirical Analysis for Germany. *EURAS Proceedings 2014*, 197-210.

Rechtbank's Gravenhage. (2010, March 17). *Philips v. SK Kassetten*, HA ZA 08-2522 en HA-ZA 08-2524.

Rechtbank's Gravenhage. (2017, February 10). *Archos v. Koninklijke Philips N.V.*, ECLI:NL:RBDHA:2017:1025.

Regulation (EU) No 1025/2012 of the European Parliament and of the Council of 25 October 2012 on European standardisation, amending Council Directives 89/686/EEC and 93/15/EEC and Directives 94/9/EC, 94/25/EC, 95/16/EC, 97/23/EC, 98/34/EC, 2004/22/EC, 2007/23/EC, 2009/23/EC and 2009/105/EC of the European Parliament and of the Council and repealing Council Decision 87/95/EEC and Decision No 1673/2006/EC of the European Parliament and of the Council.

Rennie, M. W. (1993). Global competitiveness. Born global. *The McKinsey Quarterly*, *3*, 45–52.

Riikonen, A., & Smura, T. (2013). *Mobile handset population in Finland 2005-2012*. Technical report. Retrieved from http://momie.comnet.aalto.fi/wp-uploads/Mobile_Handset_Population_2005-2012.pdf

Riikonen, A., Smura, T., Kivi, A., & Töyli, J. (2013). Diffusion of mobile handset features: Analysis of turning points and stages. *Telecommunications Policy*, *37*(6–7), 563–572. doi:10.1016/j.telpol.2012.07.011

Riillo, C. A. F. (2014). *ICT Standardization and use of ICT standards: a firm level analysis*. Available at http://mpra.ub.uni-muenchen.de/id/eprint/63436

Riillo, C. A. F. (2014). *The Engagement in Standardization Activities A Firm Level Analysis of Formal and Company Standardization*. Retrieved from http://www.statistiques.public.lu/catalogue-publications/economie-statistiques/2014/70-2014.pdf

Riquelme, H. (2001). Do consumers know what they want? *Journal of Consumer Marketing*, *18*(5), 437–448. doi:10.1108/07363760110398772

Ritter, T., Wilkinson, I. F., & Johnston, W. J. (2004). Managing in complex business networks. *Industrial Marketing Management*, *33*(3), 175–183. doi:10.1016/j.indmarman.2003.10.016

Rodon, J., Sese, F., & Christiaanse, E. (2011). Exploring users' appropriation and post-implementation managerial intervention in the context of industry IOIS. *Information Systems Journal*, *21*(3), 223–248. doi:10.1111/j.1365-2575.2009.00339.x

Rogers, E. M. (1995). *Diffusion of innovations*. New York, NY: Free Press.

Rogers, E. M. (2003). *Diffusion of Innovations* (5th ed.). New York: Free Press.

Rosenblatt, J., Womack, B., & Satariano, A. (2014). *Apple and Google declare cease-fire in legal war.* Retrieved July 29, 2017, from http://www.canada.com/technology/apple+google+declare+cease+fire+legal/9848740/story.html

Rosenbloom, R. S., & Cusumano, M. A. (1987). Technological Pioneering and Competitive Advantage: The Birth of the VCR Industry. *California Management Review*, *29*(4), 51–76. doi:10.2307/41162131

Rouhani, S., & Zare Ravasan, A. (2014). A Fuzzy TOPSIS based Approach for ITSM Software Selection. *International Journal of IT/Business Alignment and Governance*, *5*(2), 1–26. doi:10.4018/ijitbag.2014070101

Rysman, M., & Simcoe, T. (2008). Patents and the Performance of Voluntary Standard-Setting Organizations. *Management Science*, *54*(11), 1920–1934. doi:10.1287/mnsc.1080.0919

Rysman, M., & Simcoe, T. (2011). A NAASTy Alternative to RAND Pricing Commitments. *Telecommunications Policy*, *35*(11), 11. doi:10.1016/j.telpol.2011.07.003

Sahal, D. (1981). *Patterns of Technological Innovation*. Addison-Wesley.

Sales, B., Darmois, E., Papadimitriou, D., & Bourse, D. (2012). A systematic approach for closing the research to standardization gap. *The Future Internet*, 18–29.

Salle, M., & Rosenthal, S. (2005). Formulating and Implementing an HP IT program strategy using CobiT and HP ITSM. *Proceedings of the 38th Hawaii International Conference on System Sciences*. doi:10.1109/HICSS.2005.276

Salop, S. C., & Scheffman, D. T. (1983). Raising Rivals' Costs. *The American Economic Review*, *73*(2), 267–271.

Sandholtz, K. W. (2012). Making standards stick: A theory of coupled vs. decoupled compliance. *Organization Studies*, *33*(5-6), 655–679. doi:10.1177/0170840612443623

Savall, H., Zardet, V., & Bonnet, M. (2008). *Libérer les performances cachées des entreprises par un management socio-économique (2nd ed.)*. Genève: ISEOR.

Schankerman, M. (1998). How Valuable is Patent Protection? Estimates by Technology Field. *The Rand Journal of Economics*, *29*(1), 1. doi:10.2307/2555817

Schewe, G. (1992). Imitations management. Academic Press.

Schilling, M. A. (n.d.). Technological Leapfrogging: Lessons from the U.S. Video Game Console Industry. *Harvard Business Review*.

Schmalensee, R. (2009). Standard-Setting, Innovation Specialists and Competition Policy. *The Journal of Industrial Economics*, *57*(3), 526–552. doi:10.1111/j.1467-6451.2009.00388.x

Schmidt, S. K., & Werle, R. (1998). Coordinating Technology. Studies in the International Standardization of Tellecommunications. Cambridge, MA: The MIT Press.

Schnaars, S. P. (1989). *Megamistakes; Forecasting and the Myth of Rapid Technological Change*. New York: The Free Press.

Schneberger, S. L., & McLean, E. R. (2003). The Complexity Cross - Implications for Practice. *Communications of the ACM*, *46*(9), 216–225. doi:10.1145/903893.903940

Scholten, V., van de Kaa, G., & Trott, P. (2013). Riding the waves: Developing theory on start-up firm networking in the pre dominance design stage. Academic Press.

Schot, J., & Geels, F. (2007). Niches in evolutionary theories of technical change. *Journal of Evolutionary Economics*, *17*(5), 605–622. doi:10.1007/s00191-007-0057-5

Schot, J., Hoogma, R., & Elzen, B. (1994). Strategies for shifting technological systems. *Futures*, *26*(10), 1060–1076. doi:10.1016/0016-3287(94)90073-6

ScientiaMobile. (2012). *WURFL repository v. 2.3.4*. Retrieved from http://wurfl.sourceforge.net

Scott Morton, F., & Shapiro, C. (2016). Patent Assertions: Are We Any Closer to Aligning Reward to Contribution? *Innovation Policy and the Economy*, *16*(1), 89–133. doi:10.1086/684987

Scott, W. (1994). Institutional analysis: variance and process theory approaches. In Institutional Environments and Organisations: Structural Complexity and Individualism. Academic Press.

SEC. (2017). *Report of Investigation Pursuant to Section 21(a) of the Securities Exchange Act of 1934: The DAO*. Security and Exchange Commission, Release No. 81207.

Seck, M., & Evans, D. (2004). Major U.S. Cities Using National Standard Fire Hydrants, One Century After the Great Baltimore Fire. *NISTIR*, 7158. Retrieved August 30, 2017, from http://www.fire.nist.gov/bfrlpubs/fire04/PDF/f04095.pdf

Seidl, D. (2007). Standard setting and following in corporate governance: An observation-theoretical study of the effectiveness of governance codes. *Organization*, *14*(5), 705–727. doi:10.1177/1350508407080316

Senden, M. J., & Wöckel, I. (1997). Gelebte Unternehmensphilosophie contra Normierungszwaenge: ISO 9000 im DLR-Wissenschaftsmanagement'. *Wissenschaftsmanagement*, *3*, 269–272.

Seo, D. B.DongBack. (2013). Analysis of Various Structures of Standards Setting Organizations (SSOs) that Impact Tension among Members. *International Journal of IT Standards and Standardization Research*, *11*(2), 46–60. doi:10.4018/jitsr.2013070104

Shapiro, C. (2001). Navigating the Patent Thicket: Cross Licenses, Patent Pools, and Standard-Setting. In A. B. Jaffe et al. (Eds.), *Innovation Policy and the Economy I* (pp. 119–150). Cambridge, MA: MIT Press. doi:10.2139/ssrn.273550

Shapiro, C. (2001) Setting Compatibility Standards: Cooperation or Collusion? In R. Dreyfuss, D. L. Zimmerman, & H. First (Eds.), *Expanding the Boundaries of Intellectual Property Innovation Policy for the Knowledge Society* (pp. 81–101). Oxford.

Shapiro, C. (2010). Injunctions, Hold-Up, and Patent Royalties. *American Law and Economics Review*, *12*(2), 280–317. doi:10.1093/aler/ahq014

Shapiro, C., & Varian, H. R. (1998, November). Versioning: The smart way to sell information. *Harvard Business Review*, 106–114.

Shapiro, C., & Varian, H. R. (1999a). Art of Standard Wars. *California Management Review*, *41*(2), 8–32. doi:10.2307/41165984

Shapiro, C., & Varian, H. R. (1999b). *Information Rules: A Strategic Guide to the Network Economy*. Cambridge.

Sherif, M. H. (2001). A Framework for Standardization in Telecommunications and Information Technology. *IEEE Communications Magazine*, *39*(April), 94–100. doi:10.1109/35.917510

Sherif, M. H. (2003). When is Standardization Slow? *International Journal of IT Standards and Standardization Research*, *1*(1), 19–32. doi:10.4018/jitsr.2003010102

Shih, C., & Venkatesh, A. (2004). Beyond adoption: Development and application of a use-diffusion model. *Journal of Marketing*, *68*(1), 59–72. doi:10.1509/jmkg.68.1.59.24029

Shin, D. H., Kim, H., & Hwang, J. (2015). Standardization revisited: A critical literature review on standards and innovation. *Computer Standards & Interfaces*, *38*, 152–157. doi:10.1016/j.csi.2014.09.002

Sidak, G. J. (2008). Holdup, Royalty Stacking, and the Presumption of Injunctive Relief for Patent Infringement: A Reply to Lemley and Shapiro. *Minnesota Law Journal, 92.*

Sidak, G. J. (2013). The Meaning of FRAND, Part I: Royalties. *Jnl of Competition Law & Economics, 9*(4).

Sidak, G. J. (2016). Apportionment, FRAND Royalites, and Comparable Licenses after Ericsson v. D-Link. *U. Ill. L. Rev.*, 1809.

Sidak, G. (2015). The Meaning of FRAND, Part II: Injunctions. *Journal of Competition Law & Economics*, *11*(1), 1–69.

Sidak, G. J. (2015). The Antitrust Division's Devaluation of Standard-Essential Patents. *104 Georgetown L.J. Online (Bergheim)*, *48*, 61.

Siebrasse, N., & Cotter, T. F. (2015). A New Framework for Determining Reasonable Royalties in Patent Litigation. *Florida Law Review*.

Simcoe, T. (2015). Modularity and the evolution of the Internet. In Economical Analysis of the Digital Economy (pp. 21–54). Chicago: University of Chicago Press.

Simcoe, T. (2012). Standard setting committees: Consensus governance for shared technology platforms. *The American Economic Review*, *102*(1), 305–336. doi:10.1257/aer.102.1.305

Simcoe, T. S. (2006). Open Standards and intellectual property rights. In H. Chesbrough et al. (Eds.), *Open Innovation researching a new paradigm*. Oxford, UK: Oxford University Press.

Simeonova, D. (2015). *Digital Solutions: Can we 'Open Source' the Future? Commission en direct #21*. European Commission.

SIP Forum. (n.d.). Retrieved from http://www.sipforum.org/content/view/13/34/

Smura, T., Kivi, A., & Töyli, J. (2009). A framework for analysing the usage of mobile services. *Info*, *11*(4), 53–67. doi:10.1108/14636690910970973

Smura, T., Kivi, A., & Töyli, J. (2011). Mobile data services in Finland: Usage of networks, devices, applications and content. *International Journal of Electronic Business*, *9*(1), 138–157. doi:10.1504/IJEB.2011.040359

Snow, P., Deery, B., Lu, J., Johnston, D., & Kirby, P. (2014). *Factom white paper 1.0*. Retrieved September 24th, 2017, from https://www.factom.com/devs/docs/guide/factom-white-paper-1-0

Software Engineering Institute (SEI). (2012). *Change Requests*. Retrieved January 15, 2017 from http://www.sei.cmu.edu/cmmi/solutions/crs/index.cfm

Soh, P.-H., & Roberts, E. B. (2003). Networks of innovators: A longitudinal perspective. *Research Policy*, *32*(9), 1569–1588. doi:10.1016/S0048-7333(03)00065-9

Soja, P. (2008). Difficulties in enterprise system implementation in emerging economies: Insights from an exploratory study in Poland. *Information Technology for Development*, *14*(1), 31–51. doi:10.1002/itdj.20086

SOU. (2007). Standarder och tillväxt: en kommenterad forskningsöversikt – Standards and growth: a research review. SOU, Statens offentliga utredningar nr. 2007, 83, Stockholm.

SOU. (2009). *Strategi för myndigheternas arbete med e-förvaltning*. Statens Offentliga Utredningar: SOU 2009:86, e-Delegationen, Finansdepartementet, Regeringskansliet, Stockholm. (in Swedish)

Spremic, M., Zmirak, Z., & Kraljevic, K. (2008). *IT and business process performance management: Case study of ITIL implementation in finance service industry*. Paper presented at the Information Technology Interfaces, 2008. ITI 2008. 30th International Conference on.

Spring, M. B., & Weiss, M. B. (1999). Intellectual Property Issues in Standardization. In K. Jakobs (Ed.), Information Technology Standards and Standardization: A Global Perspective (pp. 63-79). Hershey, PA: IGI Global.

Spring, M. B., & Weiss, M. B. (1995). Financing the Standards Development Process. In B. Kahin & J. Abbatte (Eds.), *Standards Policy for Information Infrastructure* (pp. 289–320). Cambridge, MA: The MIT Press.

Standaardisatie. (2014). *Lijst met open standaarden, Forum Standaardisatie*. Retrieved from https://lijsten.forumstandaardisatie.nl/lijsten/open-standaarden/

Standardisation Forum. (2011). *A guide for government organisations Governance of Open Standards*. The Standardisation Forum, The Hague.

Star, S. L. (2010). This is not a boundary object: Reflections on the origin of a concept. *Science, Technology & Human Values*, *35*(5), 601–617. doi:10.1177/0162243910377624

Star, S. L., & Griesemer, J. R. (1989). Institutional Ecology, 'Translations' and Boundary Objects: Amateurs and Professionals in Berkeley's Museum of Vertebrate Zoology. *Social Studies of Science*, *19*(3), 387–420. doi:10.1177/030631289019003001

Stasik, E. (2010, September). Royalty Rates and Licensing Strategies For Essential Patents On LTE (4G) Telecommunication Standards. *les Nouvelles*.

STATEC. (2015). *Sources and methods*. Available: http://www.statistiques.public.lu/en/methodology/methodes/index.html

Steinmueller, W. E. (2003). *The role of technical standards in coordinating the division of labour in complex system industries*. Oxford, UK: Oxford University Press. doi:10.1093/0199263221.003.0008

Sterman, J. D. (2002). All models are wrong: Reflections on becoming a systems scientist. *System Dynamics Review*, *18*(4), 501–531. doi:10.1002/sdr.261

Stewart, J., Shen, X., Wang, C., & Graham, I. (2011). From 3G to 4G: Standards and the development of mobile broadband in China. *Technology Analysis and Strategic Management*, *23*(7), 773–788. doi:10.1080/09537325.2011.592284

Stinchcombe, A. L. (1965). Social structure and organizations. In J. G. March (Ed.), *Handbook of organizations*. Chicago.

Strauss, A., & Corbin, J. (1996). *Basics of Qualitative Research*. Academic Press.

Stray, S., Bridgewater, S., & Murray, G. (2001). The internationalisation process of small, technology-based firms. Market selection, mode choice and degree of internationalisation. *Journal of Global Marketing*, *25*(1), 7–29. doi:10.1300/J042v15n01_02

Strogatz, S. H. (2001). Exploring complex networks. *Nature*, *410*(6825), 268–276. doi:10.1038/35065725 PMID:11258382

Suarez, F. F. (2004). Battles for technological dominance: An integrative framework. *Research Policy*, *33*(2), 271–286. doi:10.1016/j.respol.2003.07.001

Suarez, F. F., & Utterback, J. M. (1995). Dominant Designs and the Survival of the Firms. *Strategic Management Journal*, *16*(6), 415–430. doi:10.1002/smj.4250160602

Sukarevičienė, G., & Fomin, V. V. (2013). *Modeling Dynamic Spectrum Access Infrastructure Development As a Techno-Economic System. Research in progress*. Brussels: EURAS.

Sumer, K., & Bayraktar, C. A. (2012). Business Strategies and Gaps in Porter's Typology: A Literature Review. *Journal of Management Research*, *4*(3). doi:10.5296/jmr.v4i3.1721

Swan, M. (2015). *Blockchain: blueprint for a new economy*. Sebastopol, CA: O'Reilly Media.

Swann G. M. P. (2010). *The Economics of Standardization: An Update*. Report for the UK Department of Business, Innovation and Skills.

Swann, G. M. P., & Lambert, R. (2010). Why do Standards Enable and Constrain Innovation? In *Proceedings of the 15th EURAS Conference EURAS contributions to standardization research* (vol. 2). Mainz Academic Publishers.

Swann, G. M. P., & Lambert, R. J. (2010). Why do Standards Enable *and* Constrain Innovation? *EURAS Proceedings 2010: Service Standardization*, 357-376.

Swann, P. (2000). The Economics of standardization. Final Report for Standards and Technical Regulations Directorate Department of Trade and Industry.

Swann, P. G. M. (2010). *The Economics of Standardization: An Update*. Academic Press.

Swann, G. M. P. (1990). Standards and the Growth of a Software Network: A Case Study of PC Applications Software. In J. Berg & H. Schumny (Eds.), *An Analysis of the IT Standardisation Process*. Amsterdam: Elsevier Science Publishers. doi:10.1016/B978-0-444-87390-3.50050-5

Swann, G. M. P. (1999). *The Economics of Measurement*. London: Department of Trade and Industry, Standards and Technical Regulations Directorate.

Swann, G. M. P. (2000). *The Economics of Standardization*. London: Department of Trade and Industry, Standards and Technical Regulations Directorate.

Swann, G. M. P., & Watts, T. P. (2002). Visualisation Needs Vision: The Pre-Paradigmatic Character of Virtual Reality. In S. Woolgar (Ed.). *The Virtual Society? Technology, Cyberbole, Reality*. Oxford, UK: Oxford University Press.

Swann, G. P. (1987). Industry Standard Microprocessors and the Strategy of Second Source Production. In H. L. Gabel (Ed.), (pp. 239–262). Amsterdam: Product Standardization and Competitive Strategy.

Swanson, D., & Baumol, W. (2005). Reasonable and Nondiscriminatiory (RAND) Royalties, Standards Selection, and Control of Market Power. *Antitrust Law Journal*, *73*, 1.

Tallberg, M., Hämmäinen, H., Töyli, J., Kamppari, S., & Kivi, A. (2007). Impacts of handset bundling on mobile data usage: Case Finland. *Telecommunications Policy*, *31*(10), 648–659. doi:10.1016/j.telpol.2007.08.007

Tamm Hallström, K. (2000). Organizing the process of standardization. In N. Brunsson & B. Jacobsson (Eds.), *A World of Standards* (pp. 85–100). Oxford, UK: Oxford University Press.

Tamm Hallström, K. (2004). *Organizing international standardization. ISO and the IASC in quest of authority*. Cheltenham, UK: Edward Elgar.

Tan, W.-G., Cater-Steel, A., & Toleman, M. (2007). *Implementing centralised IT service management: drawing lessons from the public sector*. Paper presented at the 18th Australasian Conference on Information Systems, Toowoomba, Australia.

Tapia, C. (2016, January). Securing a competitive future in Europe. *The Patent Lawyer*, 31-35.

Tapscott, D., & Tapscott, A. (2017). Realizing the Potential of Blockchain A Multistakeholder Approach to the Stewardship of Blockchain and Cryptocurrencies. *World Economic Forum.*

Tassey, G. (1982). Infratechnologies and the Role of Government. *Technological Forecasting and Social Change*, *21*(2), 163–180. doi:10.1016/0040-1625(82)90013-0

Taylor, A. S., & Vincent, J. (2005). An SMS history. In *Mobile World; Past, Present and Future* (pp. 75–91). London: Springer.

TBT Committee. (2000). *Decision on Principles for the Development of International Standards, Guide and Recommendations with Relation to Article 2, 5 and Annex 3 of the TBT Agreement.* Author.

Teece, D., & Sherry, E. (2016). *On the "Smallest Salable Patent Practicing Unit": An Economic and Public Policy Analysis*. Tusher Center for the Management of Intellectual Capital Working Paper Series No.11.

Teece, D. J. (1986). Profiting from technological innovation: Implications for integration, collaboration, licensing, and public policy. *Research Policy*, *15*(6), 285–305. doi:10.1016/0048-7333(86)90027-2

Teece, D. J. (2007). Explicating dynamic capabilities: The nature and microfoundations of (sustainable) enterprise performance. *Strategic Management Journal*, *28*(13), 1319–1350. doi:10.1002/smj.640

Teece, D. J., Pisano, G., & Shuen, A. (1997). Dynamic capabilities and strategic management. *Strategic Management Journal*, *18*(7), 509–533. doi:10.1002/(SICI)1097-0266(199708)18:7<509::AID-SMJ882>3.0.CO;2-Z

Tee, R., & Gawer, A. (2009). Industry architecture as a determinant of successful platform strategies: A case study of the i-mode mobile Internet service. *European Management Review*, *6*(4), 217–232. doi:10.1057/emr.2009.22

Tellis, G. J., & Golder, P. N. (1996, Winter). First to Market, First to Fail? Real Causes of Enduring Market Leadership. *Sloan Management Review*, 65–75.

Temple, P. (2005). *The Empirical Economics of Standards*. DTI economics paper, No. 12, June 2005.

Thakor, A. V. (1999). Information technology and financial services consolidation. *Journal of Banking &amp. Finance*, *23*(2–4), 697–700. doi:10.1016/s0378-4266(98)00104-6

The Boston Consulting Group. (2015). *The Mobile Revolution. How Mobile Technologies Drive a Trillion-Dollar Impact.* Retrieved July 29, 2017, from https://www.bcgperspectives.com/content/articles/telecommunications_technology_business_transformation_mobile_revolution/

Thompson, K., Miller, G. J., & Wilder, R. (1997). Wide-area Internet traffic patterns and characteristics. *IEEE Network, 11*(6), 10–23. doi:10.1109/65.642356

Tichy, G. (2004). The over-optimism among experts in assessment and foresight. *Technological Forecasting and Social Change, 71*(4), 341–363. doi:10.1016/j.techfore.2004.01.003

Tiemann, M. (2006). An objective definition of open standards. *Computer Standards & Interfaces, 28*(5), 495–507. doi:10.1016/j.csi.2004.12.003

Tilson, D., & Lyytinen, K. (2006). The 3G transition: Changes in the US wireless industry. *Telecommunications Policy, 30*(10–11), 569–586. doi:10.1016/j.telpol.2006.09.002

Timmermans, S., & Epstein, S. (2010). A World of Standards but not a Standard World: Toward a Sociology of Standards and Standardization. *Annual Review of Sociology, 36*(1), 69–89. doi:10.1146/annurev.soc.012809.102629

Tiwana, A., Konsynski, B., & Bush, A. A. (2010). Research commentary-Platform evolution: Coevolution of platform architecture, governance, and environmental dynamics. *Information Systems Research, 21*(4), 675–687. doi:10.1287/isre.1100.0323

Toppila, A., Liesiö, J., & Salo, A. (2011). A Resource Allocation Model for R&D Investments: A Case Study in Telecommunication Standardization. In A. Salo, J. Keisler, & A. Morton (Eds.), *Portfolio Decision Analysis: Improved methods for resource allocation* (pp. 241–258). New York, NY: Springer New York; doi:10.1007/978-1-4419-9943-6_11

Tornatzky, L. G., Fleischer, M., & Chakrabarti, A. K. (1990). *The processes of technological innovation* (Vol. 273). Lexington, MA: Lexington Books.

Treaty on the Functioning of the European Union, Article 102 (2008).

Treiblmaier, H., & Filzmoser, P. (2010). Exploratory factor analysis revisited: How robust methods support the detection of hidden multivariate data structures in IS research. *Information & Management, 47*(4), 197–207. doi:10.1016/j.im.2010.02.002

Trestian, R., Ormond, O., & Muntean, G. (2012). Game Theory — Based Network Selection : Solutions and Challenges. *Communications Survey & Tutorials, 14*(4), 1212–1231. doi:10.1109/SURV.2012.010912.00081

Tribunal de Grande Instance de Paris. (2013, November 29). TGI Paris, *Telefonaktiebolaget LM Ericsson vs. TCT Mobile Europe SAS, TCT Mobile International Limited, TCT Mobile Limited,* 2013, November 29, 12/14922.

Tribunale di Genoa. (2004, May 7). *Koninklijke Philips Electronics N.V. v. Computer Support Italcard s.r.l.*

Tribunale di Milano. (2012, January 5). Sezione Specializzata Proprieta' Industriale ed Intelletuale, *Samsung Electronics Co., Ltd. e Samsung Electronics Italia c. Apple Inc., Apple Italia S.R.L., Apple Retail Italia S.R.L., Apple Sales International,* R.G 59734/2011.

Tushman, M. L., & Anderson, P. (1986). Technological Discontinuities and Organizational Environments. *Administrative Science Quarterly, 31*(3), 439–465. doi:10.2307/2392832

Tushman, M. L., Anderson, P. C., & O'Reilly, C. (1997). Technology Cycles, Innovation Streams, and Ambidextrous Organizations: Organization Renewal Through Innovation Streams and Strategic Change. In M. L. Tushman & P. C. Anderson (Eds.), *Managing Strategic Innovation and Change* (pp. 3–23). Oxford, UK: Oxford University Press.

Tushman, M. L., & Rosenkopf, L. (1992). Organizational determinants of technological change. Towards a sociology of technological evolution. *Research in Organizational Behavior, 14*, 311–347.

U.S. Department of Justice. (2015). *Response to the Institute of Electrical and Electronics Engineers, Incorporated.* Retrieved July 29, 2017, from https://www.justice.gov/atr/response-institute-electrical-and-electronics-engineers-incorporated

U.S. Federal Trade Commission. (2013a, April 23). *In the Matter of Robert Bosch GmbH.* FTC File No. 121-0081.

U.S. Federal Trade Commission. (2013b, July 24). *In the Matter of Motorola Mobility and Google Inc.* FTC File No. 121-0120.

UK. (2012a). *Open Standards Principles: For software interoperability, data and document formats in government IT specifications.* HM Government, UK. Retrieved from https://www.gov.uk/government/uploads/ system/uploads/attachment_data/file/183962/Open-Standards-Principles-FINAL.pdf

UK. (2012b). *Open Standards in Government IT: A Review of the Evidence.* HM Government, UK. Retrieved from https://www.gov.uk/government/uploads/system/uploads/attachment_data/file/78891/Review-of-the-Evidence_CIPPM.pdf

UK. (2012c). *CloudStore opens for business, Cabinet Office and The Rt Hon Lord Maude of Horsham.* Retrieved from https://www.gov.uk/government/news/cloudstore-opens-for-business

UK. (2014). *Standards Hub.* CabinettOffice, UK Government. Retrieved from http://standards.data.gov.uk/

UK. (2015). *Open Standards Principles, Updated 7 September 2015.* HM Government, UK. Retrieved from https://www.gov.uk/government/uploads/system/uploads/attachment_data/file/459075/OpenStandardsPrinciples2015.pdf

Underwood, S. (2016). Blockchain beyond bitcoin. *Communications of the ACM, 59*(5), 15–17. doi:10.1145/2994581

Updegrove, A. (2009). *The EC Settlement: Rambus, Writs and the Rule of Law.* Retrieved from http://www.consortiuminfo.org/standardsblog/article.php?story=20090622043038212

Utterbach, J., & Abernathy, W. (1975). A dynamic model of process and product innovation. *Omega, 3*(6), 639–656. doi:10.1016/0305-0483(75)90068-7

Valenta, M., & Sandner, P. (2017). *Comparison of Ethereum, Hyperledger Fabric and Corda.* FSBC Working Paper.

Van Bon, J., de Jong, A., Kolthof, A., Pieper, M., Tjassing, R., van der Veen, A., & Verheijen, T. (2007). *Foundations of IT Service Management: based on ITIL.* Van Haren Publishing. Retrieved from www.vanharen.net

van de Kaa, G. (2009). *Standards Battles for Complex Systems.* Rotterdam: Empirical Research on the Home Network. ERIM PhD Series Research in Management.

van de Kaa, G., & de Bruijn, H. (2015). Platforms and incentives for consensus building on complex ICT systems: The development of WiFi. *Telecommunications Policy, 39*(7), 580–589. doi:10.1016/j.telpol.2014.12.012

van de Kaa, G., & de Vries, H. J. (2015). Factors for winning format battles: A comparative case study. *Technological Forecasting and Social Change, 91*, 222–235. doi:10.1016/j.techfore.2014.02.019

van de Kaa, G., de Vries, H. J., van Heck, E., & Van den Ende, J. (2007). The Emergence of Standards: a Meta-analysis. *Proceedings of the 40th Hawaii International Conference on Systems Sciences (HICSS'07).*

van de Kaa, G., van den Ende, J., de Vries, H. J., & van Heck, E. (2011). Factors for winning interface format battles: A review and synthesis of the literature. *Technological Forecasting and Social Change, 78*(8), 1397–1411. doi:10.1016/j.techfore.2011.03.011

Van Den Ende, J., Ravesteijn, W., & De Wit, D. (1997/1998). Shaping the Early Development of Television. *IEEE Technology and Society Magazine*, *16*(Winter), 13–26. doi:10.1109/44.642518

Varian, H. R. (2003). *Economics of Information Technology* (Unpublished paper). Retrieved September 27, 2017, from http://people.ischool.berkeley.edu/~hal/Papers/mattioli/mattioli.pdf

Vasudeva, G., Alexander, E. A., & Jones, S. L. (2015). Institutional Logics and Interorganizational Learning in Technological Arenas: Evidence from Standard-Setting Organizations in the Mobile Handset Industry. *Organization Science*, *26*(3), 830–846. doi:10.1287/orsc.2014.0940

Verkasalo, H. (2008). Dynamics of mobile service adoption. *International Journal of E-Business Research*, *4*(3), 40–63. doi:10.4018/jebr.2008070103

Vesselkov, A., & Hämmäinen, H. (2016). *Mobile handset population in Finland 2005-2015*. Technical report. Retrieved from http://metsahovi.aalto.fi/en/midcom-serveattachmentguid-1e5c68b3bf94696c68b11e5acd70192aed101490149/mobile_handset_population_2005-2015.pdf

Von Bertalanffy, L. (1972). The history and status of general systems theory. *Academy of Management Journal*, *15*(4), 407–426. doi:10.2307/255139

Vries de, H. J., Verhagen, P., & Beerepoot, M. (2008). Impact of standards on innovation - A case of energy performance standards. *EuroMOT Conference Proceedings*. Retrieved from http://webintec.ceram.fr/euromot2008/conftool/uploads/386/1-EPC_EuroMOT_2008-08-01.pdf

W3C. (2005). *World Wide Web Consortium Process Document*. Retrieved August 30, 2017, from http://www.w3.org/2005/10/Process-20051014/cover.html#toc

W3C. (n.d.). *History*. Retrieved from https://www.w3.org/History.html

Wakke, P., Blind, K., & Ramel, F. (2016, June). The impact of participation within formal standardization on firm performance. *Journal of Productivity Analysis*, *45*(3), 317–330. doi:10.1007/s11123-016-0465-3

Walden, I., & Gleeson, N. C. (2014). *'It's a Jungle Out There'? Cloud Computing, Standards and the Law*. Retrieved from https://papers.ssrn.com/sol3/Delivery.cfm/SSRN_ID2441182_code1382817.pdf?abstractid=2441182&mirid=1

Warma, H., Levä, T., Tripp, H., Ford, A., & Kostopoulos, A. (2011). Dynamics of communication protocol diffusion: The case of Multipath TCP. *NETNOMICS: Economic Research and Electronic Networking*, *12*(2), 133–159. doi:10.1007/s11066-011-9063-8

Wasserman, S., & Faust, K. (1994). *Social network analysis: Methods and applications*. Cambridge University Press. doi:10.1017/CBO9780511815478

Watts, D. J., & Strogatz, S. (1998). Collective dynamics of "small-world" networks. *Nature*, *393*(6684), 440–442. doi:10.1038/30918 PMID:9623998

Weick, K. (1990). Technology as equivoque: sensemaking in new technologies. In P. S. Goodman & L. S. Sproull (Eds.), *Technology and Organizations* (pp. 1–44). San Francisco, CA: Jossey-Bass.

Weick, K. E. (1976). Educational organizations as loosely coupled systems. *Administrative Science Quarterly*, *21*(1), 1–19. doi:10.2307/2391875

Weiss, M. B. H., & Sirbu, M. (1990). Technological choice in voluntary standards committees: An empirical analysis. *Economics of Innovation and New Technology*, *1*(1–2), 111–133. doi:10.1080/10438599000000007

Weitzel, T., Beimborn, D., & König, W. (2006). A unified economic model of standard diffusion: The impact of standardization cost, network effects, and network topology. *Management Information Systems Quarterly*, 489–514.

Wessman, R. (2013). *Upphandling av IT - inlåsningseffekter och möjligheter, Uppdragsforskningsrapport 2013:2, ISSN-nr 1401-8438*. Swedish Competition Authorities.

West, J. (2003). The role of standards in the creation and use of information systems. *MISQ special issues Workshop*.

Wiegmann, P. M., de Vries, H. J., & Blind, K. (2017). Multi-mode standardisation: A critical review and a research agenda. *Research Policy*, *46*(8), 1370–1386. doi:10.1016/j.respol.2017.06.002

Wieland, J. (2005). Corporate governance, values management and standards: A European perspective. *Business & Society*, *44*(1), 74–93. doi:10.1177/0007650305274852

Wi-Fi Alliance. (n.d.). Retrieved from http://www.wi-fi.org/

Wölker, Th. (1996). Technische Normen im Verlauf der Geschichte. *Homo Economicus*, *13*, 51–75.

Wright, A., & De Filippi, P. (2015). *Decentralized Blockchain Technology and the Rise of Lex Cryptographia*. Retrieved September 24th, 2017, https://ssrn.com/abstract=2580664

Wright, J. (2013). *SSOs, FRAND, and Antitrust: Lessons from the Economics of Incomplete Contracts*. Paper presented at the Inaugural Academic Conference of the Center for the Protection of Intellectual Property on The Commercial Function of Patents in Today's Innovation Economy, Arlington, VA.

Wright, J. D., & Ginsburg, D. H. (2014). Patent Assertion Entities and Antitrust: A Competition Cure for a Litigation Disease. *79 ANTITRUST L.J. 501, 509.*

Wurster, S., & Blind, K. (2011a). Towards a Dynamic Success Factor Model for the Establishment of De-facto-Standards by Born Global Firms and Technical Clusters. GSTF Best Paper Award. doi:10.1037/e605342012-005

Wurster, S. (2011). Leaders, Servants, Independence. A Dynamic Success Factors Model for the Establishment of De-facto Standards by Born Global Firms and Technical Clusters. *International Journal on GSTF Business Review*, *1*(2), 172–179.

Wurster, S., & Blind, K. (2011b). The Role of Standardization and Standards in the Context of Entrepreneurship: Conceptional Model and Insights form Case Studies. *Proceedings of the 6th European Conference on Innovation and Entrepreneurship*, 861-874.

Wurster, S., Blind, K., & Fischer, S. (2014). Born Global Market Dominators: Insight into a Unique Class of Young Companies and Their Environment. *International Journal of IT Standards and Standardization Research*, *12*(1), 1–16. doi:10.4018/ijitsr.2014010101

Xiaojin, R. (2017). *ICO ban seen helping improve regulation.* Retrieved September 24th, 2017, from http://www.chinadaily.com.cn/business/2017-09/09/content_31760669.htm

Yamakawa, P., Noriega, C. O., Linares, A. N., & Ramírez, W. V. (2012). Improving ITIL compliance using change management practices: A finance sector case study. *Business Process Management Journal*, *18*(6), 1020–1035. doi:10.1108/14637151211283393

Yates, J., Orlikowski, W., & Okamura, K. (1999). Explicit and implicit structuring of genres in electronic communication: Reinforcement and change of social interaction. *Organization Science*, *10*(1), 83–103. doi:10.1287/orsc.10.1.83

Yin, R. K. (2003). Case study research, design and methods (3rd ed.). Academic Press.

Yin, R. K. (2003). *Case Study Research*. Thousand Oaks, CA: Sage Publications.

Yli-Huumo, J., Ko, D., Choi, S., Park, S., & Smolander, K. (2016). Where Is Current Research on Blockchain Technology? A Systematic Review. *PLoS One*, *11*(10), e0163477. doi:10.1371/journal.pone.0163477 PMID:27695049

Yoffie, D. B., & Kwak, M. (2006). *With Friends Like These: The Art of Managing Complementors*. Boston: Harvard Business Revie.

Yoo, C. S. (2014). *Public Good Economics and Standard Essential Patents*. Faculty Scholarship. Paper 1371. Retrieved from http://scholarship.law.upenn.edu/cgi/viewcontent.cgi?article=2371&context=faculty_scholarship

Yoo, Y., Lyytinen, K., & Yang, H. (2005). The role of standards in innovation and diffusion of broadband mobile services: The case of South Korea. *The Journal of Strategic Information Systems*, *14*(3), 323–353. doi:10.1016/j.jsis.2005.07.007

Young, W. (2016, March 24). *The History of Winsock*. Retrieved from http://tangentsoft.net/wskfaq/articles/history.html

You, S., Robert, L. P. Jr, & Rieh, S. Y. (2015). The Appropriation Paradox: Benefits and Burdens of Appropriating Collaboration Technologies. *Proceedings of the 33rd Annual ACM Conference Extended Abstracts on Human Factors in Computing Systems*, 1741-1746. doi:10.1145/2702613.2732919

Yue, C., Beghin, J., & Jensen, H. (2006). Tariff Equivalent of Technical Barriers to Trade with Imperfect Substitution and Trade Costs. *American Journal of Agricultural Economics*, *88*(4), 947–960. doi:10.1111/j.1467-8276.2006.00908.x

Zahra, S. A. (2005). A Theory on International New Ventures: A Decade of Research. *Journal of International Business Studies*, *36*(1), 20–28. doi:10.1057/palgrave.jibs.8400118

Zare Ravasan, A., & Mansouri, T. (2014). A FCM-Based Dynamic Modeling of ERP Implementation Critical Failure Factors. *International Journal of Enterprise Information Systems*, *10*(1), 32–52. doi:10.4018/ijeis.2014010103

Zare Ravasan, A., & Mansouri, T. (2016). A dynamic ERP critical failure factors modelling with FCM throughout project lifecycle phases. *Production Planning and Control*, *27*(2), 65–82. doi:10.1080/09537287.2015.1064551

Zare Ravasan, A., Mansouri, T., Mohammadi, M. M., & Rouhani, S. (2014). Supporting the module sequencing decision in ITIL solution implementation: An application of the fuzzy TOPSIS approach. *International Journal of Information Technologies and Systems Approach*, *7*(2), 41–60. doi:10.4018/ijitsa.2014070103

Zeng, S., Shi, J., & Lou, G. (2007). A synergetic model for implementing an integrated management system: An empirical study in China. *Journal of Cleaner Production*, *15*(18), 1760–1767. doi:10.1016/j.jclepro.2006.03.007

Zettinig, P., & Benson-Rea, M. (2008). What becomes of international new ventures? A co-evolutionary approach. *European Management Journal*, *26*(6), 354–365. doi:10.1016/j.emj.2008.09.009

Zhao, L. J., Fan, S., & Yan, J. (2016). Overview of business innovations and research opportunities in blockchain and introduction to the special issue. *Financial Innovation*, *2*(1), 28–25. doi:10.1186/s40854-016-0049-2

Zmud, R. (1984). An examination of push-pull theory applied to process innovation in knowledge work. *Management Science*, *30*(6), 727–738. doi:10.1287/mnsc.30.6.727

Zohar, A. (2015). Bitcoin: Under the hood. *Communications of the ACM*, *58*(9), 104–113. doi:10.1145/2701411

About the Contributors

Kai Jakobs joined Aachen University's (RWTH) Computer Science Department as a member of the technical staff in 1985. Over the years, his research interests moved away from the technical nuts and bolts of communication systems to socio-economic aspects with a focus on ICT standards and the underlying standardisation process. By now, Kai has (co)-authored/edited 20+ books and published 200+ papers He has been on the program committees of numerous international conferences, and has also served as an external expert on evaluation panels of various European R&D programs, on both technical and socio-economic issues. He holds a PhD in computer science from the University of Edinburghand and is a Certified Standards Professional.

* * *

Timo Ali-Vehmas, Nokia Fellow, Head of Ecosystems Research, Nokia Technologies. Timo Ali-Vehmas was nominated in 2015 to lead an emerging ecosystem research in Nokia Technologies. Knowledge and leadership, related to ecosystems is needed to complement the competences and capabilities in technologies. The ICT driven and collaboration oriented ecosystems will drive growth within and between industries of next generation systems and services. In his previous role he was in charge of Nokia's standardization activities as the VP, Compatibility and Industry Collaboration, including Nokia's contributions in forums like the 3GPP, ETSI, CCSA, ITU, IETF, W3C, OMA, DLNA, IEEE, WFA, BT and many others. He has been working in Nokia since 1980 in a number of different areas, including R&D manager of the first Nokia GSM mobile phone and later VP, WCDMA Product Programs in Nokia Mobile Phones. He was also tasked to establish and lead the Radio Communication Laboratory in Nokia Research Centre. For several years, he has been a member of the OISPG/EU. Since 2012 Timo Ali-Vehmas has served as the chairman of the Board of Nokia Foundation and since 2008 he has been member of the Board of Walter Ahlström Foundation. He is also a member of standardization advisory board in Finnish Standardization Association (Standardointilautakunta/SFS). Timo Ali-Vehmas graduated at Helsinki University of Technology in 1982 and currently he is in the process to finalize his PhD in Aalto University.

Marie Barani is graduated in industrial property law in France and Belgium. She has complemented her studies with traineeships related to the legal aspects of standardization at an IT consultancy, a telecom operator and the European Commission. She started working as intellectual property legal counsel for public institutions in France and Belgium before performing a research on patent hold-out. She is now working for a licensing company. Views expressed are her own.

Knut Blind studied economics, political science and psychology at Freiburg University. In the course of his studies he spent one year at Brock University (Canada), where he was awarded a BA. Finally, he took his Diploma in Economics and later his doctoral degree at Freiburg University. Between 1996 and 2010 he joined the Fraunhofer Institute for Systems and Innovation Research, Karlsruhe, Germany, as a senior researcher and at last as head of the Competence Center "Regulation and Innovation". In April 2006 Knut Blind was appointed Professor of Innovation Economics at the Faculty of Economics and Management at the Berlin University of Technology. Between 2008 and 2016 he held also the endowed chair of standardisation at the Rotterdam School of Management of the Erasmus University. From April 2010 he is linked to the Fraunhofer Institute of Open Communication Systems in Berlin. In 2012, he initiated both the Berlin Innovation Panel and the German Standardization Panel. Besides numerous articles on patents he published further contributions on standardization and further innovation aspects in refereed journals.

Moritz Böhmecke-Schwafert is a PhD Candidate and research fellow at the Chair of Innovation Economics at the Institute of Technology in Berlin where he teaches the course "Digital Innovation". He holds a Master Degree in Economics and has studied in Germany, Finland and Chile. His research passion is focused on ICT innovation in developing countries and the economics of digitization. Meanwhile, he is a senior consultant for IT innovation in the sphere of AI, Blockchain and Data Analytics, contributing to the digital transformation of the German industry and working for the multinational IT company IBM.

Tineke M. Egyedi, founder of DIRoS, is policy advisor and senior researcher Standardization with a focus on policy, processes and impact issues. She has done studies and projects for the EU, standards bodies, science foundations, government agencies and international companies. She publishes in journals ranging from IEEE Communications Magazine to International Journal of Hydrogen Energy. Based on Inverse Infrastructures: Disrupting Networks from Below (2012, Edward Elgar; co-edited by Donna Mehos) and her work on 'Standards & Flexible Infrastructures' for the Next Generation Infrastructures Foundation, she was invited to develop NGI MOOC modules (2014). She initiated the game Setting Standards used by universities internationally (multiple disciplines), NIST (US policy makers) and NEN (Chinese ISO standardizers). She was president of the European Academy for Standardization (2005-2011) and chair of the International Cooperation for Education about Standardization (2009-2010). She is expert to the UNECE WP6 on 'Regulatory Cooperation and Standardization Policies' and member of IEEE committees (e.g. co-chair TPC SIIT2015). She regularly lectures at home and abroad.

Vladislav V. Fomin holds a professor's position at the department of Applied Informatics at Vytautas Magnus University and a principal researcher position at Vilnius University in Kaunas, Lithuania. Prof. Fomin also holds visiting professor's positions at the Faculty of Computing of the University of Latvia and La Rochelle Business School in France. After earning his PhD degree in 2001, Vladislav V. Fomin held academic positions at the University of Michigan in Ann Arbor, Copenhagen Business School, Delft University of Technology, Montpellier Business School, and Rotterdam School of Management, Erasmus University. Prof. Fomin has an extensive experience in international research and professional projects. He is serving on the editorial board of The International Journal of Standardization Research (IJSR) and Baltic Journal of Modern Computing (BJMC), is serving as a reviewer for many recognized journals and is regularly invited to serve as a scientific committee member to recognized international conferences. Fomin is a member of European Academy of Standardization (EURAS) and The Association for Information Systems (AIS). Prof. Fomin has over 70 scientific publications in journals, conferences, and as book chapters, including Journal of Strategic Information Systems, Communications of the Association for Information Systems, International Journal of IT Standards & Standardization Research, Telecommunications Policy, Knowledge, Technology & Policy. Conference publications and presentations include The International Conference on Information Systems, European Conference on Information systems, Hawaii International Conference on System Sciences, Academy of Management, and other.

Andrea Fried obtained a MSc (Dipl.-Kffr.) degree in business administration and vocationomics from the Friedrich Schiller University of Jena and a PhD degree from Chemnitz University of Technology in Germany. She has worked as a post-doctoral researcher and was awarded a Habilitation degree at the Chemnitz University of Technology. Before becoming an Associate Professor in business administration, in particular in management control and organization studies Andrea was a visiting international research fellow and senior lecturer at the Department of Management and Engineering, Linköping University, Sweden. Furthermore, she was a visiting research fellow at the Department of Science Research and Science Sociology, University of Basel, Switzerland, at the Institute of Management, University of St. Gallen, Switzerland, and at the Department of Computer Science and Software Engineering, Tshwane University of Technology, Pretoria, South Africa. Her earlier research looked at the philosophical foundations of knowledge and performance management. More recently, she has been actively involved in the development of the research field on management control, standards and innovation.

Jonas Gamalielsson is a researcher at the University of Skövde, Sweden. He is a member of the Software Systems Research Group, has conducted research related to free and open source software in a number of projects, and his research is reported in publications in a variety of international journals and conferences.

Ronny Gey graduated from Chemnitz University of Technology in 2009 with a masters degree in business information systems. He worked two years as a software engineer in a company for medical equipment as well as two years at the University of Leipzig as research assistant at the chair of business information systems. Currently, he is employed as a research assistant at the Friedrich-Schiller University of Jena at the Otto-Schott-Institute. Besides, he is a doctoral candidate at the Friedrich-Schiller-University Jena, School of Economics and Business Administration. Ronny is the founder and editor of the website www.sosciso.de where he talks about software tools in the social science research process.

Homa Hamidi received her M.A. degree in IT Management from MehrAlborz University, Tehran, Iran. Her research interests included Business Intelligence, Data Mining and Knowledge Management.

Heikki Hämmäinen is professor of Network Economics at Department of Communications and Networking, Aalto University, Finland. He has MSc (1984) and PhD (1991) in Computer Science from Helsinki University of Technology. His main research interests are in techno-economics and regulation of mobile services and networks. Special topics recently include measurement and analysis of mobile usage, value networks of flexible Internet access, and diffusion of Internet protocols in mobile. He is active in several journal and conference duties.

Bowman Heiden is Deputy Director of Center for Intellectual Property at Chalmers University of Technology, University of Gothenburg, and Norwegian University of Science and Technology (NTNU). He is also a Lecturer and Director of the Sahlgrenska School of Innovation and Entrepreneurship at University of Gothenburg. His background is in engineering and technology management and economics with a current focus on the intersection of law, economics, and innovation.

Frank Hofmann studied general mechanical engineering at TU Darmstadt with an exchange year at KTH in Stockholm. After graduating as Diplom-Ingenieur, he worked several years for Airbus on the digitization and harmonization of national and international development processes. Extending his academic education with an international master of business administration with focus on finance at HEC near Paris, he continued his career as a free trader on own account based on the evaluation of digital data and information flows. Recently, he became a PhD candidate at the Chair of Innovation Economics at TU Berlin. His research focuses on competitive advantages stemming from the access and the processing of digital data by organizations.

Andrew Katz is a visiting researcher at the University of Skövde, Sweden and a partner at boutique law firm Moorcrofts LLP, based in England's Thames Valley. Andrew specialises in technology law and has a particular interest in open design and development.

Heejin Lee is Professor at the Graduate School of International Studies, Yonsei University, Korea. He has written extensively on the impact of broadband, and 'time and ICT'. He is currently working on ICT standards policy in China and Korea, and ICT for development (ICT4D). He is the President of Korea Association of International Development and Cooperation.

Tapio Levä received his M.Sc. from Helsinki University of Technology, Finland in 2009 and his D.Sc. in network economics from Aalto University, Finland, in 2014. The topic of his dissertation was feasibility analysis of new Internet protocols. Tapio is currently working at Telia Company. His research interests include techno-economics of Internet architecture evolution, Internet standards adoption, information-centric networking and Internet of things.

Björn Lundell is a senior researcher at the University of Skövde, Sweden. He leads the Software Systems Research Group, has conducted research related to free and open source software in a number of projects, and his research is reported in over 100 publications in a variety of international journals and conferences.

Ivana Mijatovic is associate professor at Faculty of Organizational Sciences, University of Belgrade. She earned B.Sc in Mechanical Engineering at Faculty of Mechanical Engineering (University of Belgrade), M.Sc and PhD at Faculty of Organizational Sciences at University of Belgrade. For much of her academic career, she has focused on quality management and standardization. She is passionate teacher, on bachelor studies she teaches Standardization, Quality Engineering and Quality Planning, on master and PhD studies she teaches Standardization and ICT Standardization. Since 2015, she is member of the STARTed team (Team of Specialists on Standardization and Regulatory Techniques – education on standardization) of The United Nations Economic Commission for Europe (UNECE). She serves as a vice-president on the board of the European Academy for Standardization (EURAS) and she is a member of the Balkan Coordination Committee for Standardization, Protypes and Quality (BCC). She is a member of national technical committee KS I1/07 - Software engineering, IT for Education and Internet at Institute for Standardization of Serbia (national mirror committee in relation with ISO/IEC JTC 1/ SC 7 Software and systems engineering; CEN/TC 353 Information and Communication Technologies for Learning, Education and Training; CEN/TC 365 Project Committee - Internet Filtering; ISO/IEC JTC 1/SC 36 Information technology for learning, education and training and ISO/IEC JTC 1/SC 40 IT Service Management and IT Governance). She developed course Standardization 1 and wrote the text book with 10 case studies Standardization 1 (2015). Her current academic work addresses the question of standardization, education about standardization, teaching quality and quality aspects of technology enhanced learning. The list of her publication can be found at https://www.researchgate.net/profile/Ivana_Mijatovic3.

Mohammad Mehrabioun Mohammadi is a PhD candidate in IT Management at Allameh Tabataba'i University in Tehran, Iran. He received his M.A. in IT Management from Tehran University. He has published a number of papers in journals such as Systemic Practice and Action Research (SPAR), International Journal of Standardization Research (IJSR), International Journal of Virtual Communities and Social Networking (IJVCSN) and International Journal of Information Technologies and Systems Approach (IJITSA). His research interests include Soft Systems Methodology, business model design and development, information systems development and evaluating the diffusion of ICT in businesses and higher educations.

Roland J. Ortt (1964) is Associate Professor of Innovation management and Entrepreneurship at Delft University of Technology. Before joining the faculty of Technology Policy and Management (TPM) Roland Ortt worked as project manager and R&D manager for KPN research, the R&D lab of a Dutch telecommunication company. Roland is director of Interfaculty Education in TPM and research dean of the European NiTiM network of researchers in innovation and technology management.

Antti Riikonen, D.Sc. (Tech.), is a product lead at Verto Analytics, a consumer-centric audience measurement company. His research interests are related to the measurement of consumer behavior across all digital devices they use, as well as the diffusion of mobile devices and services. Antti Riikonen holds a D.Sc. (Technology) from Aalto University, Finland.

Cesare A. F. Riillo is a senior economist at STATEC (National Institute of Statistics and Economic Studies of the Grand Duchy of Luxembourg) and at ANEC (Agence pour la normalisation et l'économie de la connaissance). He holds a PhD, with mention of Doctor Europaeus in Human, Territory and Society Science (Dottore di ricerca in Scienze dell Uomo del Territorio e della Società), University of Trieste, Italy. His research interests include entrepreneurship, innovation and standardisation economics.

Michael Spring is an Associate Professor of Information Science at the University of Pittsburgh. Supported by NSF, he has worked on models and principles for design of human-computer interaction, standards development, visualization, virtual information spaces, and cybersecurity. With support from NCI, NIH, NIMH, Medicare and the Veterans Administration, he has developed web sites to support people with schizophrenia, and various forms of cancer as well as caregivers for individuals with cancer and traumatic brain injury. Current research efforts involve ontology mapping for semantic web applications, data mining of social web applications, secure location based services, and the automated development of secure websites to support medical research. Spring has authored more than 100 articles and book chapters in the areas of document processing, collaborative authoring, information technology standardization, visualization, virtual reality, and medical applications of web and mobile technology. He is the author of two books.

Juuso Töyli, D.Sc. (Econ. & Bus. Adm.), D.Sc. (Tech.), is Professor of Operations & Supply Chain Management at Turku School of Economics at University of Turku and Adjunct Professor of Network Economics at Aalto University. His research interest include policy issues and the diffusion of mobile technologies and their effects. He is further interested in business games and their development. Within operations and supply chain management his research is mainly focused on firm and supply chain performance (financial, logistics cost and operations performance) and how these are related to various issues like strategy choices, outsourcing, low-cost sourcing, geographic dispersion and sustainability.

Simone Wurster holds a PhD from the University of Potsdam in economics and social sciences. Besides her expertise as a project manager and consultant for technology companies, her specialisation as researcher at TU Berlin includes standardisation, entrepreneurship, civil security, privacy and data protection, blockchains, warning systems and sustainable economies. Dr. Wurster has taken part in various research projects including CRISP, ENSURE, STAR-ProBio and BlockOne among many others. She has published two books as well as numerous articles and papers, among them two award-winning contributions. Dr Wurster is also a guest member of the German mirror committee of ISO/TC 307.

Ahad Zare Ravasan is an Assistant Professor of Information Systems and Decision Sciences at MehrAlborz University, Tehran, Iran. He received his MSc and PhD in Information Technology Management at Allameh Tabataba'i University, Tehran, Iran.Ahad has published three books and a number of papers in acclaimed journals, such as the Expert Systems with Applications, Information Systems, International Journal of Production Research, Telecommunication Systems, Production Planning and Control, Scientia Iranica, International Journal of Data Warehousing and Mining, Strategic Outsourcing; An International Journal and International Journal of Enterprise Information Systems to name only a few. His key research interests include ERPs, artificial neural networks' applications in practice, business process outsourcing, business intelligence, Requirements Engineering, and IT Service management.

Hanah Zoo is Research Fellow at the Institute of International Studies, Graduate School of International Studies, Yonsei University, Korea. She received her Ph.D in international studies from Yonsei University in 2017. Her research interests include ICT for Development (ICT4D) and standards/ standardization and innovation in developing countries.

Index

T

www.ingramcontent.com/pod-product-compliance
Lightning Source LLC
LaVergne TN
LVHW060403300926
849162LV00006B/1331